P9-BZR-403

A SECOND MATHEMATICS COURSE FOR ELEMENTARY TEACHERS

A SECOND MATHEMATICS COURSE FOR ELEMENTARY TEACHERS

Eugene F. Krause
UNIVERSITY OF MICHIGAN

HARPER & ROW, PUBLISHERS
New York Evanston San Francisco London

Sponsoring Editor: George J. Telecki
Project Editor: Lois Wernick
Designer: Rita Naughton
Production Supervisor: Bernice Krawczyk

A SECOND MATHEMATICS COURSE FOR ELEMENTARY TEACHERS

Library of Congress Cataloging in Publication Data
Krause, Eugene F
A second mathematics course for elementary teachers.
1. Mathematics—1961– I. Title.
QA39.2.K7 510 74-1338
ISBN 0-06-043763-4

CONTENTS

PREFACE

One of the most obvious changes in the elementary school mathematics curriculum in recent years has been its broadening. In addition to the usual arithmetic, present day commercial and experimental textbooks include increasingly more work in such areas as geometry, measurement, algebra, sets, functions, logic, and probability. Quite naturally, the standard first course in mathematics for prospective elementary teachers is directed toward the major content area, arithmetic. Such a course inevitably includes some work with sets, number theory, and algebraic structure, but no one expects it to prepare the future teacher in all of the areas of mathematics that now appear in the elementary school curriculum.

This book is intended for a second course in which the preparation of the future teacher is rounded out. Although the main content strands are geometry and measurement, substantial work is also done in algebra, probability, functions, and applications. Lesser amounts of time are devoted to sets and logic, informal inductive and deductive reasoning, combinatorics, and graph theory.

Three main pedagogical considerations shaped the writing of this book.

1. Integration of mathematical topics often serves a useful purpose. For example, when a new topic such as probability (Chapter 6) is introduced as a first cousin of the familiar topics of length and area measurement (Chapter 2), useful analogies and a comfortable sense of familiarity come to mind. As another example, when two apparently unrelated topics such as linear equations (Chapter 2) and similarity of plane figures (Chapter 8) are brought together and seen to be two sides of a single coin (Chapter 9), one begins to look on mathematics as one discipline, not a collection of diverse studies called arithmetic, geometry, algebra, ... This point of view seems particularly important for the future elementary teacher who will be teaching a variety of topics out of a single "mathematics" text. Sometimes integration can be useful for the simple reason that it provides variety. A sprinkling of combinatorial questions through a chapter on basic geometric figures can spice up what might otherwise be rather dull. Finally, another virtue of integration is that it is often helpful to be able to look at a single problem from two different points of view. This is illustrated in Chapter 2 by dual approaches to number systems, variables, and absolute value.

Of course, integration can be overdone. Too much swinging back and forth between different mathematical areas can produce a feeling of aimlessness and can disrupt the construction of all but short trains of mathematical thought. For this reason the mathematical topics covered in this book are organized around the clearly recognizable main themes of geometry and measurement.

2. It is very important that prospective elementary teachers actually get their hands dirty doing the kind of mathematics they will be teaching. Commitment to this principle has resulted in extensive exercise lists being the most prominent feature of the book. There are routine exercises designed to clarify the basic concepts under discussion; exercises which ask the student to make original sketches, to count, to look for patterns, to offer conjectures; unusually many exercises directed toward applications; and exercises requiring informal inductive and deductive reasoning.

The goals of the exercise lists are modest. Most of the exercises are designed to be accessible to the average student. At times there will be a feeling of working through a workbook. Occasionally some students might even feel their intelligence is being insulted. Any such insult is unintentional and can be blamed on wariness about presenting mathematics at so high a level that its relationship to the elementary school is lost.

3. The approach to geometry should be broad and informal rather than narrow and axiomatic. While it might be intellectually satisfying to hew to a pure Euclid-Hilbert, or Birkhoff-SMSG, or transformational development of geometry, any one of these approaches is both too deep and too narrow for the purposes of the elementary school teacher. Perhaps because of the unsettled state of high school geometry, the elementary school is being called upon to provide a broad, rich background of geometric experiences on which any one

of several more formal developments could be based. It seems obvious that this objective implies a similar kind of training for the elementary teacher.

In this book the approach to geometry is intuitive, eclectic, and only locally deductive. That is, deduction is limited to finding local logical dependencies within a given collection of properties. Here as in the elementary school there are no axioms and theorems—only properties.

In summary, an attempt has been made to write an informal, integrated treatment of elementary mathematics, with exercises that invite active student involvement and maximize transfer from college course to classroom practice.

In trying to describe the structure of an integrated text it is convenient to think of it as a stack of horizontal content layers (chapters) woven together with various vertical threads of thought that relate and unify the material.

The overall horizontal structure is this:

Chapters 1–3:	Geometric and algebraic preliminaries
Chapters 4–6:	Theory and uses of measurement, both geometric and nongeometric
Chapters 7, 8:	Congruence, similarity, and indirect measurement
Chapter 9:	Unification of the foregoing algebra and geometry via functions
Chapter 10:	An appendixlike extension to space of the metric geometry done in the plane

More detailed descriptions of the individual horizontal layers can be found in the overviews which introduce each of the chapters.

Some of the vertical threads of thought are extensions of concepts encountered in a first course in number systems. For example, *sets* and related concepts are recalled in some detail at the beginning of Chapter 1, then reviewed and applied extensively throughout the book. Connections between sets and *logic* receive special attention in Chapters 1, 2, and 6. *Rational arithmetic* is reviewed in the new contexts of geometric measurement and probability. The view of rational numbers as *operators* runs through these uses.

Other unifying threads may be new. The major concept of *measurement* may not have been dealt with in a first course. The *function* concept appears informally in Chapters 2, 4, and 6 before coming in for detailed, explicit study in Chapter 9. *Rigid motions* (and dilations) have hazy beginnings in Chapter 2, take shape in Chapters 4, 7, and 8, and stand out clearly in Chapter 9. *Applications* are scattered through the exercise lists of every chapter and appear in heavy concentrations at the end of Chapter 2, throughout Chapter 5, and in most of Chapter 10. *Algebra* appears formally in Chapter 2, is sprinkled through ensuing chapters, and is reappraised in Chapter 9. *Combinatorial questions* are studied informally in Chapters 1 and 3, and counting techniques are formalized in Chapter 6. Questions from *graph theory* arise naturally in Chapter 3 and again in Chapter 10.

This book was written and class tested at Arizona State University in 1971–1972 during a leave of absence from the University of Michigan. I wish to express my appreciation to the Mathematics Department of Arizona State University for their hospitality during my stay there.

E. F. K.

TO THE INSTRUCTOR

This book was written for use in the second required mathematics course for prospective elementary teachers. It would be a remarkable class, though, that could cover all of the material in one semester. For an average class, two quarters or possibly two semesters would be more realistic. Since the bulk of the book consists of exercises, however, your pace through the text can be increased rather easily by omitting some of them. For certain classes it may be possible to omit many of the routine exercises, particularly in Chapters 1–4; for others, omitting the difficult starred problems might be more appropriate.

The relative independence of the individual chapters makes it possible to tailor this book to fit a variety of courses.

Chapters 1–7 by themselves constitute an integrated one-semester course with applications. Omitting Chapter 5 would lighten the emphasis on applications but permit the coverage of further basic geometry in Chapter 8. A rather pure geometry course could be obtained by using Chapters 1, 3, 4, 7, 8, and 10, with or without Chapter 5. Since this book assumes no specific prerequisites from a first course, it could also be used in conjunction with your favorite number systems text to make a two- or three-semester integrated sequence in elementary mathematics of the kind recommended recently by the CUPM. Finally, a useful course for prospective middle school or junior high school teachers could be constructed by moving quickly over routine material and concentrating on the less familiar concepts—for example, those in Chapters 3, 5, 6, and 9.

Some instructors have good success using a lecture format in their courses. My own preference is to lecture as little as possible. Consequently, a serious attempt was made to write a book that could be read by students. My students at Arizona State were expected to read all of the exposition themselves, and they did so very successfully. Class time was devoted exclusively to questions and the discussion of exercises. Depending on the size of your class and on other local factors, you might find this style of course management worth trying.

E. F. K.

A SECOND MATHEMATICS COURSE FOR ELEMENTARY TEACHERS

1

GEOMETRIC FIGURES

OVERVIEW

In this chapter we describe geometry as the study of figures in abstract space. We first introduce and illustrate the basic geometric figures (point, line, plane). Then we describe various secondary geometric figures (half-line, ray, segment, half-plane, half-space, angle) in terms of the basic figures and the set operations of union, intersection, and complement.

The obvious objective is to review the concepts, terminology, and symbolism of elementary geometry. By doing this in the language of sets, however, two other objectives are met: (1) the concepts and symbols of set theory are reviewed for use in other contexts later in the course; (2) geometry is seen to be not so very much different from the other branches of mathematics, which are also describable in the language of sets.

Further collateral objectives are attained by the use of many geometric counting problems throughout the chapter. Besides providing an interesting context in which to review geometric figures, these problems illustrate the interplay possible between simple geometry and simple whole-number arithmetic; encourage the development of systematic counting techniques and the practice of looking for number patterns; reorient the college student to the free-wheeling, pictorial, intuitive geometry of the pre-high school years.

Not much deduction is called for in this chapter, and what deduction there is, is of a localized sort. For example, many "incidence" properties are given, but they are treated as just that—properties. Although some logical dependencies among them are pointed out in the exercises, no separation of these properties into axioms and theorems is made. This approach seemed best since localized deduction is the only kind that the future elementary school teacher is likely to meet, grand axiomatic schemes for geometry being reserved for the high school level.

1.1 SPACE, POINTS, AND GEOMETRIC FIGURES

We read and hear about **physical space** almost every day. Scientists at Cape Kennedy plan and calculate with great care to ensure that two manned vehicles will "rendezvous" at a point in space. To guarantee that a rocket and the planet Venus will arrive at the same location in space at the same time is a similar problem. These problems cannot be solved by trial and error. Scientists solve these real problems about physical space by first posing and solving corresponding theoretical problems about **abstract space**. Abstract space is an idealized model of physical space which exists only in people's minds. It is not subject to rainstorms or air pollution or aurora borealis, but it is enough like physical space so that studying it leads to useful information about our real environment.

At various times in history various people have devised different abstract models for physical space. Thus there are several abstract spaces which we could study. In this book we shall study the abstract space that was first described by Euclid. This abstract space is often called **Euclidean space**, and its study is known as **Euclidean geometry**. Euclid's abstract space is an excellent model for the portion of physical space that we see around us. For this reason Euclidean geometry has been an important and useful school subject for over 2000 years. Euclidean space, which we shall refer to simply as "space" from now on, is a mathematical idea and so we describe it in mathematical terms. We describe it as the set of all points.

Space is the set of all points.

Of course this definition of space is incomplete without some clarification of the notion of **point**.

A point can be thought of as an *exact location*. If you are asked to find where you are on a globe, you might begin by pointing with your finger. But your finger is too fat. It probably only picks out the correct state. By using a pencil you might be able to indicate the city. With a sharp needle you might even be able to suggest which part of the city. But there is no hope of your *making* a mark so small that it describes your exact location: state, city, street, building, room, desk. You can *think* about this location, though, and the idea of a dot so small that no smaller dot can be drawn inside it is the idea of a point. Euclid described a point as "that which has no part."

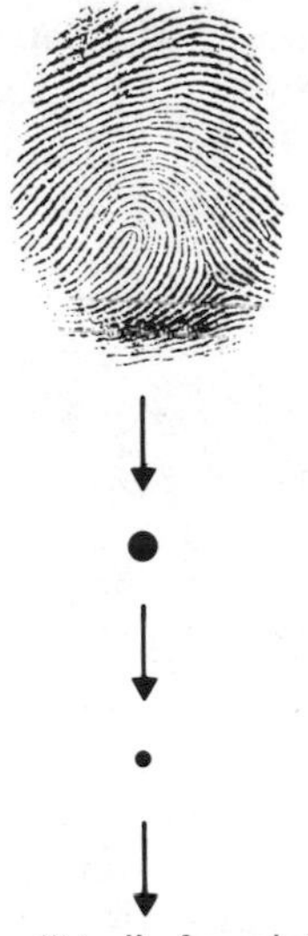

the "idea" of a point

Many things in the real world suggest the idea of point: grains of sand on a beach, specks of algae in a lake, particles of dust in the air. Entire *sets of points* are suggested by other things in the real world. A beam of light passing through a small hole in a window shade into a darkened room illuminates many dust particles. We refer to the set of points that this suggests as a *line segment*. The smoke cloud from a factory chimney is made up of many tiny smoke particles. The set of points suggested, on a calm day, is called a (solid) *cone*. A popcorn ball suggests a set of points called a (solid) *sphere*.

Real Phenomenon | Abstract Idea

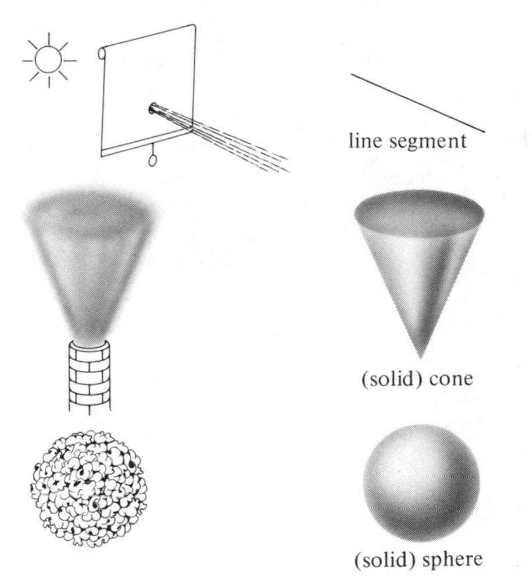

Line segments, cones, and spheres are all examples of "geometric figures." There are many other configurations of points in space that we shall want to study and refer to as geometric figures, and so we want a definition that is broad enough to include all of them. We define a geometric figure to be any set of points, or using the term "subset,"

Any subset of space is called a **geometric figure.**

All of our work in geometry can be thought of in terms of describing, classifying and studying geometric figures.

We have introduced the ideas of abstract space, point, and geometric figure very much as you might introduce them to a child—by relating them to physical reality. From a purely mathematical point of view, however, there is no need to relate these ideas to "reality." Mathematicians feel comfortable saying: "I am thinking about a set which I'll call space. Its elements I will call points. Its subsets I will call geometric figures." This use of the language of set theory in geometry will be helpful throughout the course. Exercises 5–10 below review the most common terms and symbols of set theory.

EXERCISES

Exercises 1–4 concern the connection between geometry and the physical world. They are designed to let you check how many geometric figures you remember from your school days. All of these figures will be studied in more detail later.

1. What geometric figure is suggested by
  (a) the ripples in a pond shortly after a pebble has been dropped in the middle

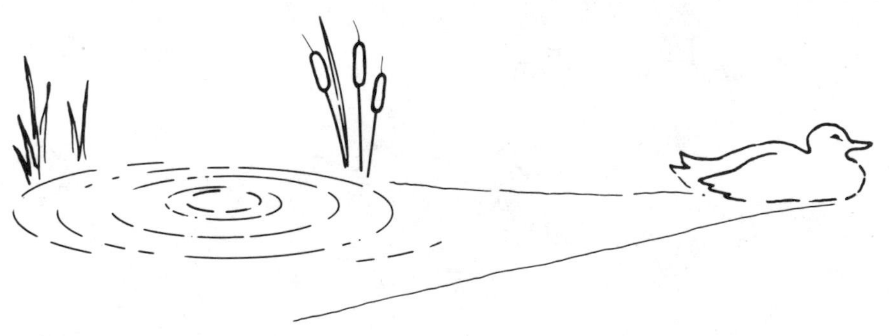

  (b) the ripples in a pond formed by a duck swimming across it
  (c) the surface of the pond on a perfectly calm day
  (d) a flock of geese in flight

(e) a log

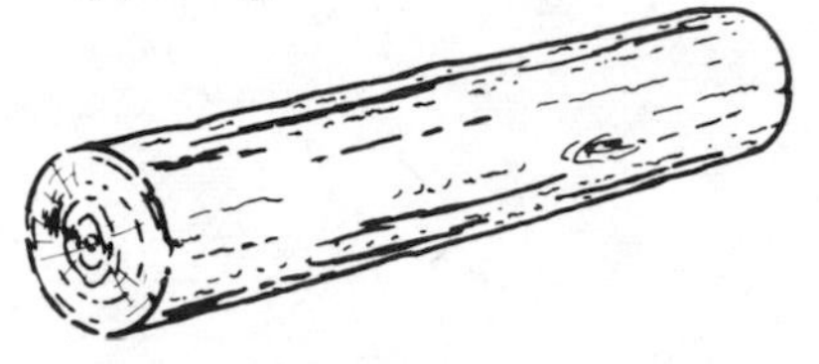

(f) the growth rings visible on the end of the log

2. Name a physical object, either natural or man-made, that suggests each of the following geometric figures:

(a) square | **(b)** cube
(c) sphere | **(d)** hemisphere
(e) parallel lines | **(f)** perpendicular lines
(g) triangle | **(h)** rectangle
(i) hexagon

3. Look for geometric figures in your room. Name both the physical object and the abstract geometric figure that it suggests.

4. **(a)** What is the smallest thing you have ever heard of? Can you imagine splitting it into pieces?

(b) Can you *draw* a point?

(c) Is the dot in the figure below a point?

(d) Is the wiggly symbol in the figure a number?

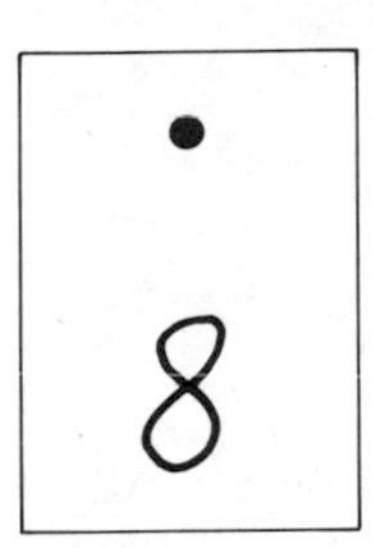

5. Below is a diagram of the set T of whole numbers less than 10.

0 2 4 6 8

1 3 5 7 9

T

(a) Draw a loop around the subset E of all even numbers.

(b) Draw a loop around the subset $A = \{1, 2, 4, 7\}$.

List (enumerate) the following subsets:

(c) $A \cup E = \{ \qquad \}$
(d) $A \cap E = \{ \qquad \}$
(e) $T - A = \{ \qquad \}$
(f) $E - A = \{ \qquad \}$
(g) $A - E = \{ \qquad \}$
(h) $T \cap A = \{ \qquad \}$

6. Match each symbol with the appropriate word or phrase:

(a) $\in$	(A) complement
(b) $\subset$	(B) empty set
(c) $\varnothing$	(C) intersection
(d) $\cap$	(D) is a member of
(e) $\cup$	(E) is a subset of
(f) $-$	(F) union

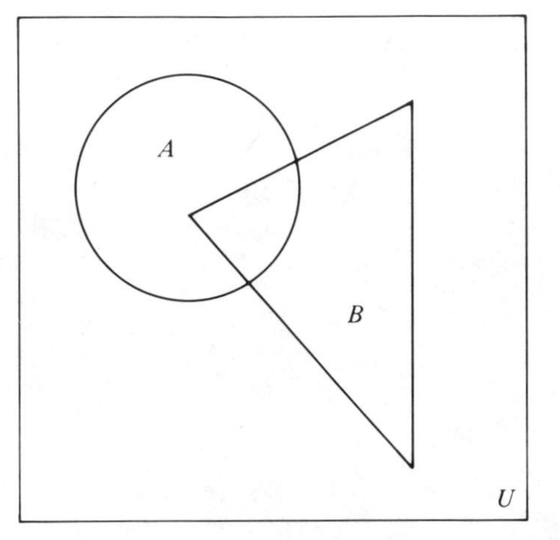

7. In the Venn diagram above, U represents the set of all points inside the rectangle, A all points inside the circle, and B all points inside the triangle. For each part below sketch a copy of the Venn diagram and then shade the indicated subset of U.

(a) $A \cup B$ **(b)** $A \cap B$
(c) $A - B$ **(d)** $U - B$
(e) $(U - A) - B$ **(f)** $A \cap (U - B)$

8. W represents the set of all whole numbers, I the set of all integers, and Q the set of all rational numbers. Mark each statement below as true or false.

(a) $6 \in W$ **(b)** $-3 \in I$
(c) $\frac{2}{3} \in I$ **(d)** $-\frac{7}{8} \in Q$
(e) $W \subset Q$ **(f)** $Q \subset I$
(g) $-5 \notin Q$ **(h)** $W - I = \varnothing$
(i) $-5 \in I - W$ **(j)** $I \not\subset W$
(k) Q and I are disjoint sets

9. Let C denote the set of all students in your math class, S the subset of all sophomores (in C), and G the subset of all students (in C) who wear glasses. Describe in words the following subsets of C.

(a) $S \cap G$ **(b)** $C - S$
(c) $S \cup G$ **(d)** $G - S$
(e) $(C - S) \cap G$

10. (Prime notation for complement in a universal set) We have used the notation $S - A$ for the complement of a set A in a set S. This notation makes sense for any two sets A and S, whether A is a subset of S or not.

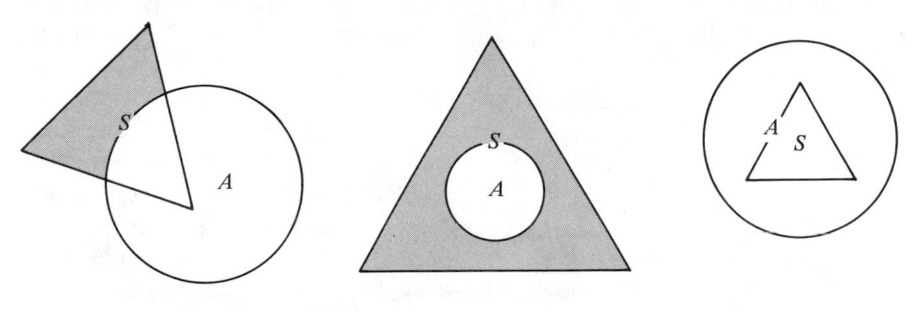

$S - A$ is the shaded set $S - A = \phi$

Very often it is clearly understood that some large set (or *universal set*), such as a plane or the set of all integers, includes every other set under discussion. When that is the case the complement of a given set A in the universal set U is written A' rather than $U - A$. For example, in the preceding exercise the large set is quite clearly C. Thus we could write $C - S$ as S' and $C - G$ as G'. We could *not* write $G - S$ as S', however, since G is not the universal set. Let us agree that in Exercise 5 the universal set is T. Enumerate

(a) A' **(b)** E'
(c) $(A \cap E)'$ **(d)** $A' \cup E'$
(e) $(A \cup E)'$ **(f)** $A' \cap E'$

1.2 LINES

Points are the most basic geometric figures. They are the material of which space consists. All other geometric figures are made up of points. After points the next most basic geometric figures are straight lines. Since most of our work will be with straight rather than curved or broken lines, *the word "line" will always mean straight* unless we specify differently.

Euclid described a line as "length without breadth." We can think of a line as the geometric figure suggested by the vapor trail of a jet plane, the horizon between sky and ocean, a telephone wire, or the juncture of the ceiling with one wall. None of these physical objects is a perfect model of the abstract idea of a line, however, because each is too "short." We should think of lines as *stretching endlessly far in each direction.* We should also think of them as being *straight* rather than being curved or having any corners. Finally, we should think of them as having *no breadth*. Another way of expressing this last property is as follows:

The removal of any point from a line "separates" the line into two pieces.

The description of a line that we have just given is an informal one. It is not a definition. Point, line, and plane are *undefined* abstract ideas. They are the basic geometric figures in terms of which other geometric figures will be defined.

Since points and lines are ideas rather than things, it is essential that we invent symbols to represent them. We can no more study points and lines without using symbols for them than we could study numbers without using some sort of numerals to represent them. (Think of two very large whole numbers and try to compute their product without writing any symbols for them.) Points are represented by dots and named by capital letters. Lines are represented by straight strokes and are usually named by lowercase script letters.

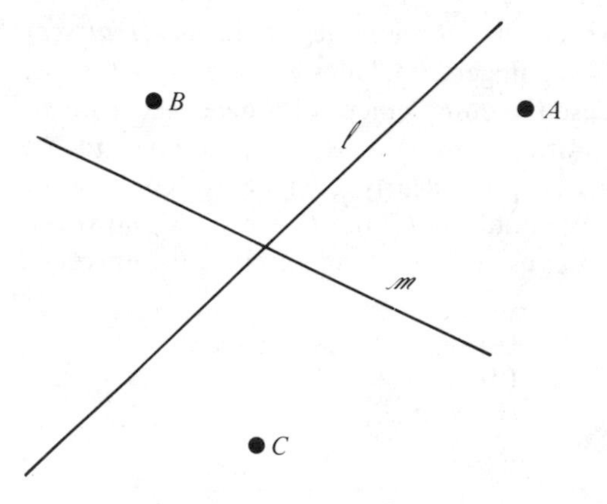

three points and two lines

We have made a decision about language. The choice was between being long-winded and logically precise, or being brief and trusting in your good judgement. We chose to be brief. Thus, instead of asking you to "make a straight stroke representing a line through the two dots representing points A and B," we shall ask you to "draw a line through points A and B." We know and you know that lines and points are really ideas, and that an idea cannot be drawn any more than it can be cut in half or painted red. We do not think it is worthwhile, however, to constantly remind each other about this distinction between ideas and the pictures which represent them.

The language of set theory is usually altered in the special case of geometry. If A is a point and ℓ is a line and $A \in \ell$, we usually say "A lies on ℓ" or "ℓ passes through A" rather than "A is a member of ℓ" or "ℓ contains A."

Just as any single line has infinitely many points on it, so any single point has infinitely many lines through it.

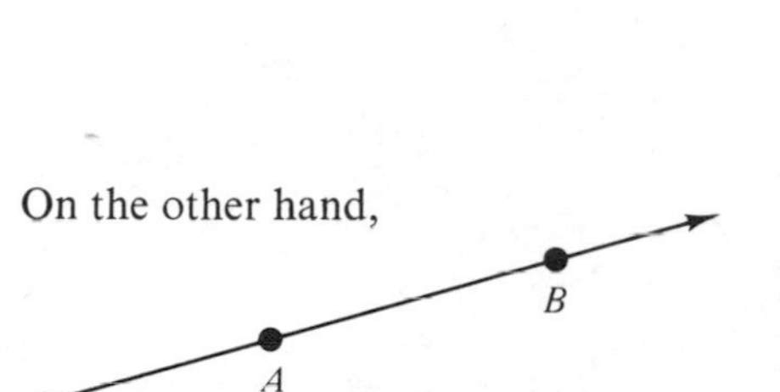

On the other hand,

Through any two points there is exactly one line.

This line is called the line **determined** by points A and B and is denoted by $\overleftrightarrow{AB}$. The two-headed arrow reminds us that a line extends endlessly far in each direction.

The corresponding situation for two lines is more complicated. Two lines may intersect in a point, or they may miss each other entirely.

The intersection of two lines is either the empty set or consists of a single point.

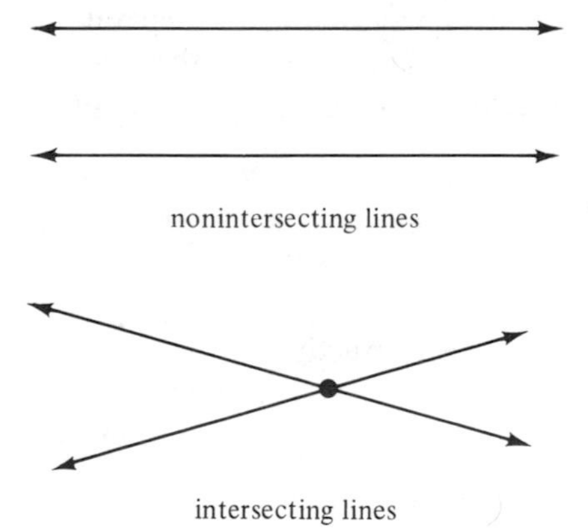

nonintersecting lines

intersecting lines

Two lines with a point in common are called **intersecting lines**. Two lines without a point in common are called **nonintersecting lines.**

Given three or more points, there may or may not be a line through all of them. If there is such a line, we say that the points are **collinear**. If there is no such line, we say that the points are **noncollinear.**

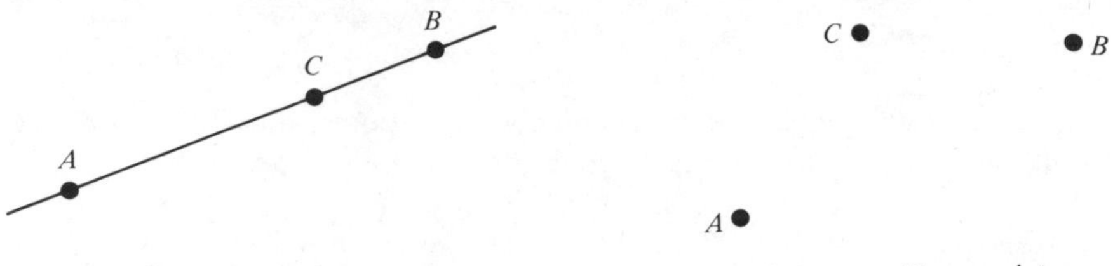

three collinear points   three noncollinear points

If three or more lines pass through a single point, they are said to be **concurrent**; otherwise, they are called **nonconcurrent**.

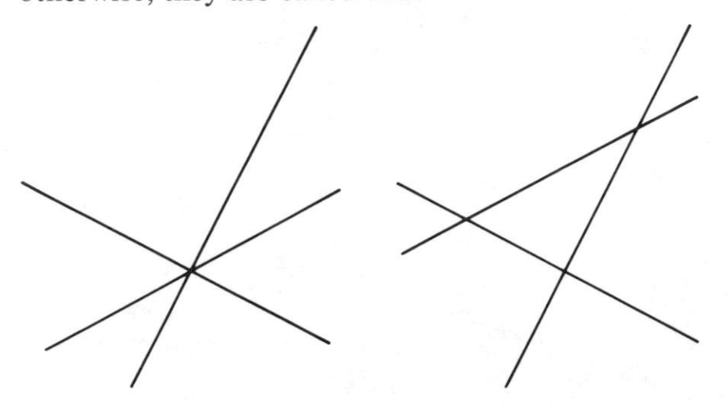

three concurrent lines   three nonconcurrent lines

It seems to be inevitable that much of the beginning work in geometry is concerned with establishing terminology. This can be quite dull unless it is done within a context in which there is room for some creativity or in which a puzzle can be solved. Many of the following exercises ask for original sketches or pose (hopefully interesting) counting questions that draw out the meanings of such terms as intersecting lines, collinearity, concurrence, and separation. The counting problems also give practice in systematic work and the recognition of patterns. Systematic counting procedures will be particularly useful when we study probability in Chapter 6.

EXERCISES

1. Decide if either of these statements

> **I.** Through any two given points there is exactly one line.
> **II.** The intersection of any two lines is either the empty set or consists of a single point.

is still true if
(a) "lines" are allowed to be curved
(b) "lines" are not allowed to be more than 3 in. long
(c) "lines" are allowed to have breadth
Draw pictures to illustrate your answers.

NOTE To say that lines are neither crooked, nor short, nor fat is not very mathematical. Exercise 1 shows that this same information can be provided by thoroughly mathematical statements. For example, statement I can be recast into the precise language of set theory as follows: Given any two points, A and B, there is exactly one line, ℓ, such that $A \in \ell$ and $B \in \ell$.

2. Can you *deduce* statement II above from statement I? That is, is II a logical consequence of I? Hint: Suppose that II were *not* true. Then there would be two lines whose intersection contained more than one point. Continue this *verbal* argument. Pictures are not permitted in a strictly logical deduction.

NOTE Exercise 2 is an example of what we mean by "localized deduction"—looking for logical relationships among just a few, here just two, geometric properties. In Exercise 5, p. 11, we look for such relationships in a somewhat larger body of properties. We make no attempt to tie *all* geometric properties together with logical threads. That is the program of an "axiomatic" geometry course such as the one you may have had in high school. No such plan is realistic in the elementary grades.

3. Try to draw three lines that have the following number of crossing points. If you claim something is impossible, defend that claim.
(a) 0 **(b)** 1 **(c)** 2 **(d)** 3 **(e)** 4

4. Try to draw four lines that have the following number of crossing points.
(a) 0 **(b)** 1 **(c)** 2 **(d)** 3 **(e)** 4 **(f)** 5 **(g)** 6 **(h)** 7

5. We say that two points "determine a line" because there is just one line containing them. The word "determine" is used in a slightly different sense when we say that "three noncollinear points determine three lines."
(a) Make a sketch to illustrate three noncollinear points "determining" three lines.
(b) How many lines are "determined" by four points no three of which are collinear?
(c) How many lines are determined by five points no three of which are collinear?
(d) Look for a pattern and decide how many lines are determined by 100 points no three of which are collinear.

6. (a) Write six other names for $\overleftrightarrow{AB}$.

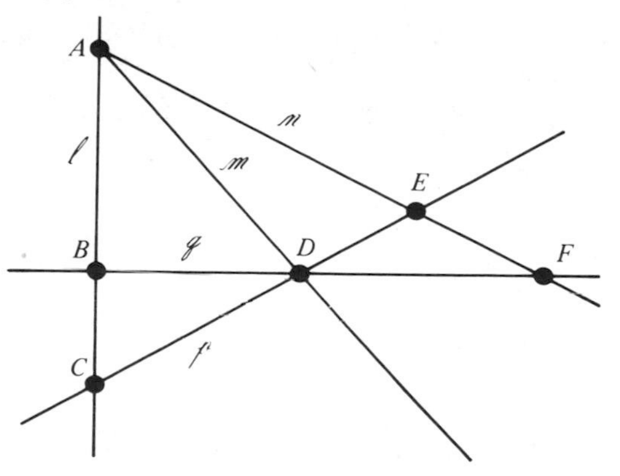

(b) Does this mean that there are seven lines, not one, through the two points A and B?

(c) Is it proper to write $\overleftrightarrow{AB} = \ell$? Why?

7. Referring to the previous figure, list the elements in each of the following sets

(a) $n \cap p$
(b) $m \cap n$
(c) $p \cap n$
(d) $\overleftrightarrow{AB} \cap \overleftrightarrow{DF}$
(e) $q \cap m \cap p$
(f) $m \cap n \cap q$

8. How many crossing points, other than A and B, will be formed by the lines in the figure below?

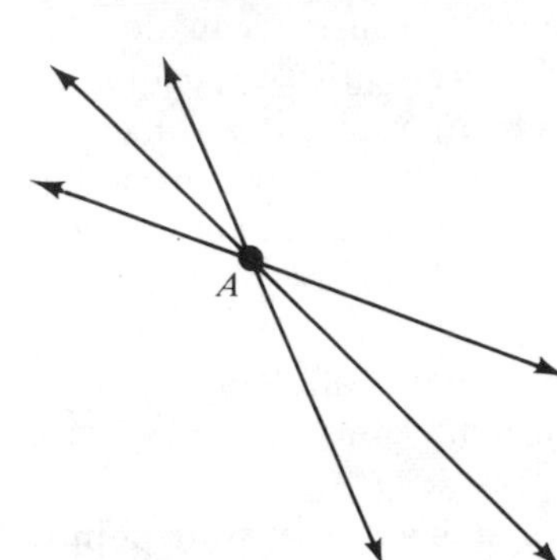

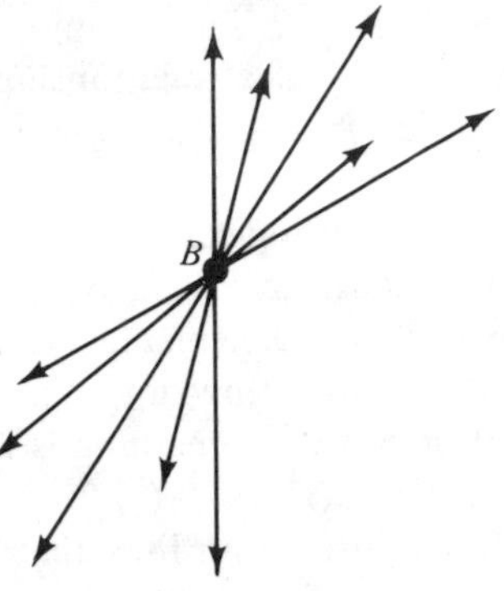

9. Decide how many points would have to be removed to "separate" the following figures into two pieces.

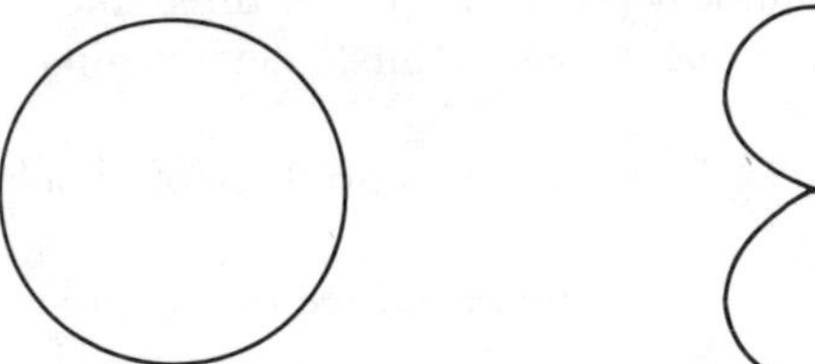

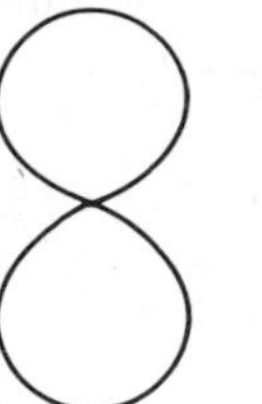

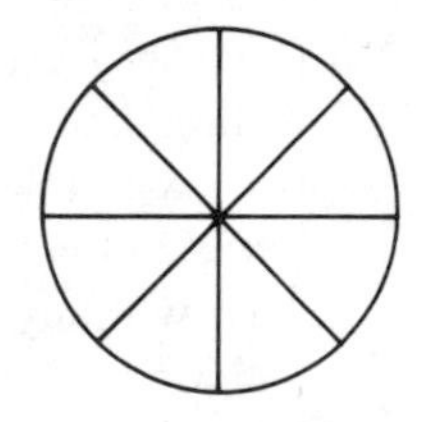

(a) the circle (rim only)
(b) the figure eight
(c) the wheel with spokes

10. Try to draw a figure "without breadth" that cannot be separated by the removal of two (well-chosen) points.

11. By removing one point at a time, could you ever separate this figure into two pieces?

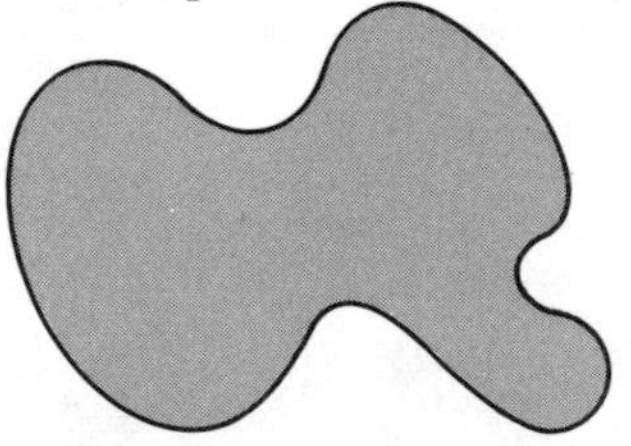

12. Removing one point from a line separates it into two pieces. Into how many pieces is a line separated by the removal of
(a) 2 points
(b) 3 points
(c) 4 points
(d) 875 points

1.3 PLANES

The third basic geometric figure is a plane. Plane, like point and line, is an undefined term, and we have to rely on our ideas about physical space to give it meaning: Physical objects that suggest the idea of plane are the surface of a pond on a calm day, the ceiling of a room, a flat sheet of paper.

Each of these objects is too " small," however, to represent a plane accurately. We should think of a plane as (1) *stretching endlessly far in all directions*, (2) being *flat*, and (3) having *no thickness*. Another way of expressing property (3) is as follows:

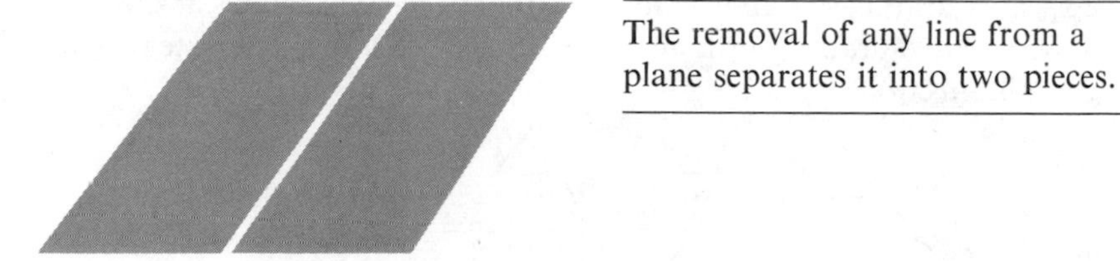

The removal of any line from a plane separates it into two pieces.

Properties (1) and (2) can also be rephrased more mathematically:

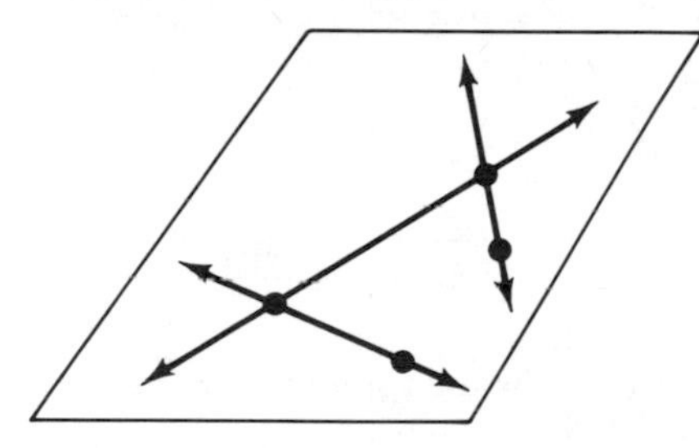

The line determined by any two points in a plane lies in that same plane.

plane

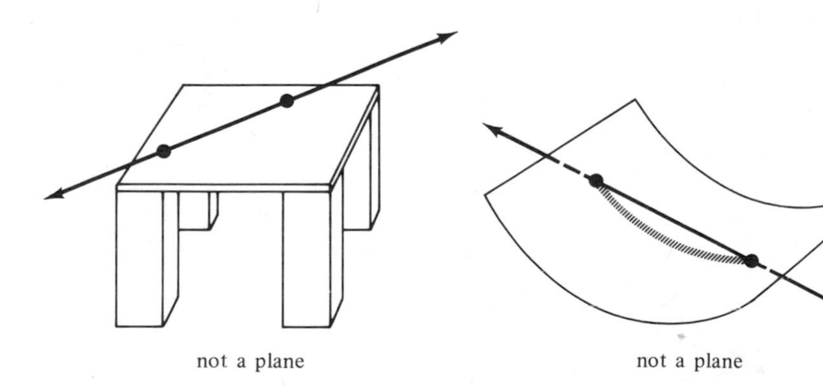

not a plane

not a plane

Planes are usually named by capital script letters such as $\mathscr{P}$, $\mathscr{R}$, $\mathscr{S}$ to distinguish them from points and lines. Again the language of set theory is abused slightly. A symbolic version of the last displayed property would be

$$\text{if } A \in \mathscr{P} \text{ and } B \in \mathscr{P}, \text{ then } \overleftrightarrow{AB} \subset \mathscr{P}$$

which would be read, in the language of set theory,

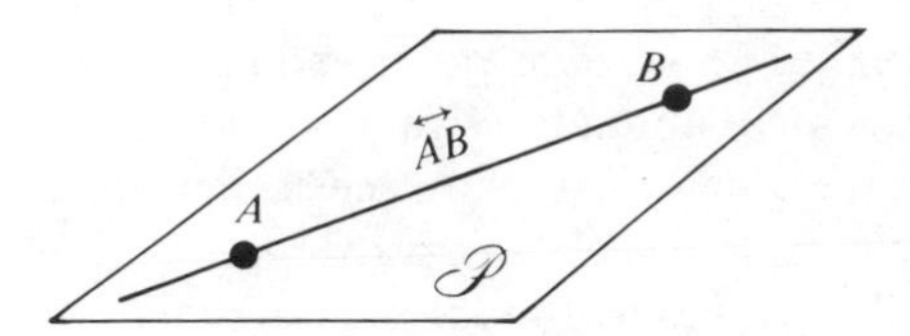

if point A is a member of plane $\mathscr{P}$ and point B is a member of plane $\mathscr{P}$, then the line determined by A and B is a subset of plane $\mathscr{P}$.

Instead of saying that a point is a member of a plane, we shall usually say that the point lies in the plane; instead of saying that a line is a subset of a plane, we shall usually say that the line lies in the plane. We represent planes by **perspective drawings** as above. Some textbooks include arrows at the edges of the perspective drawing to remind the students that a plane extends endlessly in all directions.

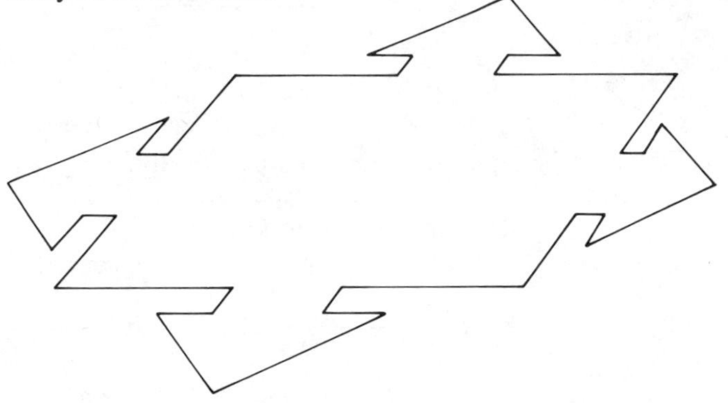

When we studied lines, we observed that through a single point there are infinitely many lines. How many planes do you think there are through a given line? How many planes do you think there are through a given point? Among all of the planes through the given line ℓ, how many do you think will also pass through the point A?

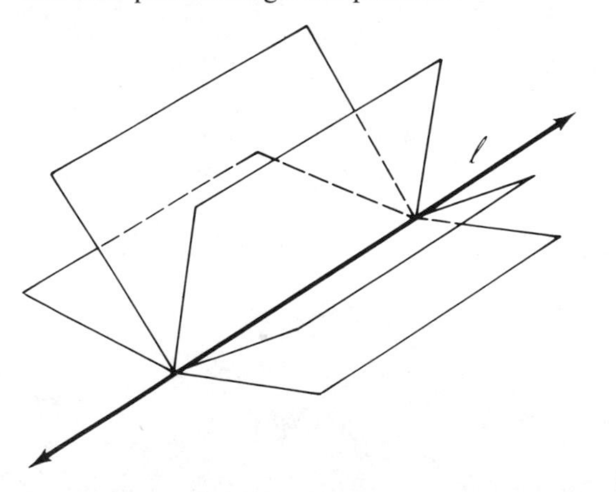

• A

Through any line and any point not on the line there is exactly one plane.

This plane is sometimes called the plane "determined" by ℓ and A. How many planes do you think there are through three given points?

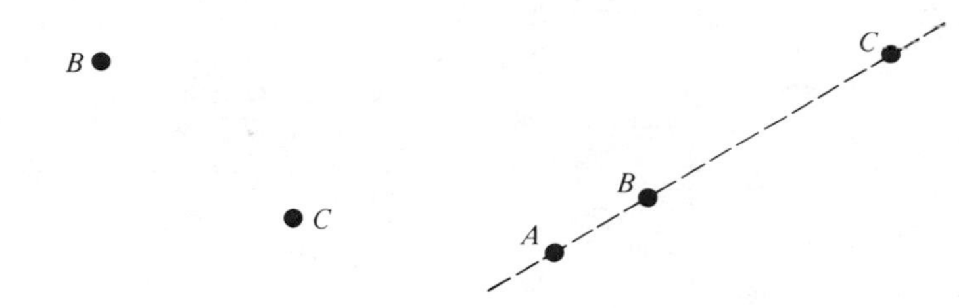

If the points are collinear there will be infinitely many, but

Through any three noncollinear points there is exactly one plane.

Do two lines "determine" a plane? The situation now is more complicated.

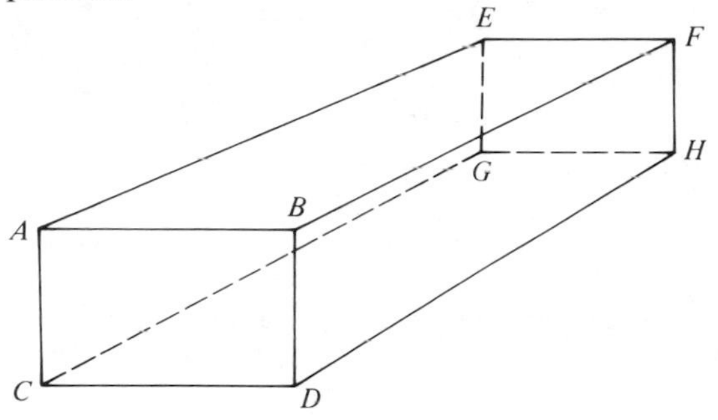

Lines $\overleftrightarrow{AB}$ and $\overleftrightarrow{BF}$ lie in a plane and intersect.
Lines $\overleftrightarrow{AB}$ and $\overleftrightarrow{CD}$ lie in a plane and do not intersect.
Lines $\overleftrightarrow{AB}$ and $\overleftrightarrow{CG}$ do not lie in a plane and do not intersect.

We can say for certain that

Two intersecting lines determine a plane.

That is, if two lines intersect, then there is exactly one plane in which they both lie. In set notation, if $\ell \cap m = \varnothing$, then there is exactly one plane $\mathscr{P}$ such that $\ell \cup m \subset \mathscr{P}$.

If two lines do not intersect, they may or may not lie in a plane. Nonintersecting lines that lie in a plane are called **parallel lines**. We write "$\ell \| m$" as shorthand for "ℓ is parallel to m." In the figure above, $\overleftrightarrow{AB} \| \overleftrightarrow{CD}$. Can you find some other pairs of parallel lines? Two lines that do not lie in a plane are called **skew lines**. In the figure above, $\overleftrightarrow{AB}$ and $\overleftrightarrow{CG}$ are skew. Can you find some other pairs of skew lines?

There is a notion of parallelism for planes as well as for lines. It is based on this property of planes in space: The intersection of two planes is either a line or is empty. If two planes have empty intersections, we say that they are **parallel** or nonintersecting. Thus:

Two planes are either parallel or intersect in a line.

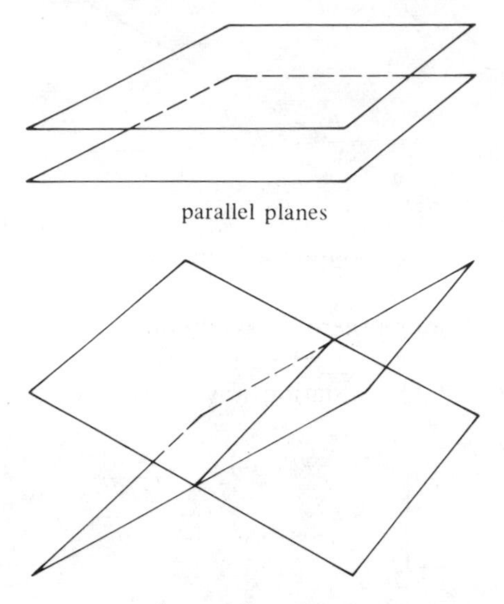

parallel planes

intersecting planes

Again we write "$\mathscr{P} \| \mathscr{R}$" as shorthand for "plane $\mathscr{P}$ is parallel to plane $\mathscr{R}$."

EXERCISES

1. (a) If $\ell \| m$ and $m \| n$, must $\ell \| n$?

(b) If $\mathscr{P} \| \mathscr{R}$ and $\mathscr{R} \| \mathscr{S}$, must $\mathscr{P} \| \mathscr{S}$?

2. (a) If point A and line ℓ (not through A) lie in plane $\mathscr{P}$, is there a line m in plane $\mathscr{P}$ such that (i) $A \in m$ and (ii) $m \| \ell$?

(b) If line ℓ does not intersect plane $\mathscr{P}$, is there a plane $\mathscr{R}$ such that (i) $\ell \subset \mathscr{R}$ and (ii) $\mathscr{R} \| \mathscr{P}$?

NOTE The assumption that there is exactly one line m having the properties of Exercise 2(a) is known to mathematicians as the Playfair postulate. This postulate plays a central role in axiomatic developments of Euclidean geometry, and altered forms of it underlie the non-Euclidean geometries known as Lobachevskian and Riemannian geometries. Despite its historical and mathematical importance, however, the Playfair property is no less (nor more) obvious to the elementary school child than any of a host of other geometric properties. There seems to be little point in emphasizing it in the elementary school.

3. Under what conditions do
   (a) three points determine a plane
   (b) three planes determine a point
4. Must there be a plane through four arbitrarily chosen points in space? (If there is, the rare case, the points are said to be **coplanar**.)
5. (a) Assuming that

   (1) through any line and any point not on it there is a plane
   (2) through any two points there is a line

   try to *deduce*

   (3) through any three noncollinear points there is a plane

   Remember that pictures and intuition are not allowed to play any essential role in a deduction. *All* that you have to work with here are properties (1) and (2).

   (b) Assuming (3) above and

   (4) on any line there are at least two points

   and

   (5) the line determined by any two points in a plane lies in that same plane

   try to *deduce* (1) above.

NOTE By now we have compiled a fairly long list of *properties* of points, lines, and planes. We have not divided them into the categories of *axioms* (assumed properties) and *theorems* (properties deduced from the assumed ones) as is done in high school geometry. Exercise 5 illustrates the arbitrary nature of such a division. Part (a) shows that if properties (1), (2) are axioms, then property (3) is a theorem; part (b) shows that if properties (3), (4), (5) are axioms, then property (1) is a theorem.

At the elementary school level all of these properties stand on an equal footing—there are no axioms and theorems—only properties. Perhaps at the junior high school level it is appropriate to begin pointing out the various logical dependencies among properties as we have done here in Exercise 5. No grand scheme of axioms seems appropriate until there has been considerable experience with this kind of localized deduction.

If you are curious, one currently popular grand axiomatic scheme can be found in *Elementary Geometry from an Advanced Standpoint* by E. E. Moise (Addison-Wesley, Reading, Mass., 1963).

6. (a) Can two skew lines ever intersect?
   (b) Are there such things as skew planes?
7. Which is easier to build, a three-legged stool that does not teeter or an ordinary four-legged chair that does not teeter? Why?
8. Drawing three-dimensional figures on a two-dimensional surface is not easy. As a teacher you will have to do it often. Here are some practice exercises. Using dotted lines and perspective, make drawings that suggest
   (a) a quarter pound of butter
   (b) a dog house

(c) one of the great pyramids of Egypt
(d) a cube
(e) a slice of pie
(f) two parallel planes
(g) two intersecting planes
(h) a line and a plane intersecting in a point
(i) a plane intersecting two other planes that are parallel to each other

9. Fill in the blanks:

GEOMETRIC FIGURE	NAMED BY	PICTURED BY
point	capital letter such as A	
line		straight stroke
plane		perspective drawing

10. Label each statement below as "reasonable" or "nonsense":
(a) $A \in \mathscr{P}$
(b) $\ell \in \mathscr{P}$
(c) $A \in \ell$
(d) $\ell \subset \mathscr{P}$
(e) $\mathscr{P} \cap \mathscr{R} = \ell$
(f) $\mathscr{P} \cap \mathscr{R} = A$
(g) $\ell \cap m = \{A\}$
(h) $\ell \cup m = \mathscr{P}$

11. Fill in the blanks with numbers:
(a) ________ points determine a line.
(b) ________ noncollinear points determine a plane.
(c) ________ intersecting lines determine a plane.
(d) ________ parallel lines determine a plane.

12. Into how many pieces is a plane separated by the removal of
(a) two parallel lines
(b) two intersecting lines
(c) three parallel lines
(d) three lines, two of which are parallel
(e) three concurrent lines
(f) three nonconcurrent lines no two of which are parallel

13. Can a plane be separated into five parts by the removal of
(a) three lines
(b) four lines

14. Into how many pieces is space separated by the removal of
(a) two parallel planes
(b) two intersecting planes
(c) three parallel planes
(d) three planes, two of which are parallel
(e) three planes that have one line in common
(f) three planes that have only a point in common
(g) three planes that cross in three parallel lines?

15. A "small piece" of a line can be isolated by removing two points.

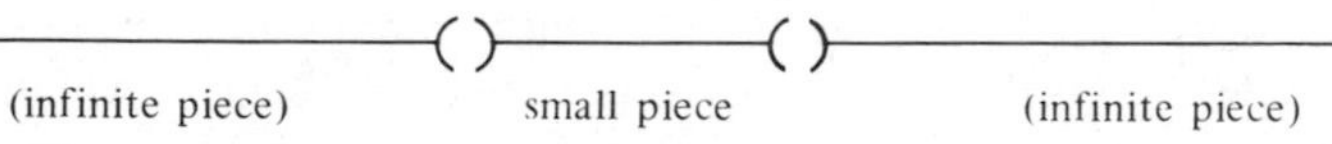

(a) How many lines must one remove from a plane to isolate a "small piece" of that plane?

(b) How many planes must one remove from space to isolate a "small piece" of space?

16. In this and the previous section we have listed many properties of points, lines, and planes. None of these properties were new or surprising, because all of them were suggested by our intuition about physical space. But we noted earlier that from a purely mathematical point of view there is no need to base geometry on physical space. For this exercise "space" consists of just four "points"—A, B, C, D—which you may want to visualize as the vertices of a tetrahedron:

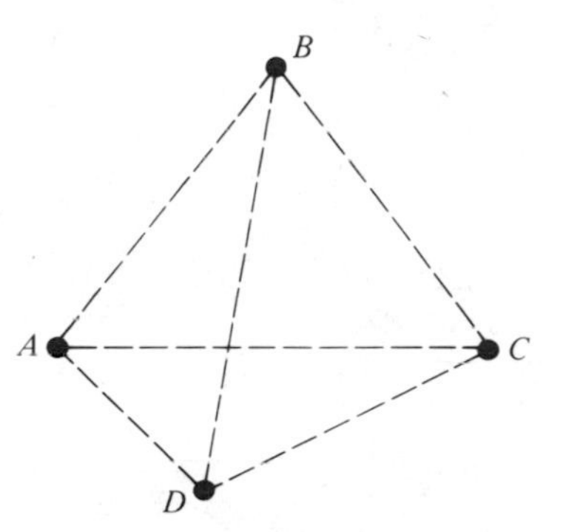

The "lines" are all possible two-element subsets of space. The "planes" are all possible three-element subsets of space.

(a) List all the "lines" in this geometry.

(b) List all the "planes" in this geometry.

(c) Nine properties of points, lines, and planes have been "displayed" in the text of Sections 1.2 and 1.3. Which of these nine properties are *not* true of the "four-point geometry"?

1.4 HALF-LINES AND RAYS

We have studied the three basic geometric figures: points, lines, and planes. These figures will now be used as the raw materials from which we shall build more geometric figures. The tools we shall use are the operations of set theory: complementation, intersection, and union. [In mathematical language, while points, lines, and planes are *un*defined, the figures we are about to study are defined.]

The first new figures we build are half-lines.

If a point is removed from a line, each of the two remaining pieces is called a **half-line**.

In the language of set theory, the complement of a point in a line (a separated or disconnected set) is the union of two connected sets, each of which is called a half-line.

Two half-lines determined by removing a point from a line.

The definition of half-line as well as several of the remarks made in previous sections depend on the intuitive notion of **connectedness**. We shall not try to define this notion,* but it might be helpful to think of it in this way. If you were a tiny bug you could walk from any point of a connected figure to any other point of the figure without ever leaving the figure.

connected figures

disconnected figures

We can use the concept of half-line to define what we mean when we say that one point is "between" two other points. *Caution: The notion of betweenness for points applies only to collinear points.*

Point A is **between** points B and C (A, B, C collinear) if B and C lie on different half-lines determined by A.

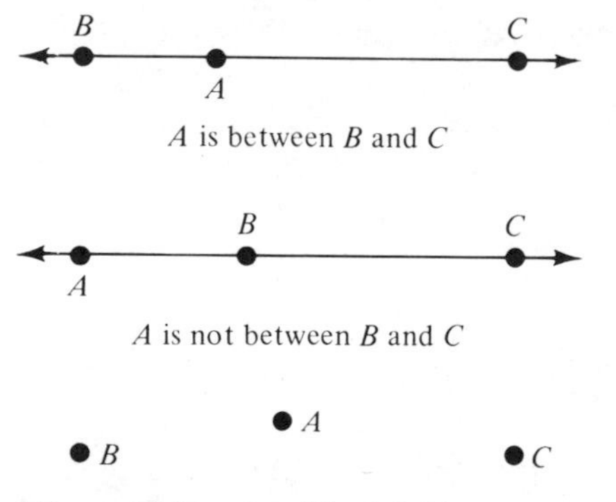

A is between B and C

A is not between B and C

* See p. 88 for a possible definition.

If A and B are two points, then the half-line determined by A that contains B, also called the half-line from A through B, is denoted by $\overset{\circ\!\!\longrightarrow}{AB}$.

half-line from A through B

The hollow circle at the tail of the arrow is to remind us that A does *not* belong to this half-line. If we unite A with this half-line, the resulting figure is called the ray from A through B, and is denoted by $\overset{\bullet\!\!\longrightarrow}{AB}$, or more commonly by $\overrightarrow{AB}$.

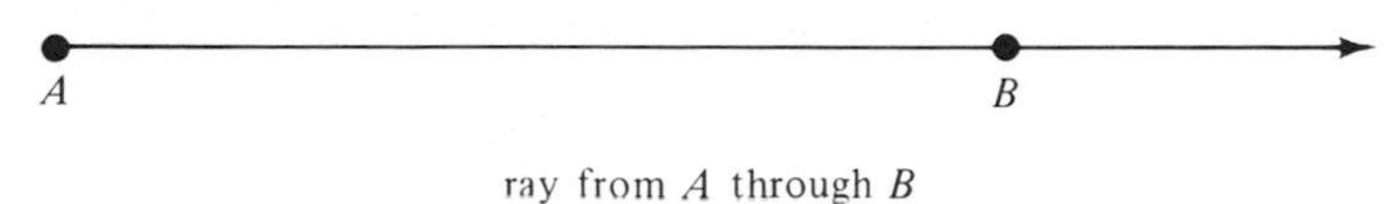

ray from A through B

The solid circle at A reminds us that A *does* belong to the ray.

The **ray from** A **through** B, $\overrightarrow{AB}$, is the union of $\{A\}$ with the half-line from A through B.

The arrow in the symbols $\overset{\circ\!\!\longrightarrow}{AB}$ and $\overrightarrow{AB}$ is always directed from left to right no matter which way the actual ray from A through B points.

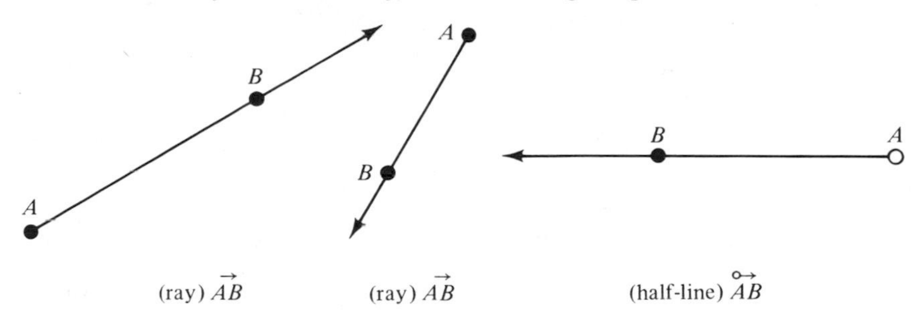

(ray) $\overrightarrow{AB}$ (ray) $\overrightarrow{AB}$ (half-line) $\overset{\circ\!\!\rightarrow}{AB}$

This is a notational convention. Of course, the symbol $\overleftarrow{BA}$ would suggest the ray from A through B as surely as $\overrightarrow{AB}$ does.

EXERCISES

1. **(a)** Name some physical "objects" that suggest the idea of ray.
   (b) Why do you think the word "ray" was chosen to name the geometric figure that it does?
2. **(a)** Do you think that between any two points on a line, no matter how close together they may be, there is another point?
   (b) Is there such a thing as a "left-most" point on the half-line $\overset{\circ\!\!\longrightarrow}{AB}$, where A and B are located as below?

(c) Is there a left-most point on the ray, $\overrightarrow{AB}$? The property of space that

between any two points there is another point

is known as the **density** property.

3. (More interpretations for the multipurpose word "determine")
 (a) Do two points determine *a* ray?
 (b) Does an ordered pair of points determine a ray?
 (c) What does it mean to say that two points A and B determine two rays?

4. Name all of the rays determined by the points shown below.

B

A C

5. Name all of the rays determined by the points A, B, and C shown below. How many are there?

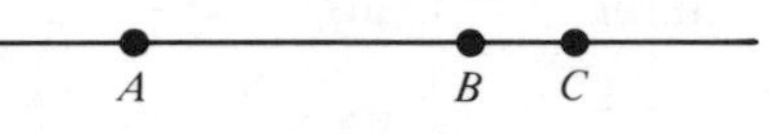

6. Sketch figures and count how many rays are determined by four points if
 (a) the four points are collinear
 (b) three of the points are collinear
 (c) no three of the points are collinear
 (Compare with Exercise 5(b), p. 11.)

7. Sketch figures and determine the maximum number of (connected) pieces into which a plane can be separated by the removal of
 (a) one ray
 (b) two rays
 (c) three rays
 (d) four rays

8. Sketch figures to illustrate all of the different ways in which two rays can intersect.

9. **(a)** Pick a point *inside* the figure above and draw a ray from it so that the ray never passes through a "corner." How many points of intersection did you make?
 (b) Repeat (a) with a different point.
 (c) Repeat (a) with still a different point.

10. Repeat Exercise 9 picking the three points *outside* the figure.

11. What do you notice about the numbers of intersection points in Exercises 9 and 10? Account for any pattern you observe.

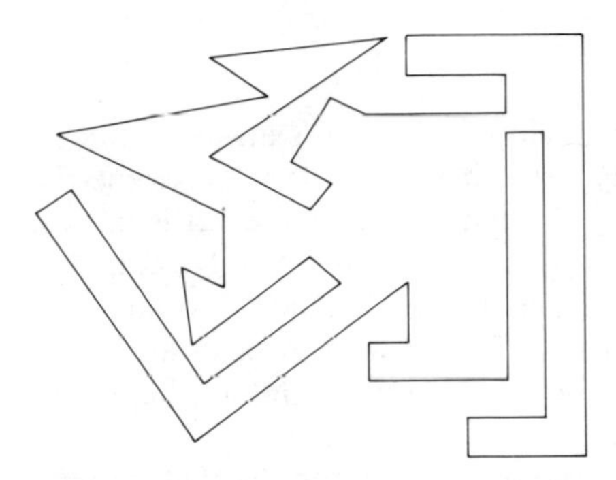

1.5 SEGMENTS

The simplest way to think of a segment is this:

A ●————————————● B

> A **segment** is a figure consisting of two points and all of the points between them.

The segment consisting of points A and B and all points between them is called "segment AB" and is written $\overline{AB}$, or sometimes $\overset{\bullet\!-\!\bullet}{AB}$. The points A and B are called its **end points.**

Strictly speaking, the figure we have just described is a **closed segment.** We can think of closed segments as being manufactured from rays by the set operation, intersection.

A B

$$\overline{AB} = \overrightarrow{AB} \cap \overrightarrow{BA}$$

By intersecting half-lines and rays in various combinations we can manufacture other types of segments. An **open segment** is produced by intersecting two half-lines.

A B

$$\overset{\circ\!-\!\circ}{AB} = \overset{\circ\!\rightarrow}{AB} \cap \overset{\circ\!\rightarrow}{BA}$$

This intersection is called "open segment AB," written $\overset{\circ\!-\!\circ}{AB}$. The easiest way to think about it is as (closed) segment $\overline{AB}$ with the end points dropped off.

Half-open segments are produced by intersecting rays and half-lines. The notation suggests which end point is missing.

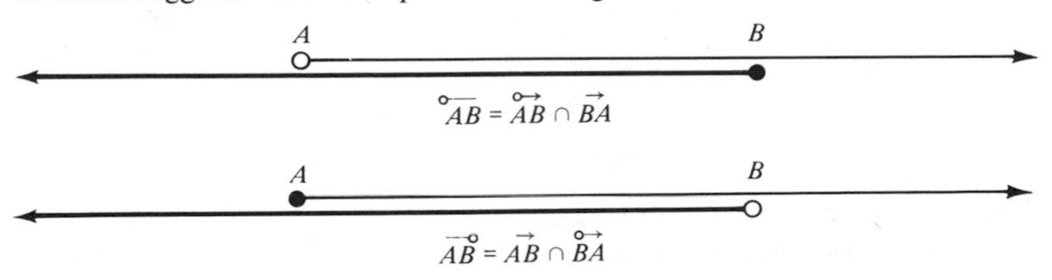

$$\overset{\circ\!-}{AB} = \overset{\circ\!\rightarrow}{AB} \cap \overrightarrow{BA}$$

$$\overset{-\!\circ}{AB} = \overrightarrow{AB} \cap \overset{\circ\!\rightarrow}{BA}$$

When the single word "segment" is used, closed segment is always meant.

By now we have built up a fairly extensive vocabulary of geometric terms, and the list will be even longer by the end of the chapter. It is fair to ask what value there is in learning so many technical words. Surely geometry for elementary school *children* should not be directed explicitly toward the learning of vocabulary. The children must be free to work with ideas. Their imagination and creativity should not be hobbled by a lot of linguistic strictures.

The reason we are looking at terminology so carefully is that correct geometric usage, like ordinary English usage, is best taught by example. If the *teacher* knows and uses the correct language, it will rub off on the children. On the other hand, a teacher who blithely refers to segments as lines is doing the students the same sort of disservice as the one who greets them in the morning with a cheery "Ain't it a nice day today?"

EXERCISES

1. Draw two points A and B.
   (a) Does $\overline{AB} = \overline{BA}$?
   (b) Does $\overrightarrow{AB} = \overrightarrow{BA}$?
   (c) Does $\overleftrightarrow{AB} = \overleftrightarrow{BA}$?
2. Interpret the diagram below, which suggests how various geometric figures can be manufactured from the raw materials of points and lines by means of set operations.

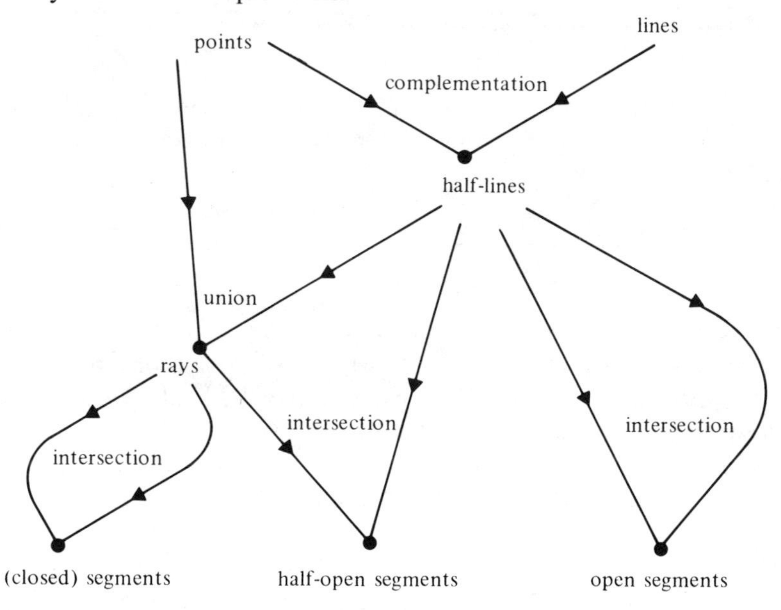

3. Two segments are said to be **parallel** if they lie on parallel lines, **skew** if they lie on skew lines, and **collinear** if they lie on a single line. In the figure that follows,

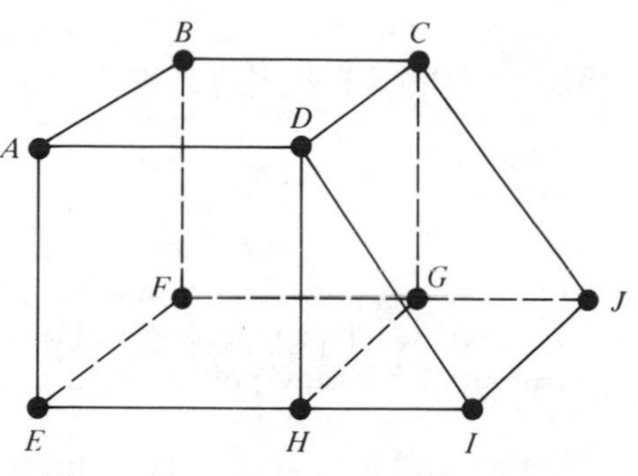

(a) name a pair of parallel segments
(b) name a pair of skew segments
(c) name a pair of collinear segments
(d) name a pair of segments that are neither parallel nor skew nor collinear.

4. Classify each of these statements about the previous figure as true or false.

(a) $J \in \overrightarrow{EI}$
(b) D is between A and I
(c) $\overline{AD} \cap \overline{DC} = \{D\}$
(d) $\overline{DC} \| \overline{IH}$
(e) $\overset{\circ\!\!\longrightarrow}{FG} = \overset{\circ\!\!\longrightarrow}{FH}$
(f) $\overline{EI} = \overline{EH} \cup \overline{HI}$
(g) $\overline{EI} = \overset{\longrightarrow\!\!\circ}{EH} \cup \overline{HI}$
(h) $\overline{FG} \subset \overset{\circ\!\!\longrightarrow}{FG}$
(i) A is between A and B
(j) $\overline{GH} = \overline{JI}$

5. Write the figure below as the union of these (pairwise) disjoint sets:

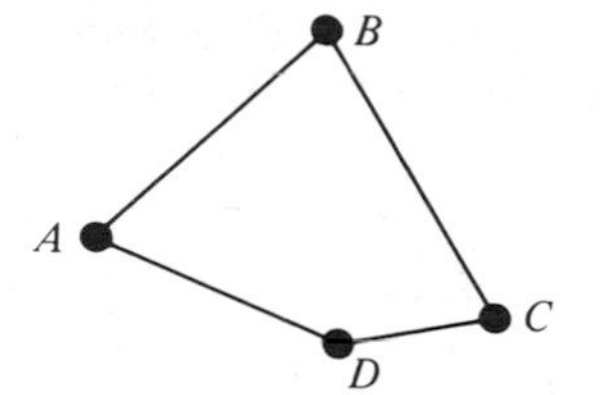

(a) four half-open segments
(b) four points and four open segments
(c) two (closed) segments and two open segments

6. (a) Write the figure below as the union of two disjoint figures of types we have already studied.
(b) Repeat (a) for three (pairwise) disjoint figures.

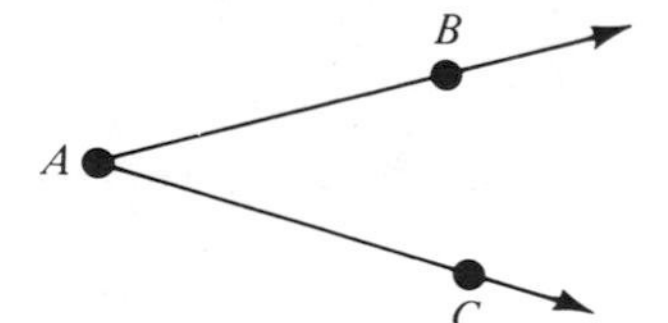

7. Draw three noncollinear points, A, B, and C, and join them by segments.
(a) How many segments did you draw?
(b) Excluding end points, how many crossing points were formed by the segments?

8. Repeat Exercise 7 for four points, A, B, C, D no three of which are collinear.

9. If in your drawing for Exercise 8 there was one crossing point, redraw the four original points so that there are no crossing points. If in your drawing for Exercise 8 there were no crossing points, redraw the four original points so that one crossing point will be formed.

10. Repeat Exercise 7 for five points, no three of which are collinear.

11. Rearrange the five points of Exercise 10 so that the connecting segments cross as few times as possible. How few crossing points could you make?

12. Rearrange the five points of Exercise 10 so that the connecting segments cross as many times as possible. How many crossing points could you make?

13. Try to draw six points, no three of which are collinear, so that the segments joining them will form only three crossing points. (You are an electrician and you want to arrange six terminals on a board in such a way that the wires joining them will cross each other a minimum number of times.)

TEACHING NOTE Mathematics is not cut and dried and neatly packaged away with all answers known. In fact, elementary school geometry abounds with questions that are easy to pose, but very difficult to answer. Exercises 7–13 suggest some such questions that could be posed to a second or third grader.

1.6 HALF-PLANES AND HALF-SPACES

Removing a point from a line separates it into two pieces each called a half-line. Removing a line from a plane separates it into two pieces* each of which is called an **open half-plane**.

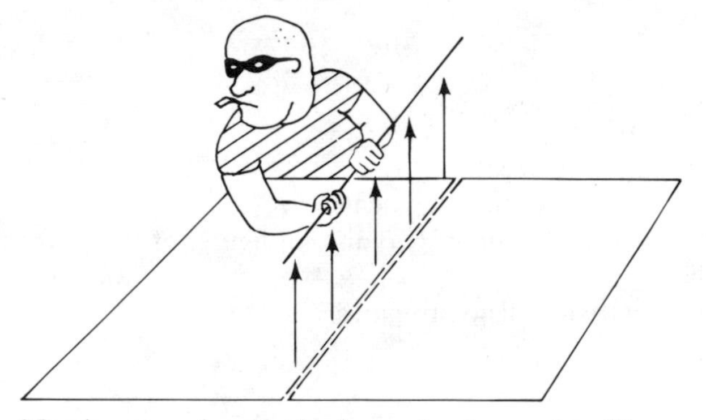

* In rigorous axiomatic developments of geometry, this property is stated as the *plane separation postulate*:

> Given a line ℓ and a plane $\mathscr{P}$ that includes it, there is a (unique) pair of sets H_1, H_2 with the properties
>
> **1.** H_1, H_2 are convex
> **2.** $H_1 \cup H_2 = \mathscr{P} - \ell$
> **3.** if $P \in H_1$ and $Q \in H_2$, then $\overline{PQ} \cap \ell \neq \varnothing$

A definition of convexity is given in Chapter 3 (p. 104).

The union of an open half-plane with the line that determined it is called a **closed half-plane**.

Closed half-planes are analogous to rays. In fact, most mathematicians would probably prefer the term "closed half-line" to the word "ray." The word "ray" is well established historically, however, and does not seem ready to disappear.

In space there are corresponding concepts, but now the pictures are harder to draw. Removing a plane from space separates it into two pieces each of which is called an **open half-space**.

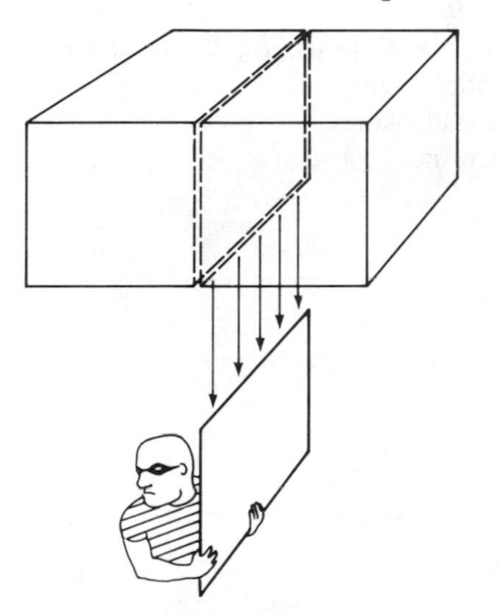

The union of an open half-space with the plane that determined it is called a **closed half-space**.

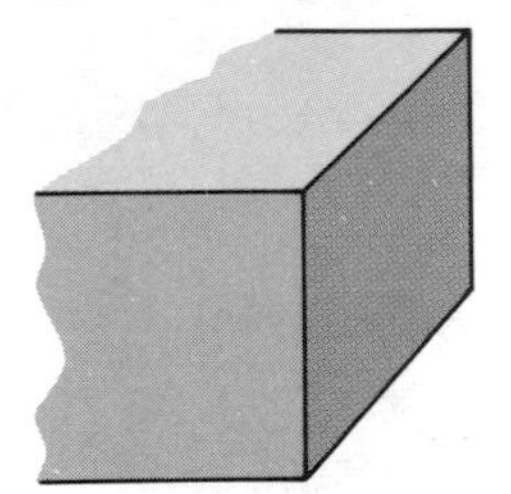

The exercises below suggest how various common geometric figures can be described in terms of half-planes and half-spaces.

EXERCISES

1. The figure below suggests several half-planes which in this exercise you may think of as either open or closed as you prefer.

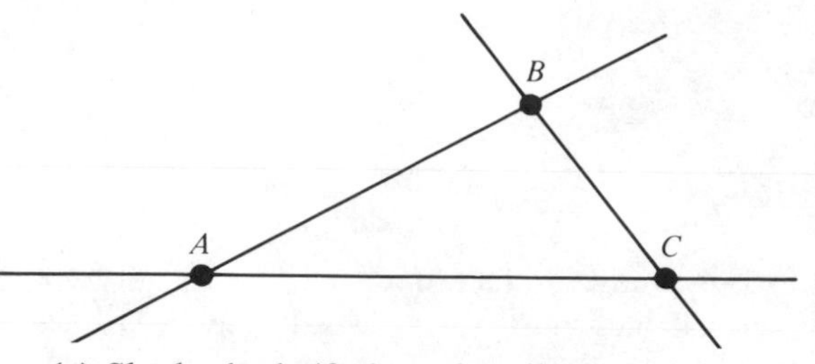

(a) Shade the half-plane described by the phrase "C's side of $\overleftrightarrow{AB}$."
(b) Describe the other five half-planes with phrases like that in part (a).
(c) Shade the intersection of "C's side of $\overleftrightarrow{AB}$" with "B's side of $\overleftrightarrow{AC}$."
(d) Copy the figure and shade the intersection of "the side of $\overleftrightarrow{AB}$ opposite C" with "the side of $\overleftrightarrow{AC}$ opposite B."
(e) Describe the **interior** of the triangle as an intersection of half-planes.
(f) Describe its **boundary** as a union of segments.
(g) Describe its **exterior** as a union of half-planes.

2. (a) Describe the "polygon" below as a union of segments.

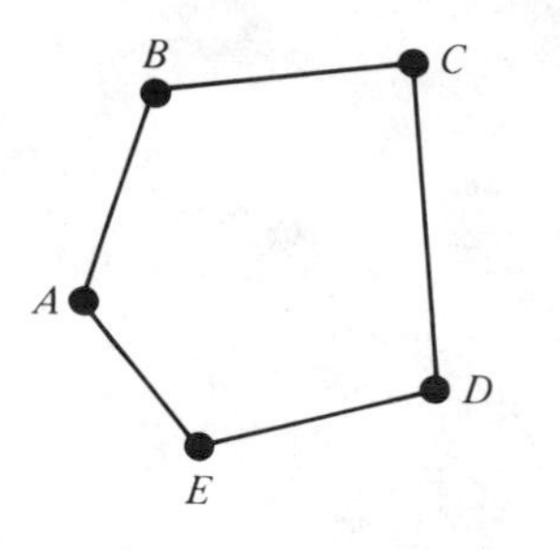

(b) Describe its interior as an intersection of half-planes.
(c) Describe its exterior as a union of half-planes.

3. Try to describe the interior of the polygon below as an intersection of half-planes.

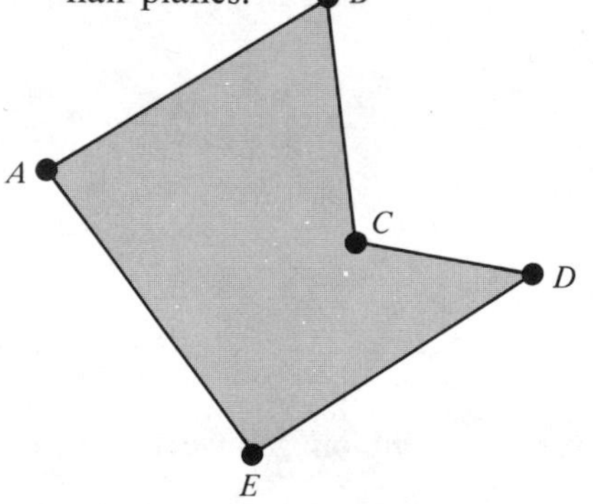

4. The figure below suggests several half-spaces which in this exercise you may think of as either open or closed as you please.

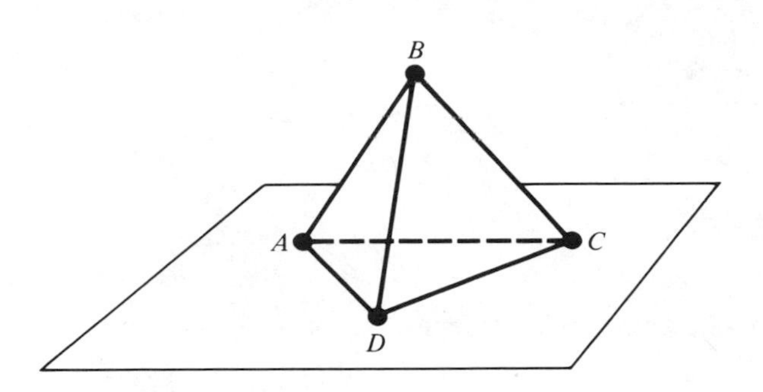

(a) Describe in words "*B*'s side of plane *ACD*."
(b) Describe the other seven half-spaces with phrases like that in (a).
(c) Describe the interior of the pyramid as an intersection of half-spaces.

5. Describe the slab of space between the parallel planes below as an intersection of half-spaces.

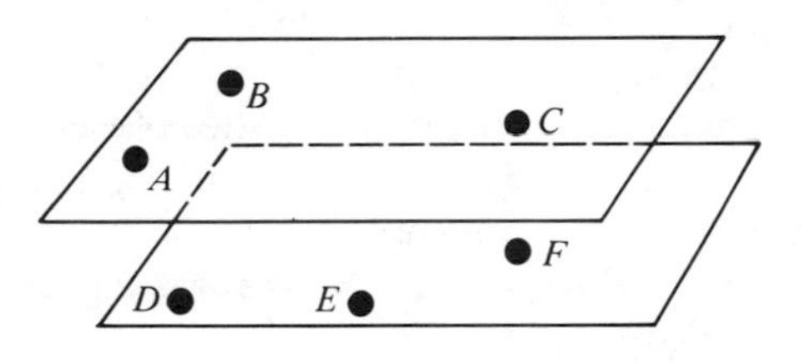

6. Describe the (solid) cube below as an intersection of half-spaces.

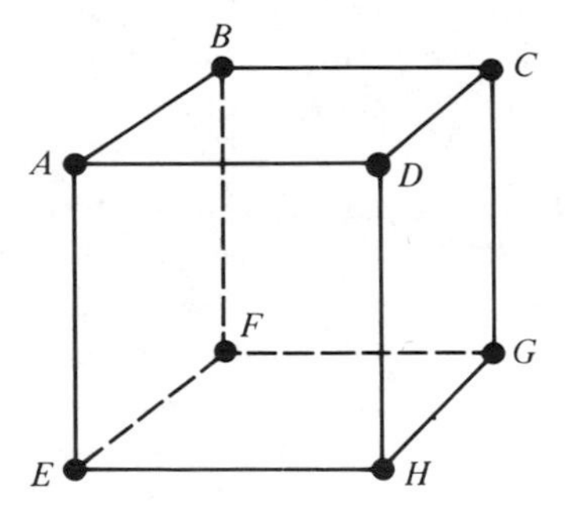

7. Describe the "upper left" infinite wedge of space between the intersecting planes (p. 30) as an intersection of half-spaces.

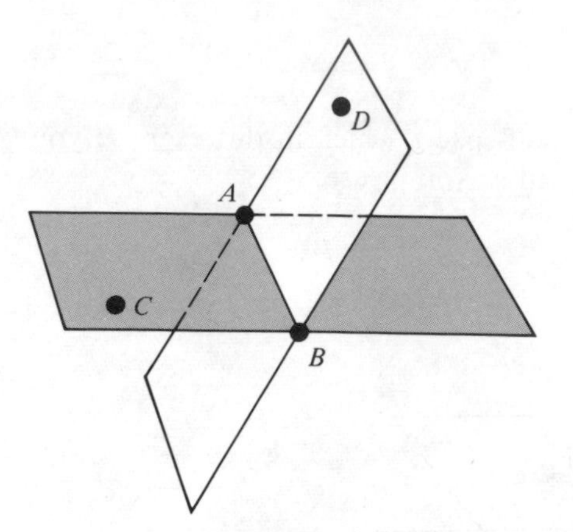

8. Describe the solid figure below as an intersection of half-spaces.

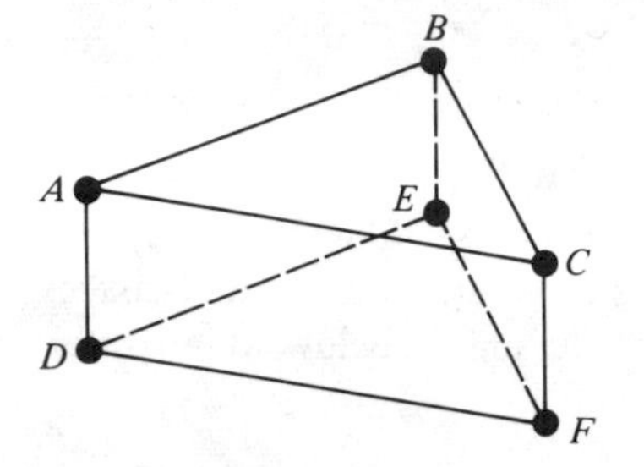

***9.** (Polya's problem)

(a) Removal of one plane from space separates it into two pieces. Into how many pieces can space be separated by the removal of two planes? three planes? four planes? five planes? six planes? seven planes? Unless a person has unusually good spatial perception, these questions become very difficult beyond about three planes. Maybe answering similar but easier questions will shed some light.

(b) Into how many pieces can a plane be separated by the removal of one line? two lines? three lines? four lines? five lines? six lines? seven lines? These questions also get quite difficult beyond about four lines. Maybe answering similar but easier questions will shed some light.

(c) Into how many pieces is a line separated by the removal of one point? two points? three points? four points? five points? six points? seven points?

(d) Look for patterns in what answers you have and extend those patterns to fill in the chart below.

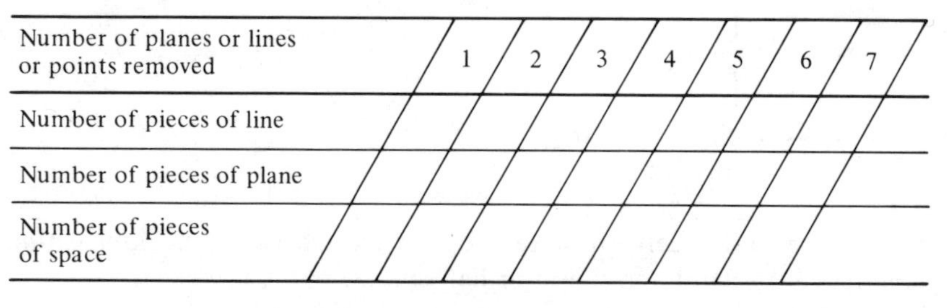

Number of planes or lines or points removed	1	2	3	4	5	6	7
Number of pieces of line							
Number of pieces of plane							
Number of pieces of space							

**(e) Try to write down formulas for the number of pieces into which space [a plane, a line] can be separated by the removal of n planes [lines, points].

1.7 ANGLES

There are two kinds of figures that are called angles. One is suggested by the hands on a clock. This figure is referred to simply as an "angle." The other is suggested by a partly opened book. This figure is called a "dihedral angle."

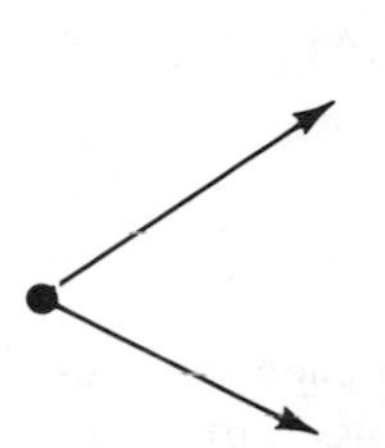

an angle

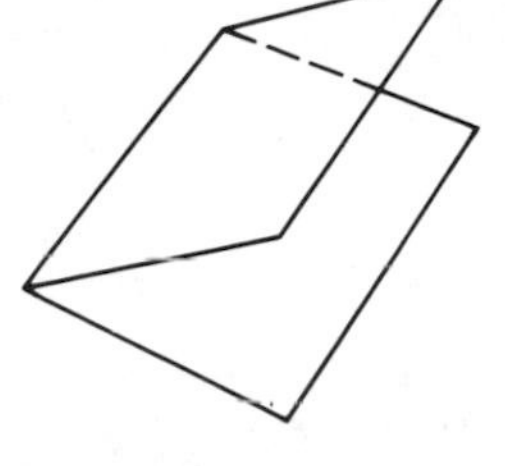

a dihedral angle

Careful definitions can be formulated in terms of rays and half-planes.

An **angle** is the union of two rays having a common end point. The common end point is called the **vertex** of the angle. The two rays are called its **sides**.

NOTE When we say "*two* rays" we really mean it. We do not recognize a "zero angle."

A **dihedral angle** is the union of two closed half-planes having a common edge.

Dihedral angles will not be of much use to us in our future work, and so we restrict our attention to (ordinary) angles.

Since an angle is the union of two rays, we could name it accordingly. For example, we could refer to the angle

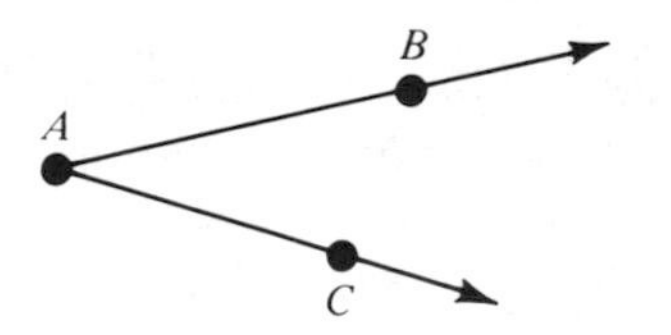

as $\overrightarrow{AB} \cup \overrightarrow{AC}$. That is not the usual way of naming angles, however. The usual way is first to make a little mark that looks like an angle ($\angle$) and then write down, in order, the names of a point on one side of the angle, the vertex, and a point on the other side of the angle. Thus the previous angle would be named

$\angle BAC$ or $\angle CAB$

The symbol "$\angle BAC$" is read "angle B, A, C."

Sometimes it is permissible simply to make the angle symbol and name the vertex. The above angle would be named "$\angle A$." At other times this abbreviated notation is ambiguous and inappropriate. In the figure below, what angle does "$\angle A$" name?

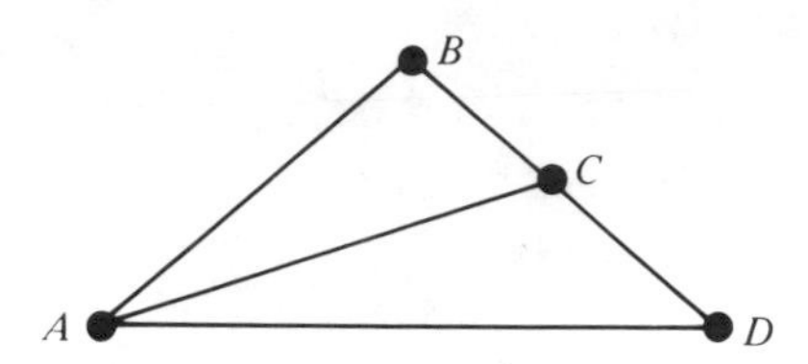

We cannot decide. If the "largest" angle with vertex A is intended, it must be named "$\angle BAD$" or "$\angle DAB$." One of the "smaller" angles with vertex A could be named "$\angle CAD$" or "$\angle DAC$." Is there any ambiguity in the symbol "$\angle B$"? "$\angle C$"? "$\angle D$"?

Removing an angle from a plane separates the plane into two pieces. [Notice that this property would be false if we recognized a "zero angle."] The "smaller" piece is called the "interior" of the angle; the "larger," the "exterior."

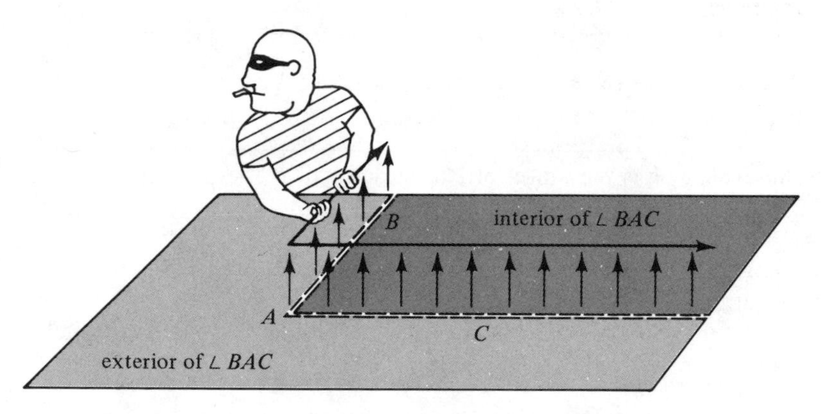

The interior and exterior of an angle need not be defined in such vague terms as "smaller" and "larger." Respectable set theoretic definitions are possible.

The **interior** of $\angle BAC$ is the intersection of the two open half-planes: B's side of $\overleftrightarrow{AC}$, and C's side of $\overleftrightarrow{AB}$. The **exterior** is the union of the two open half-planes: the side of $\overleftrightarrow{AC}$ opposite B, and the side of $\overleftrightarrow{AB}$ opposite C.

EXERCISES

1. How do you know that all points of an angle lie in the same plane?
2. Do three points determine *an* angle?
3. Do two lines determine *an* angle?
4. Does $\angle BAC$ below have an interior?

$\angle BAC$ is referred to as a **straight angle.**

5. Suppose that $\angle ABC$ lies in plane $\mathscr{P}$. Define exterior of $\angle ABC$ in terms of $\mathscr{P}$, $\angle ABC$, interior $\angle ABC$, and appropriate set operations.

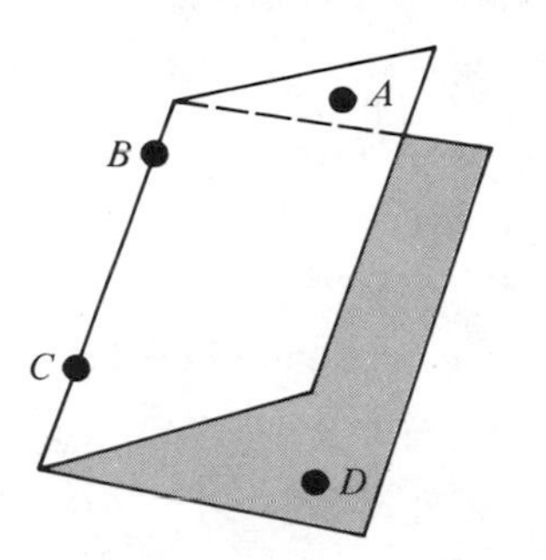

6. Define the interior of the dihedral angle below in terms of half-spaces and set operations.
7. Draw three noncollinear points A, B, and C. Name three angles "determined" by these points.
8. How many angles are determined by four points no three of which are collinear?
9. How many angles are determined by five points no three of which are collinear?

*10. Look for a pattern in your answers to Exercises 7, 8, and 9 and extend it to six points, seven points, and so on.

Certain important *pairs of angles* have been given special names. Two angles are said to be **adjacent** if

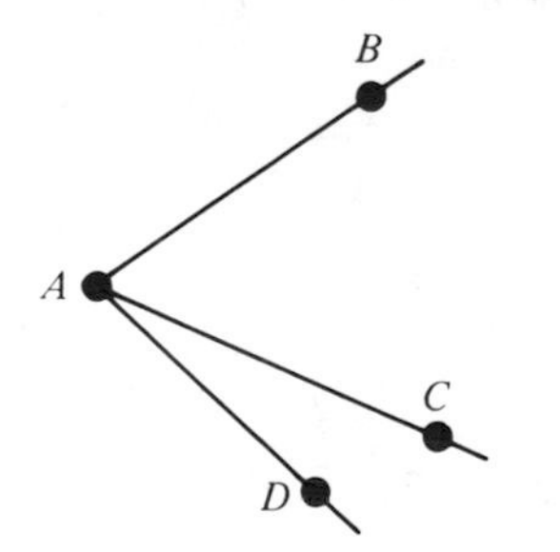

(1) they lie in the same plane
(2) they have a common vertex
(3) they have a common side
(4) their interiors do not intersect

Adjacent angles: $\angle BAC$ and $\angle CAD$.

11. Decide if each pair of angles given is a pair of adjacent angles or not. If

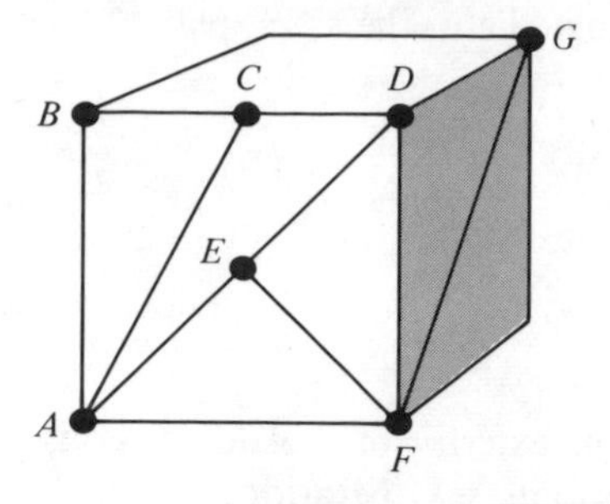

not, tell which of the four conditions above is (are) not satisfied.

(a) $\angle BAC$ and $\angle CAD$
(b) $\angle CDA$ and $\angle ADF$
(c) $\angle EFD$ and $\angle DFG$
(d) $\angle AEF$ and $\angle FED$
(e) $\angle CAD$ and $\angle DEF$
(f) $\angle FDE$ and $\angle ADC$
(g) $\angle BAC$ and $\angle DAF$
(h) $\angle DBA$ and $\angle BAF$
(i) $\angle BCA$ and $\angle ACD$
(j) $\angle BAC$ and $\angle BAD$

Two adjacent angles are said to be a **linear pair** if their noncommon sides form a line.

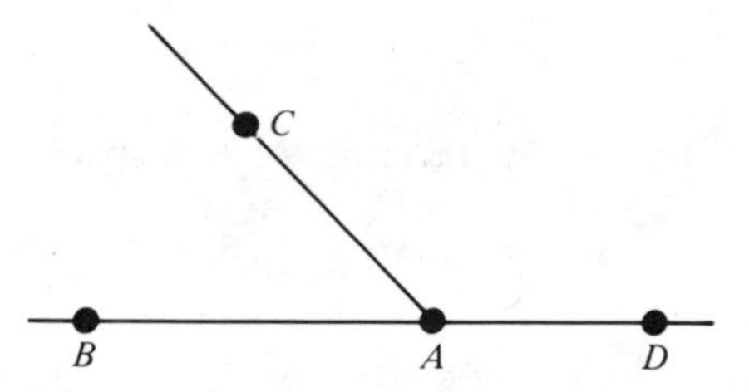

Linear pair: $\angle BAC$ and $\angle CAD$.

12. Which of the following pairs of angles are linear pairs?

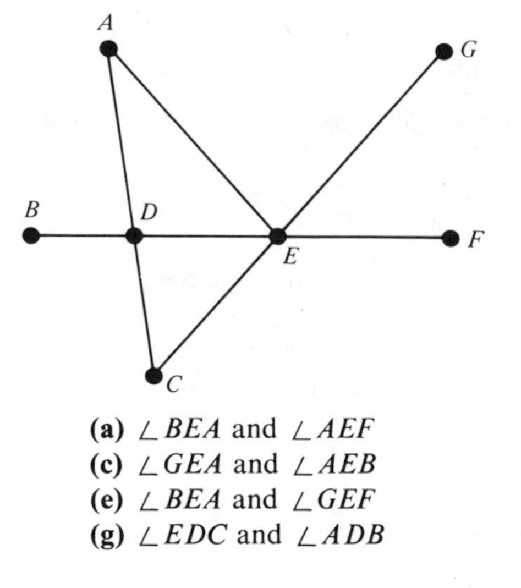

(a) $\angle BEA$ and $\angle AEF$
(b) $\angle BDC$ and $\angle BDA$
(c) $\angle GEA$ and $\angle AEB$
(d) $\angle CEF$ and $\angle FEG$
(e) $\angle BEA$ and $\angle GEF$
(f) $\angle DEC$ and $\angle GED$
(g) $\angle EDC$ and $\angle ADB$

When two lines intersect, four angles are formed. Each pair of non-adjacent angles is called a pair of **vertical angles.**

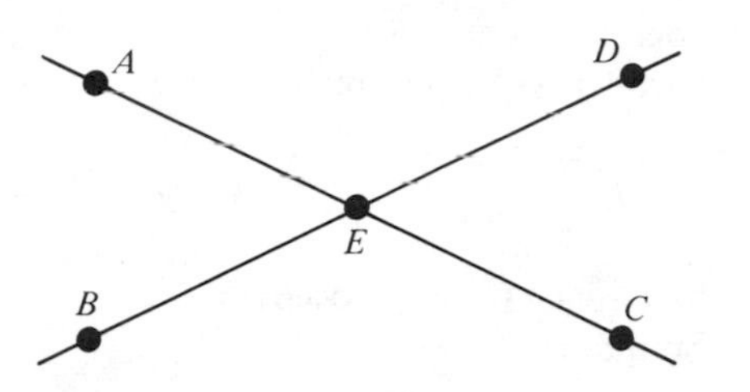

Two pairs of vertical angles: (1) $\angle AEB$ and $\angle DEC$, (2) $\angle AED$ and $\angle BEC$.

13. Find three pairs of vertical angles in the figure of Exercise 12.

VOCABULARY

1.1 Space, Points, and Geometric Figures

physical space
Euclidean space
space
geometric figure
abstract space
Euclidean geometry
point A

1.2 Lines

line determined by two points, ℓ, $\overleftrightarrow{AB}$
nonintersecting (lines)
noncollinear (points)
nonconcurrent (lines)
intersecting (lines)
collinear (points)
concurrent (lines)

1.3 Planes

perspective drawing
skew (lines)
coplanar (points)
parallel (lines)
parallel (planes)

1.4 Half-lines and Rays

half-line
between
ray from A through B, $\overrightarrow{AB}$
connectedness
half-line from A through B, $\overset{\circ\!\!\longrightarrow}{AB}$
density

1.5 Segments

segment
closed segment $\overline{AB}$ or $\overset{\bullet\!-\!-\!\bullet}{AB}$
half-open segment $\overset{\circ\!-\!-}{AB}$ or $\overset{\circ\!-\!-\!\bullet}{AB}$
endpoint
open segment $\overset{\circ\!-\!-\!\circ}{AB}$
parallel, skew, collinear segments

1.6 Half-planes and Half-spaces

half-plane (open, closed)
interior
exterior
half-space (open, closed)
boundary

1.7 Angles

angle $\angle ABC$, $\angle B$
sides
interior, of an angle
straight angle
linear pair of angles
vertex
dihedral angle
exterior, of an angle
adjacent angles
vertical angles

OPEN SENTENCES AND THEIR GRAPHS

OVERVIEW

In this chapter the geometric figures of Chapter 1 are met again, this time as graphs of open sentences. The necessary introductory work with the concepts of number line (Section 2.1) and open sentence (Sections 2.2 and 2.3) is followed by sections on solving simple and compound equations and inequalities. The final section is devoted to applications.

In the course of the chapter many ideas from previous courses are reviewed. In Section 2.1 some of the arithmetic of rational numbers is reviewed and the function concept is touched on lightly. (It will be studied intensively in Chapter 9.) Some perhaps familiar work with logic is done in Sections 2.2 and 2.3. In Section 2.4 a few of the simplest properties of equality and order are reviewed within the context of solving "linear" equations and inequalities. The relation of the logical operators—and, or, not—to the set operations—intersection, union, complement—is reemphasized strongly in Section 2.5. Sections 2.6 and 2.7 are devoted to absolute value. This may well be a review concept, but the context is new. The technique of setting up (and solving) open sentences suggested by "word problems" is the subject of Section 2.8. Like most real problems, the problems of this section are couched in decimal and percent notation. Thus, computation with decimals and percent is reviewed in a problem-solving context.

WARNING Although most of the exercises of Chapter 1 could be assigned, virtually as they stand, to a child in the early elementary grades, that is not true of this chapter. Many of the exercises of this chapter would be appropriate only at the upper elementary or junior high school levels.

2.1 THE NUMBER LINE

The numbers of elementary mathematics—the whole numbers, the integers, the rational numbers, and the real numbers—are all abstract mathematical concepts. (Remember that technically you do not write down numbers. What you write down are numerals that represent them.) To study the properties of these number systems, particularly the order properties, it is helpful to have available a visible, concrete representation of the numbers. The points of a line provide just such a representation.

Below is a typical **number line**. Notice that the integers are represented by evenly spaced points with smaller numbers to the left; rational numbers are represented by appropriate intermediate points (the point two-thirds of the way from 1 to 2 represents $1\frac{2}{3}$); irrational numbers are fit in near their rational approximations (π is a little to the left of 3.142 and $\sqrt{2}$ is a little to the right of 1.414).

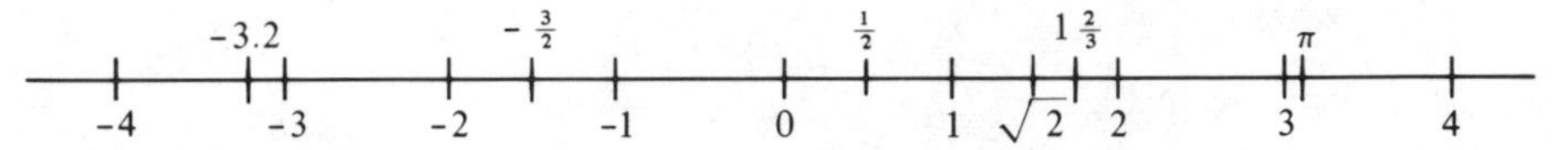

Every real number is matched with a unique point on the line, and every point corresponds to a unique real number. Furthermore, distances* between points on the line correspond to differences of the numbers matched with the points. For example, the distance from the point $-\frac{3}{2}$ to the point 2 is $2 - (-\frac{3}{2}) = 2 + \frac{3}{2} = 3\frac{1}{2}$; and the distance from the point $\frac{1}{2}$ to the point $1\frac{2}{3}$ is $1\frac{2}{3} - \frac{1}{2} = \frac{5}{3} - \frac{1}{2} = \frac{10}{6} - \frac{3}{6} = \frac{7}{6}$.

We have given a naive, intuitive description of the number line—the kind of description that is adequate in the early grades. Fundamental to a deeper understanding of the number line is the concept of **function**. A function can be thought of as a rule that assigns to each element of a first set an element of a second set. In the present context the first set, or **domain** or **input set** of the function, is the set R of real numbers. The second set, or **range** or **output set** of the function, is a line ℓ (a set of points). "The number line" is actually both of these sets together with the function that matches the numbers with the points. This function is called the **graphing function** and is denoted by g. The point corresponding to (say) the number -2 is called the **graph** of -2 and is written $g(-2)$. The number that corresponds to a point is called the **coordinate** of the point. In the figure below the coordinate of the point Q is 1.8.

* We shall examine the concept of distance closely in Chapter 4. See, in particular, the note on p. 135.

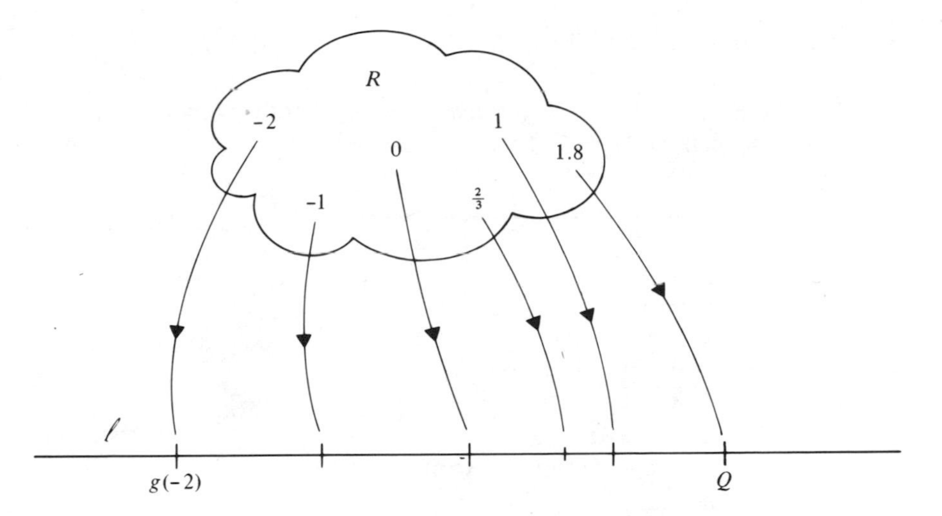

It would be well for you to retain a mental picture of the above diagram. Think of the set R as the place where arithmetic is done. This is where numbers are added, subtracted, multiplied, divided, and compared for size. Think of the graphing function as strings on puppets, the puppets being the points on the line. As arithmetic is done with the intangible numbers, certain geometric effects are visible down the line. A geometric interpretation is given to each arithmetic relation and operation. (Having two different ways of looking at a single question is very useful, as we shall soon see. Sometimes the arithmetic viewpoint is simpler, sometimes the geometric.)

The Order Relation

Arithmetic and geometric criteria for the order relation, less than ($<$), are illustrated in the following examples.

EXAMPLE *Geometrically*, $\frac{3}{2} < \frac{10}{3}$ because the point representing $\frac{3}{2}$ is obviously *to the left of* the point representing $\frac{10}{3}$. *Arithmetically*, $\frac{3}{2} < \frac{10}{3}$ because $\frac{3}{2} = \frac{9}{6}$, $\frac{10}{3} = \frac{20}{6}$, and $\frac{9}{6} < \frac{20}{6}$.

EXAMPLE *Arithmetically*, $\frac{9}{25} < \frac{3}{8}$ because $\frac{9}{25} = .36$, $\frac{3}{8} = .375$, and $.36 < .375$. To decide *geometrically* that the point corresponding to $\frac{9}{25}$ is to the left of the point corresponding to $\frac{3}{8}$ would require extremely meticulous plotting of points. For this pair of numbers the arithmetic test is simplest.

GEOMETRIC INTERPRETATION OF ORDER

For real numbers a and b, $a < b$ if and only if the point representing a on the number line lies to the left of the point representing b.

The Addition Operation

The geometric interpretation of addition (and subtraction) is quite simple.

EXAMPLE To find $-\frac{3}{2}+\frac{9}{4}$ *geometrically*, begin at the point corresponding to $-\frac{3}{2}$ and **slide** *to the right* $\frac{9}{4}$ units.

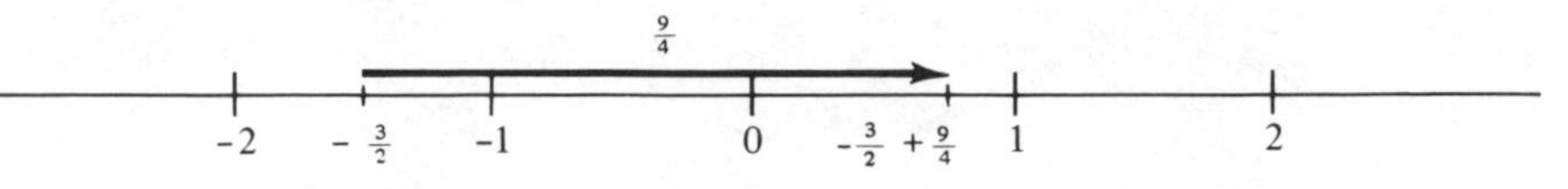

Checking *arithmetically*,

$$-\frac{3}{2}+\frac{9}{4}=-\frac{6}{4}+\frac{9}{4}=\frac{3}{4}$$

GEOMETRIC INTERPRETATION OF ADDITION AND SUBTRACTION

Adding a positive number corresponds to sliding to the right on the number line. Adding a negative number (which is the same as subtracting a positive number) corresponds to sliding to the left on the number line.

EXAMPLE To find $\frac{9}{4}-\frac{3}{2}=\frac{9}{4}+(-\frac{3}{2})$ geometrically, begin at $\frac{9}{4}$ and slide to the left $\frac{3}{2}$ units.

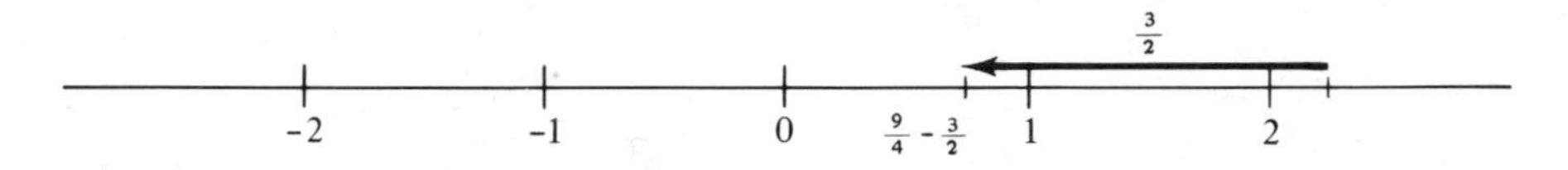

The Multiplication Operation

Multiplication can also be thought of geometrically in terms of "stretching," "shrinking," "flipping."

EXAMPLE To find $\frac{3}{2}\cdot\frac{5}{4}$ geometrically, **stretch** to $\frac{3}{2}$ its original length the arrow from 0 to $\frac{5}{4}$.

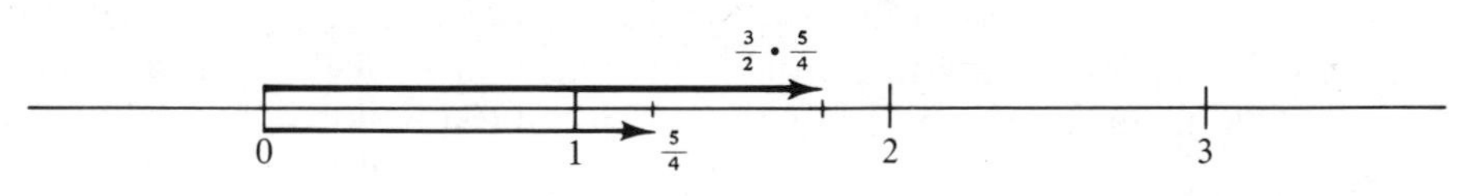

Checking arithmetically,

$$\frac{3}{2}\cdot\frac{5}{4}=\frac{3\cdot 5}{2\cdot 4}=\frac{15}{8}$$

EXAMPLE To find $\frac{1}{2}\cdot\frac{5}{3}$ geometrically, **shrink** to $\frac{1}{2}$ its original length the arrow from 0 to $\frac{5}{3}$.

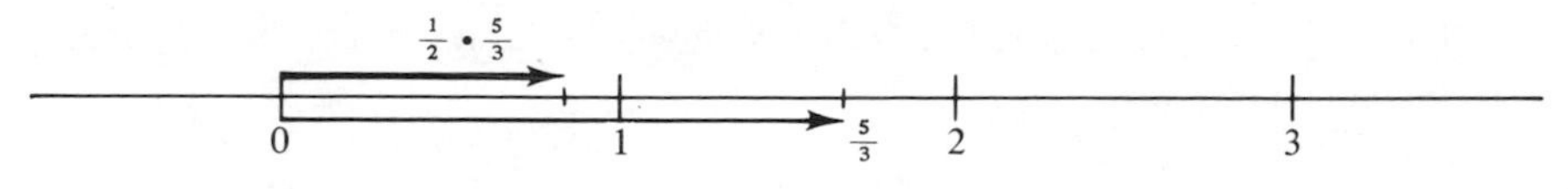

Checking arithmetically,

$$\frac{1}{2} \cdot \frac{5}{3} = \frac{5}{6}$$

EXAMPLE To find $-\frac{2}{3} \cdot \frac{6}{5}$ geometrically, shrink to $\frac{2}{3}$ its original length the arrow from 0 to $\frac{6}{5}$, and then **flip** it over so that it points the opposite way from 0.

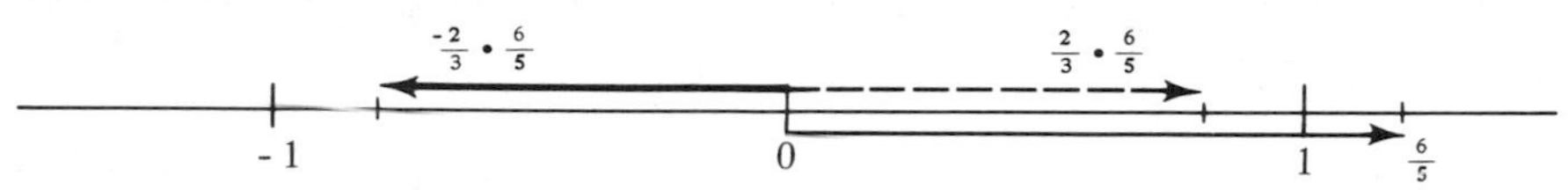

Checking arithmetically,

$$-\frac{2}{3} \cdot \frac{6}{5} = -\frac{12}{15} = -\frac{4}{5}$$

(See if you end up at the same spot by beginning with the arrow from 0 to $-\frac{2}{3}$ and stretching it by $\frac{6}{5}$.)

GEOMETRIC INTERPRETATION OF MULTIPLICATION

To multiply by a positive number greater than 1 you stretch; to multiply by a positive number less than 1 you shrink; to multiply by a negative number you stretch or shrink (depending on whether it is less than or greater than -1) and then flip.

Since any division problem can be thought of as a multiplication problem, we now have geometric interpretations for the relation $<$ and all the basic arithmetic operations: $+$, $-$, $\times$, $\div$. These geometric interpretations allow one to "see" arithmetic and can be used to good advantage in teaching. In most numerical problems, of course, working with arrows on a number line is too inaccurate to give any more than an approximation to the correct answer.

EXERCISES

1. Give the (approximate) coordinates of the points A through E shown below.

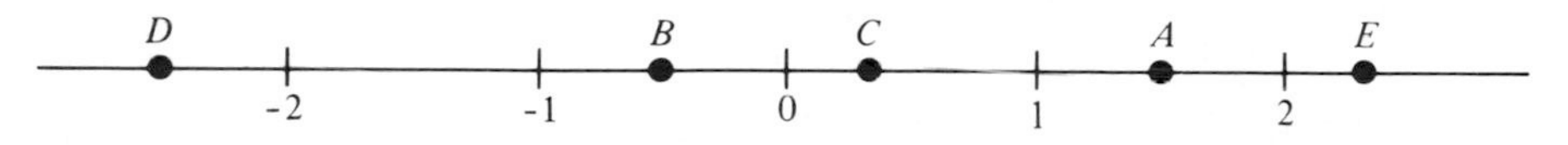

2. Using your answers to Exercise 1, give the lengths of the following segments.

(a) $\overline{CA}$ **(b)** $\overline{BE}$ **(c)** $\overline{DB}$ **(d)** $\overline{ED}$

3. (a) On a number line sketch the graphs of the numbers

$$-\frac{7}{8},\ \frac{3}{4},\ \frac{5}{-4},\ \frac{5}{16},\ \frac{-1}{2},\ \frac{5}{8},\ 0$$

(The graph of 0 is called the **origin** of the number line.)

(b) Now arrange the above numbers in order from smallest to largest.

4. On the number line below, the graphs of various numbers are shown using careful function notation.

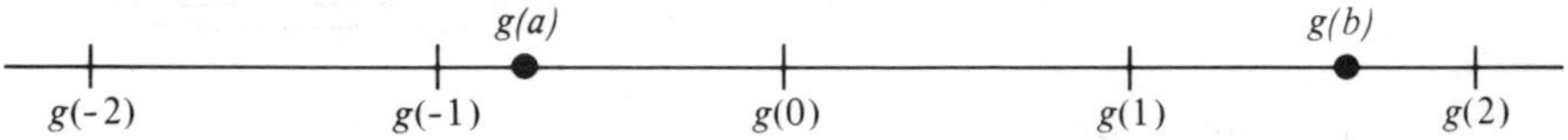

g(a) g(b)
g(-2) g(-1) g(0) g(1) g(2)

Locate the following points on this number line. Remember to simplify the inputs (perform the operations in parentheses) before plotting the outputs.

(a) $g(\frac{1}{2} \cdot 0 + \frac{1}{2} \cdot 1)$ **(b)** $g(\frac{1}{2} \cdot 1 + \frac{1}{2} \cdot 2)$
(c) $g(\frac{1}{2} \cdot -2 + \frac{1}{2} \cdot -1)$ **(d)** $g(\frac{1}{2} \cdot 0 + \frac{1}{2} \cdot 2)$
(e) $g(\frac{1}{2} \cdot -1 + \frac{1}{2} \cdot 2)$ **(f)** $g(\frac{1}{2} \cdot a + \frac{1}{2} \cdot b)$

On a fresh copy of the number line

g(a) g(b)
g(-2) g(-1) g(0) g(1) g(2)

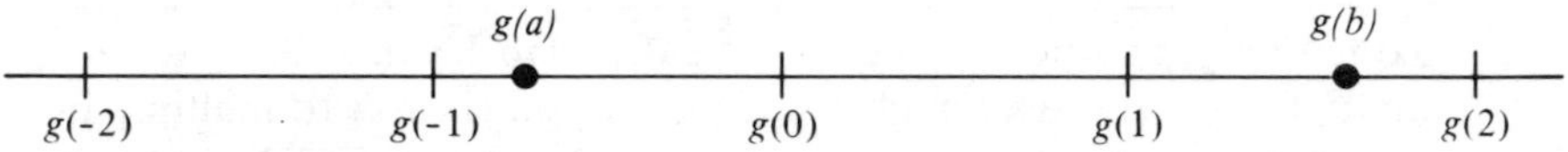

locate these points:

(g) $g(\frac{1}{3} \cdot 0 + \frac{2}{3} \cdot 1)$ **(h)** $g(\frac{1}{3} \cdot 1 + \frac{2}{3} \cdot 2)$
(i) $g(\frac{1}{3} \cdot -2 + \frac{2}{3} \cdot -1)$ **(j)** $g(\frac{1}{3} \cdot 0 + \frac{2}{3} \cdot -1)$
(k) $g(\frac{1}{3} \cdot -1 + \frac{2}{3} \cdot 2)$ **(l)** $g(\frac{2}{3} \cdot -1 + \frac{1}{3} \cdot 2)$
(m) $g(\frac{1}{3} \cdot a + \frac{2}{3} \cdot b)$ **(n)** $g(\frac{2}{3} \cdot a + \frac{1}{3} \cdot b)$

On a fresh copy of the number line

g(a) g(b)
g(-2) g(-1) g(0) g(1) g(2)

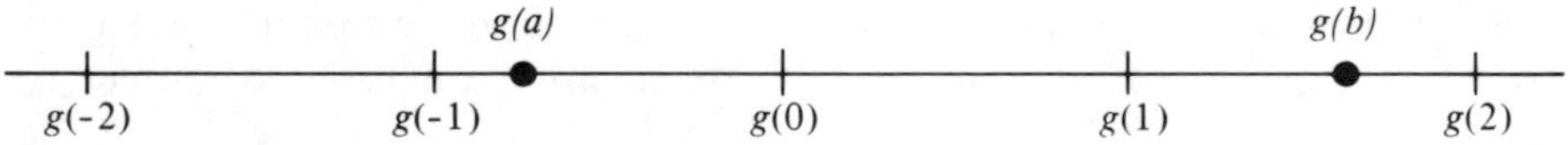

locate these points:

(o) $g(\frac{3}{4} \cdot 2 + \frac{1}{4} \cdot -2)$ **(p)** $g(\frac{1}{4} \cdot 2 + \frac{3}{4} \cdot -2)$
(q) $g(\frac{7}{8} \cdot 1 + \frac{1}{8} \cdot -1)$ **(r)** $g(\frac{3}{4} \cdot a + \frac{1}{4} \cdot b)$
(s) $g(\frac{1}{4} \cdot a + \frac{3}{4} \cdot b)$ **(t)** $g(1 \cdot a + 0 \cdot b)$
(u) $g(0 \cdot a + 1 \cdot b)$

5. What is the coordinate of the point

(a) midway between $g(-2)$ and $g(1)$

(b) $\frac{1}{4}$ of the way from $g(-1)$ to $g(1)$

(c) $\frac{2}{3}$ of the way from $g(0)$ to $g(2)$
(d) $\frac{5}{6}$ of the way from $g(1)$ to $g(-2)$
(e) $\frac{3}{8}$ of the way from $g(-2)$ to $g(2)$
(f) $\frac{1}{3}$ of the way from $g(a)$ to $g(b)$?

6. We noted in Chapter 1 that the points on a line are "dense"; that is, between any two points on a line there is another. If the points on a line accurately represent real numbers, then real numbers too should be **dense**. Find a real number "between" each two given numbers:
(a) $\frac{2}{3}, \frac{3}{4}$ **(b)** $\frac{1}{9}, \frac{1}{10}$ **(c)** $\frac{3}{8}, .37$ **(d)** $\frac{22}{7}, \pi$
Was your answer a rational number in each case?

7. (a) Is there a smallest positive integer?
(b) Is there a smallest positive rational number?

8. Write an addition or subtraction statement suggested by each diagram.

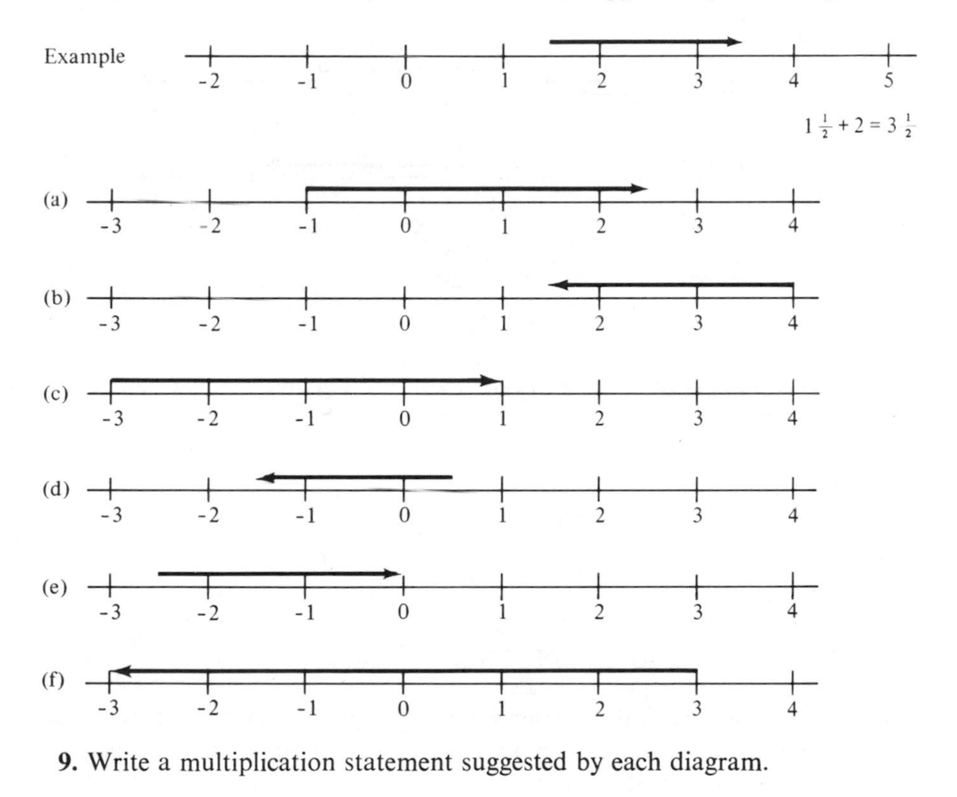

9. Write a multiplication statement suggested by each diagram.

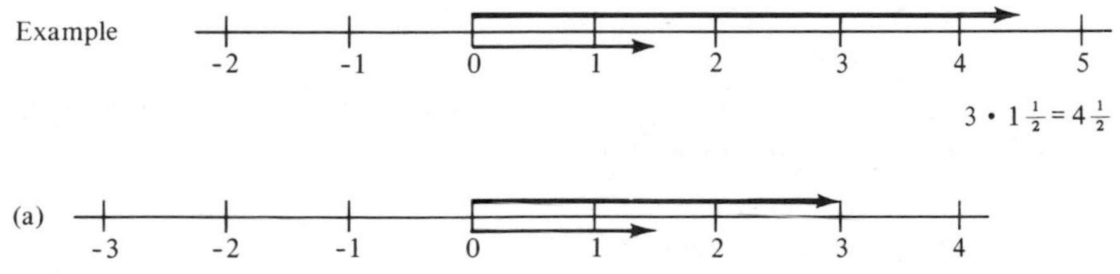

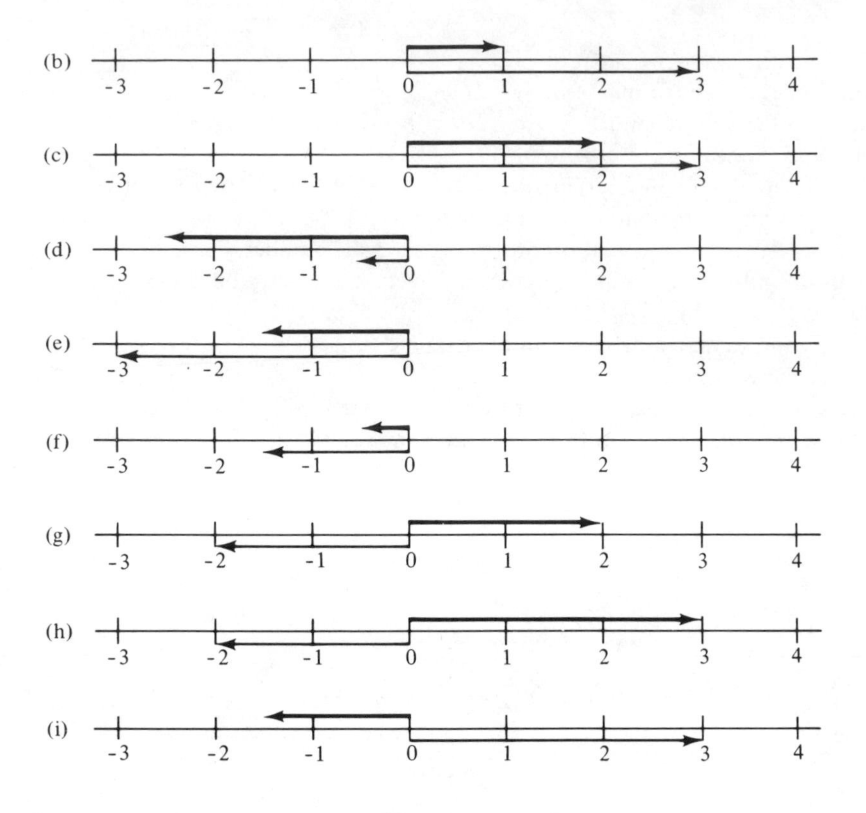

10. Find the smaller of each pair of numbers either geometrically or arithmetically, whichever is easier.

(a) $-\frac{5}{2}, -1$
(b) $-\frac{1}{64}, \frac{1}{260}$
(c) $\frac{1}{6}, .16$
(d) $\frac{5}{32}, \frac{1}{6}$

2.2 STATEMENTS AND OPEN SENTENCES

Our eventual goal in this chapter is to "graph" on the number line the "solution sets" of various "simple and compound equations and inequalities." The purpose of this and the next section is to assign precise meanings to the technical terms in the previous sentence. We begin in a nonmathematical setting.

A **statement** is a declarative sentence that is either true or false. Why is each of the following sentences a statement?

Sauerkraut contains vitamin C.
Lake Superior is the deepest lake in the world.

Osmium is heavier than lead.
Julius Caesar discovered America.

Why is each of the following sentences *not* a statement?

When does the next bus leave?
Keep off the grass.
What planets are nearer to the Sun than the Earth is?
Who was the third president of the United States?

You have probably taken examinations on which questions like the last two were disguised as expressions that look very much like statements, except that they contain a blank.

1. ________ was the third president of the United States.
2. ________ is nearer to the Sun than the Earth is.
3. ________ has a common border with Mexico.

Such expressions are called **open sentences**. An open sentence is *not* a statement, because it is neither true nor false as it stands, but it *becomes* a statement when the blank is filled. Your job on an examination is to fill the blanks so that true statements result. Sometimes it is clear what things can be tried in the blank. The set of all possible candidates for insertion in the blank is called the **replacement set** for the open sentence. The subset consisting of all members of the replacement set that yield a true statement, when inserted in the blank, is called the **truth set** of the open sentence.

For open sentence 1 the replacement set is pretty clearly the set of all U.S. presidents. The truth set is the singleton, {Jefferson}. For open sentence 2 the replacement set is pretty clearly the set of all planets in the solar system, and the truth set is {Mercury, Venus}. For open sentence 3 it is not clear what the replacement set is. This would be an unfair, ambiguous question on a geography examination. If the replacement set were the set of all nations of the world, then the truth set would be

{British Honduras, Guatemala, United States}

If the replacement set were the set of all states of the United States, then the truth set would be

{Texas, New Mexico, Arizona, California}

It is very important to remember this:

The truth set of an open sentence depends on the choice of the replacement set.

An open sentence can be thought of as a function or machine that produces statements out of raw materials fed to it from the replacement set. The outputs are also called **instances** of the open sentence.

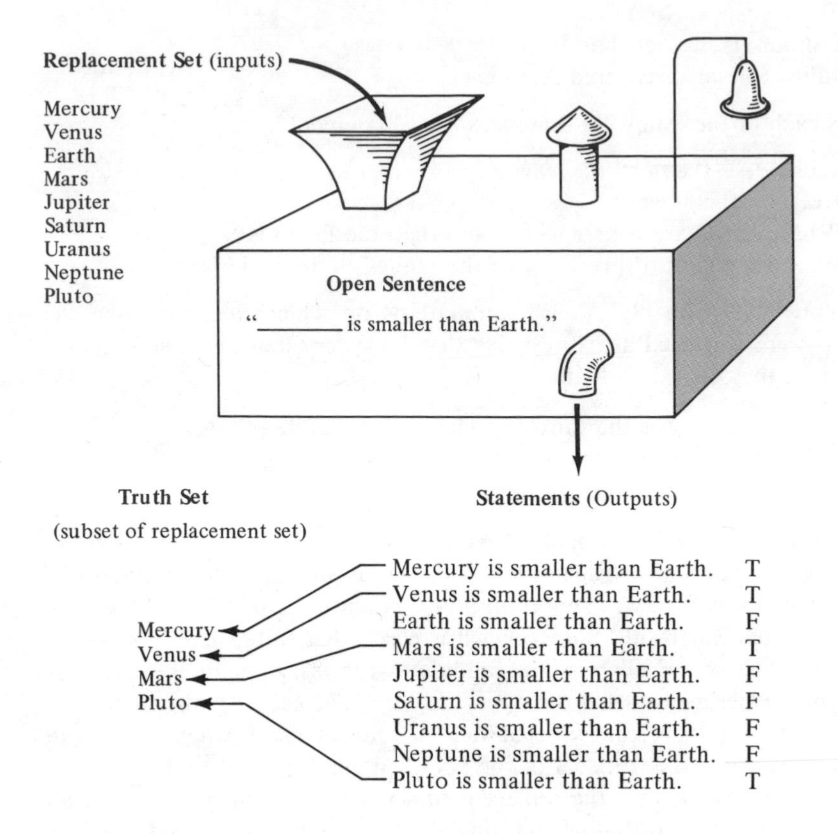

EXERCISES

1. Decide which of these sentences are statements and which are not.
   (a) The Yankees won the American League pennant in 1931.
   (b) England's Queen owns a washing machine.
   (c) Let's go play golf.
   (d) All cows eat grass.
   (e) Sit down and be quiet.
   (f) Mars has two moons.
   (g) Who's in charge here?
   (h) What is the largest prime number less than 100?
   (i) $2 + 2 = 3$
   (j) $2 + 2 = 4$
   (k) Seven is how much greater than five?
   (l) $x + 5 = 7$
   (m) $2x - 3 = 17$
2. Suggest a reasonable replacement set for each open sentence below. (Do not try to determine the truth set.)
   (a) ________ has five moons.
   (b) ________ belongs to the Security Council.

(c) ________ is a Romance Language.
(d) ________ has a population of more than 1 million people.
(e) ________ lies entirely south of the equator.
(f) ________ is a prime.
(g) ________ bisects $\angle ABC$.
(h) ________ is equidistant from A and B.

3. Questions (which, of course, are not statements) can often be restated as open sentences. Convert each question below to an open sentence and suggest a reasonable replacement set. (Do not try to determine the truth set.)

EXAMPLE 1 Who's in charge here?

Open sentence: ________ is in charge here.
Replacement set: The people in this math class.

EXAMPLE 2 What are the legal holidays in 1977?

Open sentence: ________ is a legal holiday in 1977.
Replacement set: {Jan. 1, Jan. 2, ..., Dec. 30, Dec. 31}.

(a) Who wrote *Silas Marner*?
(b) What are the vowels in our alphabet?
(c) Where do most household accidents occur?
(d) What are the prime numbers less than 20?
(e) What must you add to 17 to get 31?
(f) How much is two plus two?

4. (a) Complete this machine diagram.

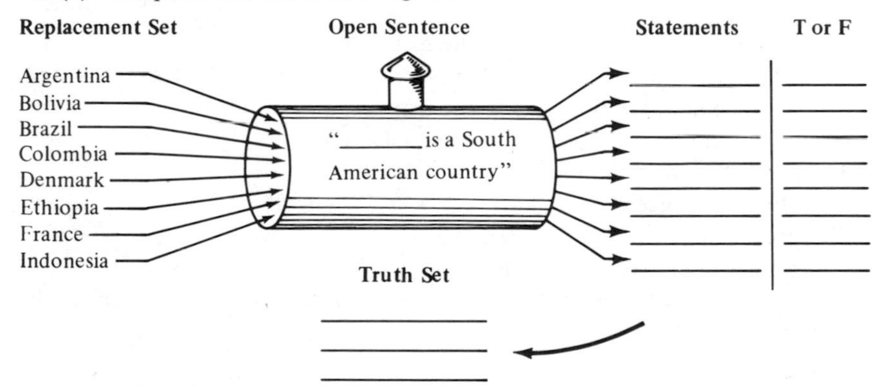

Imagine machine diagrams and list truth sets (same replacement set as above) for the following open sentences:

(b) The equator passes through ________.
(c) The equator passes through ________ or ________ is a South American country.
(*Note*: Each member of the replacement set must be put in both blanks simultaneously.)

(d) ________ is not a South American country.

5. For each open sentence and replacement set: (i) think about all statements that are formed, (ii) classify each statement as true or false, (iii) list the truth set.

EXAMPLE

Open sentence: ________ was a Republican.
Replacement set: {F. Roosevelt, Truman, Eisenhower, Kennedy, L. Johnson, Nixon}.

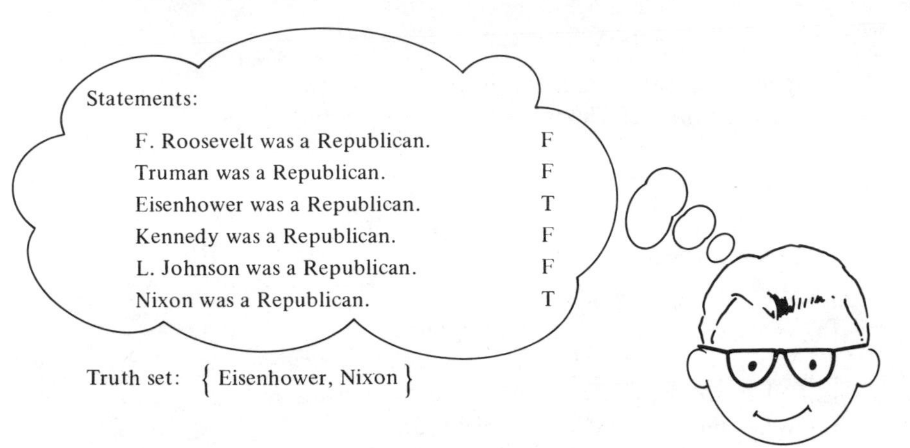

(a) Open sentence: "________ served as vice president before becoming president."
Replacement set: same as in the example.

(b) Open sentence: "________ was president during the Korean War."
Replacement set: same as in the example.

(c) Open sentence: "________ was a Democrat."
Replacement set: same as in the example.

(d) Open sentence: "________ was a Democrat and ________ was president during the Korean War."
Replacement set: same as in the example.

6. Use the given replacement set and draw a loop around the truth set of the given open sentence.

Example

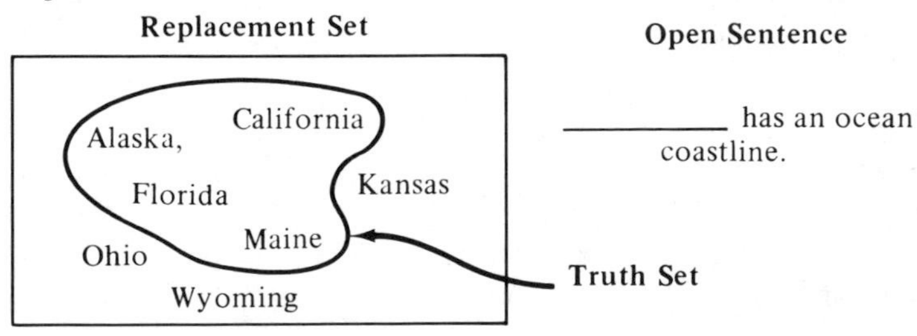

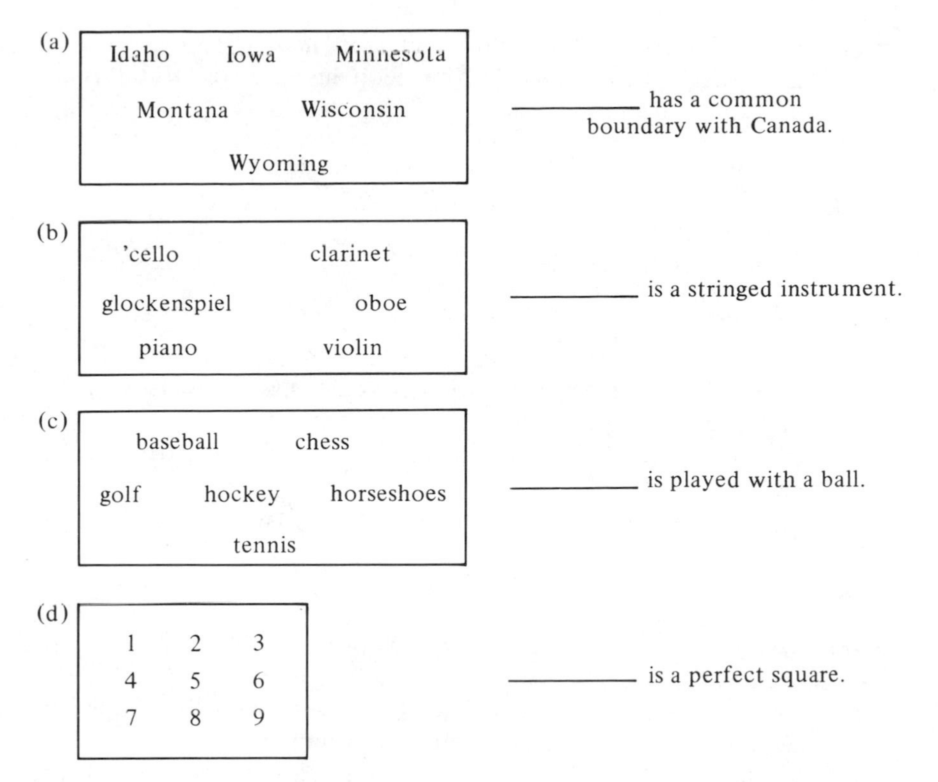

7. (a) Use the replacement set below and loop in *red* the truth set of "_________ lies west of the Mississippi River." Now loop in *black* the truth set of "_________ has a common boundary with Canada."

North Dakota
Montana Maine
Washington Michigan
Iowa Indiana
Wyoming

(b) Describe the truth set of the **compound** open sentence, "_________ lies west of the Mississippi River *and* _________ has a common boundary with Canada." [Relate this truth set to the truth sets you looped in part (a).]

(c) Describe the truth set of the compound open sentence, "_________ lies west of the Mississippi River *or* _________ has a common boundary with Canada." [Relate this truth set to the truth sets you looped in part (a).]

(d) Describe the truth set of "_________ does *not* lie west of the Mississippi River." [Relate it to a truth set you looped in part (a).]

8. (a) Use the replacement set below and loop in red the truth set of "__________ is a mammal." Now loop in black the truth set of "__________ lives in the sea."

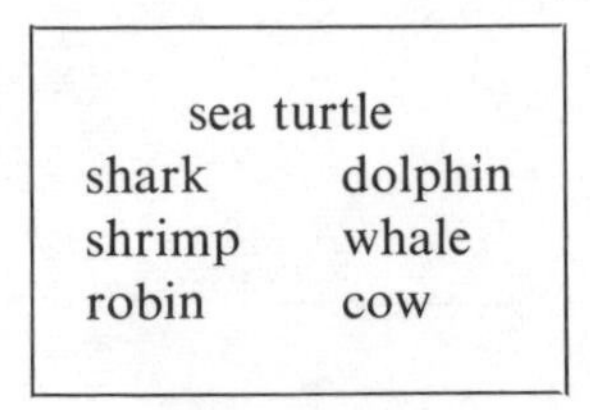

(b) The *intersection* of the two truth sets you looped in part (a) is the truth set of what open sentence?

(c) The *union* of the truth sets you looped in part (a) is the truth set of what open sentence?

(d) The *complement* of the truth set you looped in red is the truth set of what open sentence?

(e) The *complement* of the truth set you looped in black is the truth set of what open sentence?

9. (a) When two open sentences are joined by the word "and," the set operation, __________, is applied to their truth sets.

(b) When two open sentences are joined by the word "or," the set operation, __________, is applied to their truth sets.

(c) When an open sentence is denied by means of the word "not," the set operation, __________, is applied to its truth set.

10. When we describe a set verbally we often (unconsciously) describe it as the truth set of an open sentence. For example, when we speak of "the set of all award-winning movies" we are actually referring to the truth set of an open sentence (__________ won an award.) in a specific replacement set (the set of all movies). Each set described below can be thought of as the truth set, in some replacement set, of an open sentence. Give a replacement set and open sentence.

EXAMPLE. The set of all pink panthers.

One solution.
Replacement set: all panthers.
Open sentence: __________ is pink.

Another solution.
Replacement set: all pink things.
Open sentence: __________ is a panther.

(a) the set of all lonesome polecats
(b) the set of all blue-eyed babies
(c) the set of all elements that conduct electricity
(d) the Detroit Tigers' pitching staff

(e) the Senate Foreign Relations Committee
(f) the set of all four-legged, domesticated animals
(g) the set of all odd whole numbers
(h) the set of all even prime numbers

11. Use the replacement set of Exercise 8. Let M be the truth set of "________ is a mammal." let S be the truth set of "________ lives in the sea," and let L be the truth set of "________ has legs."
(a) $S \cap (L')$ is the truth set of what open sentence?
(b) $(M \cap L) \cap (S')$ is the truth set of what open sentence?
Express in terms of M, S, L, $\cup$, $\cap$, and $'$ the truth set of each open sentence below.
(c) "________ has legs or does not live in the sea."
(d) "________ lives in the sea and is not a mammal and does not have legs."
(e) "________ has legs, but is neither a mammal nor lives in the sea."
(f) "________ is a mammal, or has legs and does not live in the sea."

2.3 MATHEMATICAL SENTENCES

In mathematics much of what we do concerns statements and open sentences. The ideas of this section are very much like those of the previous section, but you will have to learn some of the special terminology that goes with mathematical sentences. Which of these mathematical sentences are statements? (That is, which are either true or false?)

1. $6 + 7 = 13$.
2. $x + 4 = 10$.
3. Every whole number is the sum of four perfect squares.
4. The intersection of two planes is always a line.
5. $3 < 1$.
6. $x < 0$.

Sentences 2 and 6 are not statements. They are open sentences. The open sentences in mathematics generally use letters instead of blanks, one reason being that blanks get mixed up with minus signs. (In the elementary school, "frames" are frequently used—for example, "$4 + \square = 7$.")

The open sentence ______ + 3 < 5 might be rewritten $x + 3 < 5$.
The open sentence 4 − ______ = 6 might be rewritten $4 - y = 6$.
The open sentence ______² − 2______ − 7 = 1 might be rewritten $t^2 - 2t - 7 = 1$.

Clearly any letter works as well as any other: $x^2 - 2x - 7 = 1$, $a^2 - 2a - 7 = 1$, $t^2 - 2t - 7 = 1$ all represent the same open sentence. Any letter that plays the role of a blank in a mathematical open sentence is called a **variable**. The replacement set for the open sentence is usually called the **domain of the**

variable, or the **domain of the open sentence**. ("Domain" and "domicile" have the same Latin root, *domus*. It might be helpful to think of the domain of the variable as the place where the variable lives.) The truth set of a mathematical open sentence is usually referred to as its **solution set**. Thus, the solution set is a subset of the domain of the variable, and the warning of the previous section can be reworded.

The solution set of a mathematical open sentence depends on the choice of the domain of the variable.

If the domain of the variable x in the open sentence $x + 3 < 5$ is the set W of all whole numbers, then the solution set is $\{0, 1\}$. If the domain of the variable is the set I of all integers, however, then the solution set is $\{\ldots -4, -3, -2, -1, 0, 1\}$.

If the domain of the open sentence $4 - y = 6$ is I, then the solution set is $\{-2\}$. If the domain is W, then the solution set is $\varnothing$. (That is what a second grader means when he says that you cannot take something away from 4 and have 6 left. There is no point in arguing with him either. For the domain he is familiar with, W, he is right.)

Again we can illustrate the relationship between a mathematical open sentence, the domain of its variable, and its solution set by means of a machine diagram.

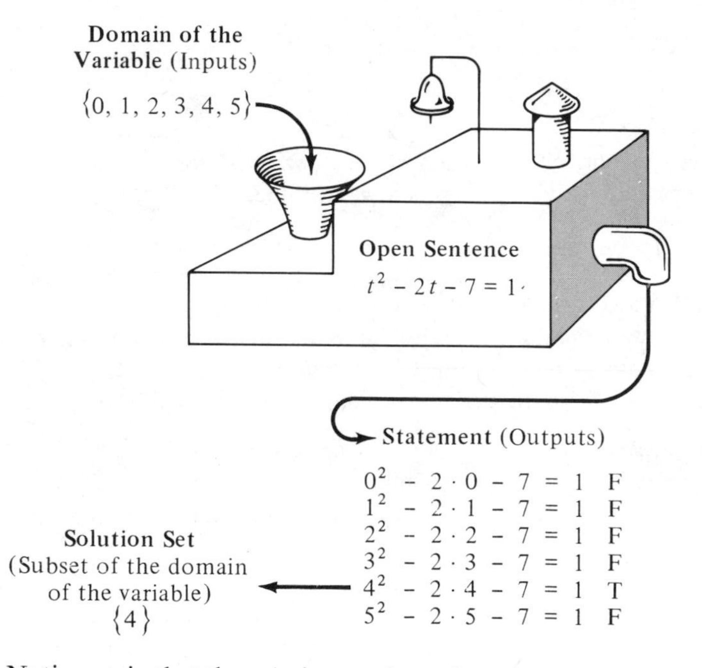

Notice again that the solution set depends on how the domain is chosen. If the domain of the variable in this example were I instead of $\{0, 1, 2, 3, 4, 5\}$, then the solution set would turn out to be $\{-2, 4\}$. (Verify that replacing t by -2 yields a true statement.)

The machine diagram illustrates nicely how easy it is, *theoretically*, to find the solution set of a mathematical open sentence: You just replace the variable by each member of the domain, in turn, and collect those members that yield true statements. At the same time, though, the machine diagram illustrates the *practical difficulty*: If the domain of the variable has many elements, then trying them all is a big job, and if the domain of the variable is *infinite* (as is the case for the most commonly used domains of whole numbers, integers, rational numbers, and real numbers), then it is *impossible* to try all members! It would certainly be hopeless to try to feed all possible fractions to this machine,

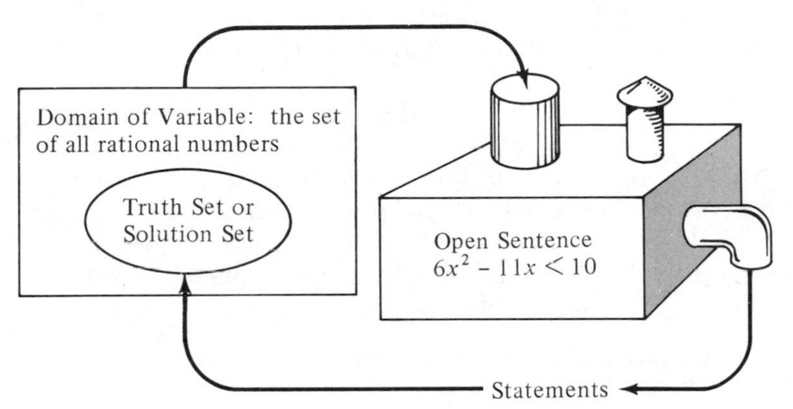

and then sort through the resulting statements to find the solution set.

In general, trying to solve an equation or an inequality is like trying to answer a multiple-choice question for which there are too many choices to permit trying them all. Techniques more efficient than trial must be used. Some of the simpler techniques will be reviewed in the next section.

EXERCISES

1. In this section we have been speaking of mathematical sentences, but we did not try to draw a careful distinction between mathematical and non-mathematical sentences. Suggest a rule of thumb for deciding if a sentence is mathematical or not.

2. Classify these mathematical sentences as statements or open sentences.

(a) $2 + 2 = 4$ **(b)** $28 = 2^2 \cdot 7$
(c) $3 < 10$ **(d)** $2x = 10$
(e) $x < 5$ or $x = 5$ **(f)** $2x - 4 > 6$
(g) $x^2 = 3x + 4$ **(h)** $x^2 = 3x + 4$ and $x < 0$
(i) $x^2 = 3x + 4$ and $2x - 3 = 5$ **(j)** $x^2 = 3x + 4$ or $x = 0$
(k) $5 = 6$ or $2 < 3$ **(l)** $2x > 5$ or $2x = 5$
(m) $5 - (3 - x) = x + 2$ **(n)** $2x - 3 < 5$ and $2x - 3 > 0$

(o) $6 < x$ or $6 = x$ **(p)** $2x - 4 > 0$ or $2x - 4 = 0$
(q) $2x - 4 > 0$ and $2x - 4 = 0$ **(r)** $x^3 = x$ and $x \neq 0$ and $x \neq 1$
(s) $3 - x > 2x - 7$ **(t)** $|x - 2| < 1$
(u) $|3 + x| = 2$ **(v)** $|-3| = 3$

3. Special names are given to various types of mathematical sentences.

(a) A *simple* (as opposed to compound) open sentence whose verb is "equals" is called an **equation**. Part (d) above is an equation. Part (a) above is *not*, although many people, including some mathematicians, would say it is. Technically, part (a) should be referred to as a **statement of equality**. Find three other equations in Exercise 2.

(b) A simple open sentence whose verb* is "is greater than" or "is less than" is called an **inequality**. Part (f) is an example of an inequality. Find two others. Technically, part (c) is a **statement of inequality**.

(c) Certain compound sentences such as (e) and (l) are also called inequalities. Do you remember a symbolic abbreviation for "$x < 5$ or $x = 5$"? Find some other inequalities of this type and abbreviate them.

4. One reason for using letters instead of blanks is that blanks look too much like minus signs or fraction symbols. Another reason is that many mathematical sentences involve more than one variable. For example, the commutative property of addition for rational numbers says that

$$a + b = b + a \text{ for all rational numbers}$$

It involves two variables, a and b. If we were to use blanks, we would have to use two different kinds.

$$\underline{\quad\quad} + \underline{\underline{\quad\quad}} = \underline{\underline{\quad\quad}} + \underline{\quad\quad} \text{ for all rational numbers}$$

suggests the commutative property of addition, but

$$\underline{\quad\quad} + \underline{\quad\quad} = \underline{\quad\quad} + \underline{\quad\quad} \text{ for all rational numbers}$$

does not. State some other properties that require more than one variable for their statement.

5. Complete the machine diagram

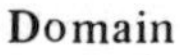

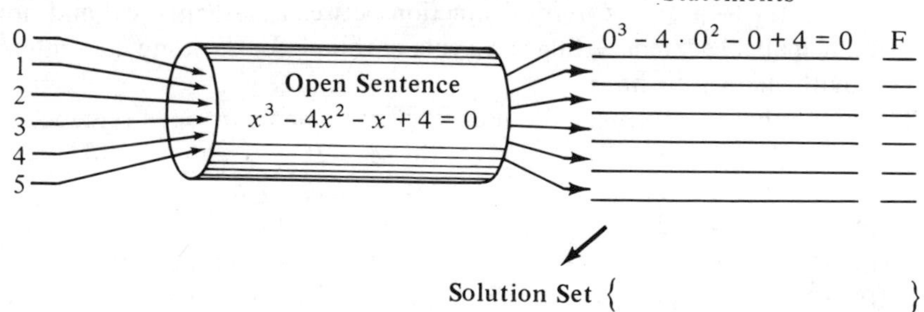

* If you do not like to call "is greater than" a verb, call it a verb phrase. Its synonym, "exceeds," is a verb.

6. **(a)** Repeat Exercise 5 but change the domain to $\{-2, -1, 0, 1, 2, 3\}$.
(b) Compare the solution sets in Exercises 5 and 6(a).
(c) Comment on the expression, "The solution set of $x^3 - 4x^2 - x + 4 = 0$."

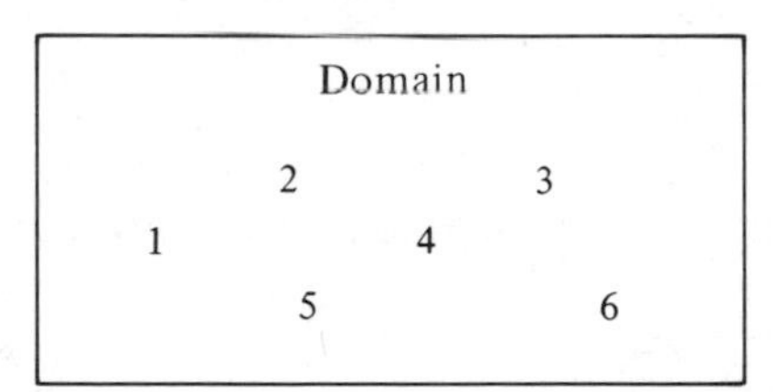

7. **(a)** In the figure above, loop the solution set of "$x < 4$."
(b) Loop the solution set of "x is odd."
(c) Relate the solution set of "$x < 4$ and x is odd" to the solution sets in (a) and (b).
(d) Relate the solution set of "$x < 4$ or x is odd" to the solution sets in (a) and (b).
(e) Describe how to obtain from one of the solution sets above, the solution set of "$x \geq 4$."

8. For this exercise the domain is $\{-3, -2, -1, 0, 1, 2, 3, 4, 5\}$, but these numbers are represented, as usual, as points on a number line.

-3 -2 -1 0 1 2 3 4 5

(a) Draw a black circle around each point that represents a solution of "$2x < 5$."
(b) Draw a red circle around each point that represents a solution of "$x > -2$."
(c) The doubly circled points represent the solution set of what open sentence?
(d) The points that are circled at least once represent the solution set of what open sentence?
(e) The points not circled in black represent the solution set of what open sentence?
(f) The points not circled in red represent the solution set of what open sentence?

9. The set $\{0, 1, 2, 3, 4\}$ can be thought of as the solution set, in an appropriate domain, of an open sentence. For example,

domain, W open sentence, $x < 5$

Using the set W of all whole numbers as domain, write an open sentence whose solution set is
(a) $\{0, 2, 4, 6, 8, 10, \ldots\}$
(b) $\{6, 7, 8, 9, 10, 11, \ldots\}$

(c) $\{2, 3, 5, 7, 11, 13, 17, 19, 23, \ldots\}$
(d) $\{4, 5, 6, 7, 8\}$
(*Hint*: $\{4, 5, 6, 7, 8\} = \{4, 5, 6, 7, 8, 9, 10, \ldots\} \cap \{0, 1, 2, 3, 4, 5, 6, 7, 8\}$.)
(e) $\{0, 1, 2, 3; 18, 19, 20, 21, \ldots\}$
(f) $\{4\}$
(g) $\{4, 7\}$
(h) $\varnothing$

Set builder notation is used to describe a set as the solution set of an open sentence. We write

$$\{0, 1, 2, 3, 4\} = \{x \in W \mid x < 5\}$$

the letter used as variable — the domain of the variable — the open sentence

and we read the symbol as follows:

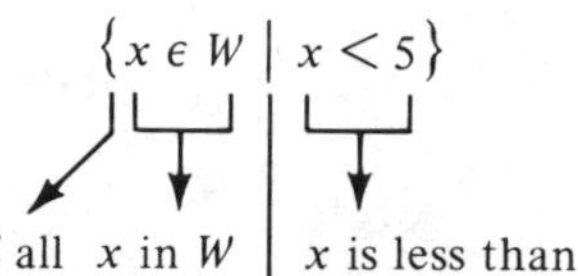

10. "Read" each set builder symbol and then enumerate the set described.
(a) $\{x \in W \mid x < 2\}$
(b) $\{x \in W \mid 2x < 7\}$
(c) $\{x \in W \mid x^2 = 4\}$
(d) $\{t \in W \mid 3t - 2 = 7\}$
(e) $\{x \in W \mid x^2 < 8\}$
(f) $\{x \in W \mid x^2 < 0\}$

11. Describe each set in Exercise 9 in set builder notation.

12. For this exercise $D = \{0, 1, 2, 3, 4, 5, 6, 7\}$. List the following sets. (*Hint*: First "read" the set builder notation.)
(a) $\{x \in D \mid x \text{ is even}\}$
(b) $\{x \in D \mid 5 - x < 0\}$
(c) $\{x \in D \mid x = x\}$
(d) $\{x \in D \mid 2x - 5 < 1\}$
(e) $\{x \in D \mid x^2 \in D\}$
(f) $\{x \in D \mid 2x \notin D\}$
(g) $\{x \in D \mid x^2 - 5x + 4 \in D\}$
(h) $\{x \in W \mid \sqrt{x} \in D\}$

13. Describe each set below in set builder notation.
(a) $\{1, 3, 5, 7, 9, 11, \ldots\}$
(b) $\{1, 2, 3, 4, 5, 6, 7, 8, 9\}$
(c) $\{1, 3, 5, 7, 9\}$
(d) $\{1, 4, 9, 16, 25, 36, \ldots\}$
(e) $\{1, 9, 25, 49, 81, 121, \ldots\}$
(f) $\{\ldots, -3, -2, -1, 0, 1\}$
(g) $\{-3, -2, -1, 0, 1, 2\}$
(h) $\overline{AB}$
(i) $\overrightarrow{AB}$

A B

(j) Is it possible to describe an infinite set in set builder notation?
(k) Is it possible to list an infinite set?

14. When you are asked to **solve an equation**, you are really being asked to *describe its solution set*. Using $\{-5, -4, -3, -2, -1, 0, 1, 2, 3, 4, 5\}$ as domain, solve these equations by enumerating (listing) the solution set.
(a) $x = 5$ (*Note*: This is an equation, not a solution set.)
(b) $x + 4 = 7$
(c) $x - 4 = 7$
(d) $x^2 = 4$
(e) $2x - 4 = 4$
(f) $x(3 - x) = 0$
(g) $8 - x = 5$
(h) $4 - x = 5$
(i) $6 - 5x = 21$

15. When you are asked to **solve an inequality**, you are really being asked to *describe its solution set*. Using $\{0, 1, 2, 3, 4, 5, 6, 7\}$ as domain, solve these inequalities by listing solution sets.
(a) $x < 5$ **(b)** $x > 5$
(c) $2x - 3 < 0$ **(d)** $3 - x < 1$
(e) $4 \geq x$ **(f)** $2 - x < 4x + 5$
(g) $x^2 - 3x < 6$ **(h)** $x^2 < 2x - 1$

16. If the domain in Exercise 15 were the set of all real numbers, could you describe the solution set of inequality (c) by listing it? How might you describe it?

17. When you are asked to **graph** the solution set of an open sentence, you are being asked to describe its solution set in a geometric or schematic way. For each open sentence below, the domain is $\{1, 2, 3, 4, 5, 6, 7, 8, 9, 10\}$. These domain elements are represented by points on a number line. Graph the solution set of each open sentence by darkening the points that represent its solution set.

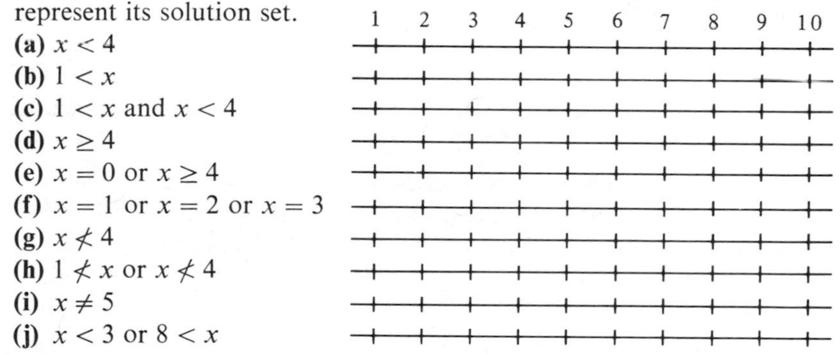

(a) $x < 4$
(b) $1 < x$
(c) $1 < x$ and $x < 4$
(d) $x \geq 4$
(e) $x = 0$ or $x \geq 4$
(f) $x = 1$ or $x = 2$ or $x = 3$
(g) $x \not< 4$
(h) $1 \not< x$ or $x \not< 4$
(i) $x \neq 5$
(j) $x < 3$ or $8 < x$

18. Without trying to solve any open sentences, fill in the blanks.
(a) $\{x \in Q \mid 3x^4 - 2x^3 + 7x = 8 \text{ or } 5x^3 - 2x^2 + 3x = 7\} =$
$\{x \in Q \mid 3x^4 - 2x^3 + 7x = 8\}$______$\{x \in Q \mid 5x^3 - 2x^2 + 3x = 7\}$
(b) $\{x \in Q \mid 5x^5 - 3x < 4 \text{ and } 3x^3 - 2x^2 > 2\} =$
$\{x \in Q \mid 5x^5 - 3x < 4\}$______$\{x \in Q \mid 3x^3 - 2x^2 > 2\}$
(c) $\{x \in Q \mid x > 0\} \cap \{x \in Q \mid 6x^4 - 3x^3 + 2x = 12\} =$
$\{x \in Q \mid x > 0$______$6x^4 - 3x^3 + 2x = 12\}$

(d) $\{x \in Q \mid 2x - 5 < 0\} \cup \{x \in Q \mid x^3 - 2x^2 - x + 2 = 0\} =$
$\{x \in Q \mid \_\_\_\_\_\_\_\_\}$
(e) $\{x \in Q \mid 3x^3 - 2x^2 > 2\}' = \{x \in Q \mid \_\_\_\_\_\_\_\_\}$
(f) $\{x \in Q \mid 6x^4 - 3x^3 + 2x \neq 12\} = (x \in Q \mid \_\_\_\_\_\_\_\_\}'$

2.4 SOLVING SIMPLE OPEN SENTENCES

In the previous section we decided that to "solve" an open sentence is to "describe its solution set." We also learned two ways of giving such a description.

1. by listing, or enumerating, the solution set
2. by graphing the solution set

EXAMPLE 1

Open sentence: $2x - 1 < 6$
Domain of variable: W

Solution Set (listed): $\{0, 1, 2, 3\}$

Graph

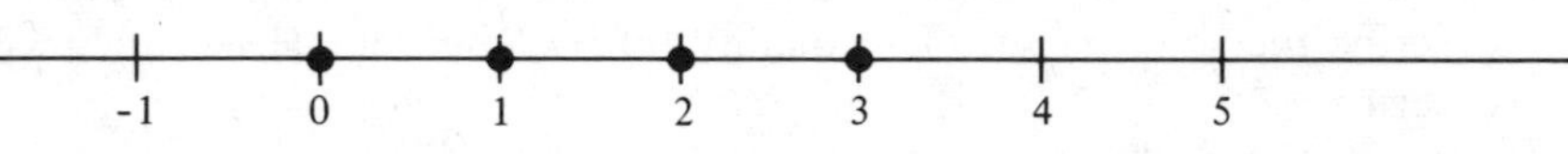

EXAMPLE 2

Open sentence: $4x + 5 = 8$
Domain of variable: R

Solution Set (listed): $\{\frac{3}{4}\}$

Graph

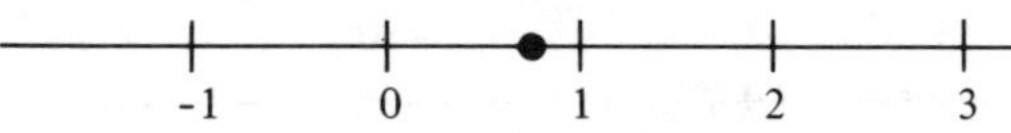

It is not always possible, however, to list the solution set. In such cases the geometric technique of graphing still works, and a nongeometric means of describing the solution set is provided by set builder notation.

EXAMPLE 3

Open sentence: $x - 2 < 1$
Domain of variable: R

Solution Set (in set builder notation): $\{x \in R \mid x < 3\}$

Graph

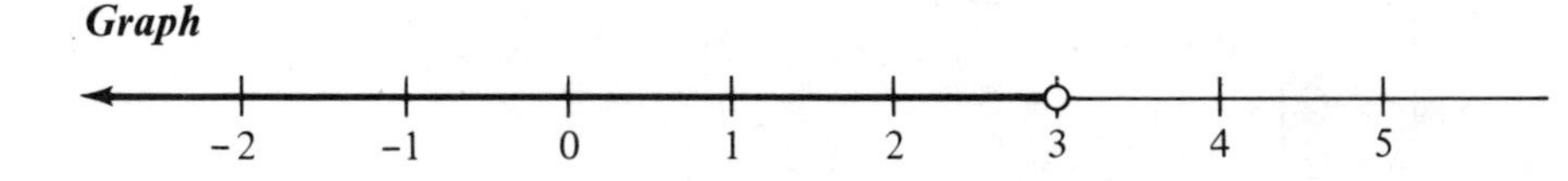

EXAMPLE 4

Open sentence: $2 < 3 + 2x$
Domain of variable: R

Solution Set (in set builder notation): $\{x \in R \mid -\frac{1}{2} < x\}$

Graph

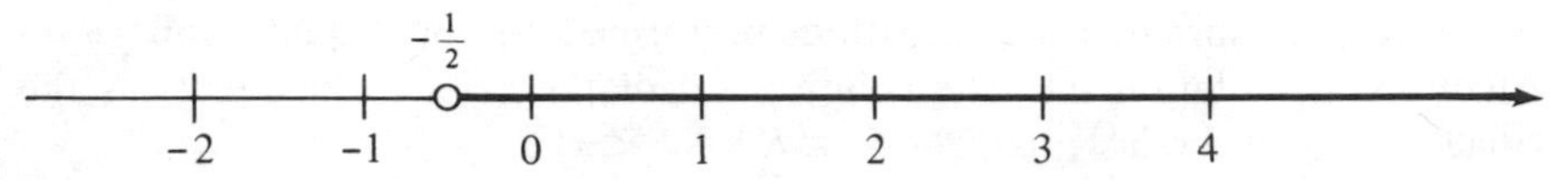

Several observations are suggested by these last two examples. One is that the figure called a half-line which we studied in Chapter 1 is not just a geometric curiosity; it arises naturally in the algebraic context of solving inequalities. Another is that set builder notation needs to be used with good taste. If you are asked to solve the inequality $2 < 3 + 2x$ of Example 4, it is not enough simply to write

$$\text{solution set} = \{x \in R \mid 2 < 3 + 2x\}$$

That is a true statement, but it is begging the question. It is very difficult to look at this description and "see" the solution set. A more useful statement, but still not the answer, is this one:

$$\text{solution set} = \{x \in R \mid -1 < 2x\}$$

It is still not easy to see the solution set. The answer is the one given in the example:

$$\text{solution set} = \{x \in R \mid -\tfrac{1}{2} < x\}$$

From this description it is obvious which real numbers are in the solution set and which are not. When you are asked to solve an open sentence, you are always expected to "simplify" the set builder description as much as possible.

The most important observation to be made from these examples, however, is that the tedious process of—(1) replacing the variable by each domain element in turn, (2) classifying the resulting statements as true or false, and thus (3) identifying the truth set—has been superseded by some powerful algebraic shortcuts. In Example 2 (open sentence, $4x + 5 = 8$; domain of variable, R) you probably arrived at the solution set $\{\frac{3}{4}\}$ as follows:

1. The given open sentence (an equation): $4x + 5 = 8$
2. Add -5 to both members ("sides"): $4x = 3$
3. Multiply both members by $\frac{1}{4}$: $x = \frac{3}{4}$

You feel certain that all three equations have the same solution set, and since the solution set of the last equation is obviously $\{\frac{3}{4}\}$, that is the solution set of the first equation as well.

In Example 4 (open sentence, $2 < 3 + 2x$; domain of variable, R) the procedure was identical.

1. The given open sentence (an inequality): $2 < 3 + 2x$
2. Add -3 to both members: $-1 < 2x$
3. Multiply both members by $\frac{1}{2}$: $-\frac{1}{2} < x$

Again you feel sure that the operations performed on the inequalities have no effect on the solution set. The solution set of the original inequality is the same as that of the last; namely, $\{x \in R \mid -\frac{1}{2} < x\}$.

Let us look more closely at these solution techniques to see *why* they work, first in the context of equations. Underlying the procedure, illustrated above, for solving equations are two obvious, but fundamental, principles of arithmetic.

---

1. The addition principle of equality: For any real numbers a, b, and c,

$$\text{if } a = b, \text{ then } a + c = b + c$$

For example, since $14 = 2 \cdot 7$, we can conclude that $14 + 7 = 2 \cdot 7 + 7$.

---

2. The multiplication principle of equality: For any real numbers a, b, and c,

$$\text{if } a = b, \text{ then } a \cdot c = b \cdot c$$

For example, since $14 = 10 + 4$, we can conclude that $14 \cdot 7 = (10 + 4) \cdot 7$. We return to Example 2 to illustrate the use of these principles.

The original equation to be solved is $4x + 5 = 8$. For the moment think of x not as a blank, but as a particular real number for which the statement, $4x + 5 = 8$, is true. Then, by the addition principle of equality, the statement, $4x + 5 + (-5) = 8 + (-5)$, is also true; that is, the statement, $4x = 3$, is true. In other words, any element of the solution set of the first equation, $4 \cdot \square + 5 = 8$, is also an element of the solution set of the second equation, $4 \cdot \square = 3$. In subset notation,

$$\text{solution set of Equation 1} \subset \text{solution set of Equation 2}$$

Conversely, if x is any real number for which the statement, $4 \cdot x = 3$, is true then, by the addition principle of equality, the statement, $4 \cdot x + 5 = 3 + 5$, is also true; that is, the statement, $4x + 5 = 8$, is true. In other words, any element of the solution set of the second equation, $4 \cdot \square = 3$, is also an element of the solution set of the first equation, $4 \cdot \square + 5 = 8$. In subset notation

$$\text{solution set of Equation 2} \subset \text{solution set of Equation 1}$$

We have proved, on the basis of the addition principle of equality, that Equations 1 and 2 have the same solution set.

A similar kind of argument, based on the multiplication principle of equality, can be given to show that Equation 2, $4x = 3$, and Equation 3, $x = \frac{3}{4}$, have the same solution set. By using the arrow "$\Rightarrow$" as an abbreviation for "implies," the argument can be given quite briefly in symbolic form:

$$4x = 3 \Rightarrow \tfrac{1}{4} \cdot 4x = \tfrac{1}{4} \cdot 3 \Rightarrow x = \tfrac{3}{4} \Rightarrow$$

solution set of Equation 2 $\subset$ solution set of Equation 3

Conversely,

$$x = \tfrac{3}{4} \Rightarrow 4 \cdot x = 4 \cdot \tfrac{3}{4} \Rightarrow 4x = 3 \Rightarrow$$

solution set of Equation 3 $\subset$ solution set of Equation 2

Thus

solution set of Equation 2 = solution set of Equation 3

In proving that two equations have the same solution set, it is crucial that you be able to "get" from either one to the other. We got from the first equation, $4x + 5 = 8$, to the second, $4x = 3$, by adding -5 to both sides. To get back from the second to the first we needed to add the opposite of -5, namely 5. Since every real number (not just -5) has an opposite (or additive inverse), it is clear that we can always get back and forth by addition.

The addition technique for solving equations: Adding the same number to both members of an equation does not change the solution set.

For multiplication the situation is similar except for one danger spot. We got from the second equation, $4x = 3$, to the third, $x = \frac{3}{4}$, by multiplying both sides by $\frac{1}{4}$. To get back from the third to the second we needed to multiply by the reciprocal of $\frac{1}{4}$, namely 4. Since every real number *except zero* has a reciprocal (or multiplicative inverse), we can get back and forth as long as we avoid using zero as a multiplier.

The multiplication technique for solving equations: Multiplying both members of an equation by the same *nonzero* number does not change the solution set.

The procedure, then, for solving equations of the type we have studied is clearly this: Use the addition and multiplication techniques to replace the original equation by successively simpler equations until an equation is reached whose solution set is obvious. This solution set will also be the solution set of the original equation.

In the exercises below you are asked to formulate addition and multiplication principles and techniques for inequalities.

EXERCISES

1. (a) Graph the points corresponding to $-\frac{3}{2}$ and 2 on the number line

-5 -4 -3 -2 -1 0 1 2 3 4 5 6

and then enter these two numbers in the blanks:

________ < ________.

(b) Starting at the points above, slide right three units.
Fill in the blanks: $-\frac{3}{2}$ + ________ < ________ + ________.

(c) Starting at the points in part (a), slide left $1\frac{1}{2}$ units.
Fill in the blanks: ________ + ________ < ________ + ________.

2. Complete this statement of the **addition principle of inequality**:* For any real numbers a, b, and c, if $a < b$, then ________.

3. (a) Graph the arrows corresponding to $\frac{2}{3}$ and 2 on the number line

-5 -4 -3 -2 -1 0 1 2 3 4 5 6 7

and then fill in the blanks: ________ < ________.

(b) Stretch the above arrows by 3 and fill in the blanks:

________ · ________ < ________ · ________.

(c) Shrink the arrows in part (a) by $\frac{1}{2}$ and fill in the blanks:

________ · ________ < ________ · ________.

(d) Flip the arrows in part (a) and fill in the blanks:

________ < ________.

4. Fill in the correct symbol: <, =, or >.

(a) $\frac{3}{4} \square 3$ **(b)** $2 \cdot \frac{3}{4} \square 2 \cdot 3$

(c) $\frac{1}{6} \cdot \frac{3}{4} \square \frac{1}{6} \cdot 3$ **(d)** $-1 \cdot \frac{3}{4} \square -1 \cdot 3$

(e) $-\frac{2}{3} \cdot \frac{3}{4} \square -\frac{2}{3} \cdot 3$ **(f)** $-2 \cdot \frac{3}{4} \square -2 \cdot 3$

(g) $0 \cdot \frac{3}{4} \square 0 \cdot 3$

5. Complete this statement of the **multiplication principles of inequality**.** For any real numbers a, b, c:

If $a < b$ and $c > 0$, then ________.
If $a < b$ and $c = 0$, then ________.
If $a < b$ and $c < 0$, then ________.

6. The addition and multiplication principles of inequality play the same role in the solution of inequalities as the addition and multiplication

* Also sometimes called the *monotony law of addition*.

** Also called the *monotony laws of multiplication*.

principles of equality play in the solution of equations. Fill in the blanks below. (Think of x as some real number, not as a blank.)

(a) $2x - 3 < 4 \Rightarrow 2x - 3 + 3 < 4 + 3$ (i.e., $2x < 7$) by the ________ principle of inequality

(b) $2x < 7 \Rightarrow \frac{1}{2} \cdot 2x < \frac{1}{2} \cdot 7$ (i.e., $x < \frac{7}{2}$) by the ________ principle of inequality

(c) $4 - 2x < 1 \Rightarrow -4 + 4 - 2x < -4 + 1$ (i.e., $-2x < -3$) by the ________ principle of inequality

(d) $-2x < -3 \Rightarrow -\frac{1}{2} \cdot -2x > -\frac{1}{2} \cdot -3$ (i.e., $x > \frac{3}{2}$) by the ________ principle of inequality

(e) $5x + 2 < 7 \Rightarrow 5x < 5$ by the ________ principle of inequality

(f) $4 < 6 - 3x \Rightarrow -2 < -3x$ by the ________ principle of inequality

(g) $\frac{3}{4}x < 6 \Rightarrow x < 8$ by the ________ principle of inequality

(h) $\frac{2}{3}x < 5 \Rightarrow x <$ ________

(i) $x + 2 < \frac{3}{4} \Rightarrow x <$ ________

(j) $-6x < 5 \Rightarrow x >$ ________

In Exercises 7–10 justify each implication arrow with the words "addition principle of inequality" or "multiplication principle of inequality." Also state what number has been added or used to multiply.

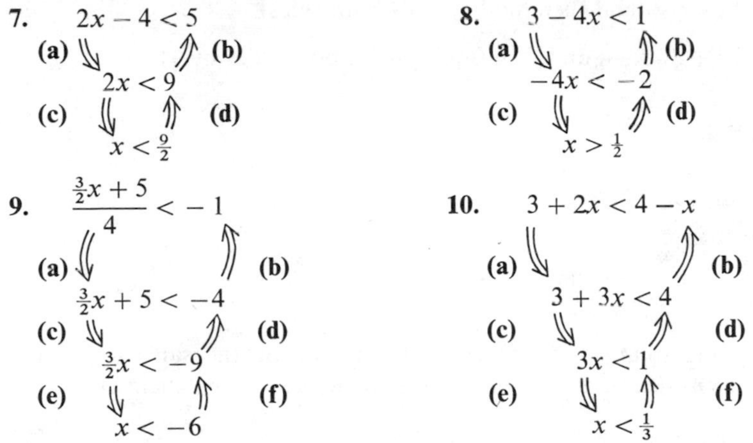

11. The reversibility of the implication arrows in Exercises 7–10 suggests addition and multiplication techniques for solving inequalities similar to those for solving equations.

(a) State the **addition technique for solving inequalities**.

(b) State the **multiplication techniques for solving inequalities**.

In actually writing out the steps in solving an inequality (or equation), it is customary to use the **double implications** symbol, "$\Leftrightarrow$," which you can pronounce as "if and only if." For example, in solving the inequality

$$\frac{\frac{3}{2}x + 5}{4} < -1$$

one could write:

$$\frac{\frac{3}{2}x + 5}{4} < -1 \Leftrightarrow \tfrac{3}{2}x + 5 < -4 \Leftrightarrow \tfrac{3}{2}x < -9 \Leftrightarrow x < -6$$

The double arrows mean that you can read from left to right *and* from right to left. Reading from left to right the reasoning is this:

If $\frac{\frac{3}{2}x + 5}{4} < -1$, then (multiplying both sides by 4)

$\frac{3}{2}x + 5 < -4$ and then (adding -5 to both sides)

$\frac{3}{2}x < -9$ and *then* (multiplying both sides by $\frac{2}{3}$)

$x < -6$

Reading from right to left the reasoning is this:

If $x < -6$, then (multiplying both sides by $\frac{3}{2}$)

$\frac{3}{2}x < -9$ and then (adding 5 to both sides)

$\frac{3}{2}x + 5 < -4$ and *then* (multiplying both sides by $\frac{1}{4}$)

$\frac{\frac{3}{2}x + 5}{4} < -1$

Thus the open sentences,

$$\frac{\frac{3}{2}x + 5}{4} < -1 \quad \text{and} \quad x < -6$$

have the same solution set. Open sentences having the same solution are said to be **equivalent**. All four of the open sentences in the chain

$$\frac{\frac{3}{2}x + 5}{4} < -1 \Leftrightarrow \tfrac{3}{2}x + 5 < -4 \Leftrightarrow \tfrac{3}{2}x < -9 \Leftrightarrow x < -6$$

are equivalent.

In working Exercise 12 below, use double-arrow notation and produce a chain of simpler and simpler equivalent open sentences.

12. For each open sentence below, (i) describe its solution set in (simplest possible) set builder notation, and (ii) graph its solution set on the number line.

(a) $\frac{1}{2}x - 3 < -1$
(b) $\frac{2}{3}x + 4 < 6$
(c) $2 - 3x < 6$

(d) $-(4 + x) < 1$
(e) $3x - 7 = 1$
(f) $2 < x + 5$
(g) $3 - (x - 1) < 4$
(h) $-2 < 1 - x$
(i) $(4 - x)/3 < 1$
(j) $2x + 5 < 2$
(k) $2x + 5 = 2$
(l) $2x + 5 > 2$
(m) $(5 - 3x) - (x + 2) > 4(2 - x) - (5 - x)$
(n) $2x(5 - x) + 4 < x(5 - 2x) + (3 - x)$
(o) $2(3 - x) + x < 5 - (x - 3)$
(p) $-3(4 - x) > 6 - (5 - 3x)$

13. The half-lines (or open rays) below represent sets of real numbers. Describe each set in set builder notation.

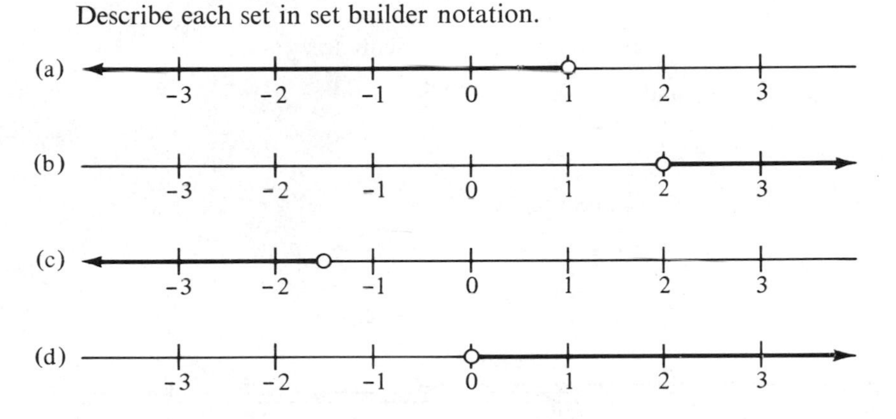

2.5 SOLVING COMPOUND OPEN SENTENCES

In the previous section we solved "simple" equations and inequalities. In this section we study "compound" open sentences. "Simple" and "compound" have nearly the same meaning in mathematics as they do in grammar.

EXAMPLES

1. "Eddie likes spaghetti" is a simple sentence.
2. "Eddie likes spaghetti and Tommy likes salami" is a compound sentence.
3. "$2x + 3 < 4$" is a simple (open) sentence.
4. "$2x + 3 < 4$ and $0 < 2x + 3$" is a compound (open) sentence.
5. "$2x + 3 < 4$ or $2x + 3 = 4$" is a compound (open) sentence.

The compound sentences above are made up of simple sentences joined by the **logical operators**: "and," "or" The word "not" is the third basic logical operator. While a grammarian would not classify the sentence

Eddie does not like spaghetti

as compound, mathematicians sometimes do refer to an open sentence such as

$2x + 3$ is not less than 4

as compound because it is derived from a simple open sentence ($2x + 3 < 4$) by means of a logical operator (not).

To solve a compound open sentence one must be able to do two things:

I. Determine the solution sets of the simple open sentences that appear in it.

II. Apply to these solution sets the set operations that correspond to the logical operators in the compound sentence.

In the examples that follow, and for the rest of the chapter, the domain of the variable is the set R of all real numbers.

EXAMPLE Solve the compound open sentence,

$$2x + 3 < 4 \quad \text{and} \quad 0 < 2x + 3$$

Solution Set

$$\{x \in R \mid 2x + 3 < 4 \text{ and } 0 < 2x + 3\}$$
$$= \{x \in R \mid 2x + 3 < 4\} \cap \{x \in R \mid 0 < 2x + 3\}$$
$$= \{x \in R \mid x < \tfrac{1}{2}\} \cap \{x \in R \mid -\tfrac{3}{2} < x\}$$

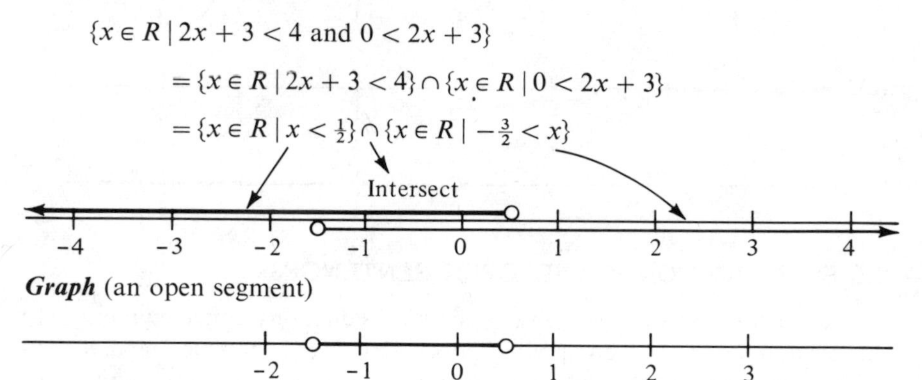

Graph (an open segment)

EXAMPLE Solve the compound open sentence,

$$2x + 3 < 4 \quad \text{or} \quad 2x + 3 = 4$$

Solution Set

$$\{x \in R \mid 2x + 3 < 4 \quad \text{or} \quad 2x + 3 = 4\}$$
$$= \{x \in R \mid 2x + 3 < 4\} \cup \{x \in R \mid 2x + 3 = 4\}$$
$$= \{x \in R \mid x < \tfrac{1}{2}\} \cup \{x \in R \mid x = \tfrac{1}{2}\}$$

Unite

-4 -3 -2 -1 0 1 2 3

Graph (a ray)

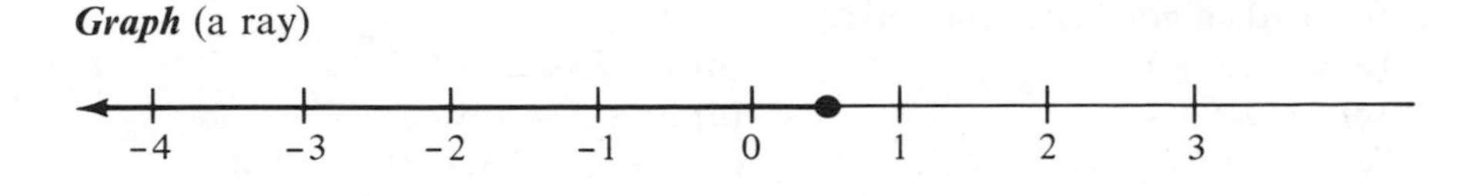

E X A M P L E Solve the compound open sentence,

$2x + 3$ is not less than 4

Solution Set

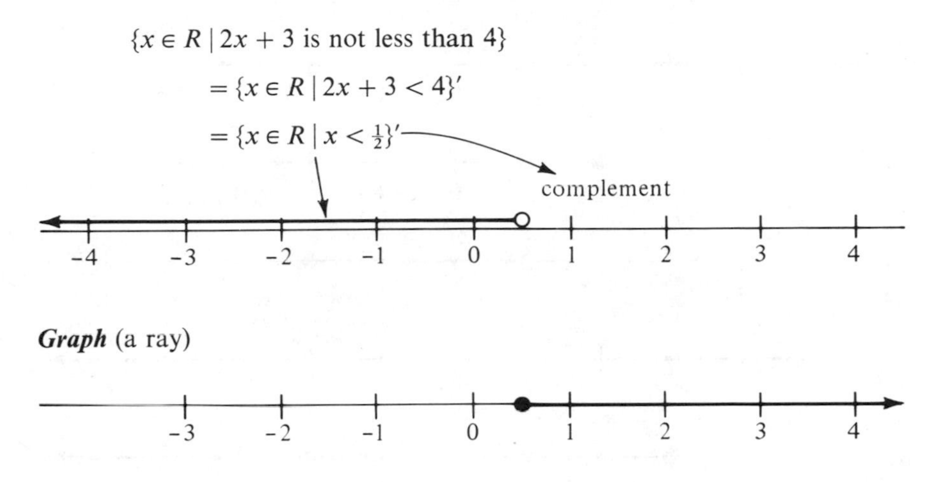

$$\{x \in R \mid 2x + 3 \text{ is not less than } 4\}$$
$$= \{x \in R \mid 2x + 3 < 4\}'$$
$$= \{x \in R \mid x < \tfrac{1}{2}\}'$$

Graph (a ray)

EXERCISES

1. The presence of logical operators is often obscured by the brevity of mathematical symbolism. For example, the symbols "$-3 < x < 1$" mean "$-3 < x$ *and* $x < 1$"; the symbols "$4 \leq x$" mean "$4 < x$ *or* $4 = x$"; the symbols "$x \nless 2$" mean "x is *not* less than 2." Write out the meaning of each open sentence below showing the logical connectives used. Then graph the solution set and name it in geometric terminology.

(a) $0 < x < 3$
(b) $-1 \leq x$
(c) $x \nless 2$
(d) $x \neq 0$
(e) $-3 < x \leq 1$
(f) $1 \leq x < 4$
(g) $-1 \leq x \leq 2$
(h) $x \nleq 1$

2. Graph the solution sets of these compound open sentences and describe the graphs in geometric terminology.

(a) $x < 0$ or $x > 2$
(b) $x \leq 1$ or $x > 2$
(c) $x < 0$ or $x = 3$
(d) $x > 0$ and $x > 3$
(e) $x < 0$ and $x > 1$
(f) $x < 0$ and $x > -2$
(g) not $(x < 0$ and $x > -2)$
(h) $x \nless 0$ or $x \ngtr -2$
(i) $2 \leq x$ and $x \leq 2$
(j) $x = 3$ and $x \ngtr 1$

3. Solve, then graph solution sets:

(a) $0 < 2x + 1 < 4$

(b) $1 < 3 - 2x < 3$

(c) $5 - 2x \leq 4$

(d) $3x + 5 \not> -1$

(e) $4 - x < 2$ or $2x + 1 > 2$

(f) $4 - x < 2$ and $2x + 1 > 2$

(g) $2x - 3 < 4$ and $-(2x - 3) < 4$

(h) $3 - 2x > 1$ or $-(3 - 2x) > 1$

(i) $2 < 1 - 3x$

(j) $2 \not< 1 - 3x$

4. Each geometric figure below is the graph of some open sentence. Write the open sentence.

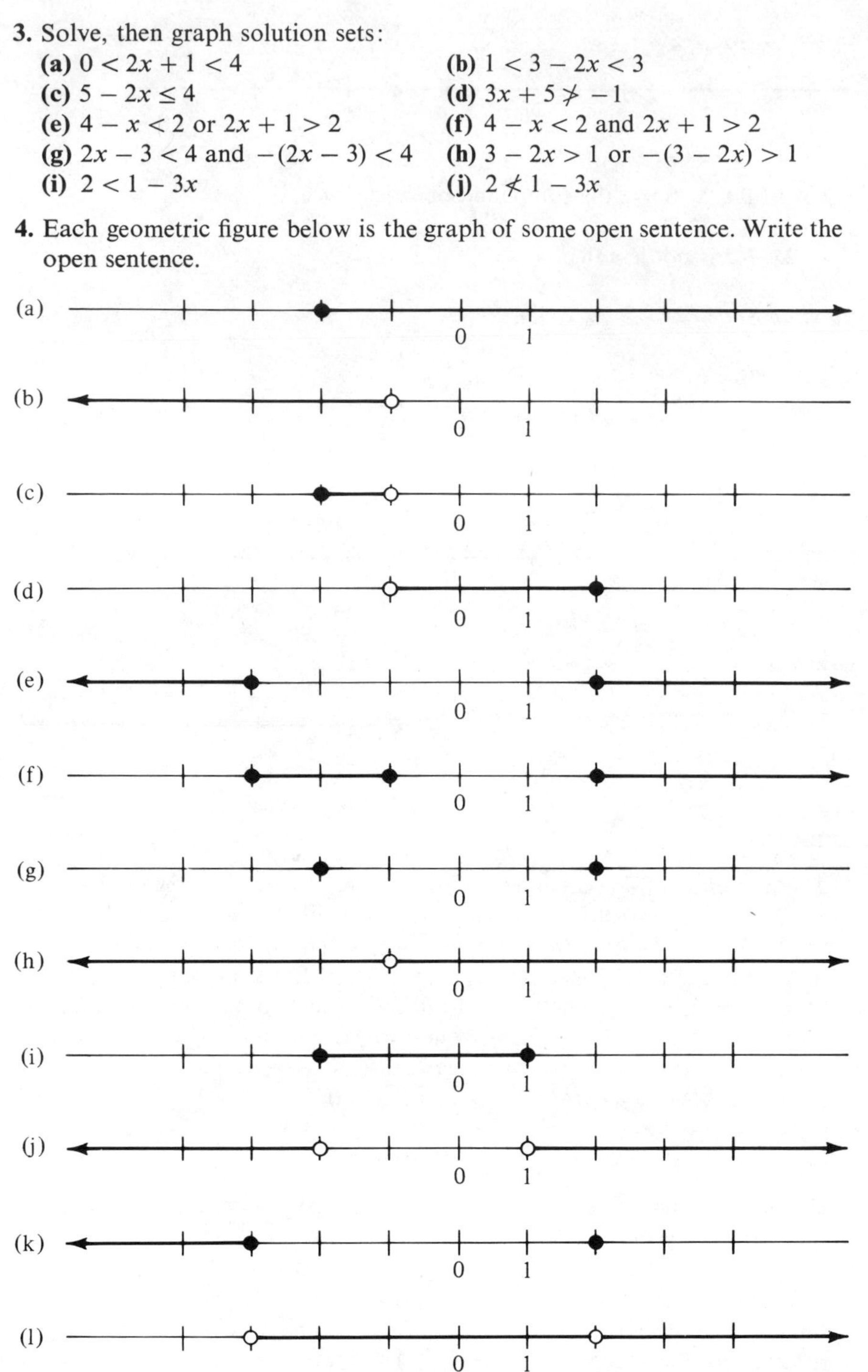

5. A simple inequality such as $\frac{1}{3}x + 2 < 1$ can be solved very quickly using addition and multiplication techniques:

$$\tfrac{1}{3}x + 2 < 1 \Leftrightarrow \tfrac{1}{3}x < -1 \Leftrightarrow x < -3$$

Can the very same techniques be used to solve the inequality $\frac{1}{3}x + 2 \leq 1$ which, technically, is compound? Be prepared to justify each equals sign in this illustration that they can.

$$\begin{aligned}
&\{x \in R \mid \tfrac{1}{3}x + 2 \leq 1\} \\
&\quad = \{x \in R \mid \tfrac{1}{3}x + 2 < 1 \text{ or } \tfrac{1}{3}x + 2 = 1\} \\
&\quad = \{x \in R \mid \tfrac{1}{3}x + 2 < 1\} \cup \{x \in R \mid \tfrac{1}{3}x + 2 = 1\} \\
&\quad = \{x \in R \mid \tfrac{1}{3}x < -1\} \cup \{x \in R \mid \tfrac{1}{3}x = -1\} && \text{(the "add } -2\text{" step)} \\
&\quad = \{x \in R \mid x < -3\} \cup \{x \in R \mid x = -3\} && \text{(the "multiply by 3" step)} \\
&\quad = \{x \in R \mid x < -3 \text{ or } x = -3\} \\
&\quad = \{x \in R \mid x \leq -3\}
\end{aligned}$$

2.6 ABSOLUTE VALUE AND INEQUALITIES —A GEOMETRIC VIEW

The dual, arithmetic–geometric interpretation of real numbers that we introduced in Section 2.1 is especially useful in solving inequalities that involve absolute value. In this section we take a geometric view. In the next section we take an arithmetic view. For some inequalities the geometric approach is easier, for others the arithmetic is easier.

In geometric terms, the absolute value of any number x, written $|x|$, is the distance from its point on the number line to the origin.

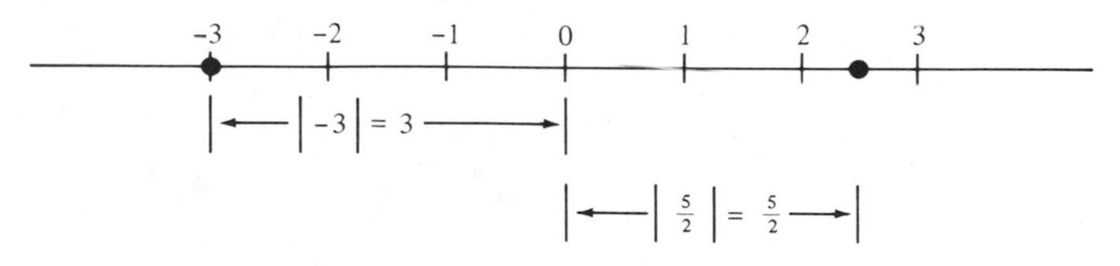

GEOMETRIC DEFINITION OF ABSOLUTE VALUE

$|x|$ = distance from x to 0 on number line

Why is it true that $|x| \geq 0$ for every real number x?

The absolute value symbol is often used in writing inequalities. As a rule you can expect that a "simple" looking inequality involving an absolute value is actually a compound inequality in disguise.

EXAMPLE Solve the inequality, $|x| < 5$.

Solution $|x| < 5 \Leftrightarrow$ the distance from x to the origin is less than $5x \Leftrightarrow > -5$ and $x < 5$; that is $-5 < x < 5$.

Graph

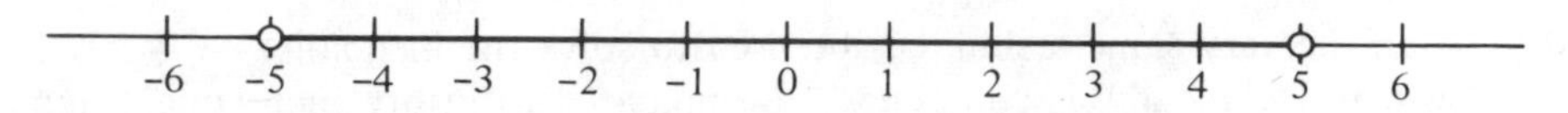

EXAMPLE Solve the inequality, $|x| > 2$.

Solution $|x| > 2 \Leftrightarrow$ the distance from x to the origin is greater than $2 \Leftrightarrow x < -2$ *or* $x > 2$.

Graph

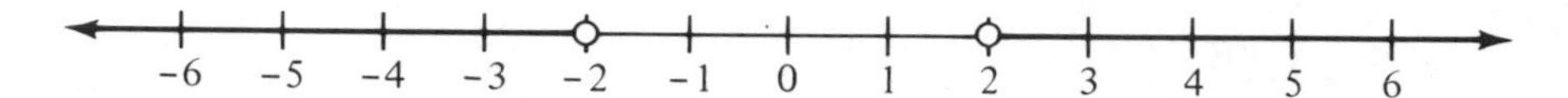

EXERCISES

1. Simplify as much as possible:

(a) $|-6|$ **(b)** $|3|$
(c) $|5 - 8|$ **(d)** $|4 - (-1)|$
(e) $|(-1) - 4|$ **(f)** $|8 - 5|$
(g) $|-3/4|$ **(h)** $|-3/-4|$

2. For each pair of numbers do two things: (i) Write a statement of inequality telling which is larger; (ii) write a statement telling which is larger in absolute value.

EXAMPLE $-3, 1$

(i) $-3 < 1$
(ii) $|1| < |-3|$

(a) $-4, 0$ **(b)** $-5, -2$
(c) $-2, 3$ **(d)** $-4, 3$
(e) $0, 5$ **(f)** $-50, -40$
(g) $50, 40$ **(h)** $-40, 40$

3. For each pair of numbers do three things: (i) Graph them on the number line; (ii) determine the distance between their graphs; (iii) find the absolute value of the difference of the two numbers. (Will it matter which number is subtracted from which?)

EXAMPLE $-2, 3$

(i)

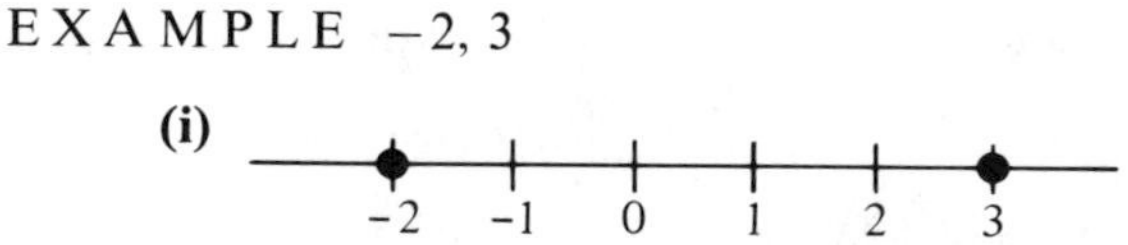

(ii) 5
(iii) $|-2-3| = |-5| = 5$

(a) $-3, -1$ **(b)** $2, -3$
(c) $-2, 0$ **(d)** $1, 5$
(e) $1, -1$ **(f)** $-\frac{1}{2}, 2\frac{1}{4}$

4. "For any two real numbers a and b, $|a-b|$ represents the distance between point ________ and point ________ on the number line."

5. "If x is a real number for which $|x-4|<2$, then the distance from point x to point ________ is less than ________." Now graph $\{x \in R \mid |x-4|<2\}$.

6. "If x is a real number for which $|x-1|>3$, then the distance from point ________ to point ________ is greater than ________." Now graph $\{x \in R \mid |x-1|>3\}$.

7. Translate each open sentence below into a (English) sentence about distance. Then graph the solution set on a number line.

EXAMPLE $|x-2|<4$

The distance from x to 2 is less than 4.

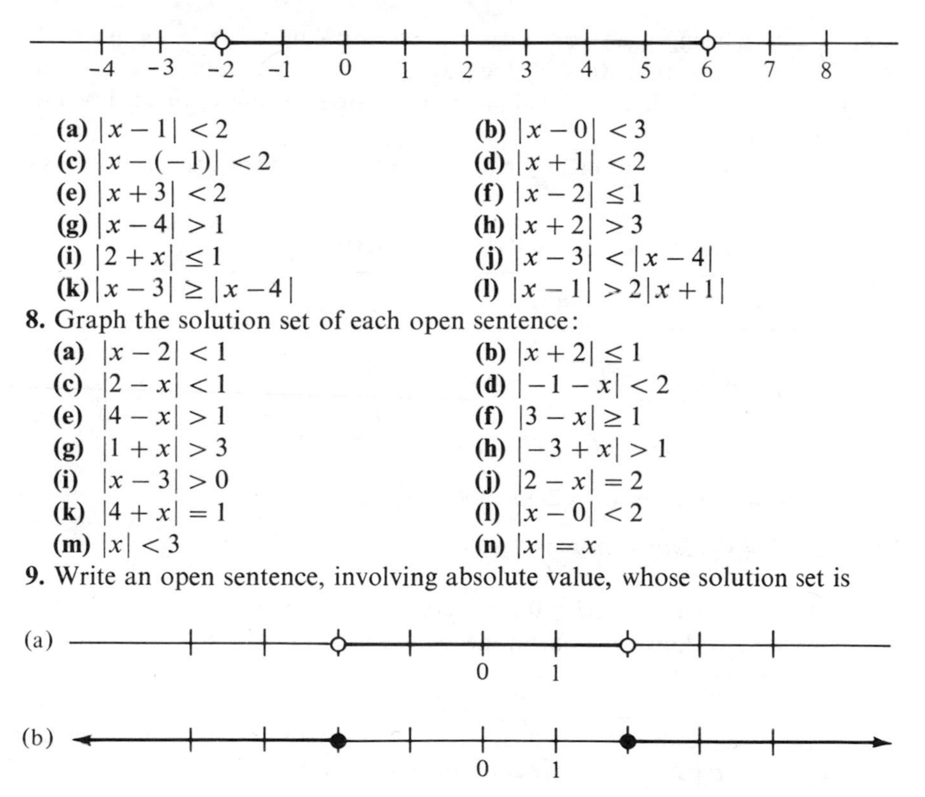

(a) $|x-1|<2$ **(b)** $|x-0|<3$
(c) $|x-(-1)|<2$ **(d)** $|x+1|<2$
(e) $|x+3|<2$ **(f)** $|x-2| \le 1$
(g) $|x-4|>1$ **(h)** $|x+2|>3$
(i) $|2+x| \le 1$ **(j)** $|x-3|<|x-4|$
(k) $|x-3| \ge |x-4|$ **(l)** $|x-1|>2|x+1|$

8. Graph the solution set of each open sentence:

(a) $|x-2|<1$ **(b)** $|x+2| \le 1$
(c) $|2-x|<1$ **(d)** $|-1-x|<2$
(e) $|4-x|>1$ **(f)** $|3-x| \ge 1$
(g) $|1+x|>3$ **(h)** $|-3+x|>1$
(i) $|x-3|>0$ **(j)** $|2-x|=2$
(k) $|4+x|=1$ **(l)** $|x-0|<2$
(m) $|x|<3$ **(n)** $|x|=x$

9. Write an open sentence, involving absolute value, whose solution set is

(a) 0 1

(b) 0 1

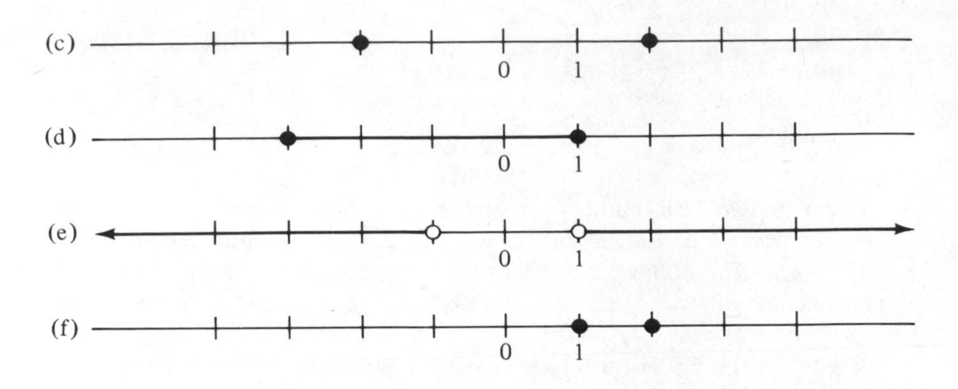

2.7 ABSOLUTE VALUE AND INEQUALITIES —AN ARITHMETIC VIEW

It is usually useful to know more than one method of solving problems. Then if one method fails another can be tried. Consider the problem of finding the solution set of the inequality

$$|4 - 3x| > x + 2$$

Can you do it using what you learned about absolute value in the previous section? In this section we shall look at absolute value from an arithmetic point of view. This viewpoint will allow us to solve the above inequality and many others.

The absolute value of any real number x is defined arithmetically as follows:

ARITHMETIC DEFINITION OF ABSOLUTE VALUE

$|x|$ = the larger of x and $-x$

EXAMPLES

$|4|$ = the larger of 4 and $-4 = 4$

$|-2|$ = the larger of -2 and $-(-2) = 2$

$|0|$ = the larger of 0 and $-0 = 0$

Why is it true that $|x| \geq 0$ for every real number x?

Before attacking the difficult inequality, $|4 - 3x| > x + 2$, we warm up with some easier ones.

EXAMPLE Solve the inequality, $|x| < 5$. [Recall that to solve an inequality is to replace it by an equivalent inequality whose solution set is "obvious." In this case, solving amounts to getting rid of the absolute value symbol.]

(Old) Geometric Solution $|x| < 5 \Leftrightarrow$ the distance from point x to the origin is less than $5 \Leftrightarrow -5 < x$ *and* $x < 5$; that is, $-5 < x < 5$.

(New) Arithmetic Solution $|x| < 5 \Leftrightarrow$ the larger of x and $-x$ is less than $5 \Leftrightarrow$ (both) $x < 5$ *and* $-x < 5 \Leftrightarrow x < 5$ and $x > -5$; that is, $-5 < x < 5$.

Graph

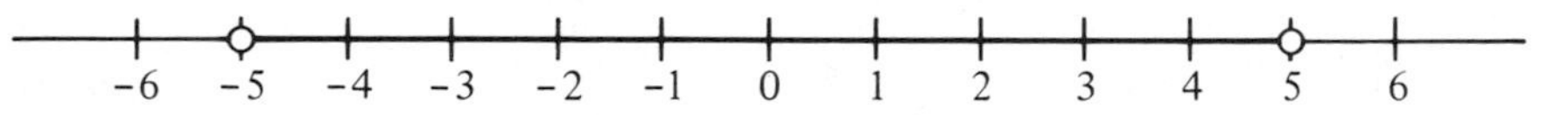

EXAMPLE Solve the inequality, $|x - 1| > 2$.

(Old) Geometric Solution $|x - 1| > 2 \Leftrightarrow$ the distance from x to 1 is greater than $2 \Leftrightarrow x < -1$ or $x > 3$.

(New) Arithmetic Solution $|x - 1| > 2 \Leftrightarrow$ the larger of $(x - 1)$ and $-(x - 1)$ is greater than $2 \Leftrightarrow$ (either) $x - 1 > 2$ *or* $-x + 1 > 2 \Leftrightarrow x > 3$ or $x < -1$.

Graph

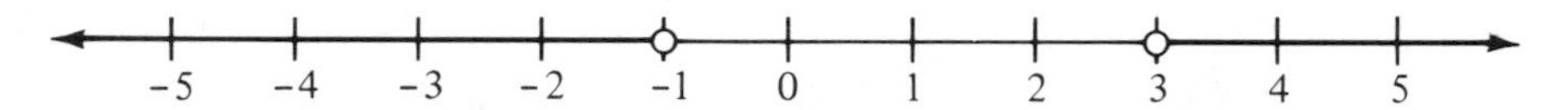

If you found the geometric solutions in the preceding examples easier to follow than the arithmetic solutions, that is because the inequalities involved ($|x| < 5$, $|x - 1| > 2$) were so uncomplicated. In more complicated situations the arithmetic technique is often preferable.

EXAMPLE Solve the inequality, $|4 - 3x| > x + 2$.

Solution (Arithmetic)

$$|4 - 3x| > x + 2$$

$\Leftrightarrow$ the larger of $(4 - 3x)$ and $-(4-3x)$ is greater than $x + 2$

$\Leftrightarrow$ (either) $4 - 3x > x + 2$ or $-(4 - 3x) > x + 2$

$\Leftrightarrow 4 - 3x > x + 2$ or $-4 + 3x > x + 2$

$\Leftrightarrow 4 - 2 > x + 3x$ or $3x - x > 2 + 4$

$\Leftrightarrow 2 > 4x$ or $2x > 6$

$\Leftrightarrow x < \frac{1}{2}$ or $x > 3$

Graph

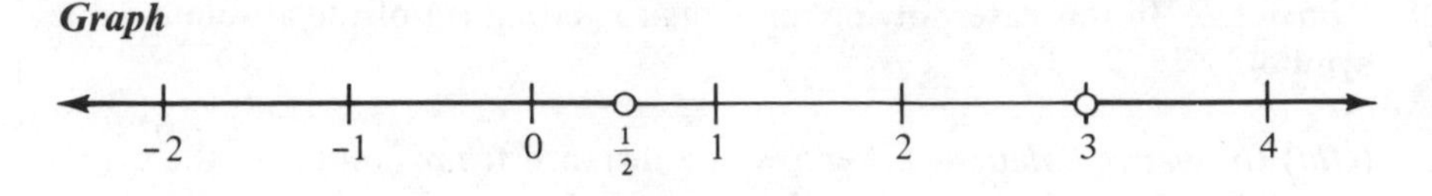

EXERCISES

1. In solving an inequality by the arithmetic method, it is crucial that the proper logical operator be used. Remembering that $|x - 4|$ is the *larger* of $(x - 4)$ and $-(x - 4)$, fill in the boxes with "and" or "or."

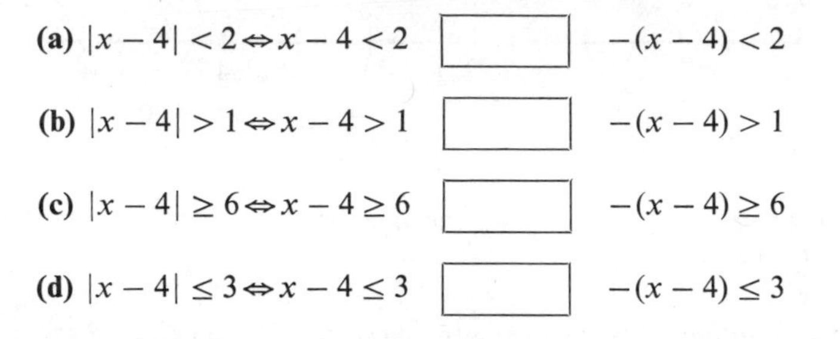

(a) $|x - 4| < 2 \Leftrightarrow x - 4 < 2$ ☐ $-(x - 4) < 2$

(b) $|x - 4| > 1 \Leftrightarrow x - 4 > 1$ ☐ $-(x - 4) > 1$

(c) $|x - 4| \geq 6 \Leftrightarrow x - 4 \geq 6$ ☐ $-(x - 4) \geq 6$

(d) $|x - 4| \leq 3 \Leftrightarrow x - 4 \leq 3$ ☐ $-(x - 4) \leq 3$

2. Fill in the blanks as in Exercise 1:
(a) $|x - 4| < 4 \Leftrightarrow$ ______________________________
(b) $|x - 4| \geq 2 \Leftrightarrow$ ______________________________

3. Complete the solution of the inequality, $|2x - 1| < 5$, begun below. Then graph the solution set.

$|2x - 1| < 5 \Leftrightarrow$ the larger of $2x - 1$ and $-(2x - 1)$ is less than 5

$\Leftrightarrow 2x - 1 < 5$ ☐ ? ☐ $-(2x - 1) < 5$

$\Leftrightarrow$ ____________

$\vdots \qquad \vdots$

4. Solve each inequality below by the careful arithmetic technique suggested in Exercise 3 and the last example on p. 73. Graph each solution set.
(a) $|4 + 3x| \geq 1$
(b) $|5 - 2x| \leq 3$
(c) $|2x + 3| > 5 - x$

5. Solve geometrically or arithmetically, whichever is easier. Graph each solution set.

(a) $|x - 2| < 2$
(b) $|x + 3| < \frac{3}{2}$
(c) $|3x + 2| < 4$
(d) $|5 - x| > 1$
(e) $x = |x|$
(f) $x \cdot |x| > 0$
(g) $|3 - 4x| \geq 1 + x$
(h) $|x - 2| = |x - 3|$
(i) $|-x| = |x|$
(j) $|x - 4| < -2$
(k) $3|x - 2| < 6$
(l) $4 + x = |4 + x|$

(m) $|x - 4| = \frac{1}{2}|x - 1|$ **(n)** $|2x - 3| < 1$
(o) $|2x - 3| \geq 1$ **(p)** $2 \leq |x - 4| \leq 3$
(q) $0 < |x + 2| < 1$ **(r)** $2 \leq |x - 1| \leq 1$

6. Explain geometrically and prove arithmetically that $|a - b| = |b - a|$ for all real numbers a and b.

7. In describing the number line on p. 38 we stated, rather loosely, that "distances between points on the line correspond to differences of the numbers matched with the points." Using the concept of absolute value, rewrite more precisely what is really meant by the statement in quotation marks.

2.8 APPLICATIONS

In this chapter two different views of the symbol x have emerged. One is that x is a *blank* and that an expression such as

$$2x + 3 = 4$$

is an *open sentence*, being neither true nor false until x is replaced by a member of its domain (replacement set). From this point of view, solving the above equation amounts to replacing it by successively simpler equivalent open sentences until one is reached whose solution set is obvious.

$$2x + 3 = 4 \Leftrightarrow \qquad \text{(read "}\Leftrightarrow\text{" as "is equivalent to")}$$
$$2x = 1 \Leftrightarrow$$
$$x = \tfrac{1}{2}$$

Answer $\{\frac{1}{2}\}$ because solution set of first open sentence = solution set of last open sentence = $\{\frac{1}{2}\}$ (obviously).

The other view of x is that x is a (any) real *number* for which

$$2x + 3 = 4$$

is a true *statement*. From this point of view, "solving" involves the same steps, but slightly different language:

$$2x + 3 = 4 \Leftrightarrow \qquad \text{(read "}\Leftrightarrow\text{" as "if and only if")}$$
$$2x = 1 \Leftrightarrow$$
$$x = \tfrac{1}{2}$$

Now the final line is the answer: The number x, for which $2x + 3 = 4$ is true, is $\frac{1}{2}$.

You can expect to meet both viewpoints in your teaching. Each has its virtues. The first is generally more satisfying when mathematics is being studied for its own sake. The second is often more natural when mathematics is being applied to real problems. The differences are psychological and stylistic rather than substantive, and it would not be advisable, in your teaching, to insist that one is "better" than the other. In this section, on applications, you will probably be most comfortable thinking of x as an **unknown** number rather than as a blank.

When we studied simple and compound open sentences it became clear that mathematics is, among other things, a language. Its vocabulary consists of many letters and numerals, some special symbols ($=$, $<$, $+$, $-$, $\times$, ...), and a few common words (and, or, not, if, then, ...). It is not a very colorful language, but it is generally unambiguous and concise. Those are its major virtues.

In solving a real problem there are three main steps: (1) Translate the problem from English into Mathematics, (2) solve the mathematical problem, and (3) translate the answer back into English. We practiced step (2) when we studied techniques for solving equations and inequalities. Step (3) is easy once step (1) has been taken. Only step (1) needs further explanation.

Step (1) is actually a two-phase procedure. The first thing one usually does is introduce letters to represent unknown quantities. Only then can the English phrases and sentences be translated into Mathematics. The diagram suggests this sequence of steps to follow in solving a real problem.

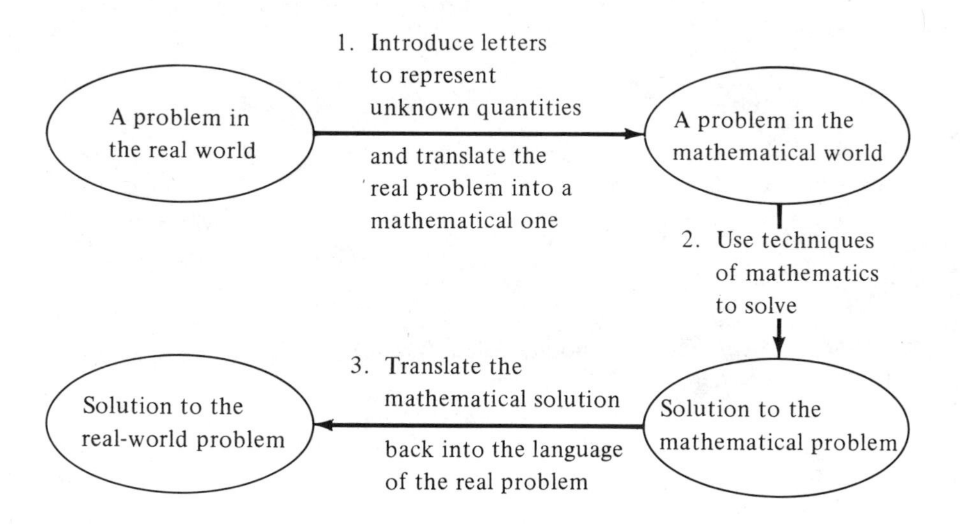

EXAMPLE (real problem) A nursery guarantees that 85 percent of its apple tree seedlings will survive. A farmer wants to have an apple orchard with 306 trees. How many seedlings should he buy?

Step 1 Let x = number of seedlings he should buy. Then $.85x$ is the number that will survive. He wants to have $.85x = 306$ (mathematical problem).

Step 2

$$.85x = 306 \Leftrightarrow$$

$$85x = 30{,}600 \Leftrightarrow$$

$$x = \frac{30{,}600}{85} \Leftrightarrow$$

$$x = 360 \quad \text{(solution to mathematical problem)}$$

Step 3 He should buy 360 seedlings (solution to real problem).

EXAMPLE (real problem) Mr. Dodge lives in a state with a 4 percent tax on taxable income. (Taxable income is total income minus tax-exempt income.) If his total income was $8600 and he paid a tax of $228, how much of his income was tax-exempt?

Step 1 Let x = his tax-exempt income (in dollars). Then $8600 - x$ is his taxable income, and $.04(8600 - x)$ is his tax (on taxable income). Thus, $.04(8600 - x) = 228$ (mathematical problem).

Step 2

$$.04(8600 - x) = 228 \Leftrightarrow$$

$$4(8600 - x) = 22{,}800 \Leftrightarrow$$

$$8600 - x = \frac{22{,}800}{4} \Leftrightarrow$$

$$8600 - x = 5700 \Leftrightarrow$$

$$8600 - 5700 = x \Leftrightarrow$$

$$x = 2900 \quad \text{(solution to mathematical problem)}$$

Step 3 Mr. Dodge's tax-exempt income was $2900 (solution to real problem).

EXERCISES

1. In translating a real problem into a mathematical problem we eventually want to arrive at an open *sentence*. On the way to getting these sentences it is usually necessary to translate English *phrases* into mathematical phrases. In the apple tree example the English phrase,

the number that will survive

translated into the mathematical phrase, $.85x$. In the income-tax example the English phrases

taxable income and tax (on taxable income)

became $8600 - x$ and $.04(8600 - x)$, respectively.

For each of the following English phrases, (i) introduce an unknown, and (ii) write a corresponding mathematical phrase.

EXAMPLE 5 more than Bob's age

(i) Let b = Bob's age (x would be just as good)
(ii) $b + 5$

(a) half of Carol's money
(b) 9 in. less than Walter's height
(c) ten times as expensive as the hat
(d) six more than a third of the class
(e) \$10 less than twice Ann's savings

2. Mary is x years old. Sue is 5 years less than twice as old as Mary.
(a) Write an expression for Sue's age.
Sue is also 10 years older than Mary.
(b) Write an appropriate equation.
(c) How old is Mary?
(d) How old is Sue?

The previous exercise is a typical "textbook" problem. It illustrates a method of setting up and solving equations, but is not a very natural, real-life problem. At present many mathematics educators are calling for more "relevant" real-world problems. An attempt has been made, in the apple tree and income-tax examples and in the following exercises, to avoid artificiality.

3. In one month the New York Yankees won 60 percent of their games. During that month they won 18 times. How many games did they play?

4. Since John began delivering newspapers he has saved 40 percent of his earnings. His parents have also contributed \$24 to his savings. His total savings amount to \$130. How much has he earned?

5. The passenger capacity of a 707 airplane is about 40 percent that of the 747. If the 707 can carry about 145 passengers, about how many can the 747 carry?

6. A 6 percent vinegar (and 94 percent water, by volume) solution is used for washing windows. How many gallons of this solution can be made up using 1 quart of vinegar?

7. Mr. Wheeler needs \$1000 in a hurry. The bank will loan him money at 5 percent. How much should he borrow so that he can set aside the interest and still have \$1000 left? (Round off your answer to the nearest penny.)

8. Joe plans to retire as soon as he has saved enough money so that he will get \$365 per year in interest. He keeps his money in a bank that pays

5 percent interest. How big will his savings account have to be before he can retire?

9. Bob has twice as much money as Dan. Bob contributes 20 percent of his money and Dan contributes 35 percent of his money to buy a $30 bicycle that they will share. How much does each boy contribute?

10. There are 15 more seventh graders than eighth graders. Eight percent of the seventh graders made the honor roll, and 10 percent of the eighth graders made it. A total of 30 seventh and eighth graders made the honor roll. How many students are in each grade?

11. A worker earns $4.60 per hour. For any hours over 40 that he works during a week he gets "time-and-a-half." One week his pay was $235.75. How many hours had he worked?

12. The total pressure $P(x)$ exerted on an object x ft below the surface of the sea is given by

$$P(x) = 14.70 + .43x \text{ pounds per square inch.}$$

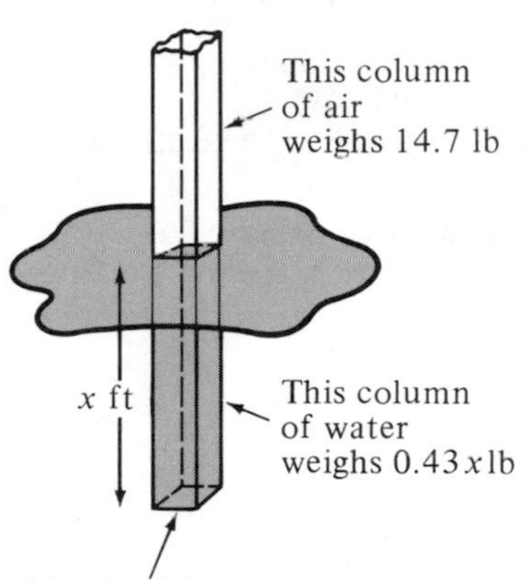

The 14.70 represents the weight in pounds of the column of air (from sea level to the top of the atmosphere) standing over a 1 in. by 1 in. square of sea water. The $.43x$ represents the weight in pounds of a column of water, 1 in. by 1 in. by x ft.

(a) A submarine is built to withstand a total pressure of 148 pounds per square inch. How deep can that submarine go?

(b) If the pressure gauge in the submarine registers a total pressure of 128.65 pounds per square inch, how deep is the submarine?

13. A gardener has 100 ft of fence. He wants to fence off a rectangular plot of ground 6 ft longer than it is wide. What should its length and width be?

14. In 5 years Joe will be more than 13 years old. Write an inequality and point out what can be concluded about Joe's (present) age.

15. The inequality, $32 < x < 212$, describes the range of Fahrenheit temperatures at which water remains in a liquid state (under standard pressure). Kelvin temperature, K, is related to Fahrenheit temperature, x, by the equation

$$x = \tfrac{9}{5}(\mathrm{K} - 273) + 32$$

Give the range of Kelvin temperatures at which water remains in a liquid state.

16. The inequality, $45 \le x \le 70$, describes the range of permissible speeds in miles per hour on a certain freeway. Your foreign sports car has a speedometer that reports speed in kilometers (km) per hour. What range of speeds is permissible for your car? (Use 1 mile $\doteq$ 1.6 km.)

17. In order to join the Chelsea fire department, one must be within 6 in. of being 72 in. tall, within 40 lb of weighing 140 lb, and within 8 years of being 28 years old. Introduce three unknowns and write absolute value inequalities that describe these specifications.

18. The length, ℓ, of a rectangle is within 1 ft of being 20 ft, and the width, w, is within 1 ft of being 10 ft.

(a) Write absolute value inequalities for these statements.

(b) Write an absolute value inequality for the perimeter, P (feet), of the rectangle.

(c) Write an absolute value inequality for the area, A (square feet), of the rectangle.

VOCABULARY

2.1 The Number Line

number line
function
domain, input set
range, output set
graphing function, g
graph (of a number)
coordinate (of a point)
slide
stretch
shrink
flip
origin
density (of real numbers)

2.2 Statements and Open Sentences

statement
open sentence
replacement set
truth set
instance (of an open sentence)
compound (sentence)

2.3 Mathematical Sentences

variable
domain of a variable
domain of an open sentence
solution set
equation
statement of equality
inequality
statement of inequality
set builder notation
solve (an equation or inequality)
graph (the solution set of an open sentence)

2.4 Solving Simple Open Sentences

addition principle of equality (inequality)
multiplication principle of equality (inequality)
implies, $\Rightarrow$
addition technique for solving equations (inequalities)
multiplication technique for solving equations (inequalities)
double implication, $\Leftrightarrow$, if and only if
equivalent (open sentences)

2.5 Solving Compound Open Sentences

simple (sentence)
compound (sentence)
logical operators

2.6 Absolute Value and Inequalities—A Geometric View

absolute value, $|\quad|$ (defined geometrically)

2.7 Absolute Value and Inequalities—An Arithmetic View

absolute value (defined arithmetically)

2.8 Applications

unknown
phrase (mathematical)

CLASSIFICATION OF GEOMETRIC FIGURES—NONMETRIC

OVERVIEW

In Chapter 1 we studied the most basic geometric figures: points, lines, planes, spaces, half-lines, rays, segments, angles, half-planes, dihedral angles, half-spaces. In this chapter we shall study more geometric figures and we shall begin to classify, or catalog, them. In the next chapter we shall continue this classification. There we shall use the powerful *metric* concepts of congruence and measurement. Here we shall classify figures on the basis of **nonmetric** properties. These you can think of as qualitative rather than quantitative properties.

Chapter 1 was probably a review of concepts, terminology, and symbols already familiar to you from your school days. In this chapter you will meet some ideas (such as connectivity, dimension, and convexity) that are quite new to the school curriculum and that may well be new to you too. Further traditional terminology is also reviewed in what we hope is an interesting, nonroutine context. Many of the exercises require originality and ingenuity. They are meant to be enjoyed. No one is expected to be able to do every one of them.

3.1 INTRODUCTION

The following exercises are to give you a feeling for the kind of figures we shall study in this chapter, and for what we mean by their nonmetric classification. There is no single, "correct" answer to each exercise. Any answer that can be defended reasonably is correct.

EXERCISES

From each set of figures below, pick out one figure that does not belong, and explain why it does not.

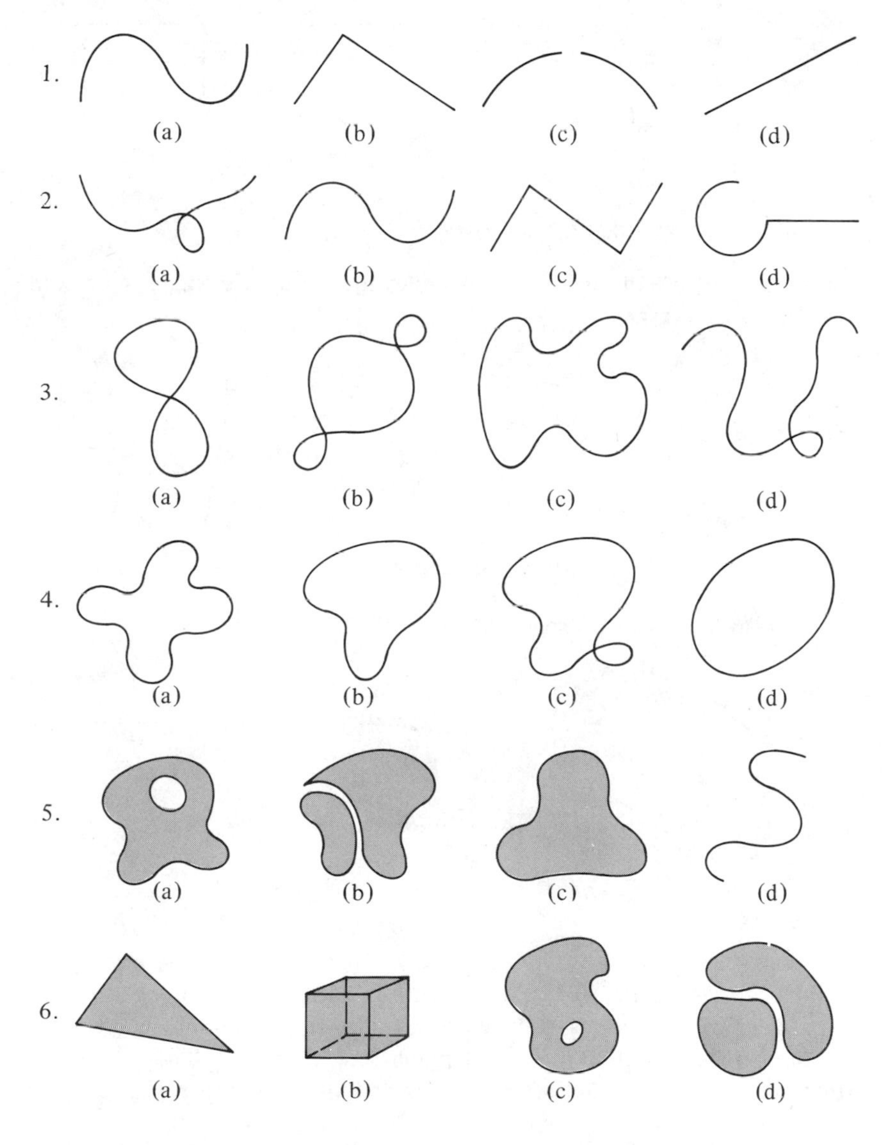

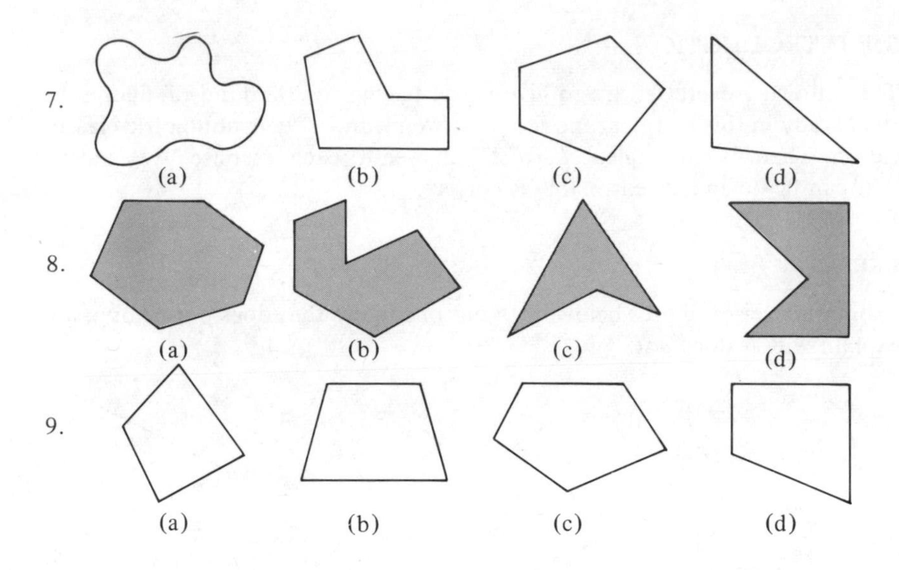

3.2 CURVES AND SIMPLE CURVES

If you drop a length of string onto a table, the geometric figure suggested is called a **planar curve**.

planar curves

There are also figures called **space curves**.

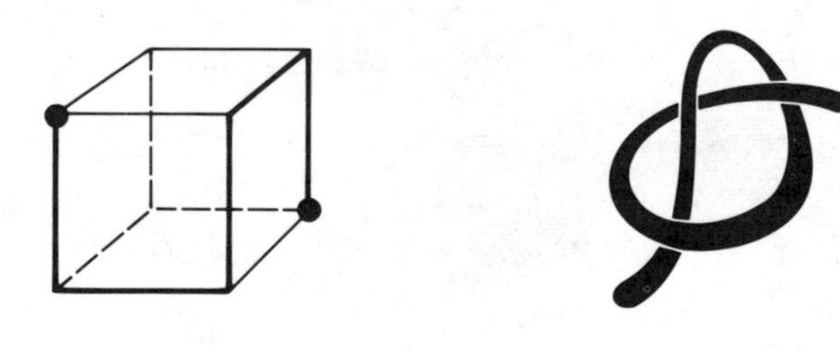

space curves

The single word "curve" (or "path") will be used to describe both planar and space curves. Notice that a curve need not have any smooth bends in it.

Another way of thinking about planar curves is this. They are those figures you can draw without lifting your pencil from the paper. Curves have a beginning and an end point. Thus (closed) segments are curves, but lines, half-lines, rays, half open segments, open segments, and angles are not.* Curves that never intersect themselves are called **simple curves**.

simple curves nonsimple curves

(In Section 3.4 we shall use the word "simple" in a slightly broader sense.) The most familiar simple curve is a segment.

EXERCISES

1. Find as many curves as you can among the figures in the preceding set of Exercises, and decide which are simple.
2. Beginning outside each figure, draw a simple curve as in part (a) that cuts each side exactly once.

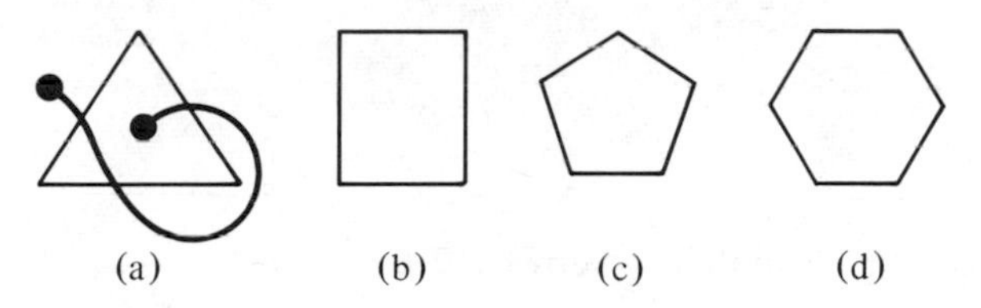
(a) (b) (c) (d)

3. Try to draw a simple curve that cuts across each of the sides of the figures below exactly once. (Do not go through any corners.)

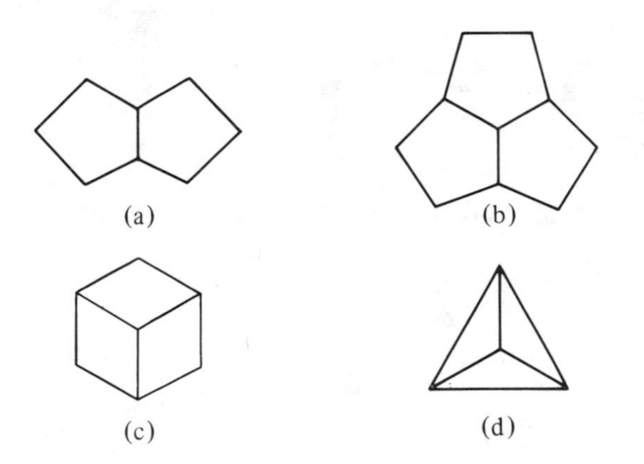
(a) (b) (c) (d)

* Unfortunately, there is some disagreement among mathematicians as to what type of figures should be called curves. We have accepted the definition that requires a curve to have a beginning and an end point. Another definition does not make that demand, so that to some mathematicians lines, rays, open segments, etc., qualify as curves.

4. If a figure, such as those in Exercise 2, has an odd number of sides and if you begin outside and draw a simple curve cutting each side once, will you finish up inside or outside? Reconsider Exercises 3(b) and 3(d). [*Proofs* that certain tasks are impossible play an important role in mathematics. See p. 281 and Exercise 10, p. 288.]

To **trace** a "graph" (or network) is to begin at a "vertex" and traverse each "edge" exactly once without lifting your pencil—that is, the tracing is a curve that passes over each edge exactly once. (It may well pass through a vertex more than once.)

5. Can you trace this graph beginning at vertex A? B? C?

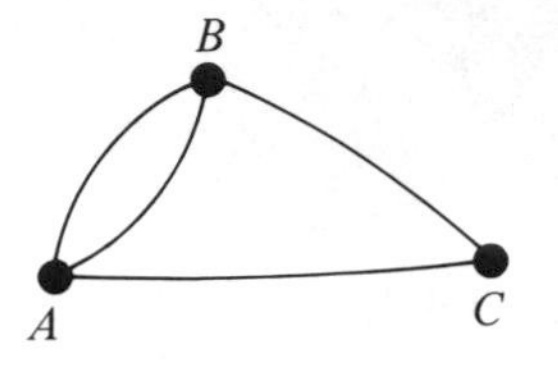

6. Can you trace this graph beginning at vertex A? B? C?

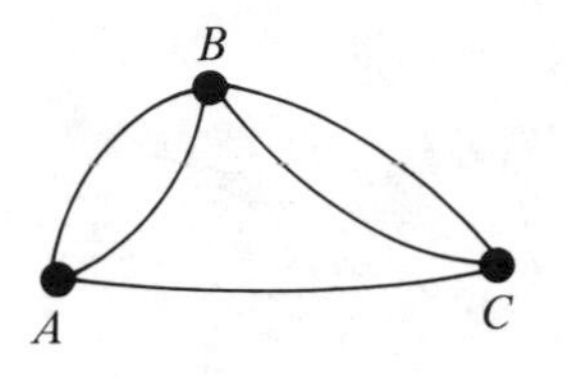

7. Can you trace these graphs beginning at vertex A? B? C? D?

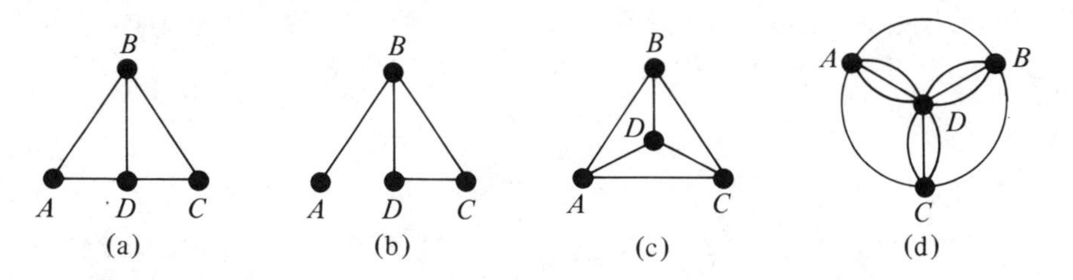

8. Can you trace these graphs at all? (You get to pick the vertex at which to begin.)

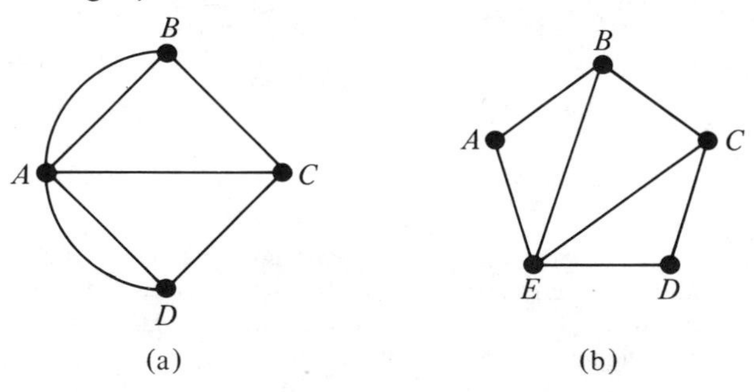

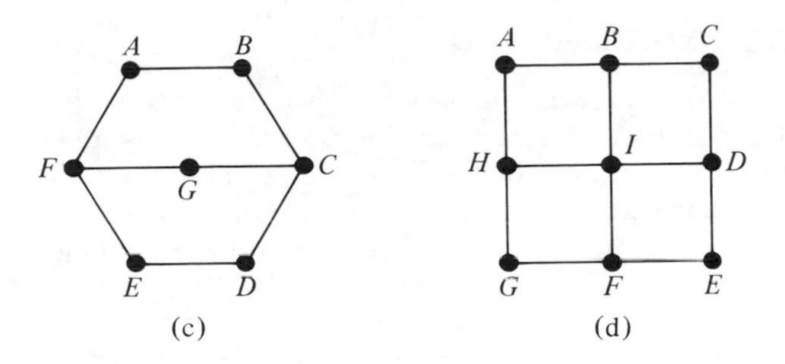

9. A famous mathematical problem is known as "The seven bridges of Königsberg" problem. Try to draw a simple curve that passes over each of the seven bridges exactly once (and does not pass through the water).

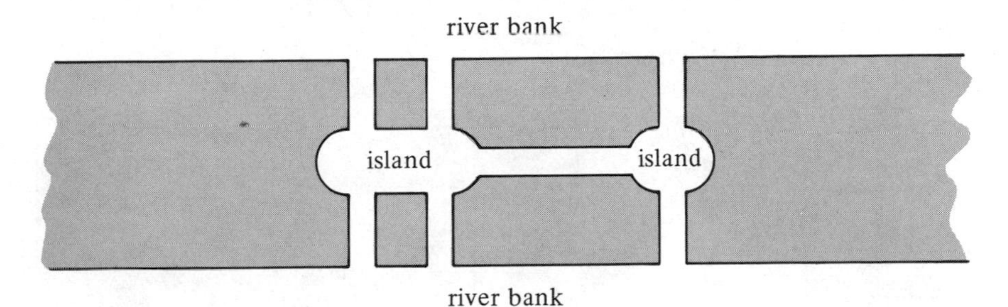

10. Can a rat walk through each doorway in this four-room cage exactly once?

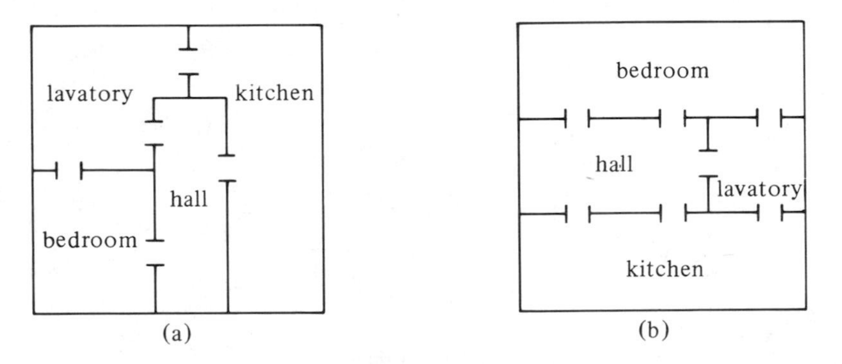

NOTE The "definition" of a planar curve as the sort of figure "you can draw without lifting your pencil from the paper" probably strikes you as extremely imprecise and intuitive. Unfortunately, the accepted mathematical definition of a curve as a "continuous function" with domain $\{x \in R \mid 0 \leq x \leq 1\}$ and range included in the plane (or in space for space curves) is not easily accessible. The concept of continuous function stands at the end of a rather long chain of prerequisite concepts, most of which are outside the boundaries of the elementary school mathematics curriculum.

3.3 CONNECTIVITY AND DIMENSION

When is a geometric figure connected? The simplest answer is: "When it is just one piece." In Chapter 1 we suggested another way of thinking about connectivity. A figure is connected if a tiny bug can walk from any point in the figure to any other point in the figure without stepping out of the figure. If the bug has dirty feet, his path will be a curve. This suggests defining connectivity in terms of curves.

A geometric figure is **connected** if any two points in it can be joined by a curve that lies entirely within it.

A figure that is not connected is said to be **disconnected** or **separated**.

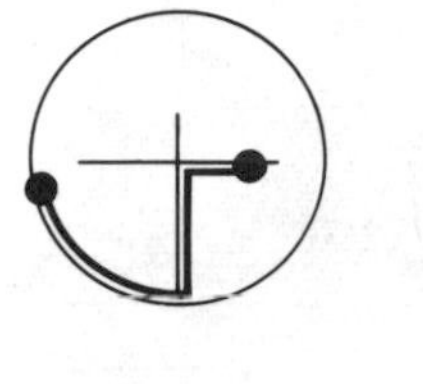
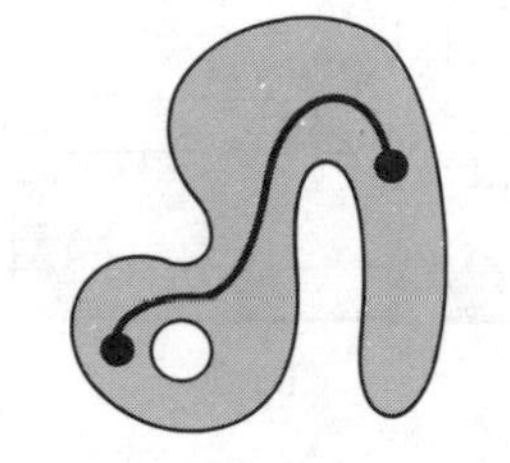
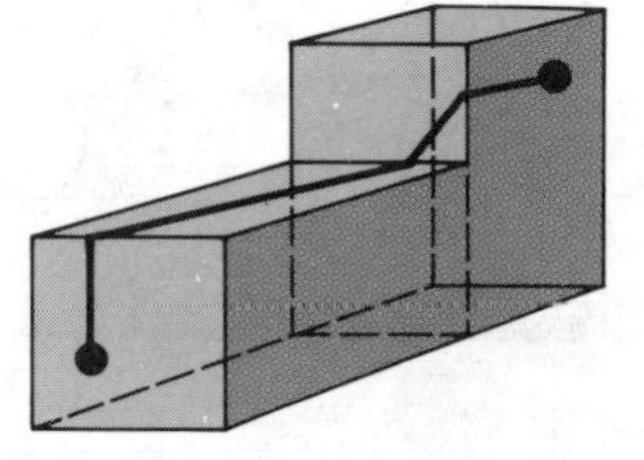

connected figures

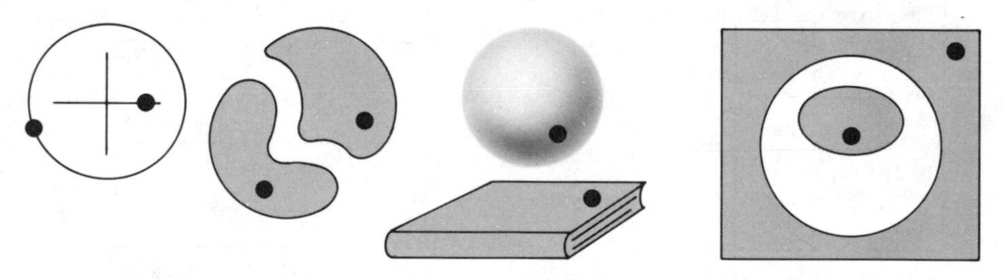

disconnected figures

The concept of connectivity can be used to make somewhat more precise our idea of the "dimension" of a figure. (A completely rigorous treatment of dimension is beyond the scope of this text.) We think of a point as having no size at all, so it is natural to agree that a point has dimension 0. We further agree that:

Any set of "isolated" points has **dimension 0**.

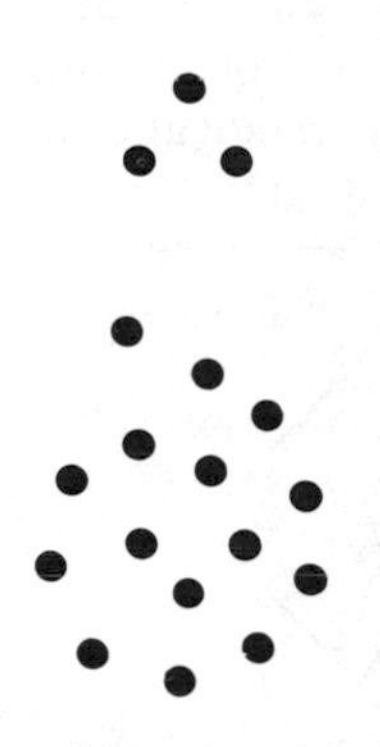

zero-dimensional figures

Intuitively, a figure is one-dimensional if it can be made of wire. In terms of connectivity,

A **one-dimensional** figure is one that is not zero-dimensional but can be disconnected by the removal of a zero-dimensional subset.

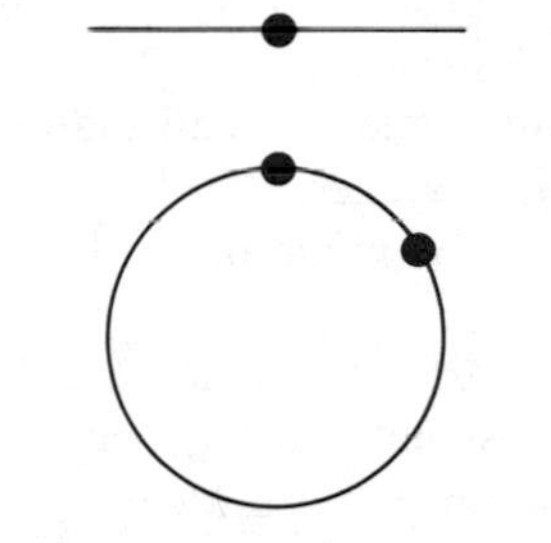

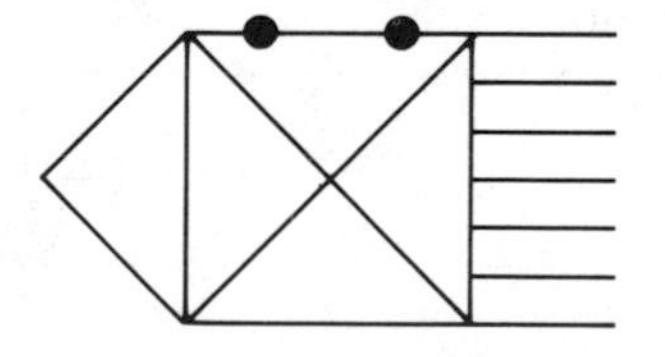

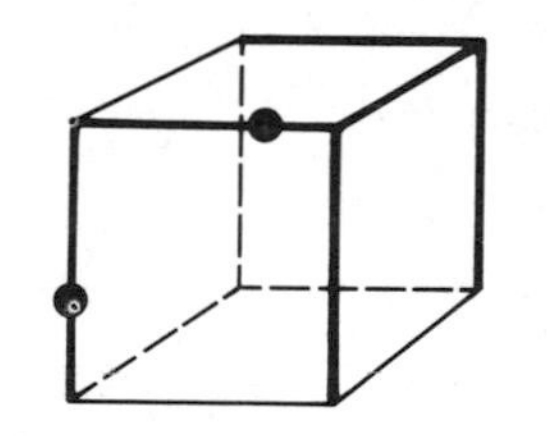

one-dimensional figures

Familiar examples of one-dimensional figures include lines, half-lines, rays, and angles.

Intuitively, a figure is two-dimensional if it can be made of paper. In terms of connectivity,

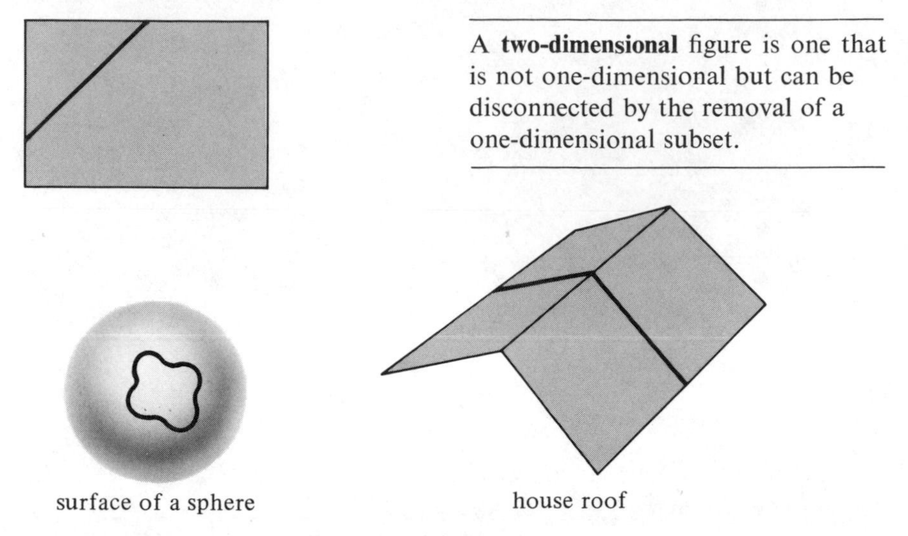

A **two-dimensional** figure is one that is not one-dimensional but can be disconnected by the removal of a one-dimensional subset.

surface of a sphere

house roof

two-dimensional figures

Familiar examples of two-dimensional figures include planes, **half-planes**, dihedral angles, and interiors and exteriors of ordinary angles.

Intuitively, a figure is three-dimensional if it can be made of modeling clay. In terms of connectivity,

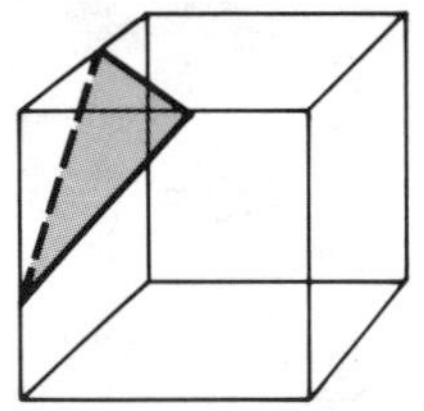

solid cube

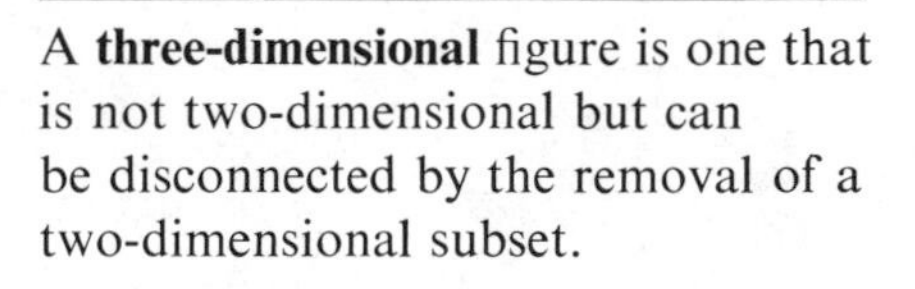

A **three-dimensional** figure is one that is not two-dimensional but can be disconnected by the removal of a two-dimensional subset.

solid sphere

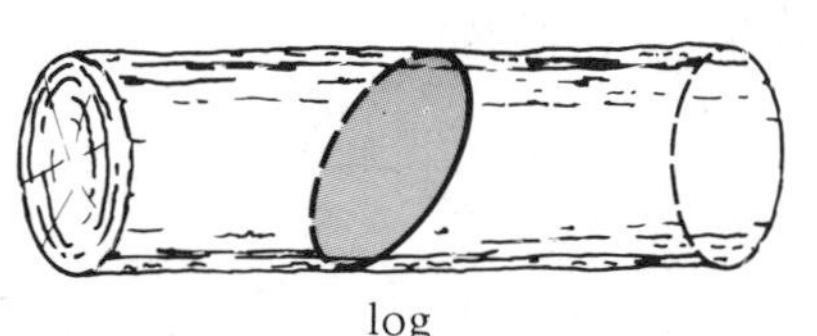

log

three-dimensional figures

Familiar three-dimensional figures include space itself, half-spaces, and interiors of dihedral angles.

Sometimes the dimension of a figure seems to depend on what part of it you happen to be looking at.

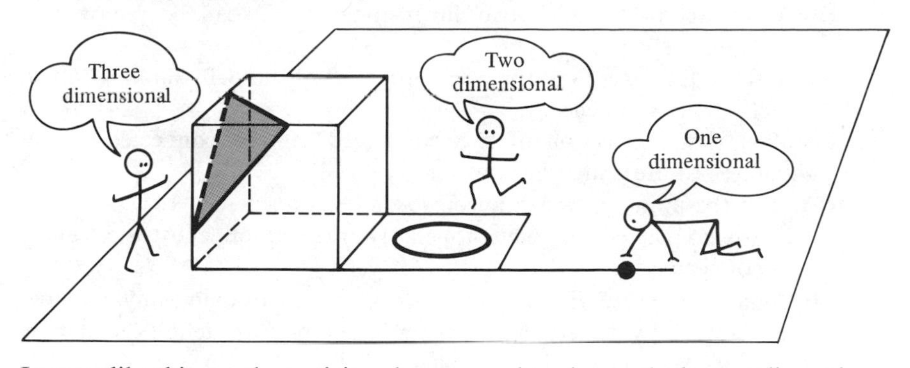

In cases like this, mathematicians have agreed to choose the largest dimension. Thus we would say that the figure above has dimension 3. Fortunately, ambiguous figures like this one will not turn up very often in our work with geometry.

EXERCISES

1. Name some physical objects that suggest geometric figures having
   (a) dimension 0
   (b) dimension 1
   (c) dimension 2
   (d) dimension 3
   (e) "mixed" dimension

2. The word "dimension" is used by different people in different ways. You might say that a sheet of paper has dimension 2. A draftsman might say it has dimensions "$8\frac{1}{2}$ by 11." If you could get him to agree that it had dimension 2, how would he probably think about the "2"?

3. An inner tube can be disconnected (separated) by the removal of a "bent" circle.

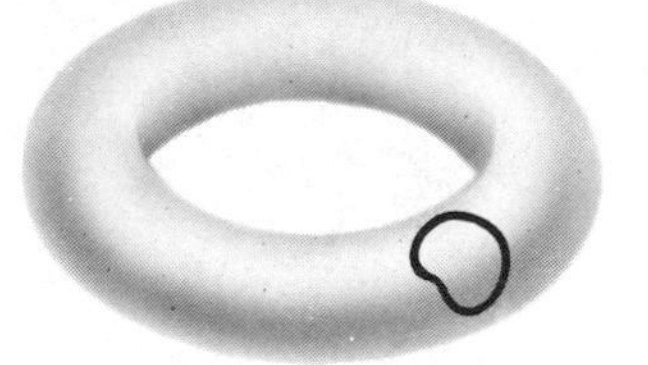

   (a) Can it be disconnected by the removal of a "flat" (planar) circle?
   (b) Can it be disconnected by the removal of two flat circles?
   (c) Can you find two flat circles whose simultaneous removal will *not* disconnect the inner tube?

4. **(a)** We know that removal of a line disconnects a plane. But removing an "infinitely long" figure sounds like hard work. Can you think of a "small" one-dimensional figure whose removal will disconnect a plane?
   (b) What sort of "small" two-dimensional figure can be removed to disconnect space?
5. Reconsider the exercises (pp. 83–84) of Section 3.1 thinking about connectivity and dimension.
6. **(a)** Must the intersection of two connected sets be connected? Draw pictures to illustrate your answer.
   (b) Must the union of two connected sets be connected?
   (c) If two connected sets have nonempty intersection, must their union be connected?
   (d) Could you *prove* the answer to (c) without drawing any pictures, but simply by a verbal argument based on the definition of connectivity?
7. What is the dimension of the geometric figure suggested by
   (a) a square of chicken wire **(b)** a basketball
   (c) a baseball **(d)** the highways on a map
   (e) the towns on a map **(f)** the map
   (g) an empty milk carton **(h)** a Yo-Yo on a string
   (i) an ice cream cone **(j)** a balloon on a string
8. Draw two two-dimensional figures whose intersection is
   (a) a two-dimensional figure **(b)** a one-dimensional figure
   (c) a zero-dimensional figure **(d)** the empty set
9. Sketch a cube. It has ________ faces, ________ edges, and ________ vertices. Each face is a ________ dimensional figure, each edge is a ________ dimensional figure, each vertex is a ________ dimensional figure and the solid cube itself is a ________ dimensional figure.
10. **(a)** What is the dimension of the surface of a cube?
    (b) Is the surface of a cube a connected set?
    (c) Does the surface of a cube separate space?

3.4 CLOSED CURVES AND SIMPLE CLOSED CURVES

A curve, you remember, has a beginning point and an end point.

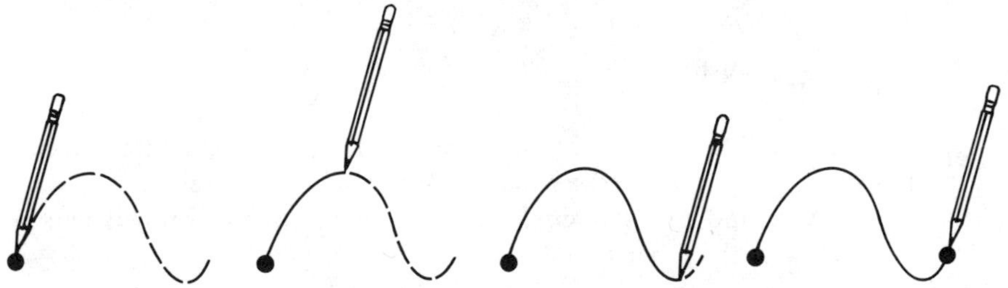

If the beginning and end points of a curve coincide, then the curve is called a **closed curve**. Closed curves, like ordinary curves, may lie in a plane or may wander around in space.

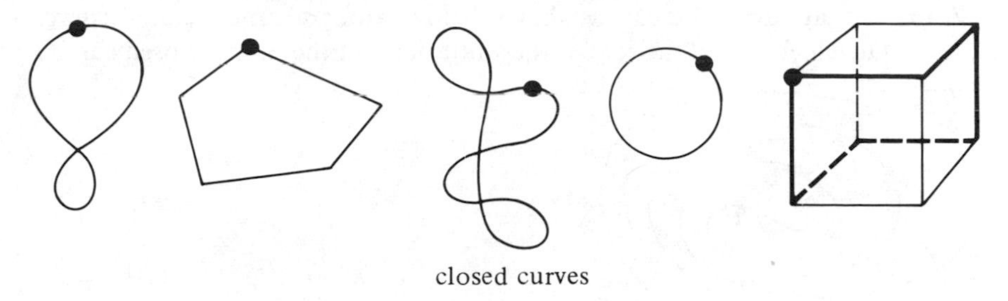

closed curves

Technically no closed curve is simple, since every closed curve intersects itself at its common beginning–end point. Nevertheless, certain closed curves are called simple. A closed curve is called a **simple closed curve** if, in drawing it, no intersections are made until the end point is drawn.

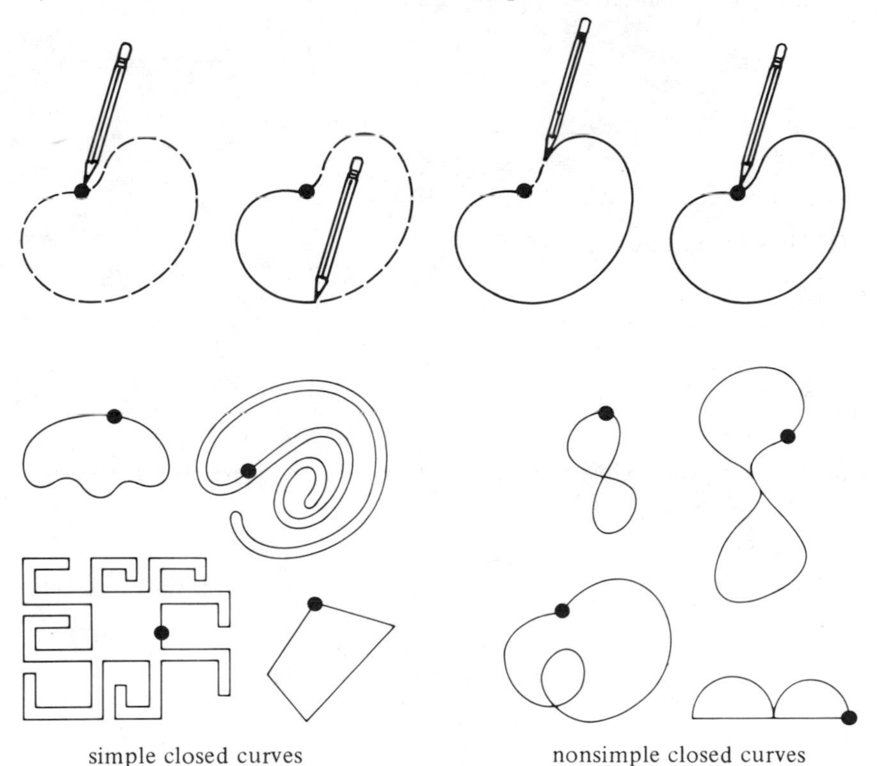

simple closed curves

nonsimple closed curves

Many of the geometric figures that are most familiar to us are examples of simple closed curves: triangles, squares, rectangles, trapezoids, pentagons, hexagons, circles, ovals. It is exactly because there are so many useful special examples that we study the general class of figures called simple closed curves.

EXERCISES

1. If a loop of string is dropped onto a table, what sort of curve will be formed?

2. (a) You are driving a car, as shown below, along a simple closed curve. On which side of the road is the "interior" of the simple closed curve?

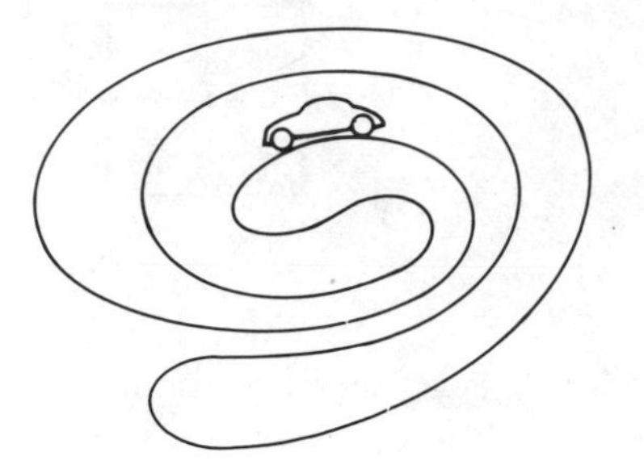

(b) Suppose that you drive this nonsimple closed curve. Is its "interior" always on your left?

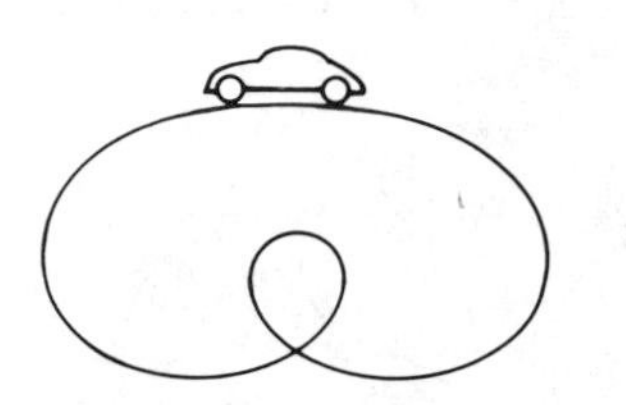

(c) How about this one?

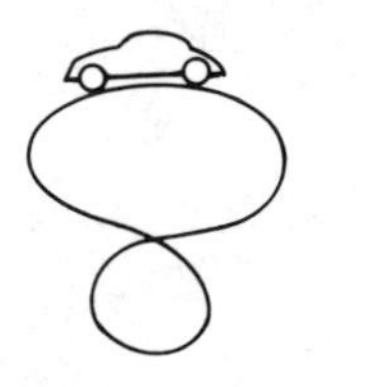

3. (a) Draw a simple closed curve that fences in the sheep and fences out the wolves.

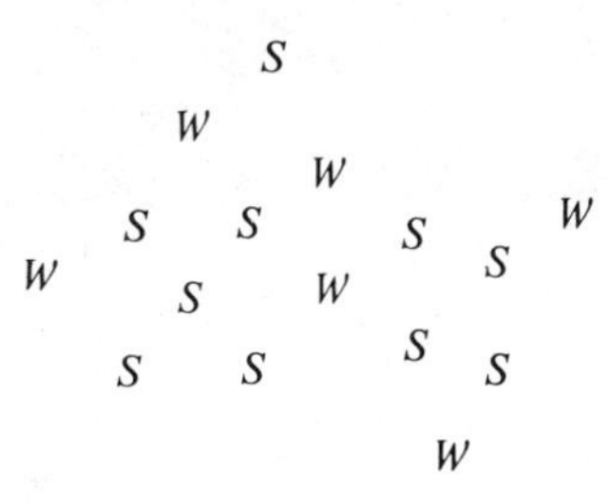

(b) Draw a simple closed curve that fences in the wolves and fences out the sheep.

(c) Try writing a definition of "the exterior of a simple closed (planar) curve."

4. Try to draw a simple closed curve that cuts each side of the figure below exactly once.

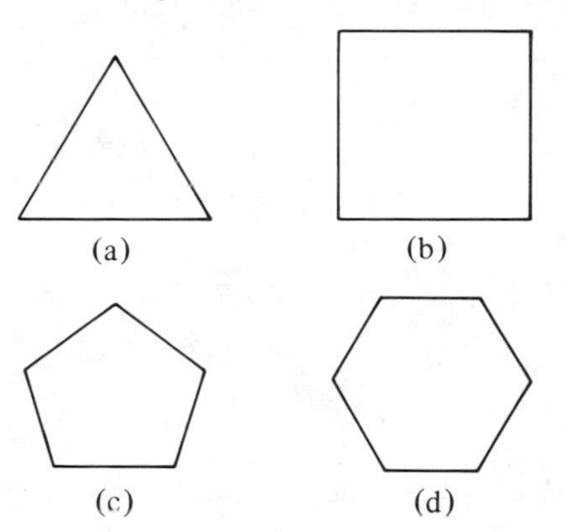

5. Removing a simple closed curve from a plane separates the plane into two (connected) pieces. (This rather obvious statement is very difficult to prove. It is a famous mathematical theorem known as the "Jordan curve theorem.") Draw a closed curve, which is not simple, whose removal separates the plane into
  (a) three pieces (Ignoring end points, how many times did your curve intersect itself?)
  (b) four pieces (same question)
  (c) five pieces (same question)

6. **(a)** How many points must be removed to disconnect a simple closed curve?
  (b) Draw a closed curve that is not simple and that can be disconnected by the removal of a single (carefully chosen) point.
  (c) Draw a closed curve that is not simple and cannot be disconnected by the removal of a single point, no matter how carefully chosen.

7. Reconsider the exercises on pp. 83–84.

3.5 POLYGONS

A curve that is made up of segments joined at their end points is called a **polygonal curve**.

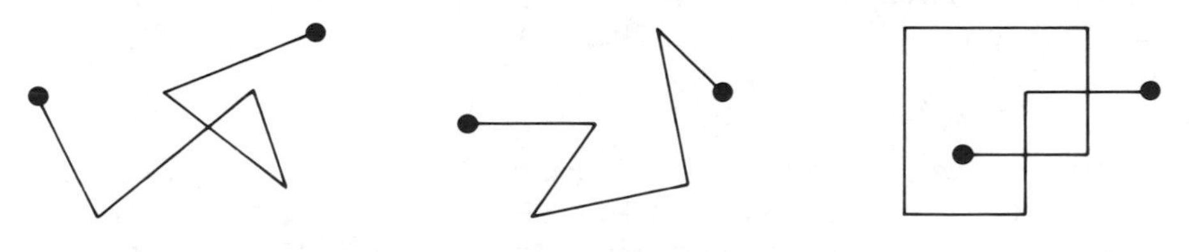

polygonal curves

A closed curve made up of segments joined at their end points probably should be called a polygonal closed curve, but in fact is called just a **polygon**.

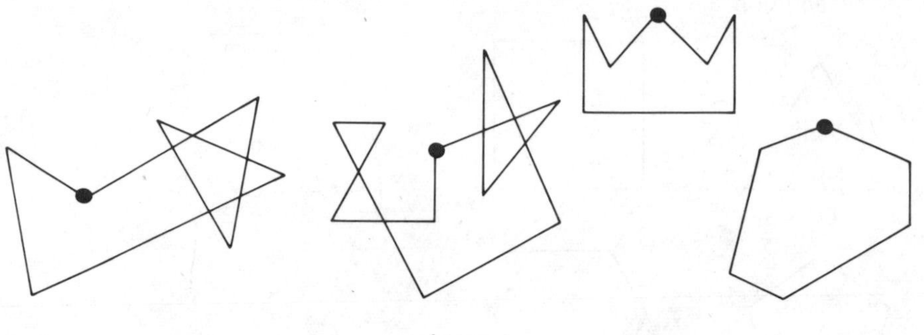

polygons

A polygon that is also a simple closed curve is called a **simple polygon**. Among the figures above, the last two polygons are simple polygons while the first two are not. The segments of which a polygon is constructed are called its **sides**; the end points of these segments are called its **vertices**.

All the figures we have drawn are planar. There are, of course, polygonal space curves and polygonal closed space curves, but we shall have little to do with them. Unless specified otherwise, you can assume that all polygonal curves are planar.

One source of difficulty in studying polygons is the language. Over the many centuries during which polygons have been investigated, special names have arisen. These names refer to the number of sides of the polygon. Etymologically, most of them refer to the number of angles, not sides, but this distinction is not important: A (simple) polygon has exactly as many sides as angles.

polygon	poly- (Gk., *many*)	+ gon (Gk., *angle*)
triangle	tri- (L., *three*)	+ angle
quadrilateral	quadri- (L., *four*)	+ laterus (L., *side*)
pentagon	penta- (Gk., *five*)	+ gon
hexagon	hexa- (Gk., *six*)	+ gon
heptagon	hepta- (Gk., *seven*)	+ gon
octagon	octa- (Gk., *eight*)	+ gon
nonagon	nona- (L., *nine*)	+ gon
decagon	deca- (Gk., *ten*)	+ gon

Occasionally, mathematicians find themselves studying polygons having a large number of sides. For example, the great mathematician Gauss proved an interesting theorem about polygons that have 3, 5, 17, 257, and 65,537 sides. If a mathematician needs to talk about a 17-sided polygon, he will not look for an appropriate Greek prefix, but will call it simply a 17-gon.

EXERCISES

1. Draw a simple *nonclosed* polygonal curve made up of six segments.
   (a) How many angles do you see?
   (b) How many vertices?
2. Draw a simple *closed* polygonal curve made up of six segments.
   (a) How many angles do you see?
   (b) How many vertices?
   (c) In drawing one of the segments you created two angles. When did this happen?
   (d) In drawing one of the segments you created no new vertices. When did this happen?
3. Explain why a simple polygon has
   (a) exactly as many sides as angles
   (b) exactly as many sides as vertices
4. Polygons can be named by naming their vertices in the order in which they are to be traced. It is understood that the last vertex named is to be joined to the first vertex named. For each polygon named below, first sketch a copy of this set of points and then sketch the polygon.

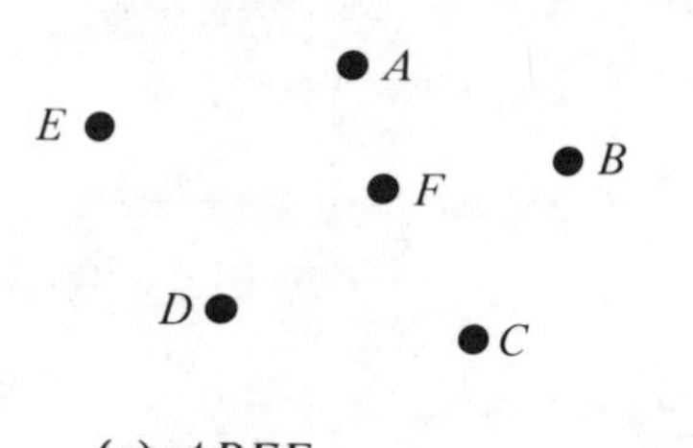

(a) *ABFE* **(b)** *AFCB*
(c) *ACDB* **(d)** *AED*
(e) *ACE* **(f)** *ABCDE*
(g) *ABCFDE* **(h)** *DEABCF*

5. Name each polygon in Exercise 4 with a different name.
6. **(a)** How many polygons can you draw using the four points shown below as vertices?
   (b) How many are simple?

D● ●C

7. Repeat Exercise 6 for these four points.

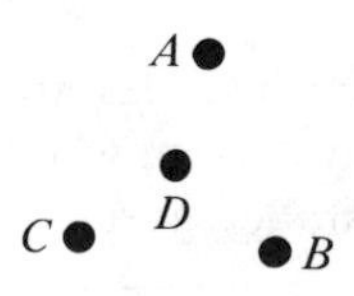

8. Repeat Exercise 6 for these five points.

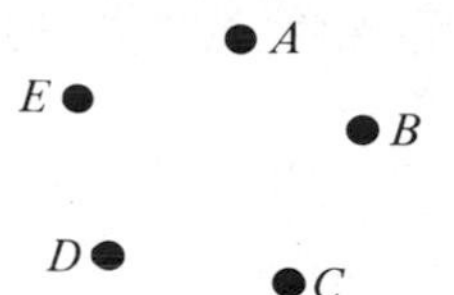

(*Hint*: Name each polygon using A as the first letter. How many five-letter names beginning with A are possible? Each polygon has how many five-letter names beginning with A?)

9. Repeat Exercise 6 for these five points.

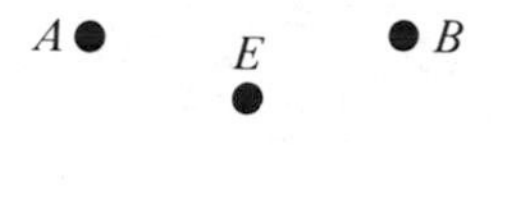

10. **(a)** Draw as many polygonal curves as you can beginning at A, ending at E, and using all of the other points (B, C, D) as vertices.

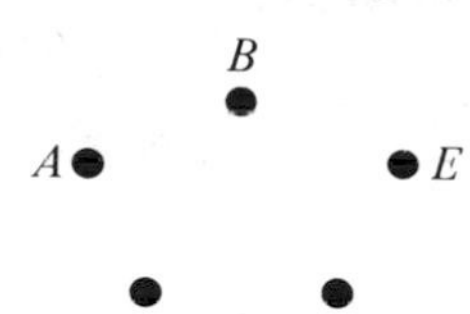

(b) How many such curves did you find?

(c) How many are simple?

(d) Try to relocate the points B, C, D so that all of the polygonal curves [drawn as in (a)] will be simple.

11. A quadrilateral can bc **triangulated** by drawing one segment.

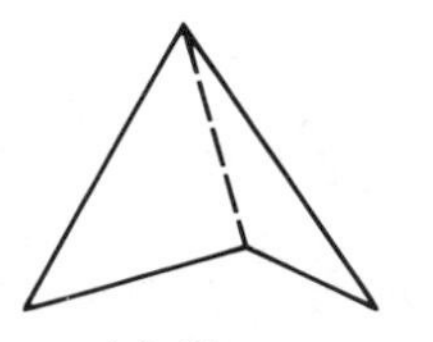

(a) Draw a pentagon and triangulate it. How many segments did you have to draw to triangulate it?

(b) Repeat (a) for a hexagon, a heptagon, a nonagon, a decagon, an n-gon.

(In Chapter 4 we shall review the technique for finding the area of a triangle. The technique of triangulation will then allow us to find the area of any kind of polygon.)

12. A quadrilateral with the property that both pairs of opposite sides are parallel is called a **parallelogram**. (Do you remember what we mean when we say that two segments are parallel?) Name each parallelogram in this figure by naming its vertices.

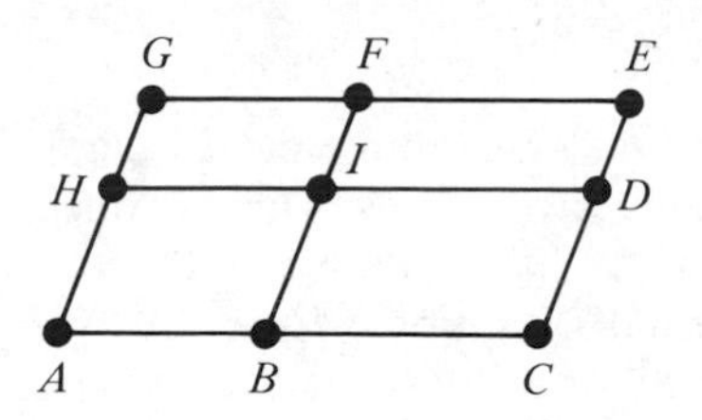

13. **(a)** How many simple polygonal space curves are there from A to H in the figure below that consist of three edges of the cube? (One, $ABCH$, is emphasized in the figure.)

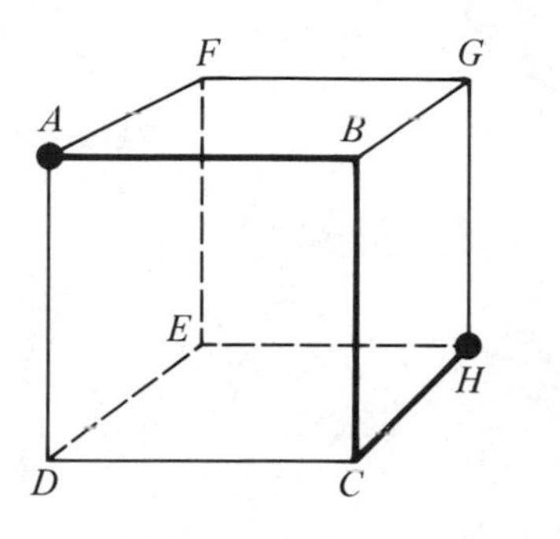

(b) Repeat (a) for four edges instead of three.
(c) Repeat (a) for five edges.
(d) Repeat (a) for six edges.
(e) Repeat (a) for seven edges.

14. Explain why
(a) every "edge-path" from A to B must consist of an odd number of edges.
(b) every edge-path from A to C must consist of an even number of edges.
(c) every edge-path from A to H must consist of an odd number of edges.

15. Reconsider the exercises on pp. 83–84.

TEACHING NOTE A multipurpose teaching device known as the "geoboard" is particularly useful in connection with the study of polygons. A five-by-five geoboard is made from a piece of wood and 25 nails as shown. Rubber bands can then be stretched over the nails to form a variety of polygons. How many squares could you make on this geoboard?

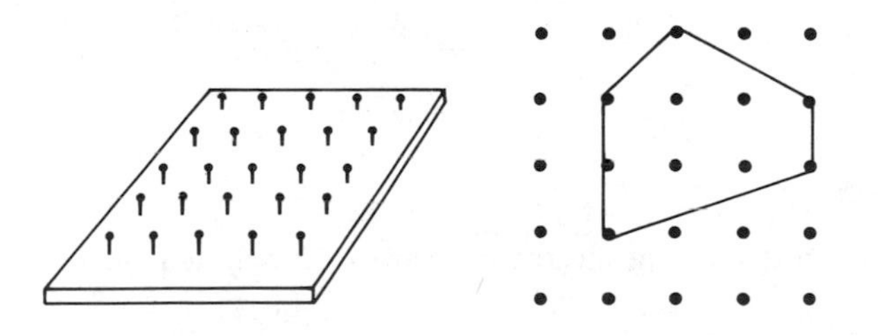

3.6 REGIONS

According to the Jordan curve theorem, removal of a simple closed curve from a plane separates the plane into two connected pieces. One is called the interior of the curve, the other the exterior. In any picture it is obvious which piece is which. We can also write formal definitions. Here are two possible definitions. (Perhaps you invented these or some others in Exercise 3(c), p. 94.) The **exterior** of a simple closed curve is that connected piece of its complement which contains at least one ray. The **interior** of a simple closed curve is that connected piece of its complement which contains no ray.

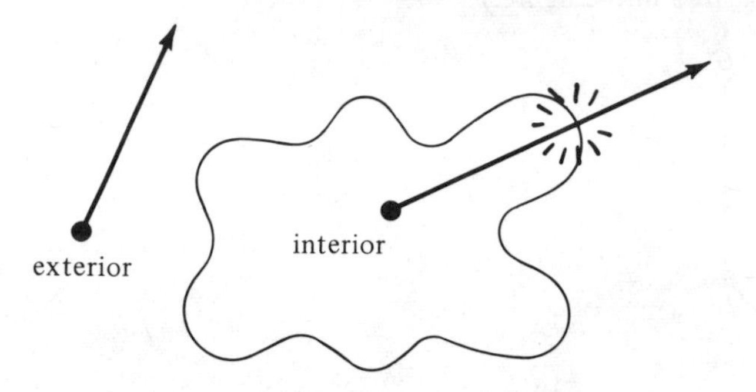

Neither the interior nor the exterior contains points of the curve itself. Thus, neither the interior nor the exterior contains any of its boundary points. Each is an example of a general type of figure called a (plane) "region."

A figure in a plane is called a (plane) **region** if (1) it is connected, and (2) it contains none of its boundary points.

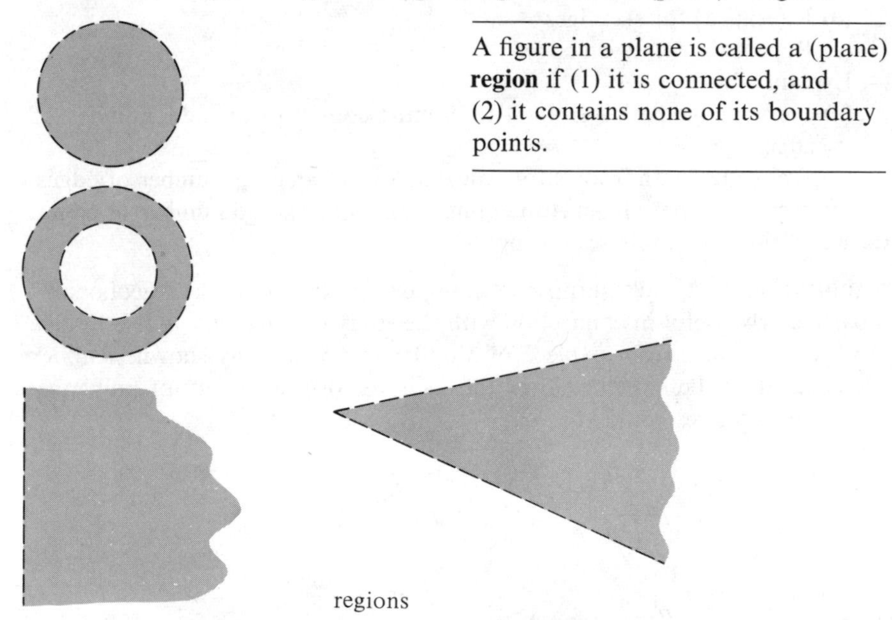

regions

The dotted lines are to suggest that the boundaries are not included in the figures.

In space there is a similar concept:

A figure in space is called a **region** (in space) if (1) it is connected, and (2) it contains none of its boundary (surface) points.

Some examples of regions in space are the inside of a ball or cube, an open half-space, the interior or exterior of a dihedral angle.

EXERCISES

1. (a) Which part of the surface of the sphere below would you call the "interior" of the simple closed space curve drawn on it?

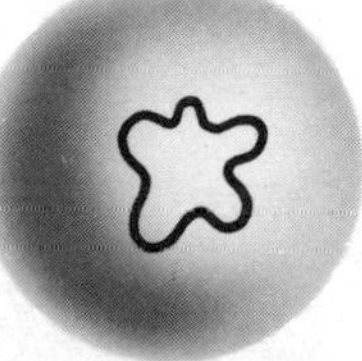

(b) The stitching on a baseball (or the groove on a tennis ball) suggests a simple closed space curve on a sphere. Which part of the surface of the ball would you call the "interior" of this curve?

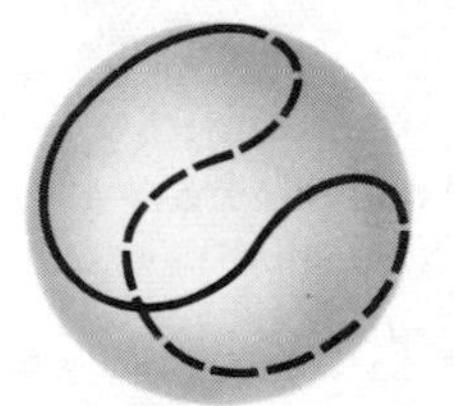

2. (a) Is the point P an interior point or an exterior point of the simple closed curve below?

(b) Try to devise a quick test for deciding whether a point is in the interior or the exterior of a simple closed curve.

(c) Does your test work for a simple closed curve on the surface of a sphere?

3. Explain why each figure below is *not* a region.

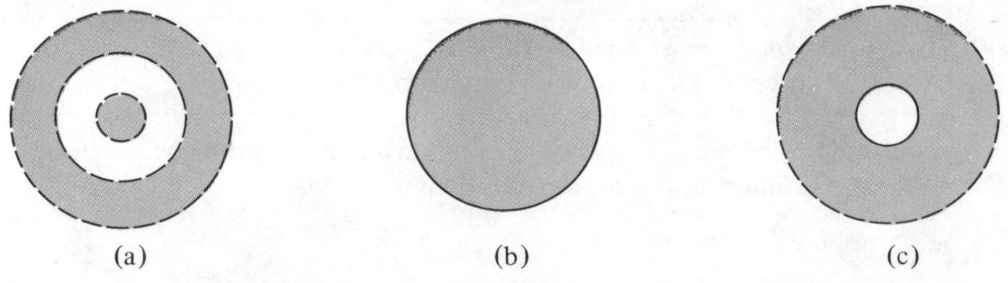

4. The idea of a "boundary point," on which the concept of region depends, has been left to your intuition until now. Here is a precise definition.

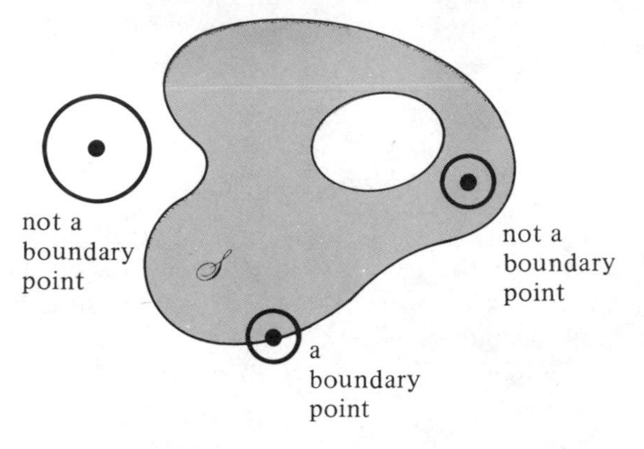

A point P (of the plane) is a **boundary point** of a (planar) figure $\mathscr{S}$ if and only if *every* circle with center at P (no matter how small) encloses two kinds of points: points of $\mathscr{S}$ and points of $\mathscr{S}'$.

The definition of the boundary of a figure is the obvious one:

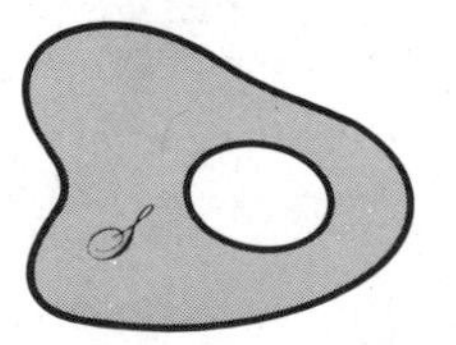

The **boundary** of a figure is the set of all of its boundary points. (The boundary of the figure $\mathscr{S}$ is the disconnected one-dimensional figure emphasized in the drawing.)

(a) Use the definitions to determine the boundaries of these (plane) figures. Sketch those boundaries.

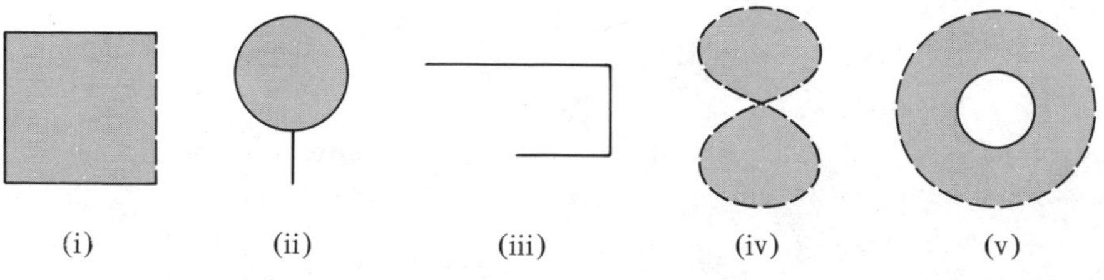

(b) Must a boundary point of a figure belong to that figure?
(c) Must the boundary of a connected set be connected?
(d) Must the boundary of a disconnected set be disconnected?

5. The word "interior" is used in several distinct senses. We encountered one usage in Chapter 1 when we discussed the *interior of an angle.* (Do you remember that definition?) We used the word somewhat differently in this section when we referred to the *interior of a simple closed curve.* Still a third meaning is intended when a person refers to the *interior of an arbitrary* (planar) *figure.*

> The **interior** of a figure, $\mathscr{S}$, is what remains when the boundary is is removed.

We can state this definition in the notation of set theory if we denote the set operation complementation by the usual minus sign:

interior of $\mathscr{S} = \mathscr{S}$ − boundary of $\mathscr{S}$

EXAMPLE

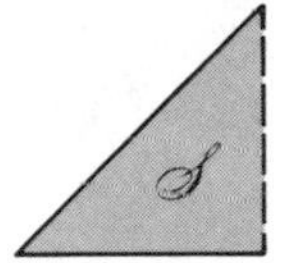

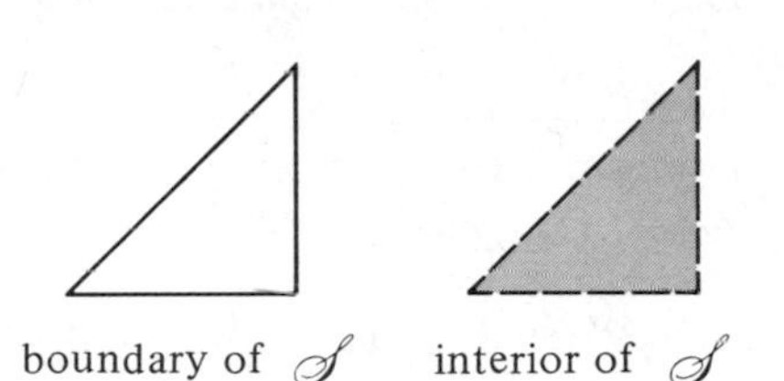

boundary of $\mathscr{S}$ interior of $\mathscr{S}$

(a) Sketch the interior of each figure in Exercise 4(a).
(b) Using the third meaning of interior, what is the interior of any angle? What is the interior of any simple closed curve?

TEACHING NOTE While it is certainly important that an elementary teacher know what is meant by the boundary of a figure, and that she be able to arbitrate disputes on the meaning of the equivocal word "interior," in all honesty it must be admitted that the precise mathematical definition of region that we have given here is widely ignored without disastrous consequences.

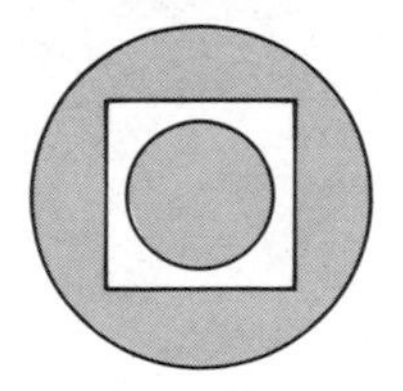

A teacher or textbook that asks a child to "find the area of the region shown" will be understood even though, technically, the figure in question is not a region. The fact that it is disconnected might cause some child to ask "which region?" But the fact that it contains boundary points will go unnoticed, as well it should since the area of the figure with its boundary points is the same as the area of the figure without the boundary points.

3.7 CONVEX FIGURES

An ordinary little bug can walk from any point in a connected figure to any other point in that figure without ever leaving the figure. Suppose now that the bug is not ordinary. Suppose he refuses to walk along anything but a straight line to his goal. Then in some connected figures he can still get around, but in others he cannot.

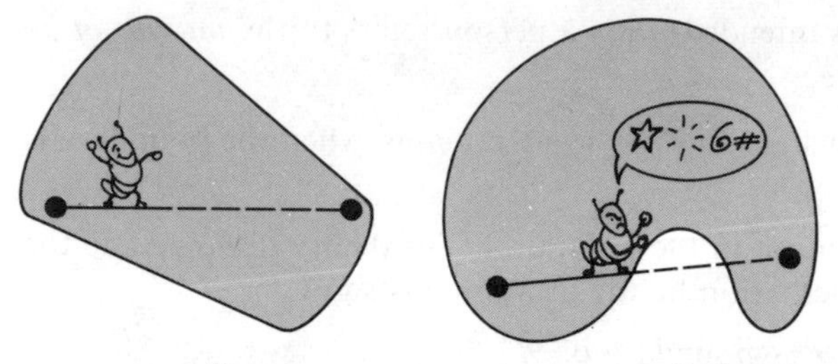

convex figure nonconvex figure

Figures in which this stubborn bug can get around are called convex figures. In mathematical language,

> A figure is **convex** if whenever two points are in it so is the segment joining them.

Convex figures, then, are very special kinds of connected figures. If a figure is convex it must be connected, but if it is connected it may or may not be convex.

The definition of convexity applies as well to figures in space as to figures in a plane.

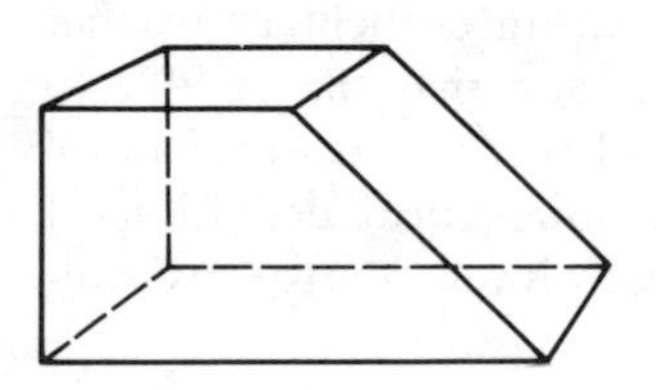

(solid) convex space figure

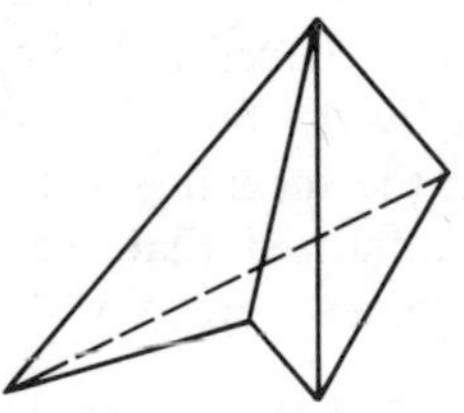

(solid) nonconvex space figure

EXERCISES

(a) (b)

side view of two lenses

1. Which lens above is called convex? What is the other one called?

2. If a polygonal figure includes all of its "diagonals," must it be convex?

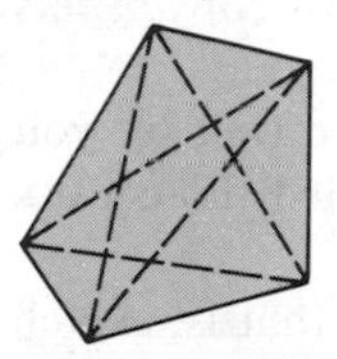

polygonal figure
with diagonals

3. Which of the following figures are convex?

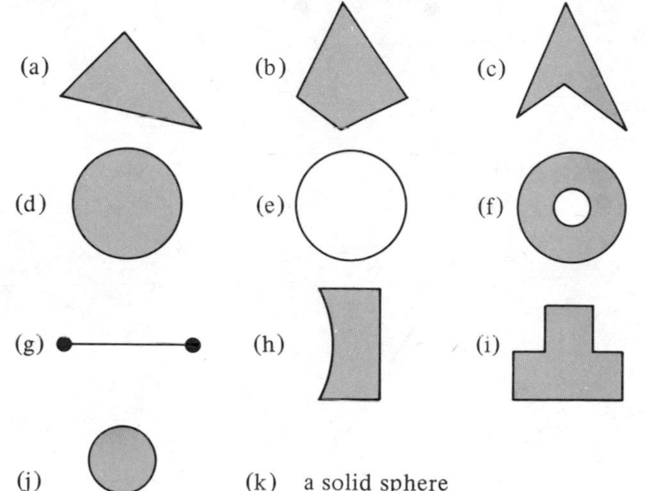

(l) the surface of a sphere
(m) a solid cube
(n) the surface of a solid cube
(o) a point

4. **(a)** A figure $\mathscr{S}$ has this property: *For all* points A and B, if $A \in \mathscr{S}$ and $B \in \mathscr{S}$ then $\overline{AB} \subset \mathscr{S}$. What kind of figure is $\mathscr{S}$?

(b) A figure $\mathscr{T}$ has this property: *There are* points A and B such that $A \in \mathscr{T}$, $B \in \mathscr{T}$, and $\overline{AB} \not\subset \mathscr{T}$. What kind of figure is $\mathscr{T}$?

5. **(a)** If a figure is disconnected, could it be convex?

(b) If a figure is not convex, could it be connected?

(c) If a figure is not convex, could it be disconnected?

6. Draw two convex figures which contain all four points A, B, C, D.

A ●

B ● C ●

D ●

Shade the intersection of your two figures. Is the intersection convex?

7. Try to prove, without pictures, using only a verbal argument based on the definition of convexity, that the intersection of any two convex sets is convex.

8. Copy the points of Exercise 6 and draw the "smallest" convex set you can that contains A, B, C, and D. This smallest set is called the **convex hull** of A, B, C, and D.

9. Sketch the smallest convex set that includes the given set; that is, sketch their convex hulls.

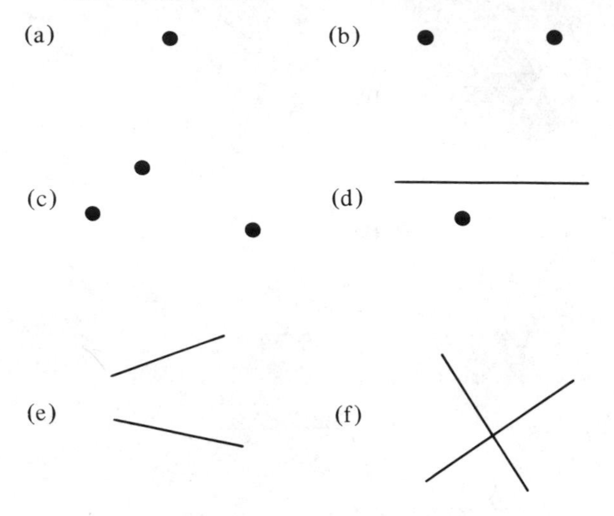

10. Sketch or describe in words the convex hull of

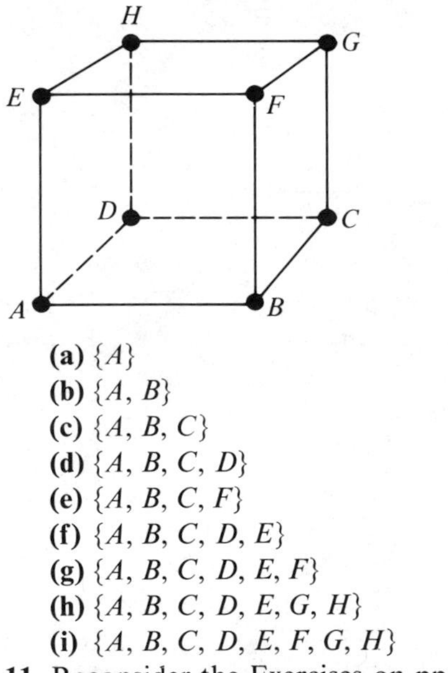

(a) $\{A\}$
(b) $\{A, B\}$
(c) $\{A, B, C\}$
(d) $\{A, B, C, D\}$
(e) $\{A, B, C, F\}$
(f) $\{A, B, C, D, E\}$
(g) $\{A, B, C, D, E, F\}$
(h) $\{A, B, C, D, E, G, H\}$
(i) $\{A, B, C, D, E, F, G, H\}$

11. Reconsider the Exercises on pp. 83–84.

VOCABULARY

3.1 Introduction

nonmetric

3.2 Curves and Simple Curves

planar curve
space curve
curve
simple curve
trace (a graph)

3.3 Connectivity and Dimension

connected
disconnected
separated
dimension: zero-, one-, two-, three-dimensional

3.4 Closed Curves and Simple Closed Curves

closed curve
simple closed curve

3.5 Polygons

polygonal curve
polygon: triangle, quadrilateral, pentagon, hexa-, hepta-, . . .
simple polygon
sides
vertices, convention for naming a polygon
triangulate
parallelogram

3.6 Regions

exterior (of simple closed curve)
interior (of simple closed curve)
region
boundary
interior (of a plane figure)

3.7 Convex Figures

convex
convex hull

CLASSIFICATION OF GEOMETRIC FIGURES—METRIC

OVERVIEW

In the previous chapter we classified geometric figures according to a variety of more or less exotic properties. The familiar properties of size and shape were not considered at all. In this chapter we shall study size and, to a lesser degree, shape. (Shape will be investigated more fully in Chapter 8.) These ideas are closely linked to ideas of measurement. The Greek word for measure is *métron*. That is why the word "**metric**" appears in the title of this chapter.

The ideas of congruence and counting underlie the common processes of measurement. To measure a stick you set down beside it, end-to-end, congruent copies of a unit segment, such as an inch or a centimeter, and count how many are needed. In this chapter, then, the basic concept of congruence is developed from an initial, rather vague same-size-and-shape criterion, through the less vague superposition test, to a definition in terms of rigid motions. Then the process of measurement is described, as above, common properties of measure functions are abstracted, and basic information about length, area, volume, and angle measure is deduced.

For this chapter you will need a compass, a protractor, and an inch–centimeter ruler.

4.1 INTRODUCTION

For a taste of what is meant by the metric classification of figures, try these exercises.

EXERCISES

From each set of figures below, pick out one that does not belong and explain why it does not.

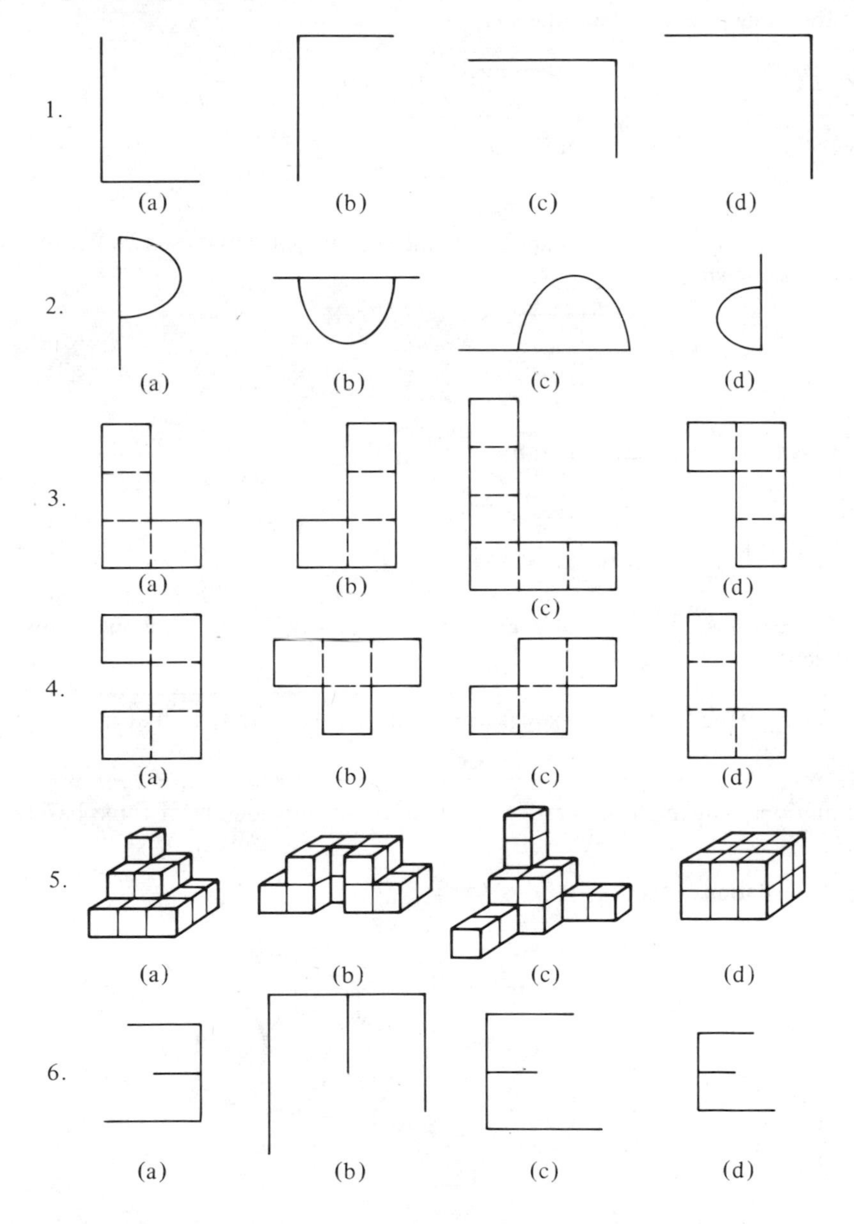

7. 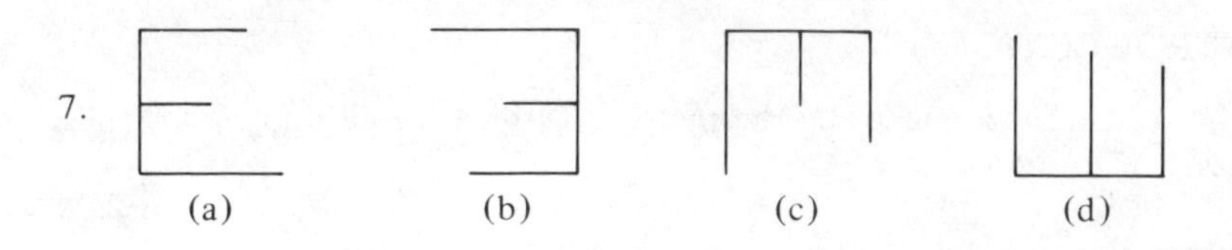

4.2 CONGRUENCE OF PLANE FIGURES

We would like to say that two geometric figures are "congruent" if they have both the same size and the same shape.

congruent figures

But the ideas of size and shape are somewhat vague. Do these two figures have the same shape?

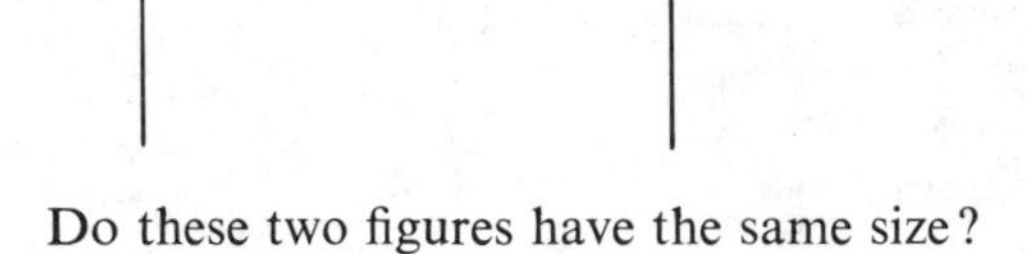

Do these two figures have the same size?

For figures in a plane there is another way of saying what we mean by congruence.

Two (plane) figures are **congruent** if one can be moved so that it coincides with the other.

The motions that might be needed to make them coincide are of three basic types:

Translations ("slides")

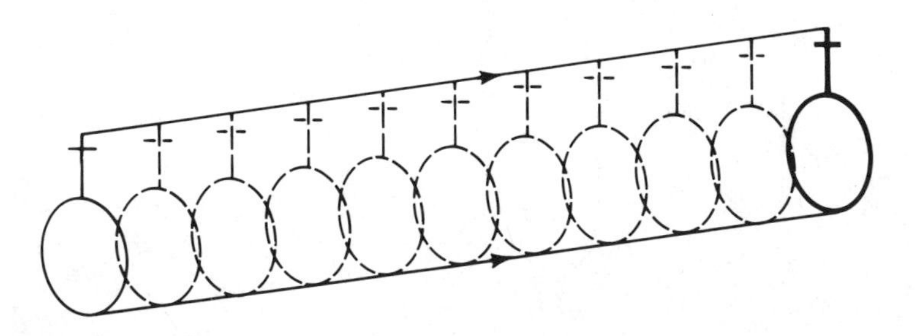

Rotations ("turns")

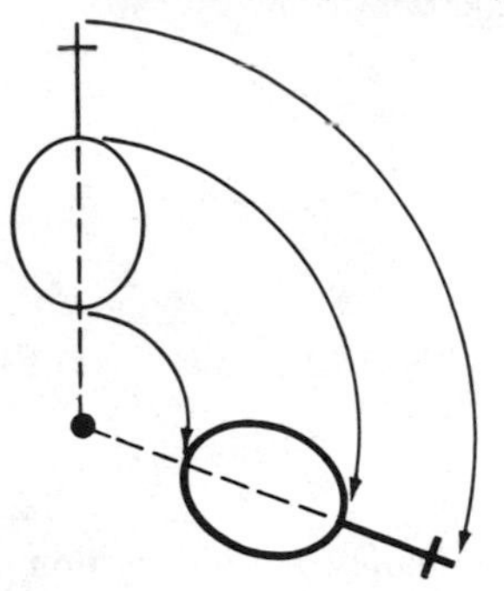

Reflections ("flips"). The word "reflection" is suggested by the relation of a figure to its image in a mirror.

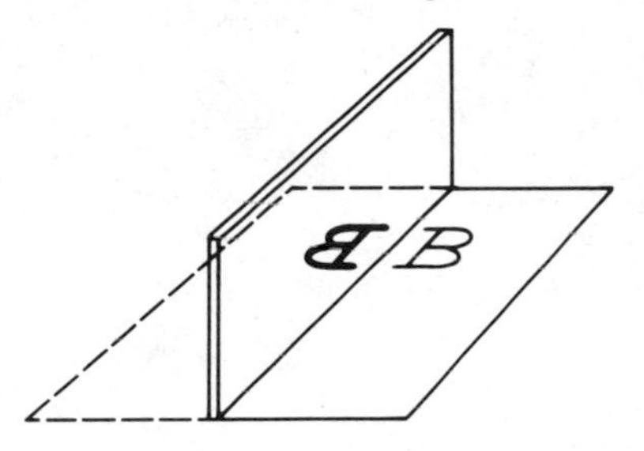

A reflection can also be thought of as a flip.

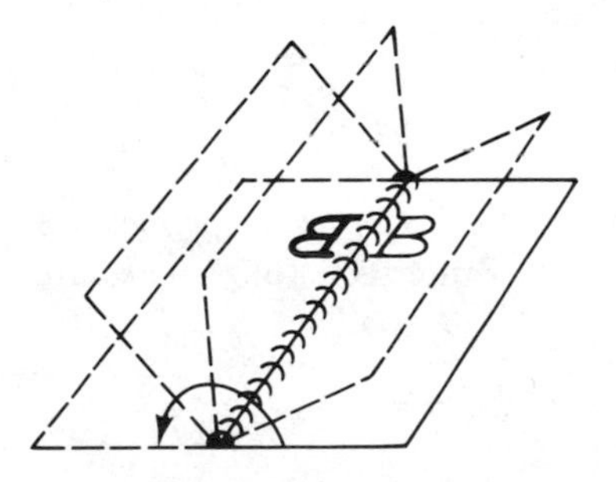

More than one of these motions might be needed to bring two congruent figures into coincidence,

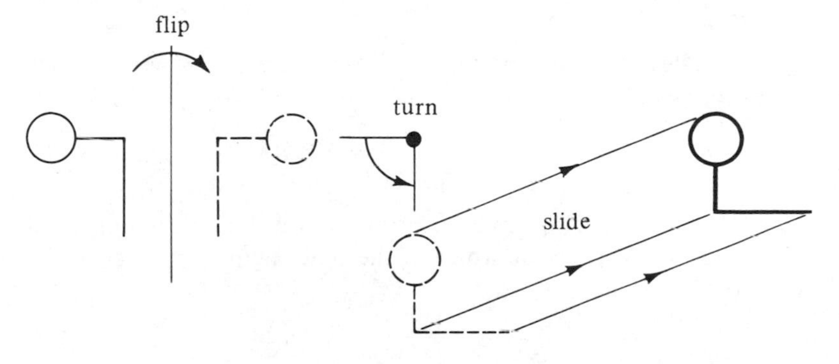

and there will always be more than one sequence of these motions that works.

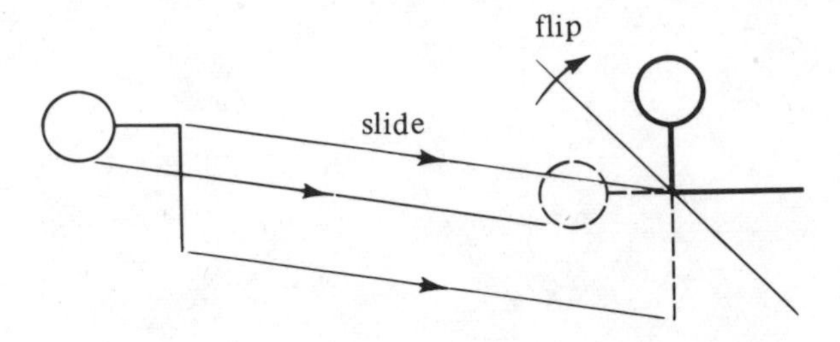

Any sequence of motions of the above three types is called a **rigid motion**. The figure below suggests a rigid motion that makes $\angle ABC$ coincide with $\angle DEF$. Thus, $\angle ABC$ is congruent to $\angle DEF$.

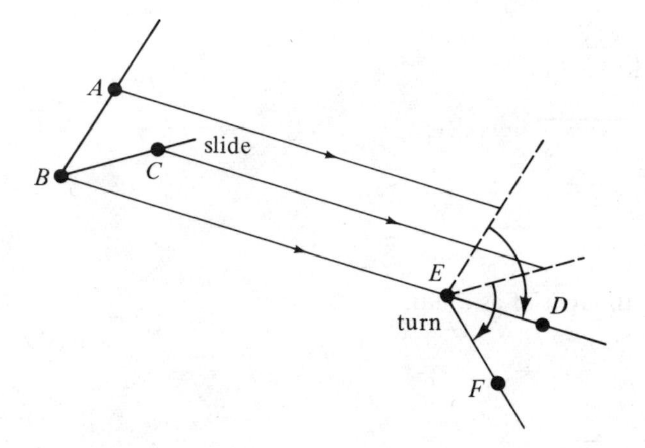

The idea of congruence of angles leads to the very important ideas of perpendicularity and right angles. When two lines intersect, four angles are formed.

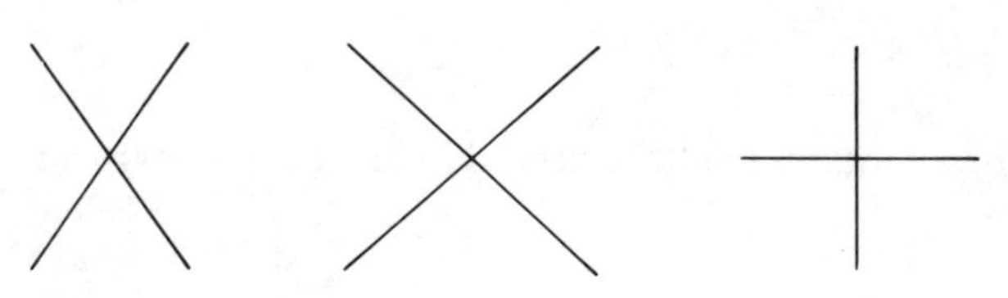

If two adjacent angles are congruent, then the lines are said to be **perpendicular** and the angles are called **right angles**.

EXERCISES

Trace figure (a) onto a sheet of translucent paper. By sliding, rotating, and flipping the sheet of paper, decide which of the other figures (b), (c), (d) are congruent to the original one.

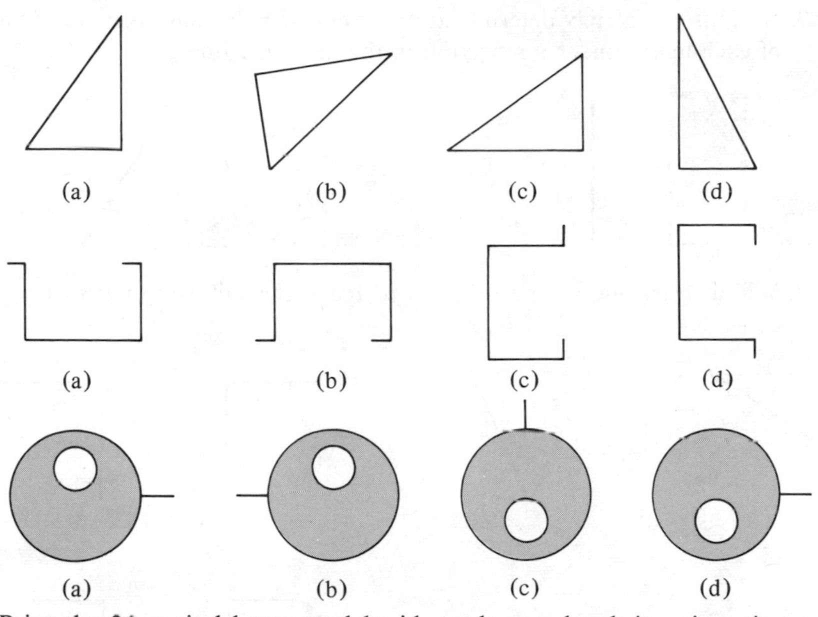

4. Print the 26 capital letters and beside each one sketch its mirror image in a vertical line. Which letters are indistinguishable from their mirror images?

A|A, B|B, . . .

5. A slide has both length and direction. It can be specified by an arrow, or "vector." Sketch the "image" of each figure under the indicated slide.

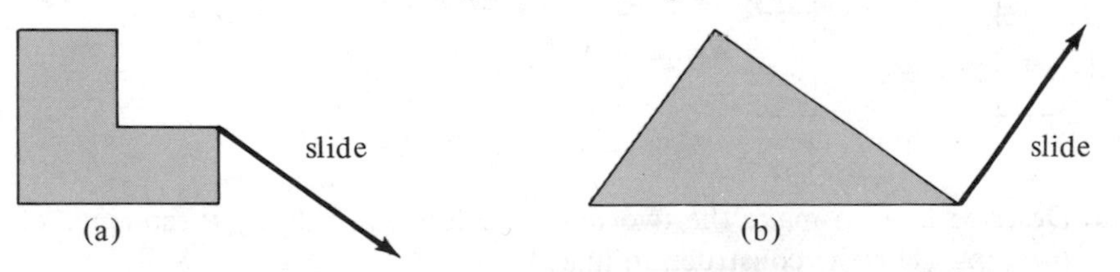

6. A turn has both a center and an angle. Sketch the image of each figure under the indicated turn.

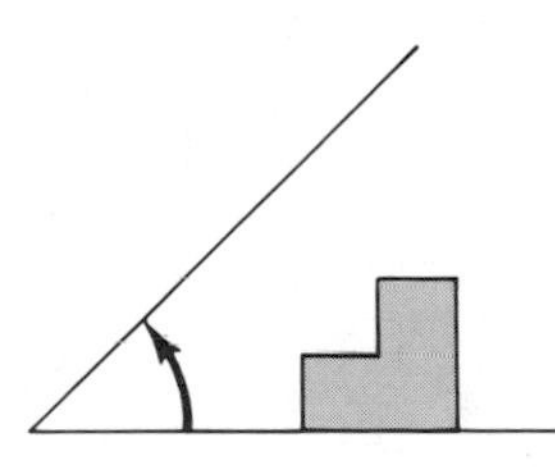

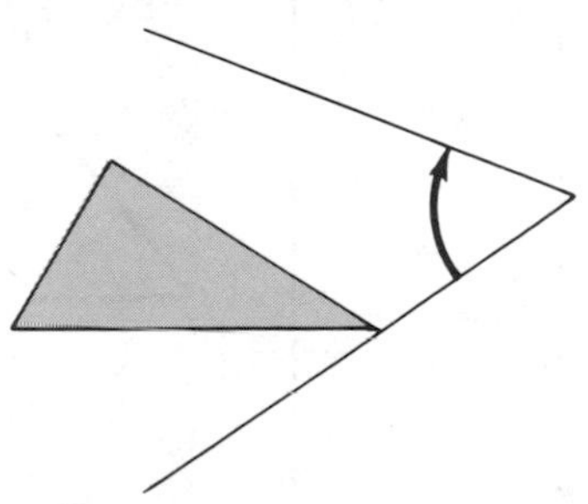

7. A flip is completely determined by its **line of reflection**. Sketch the image of each figure under a reflection in the indicated line.

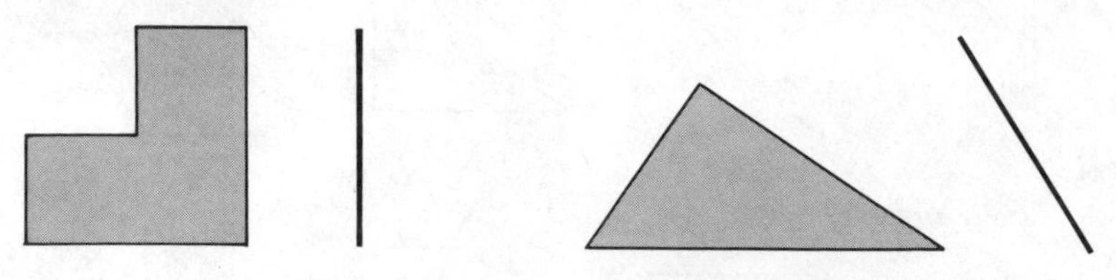

8. Will at least one flip be required to make the following pairs of figures coincide?

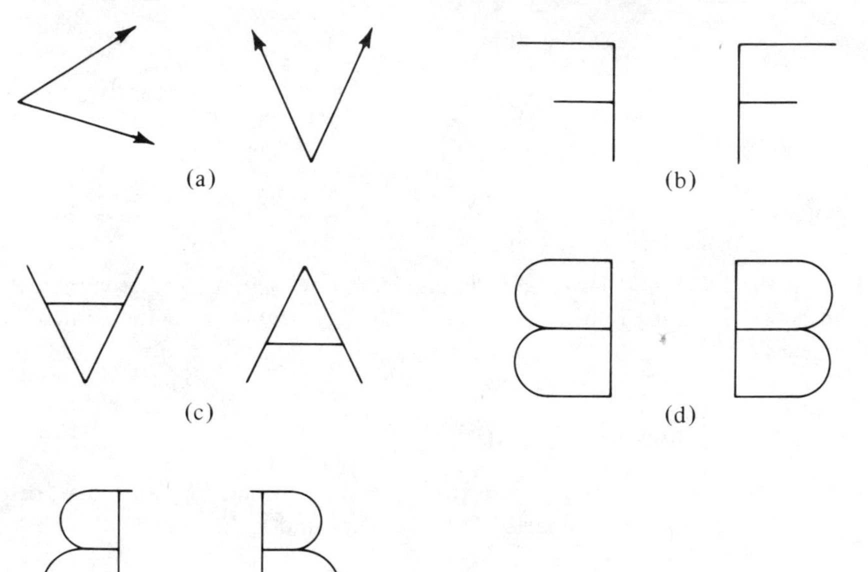

9. Describe how to make the two figures below coincide by a sequence of *two* flips. (Use the construction lines.)

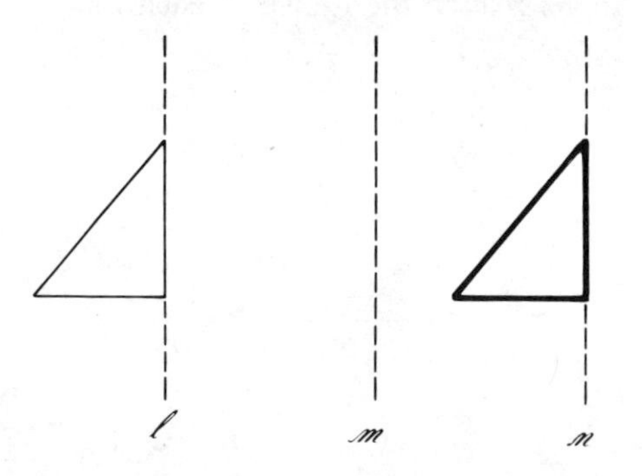

10. Repeat for the two figures below.

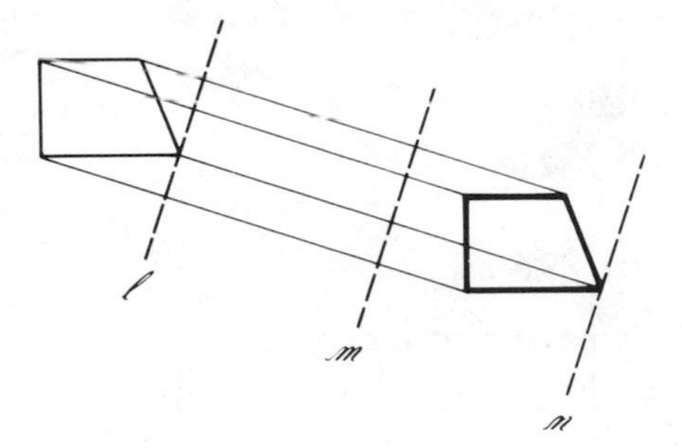

Would you guess that every translation can be accomplished by two flips?

11. Describe how to make the two figures below coincide by a sequence of *two* flips.

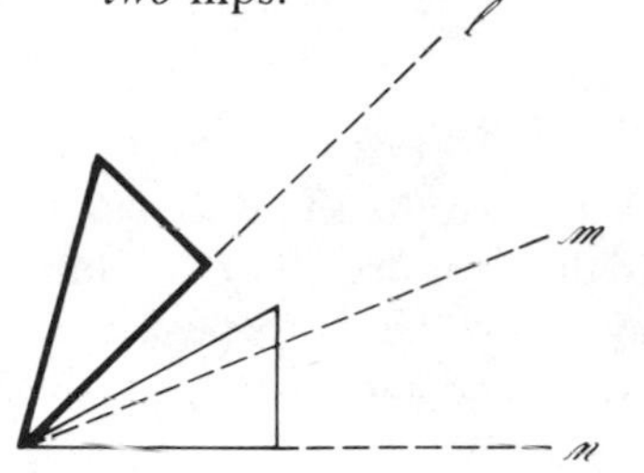

12. Repeat for the two figures below.

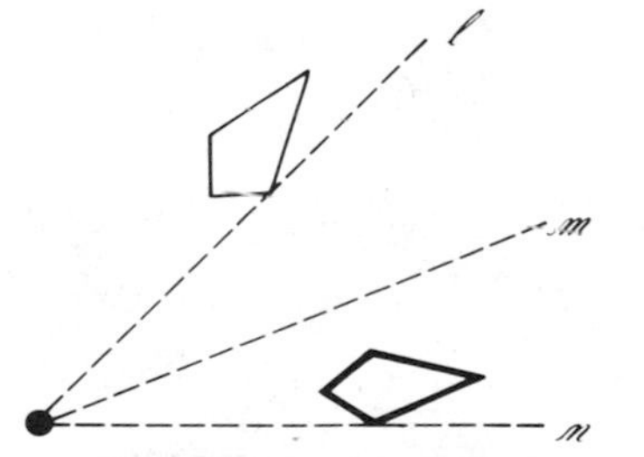

Would you guess that every rotation can be accomplished by two flips?

13. Would you guess that every rigid motion can be accomplished by a sufficient number of flips?

TEACHING NOTE "Mirror card" kits are available commercially and provide children with manipulative experience in flipping geometric figures. These kits even include hinged mirrors to illustrate the idea developed in Exercises 11 and 12.

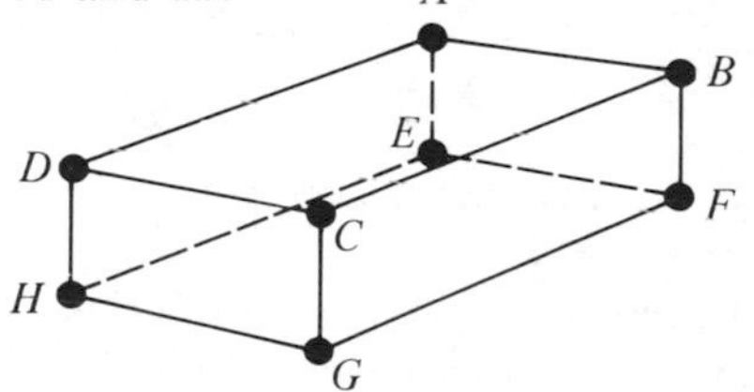

Figure for Exercise 14, p. 116

14. In the figure:

(a) Name six right angles.

(b) How many right angles are shown?

(c) Name several pairs of congruent edges.

(d) Name all pairs of congruent faces.

4.3 CONGRUENCE OF FIGURES IN SPACE

For figures in space, the problem of checking for congruence is more difficult. It is not possible to move two wooden blocks until they "coincide." How would you decide if the two blocks are congruent?

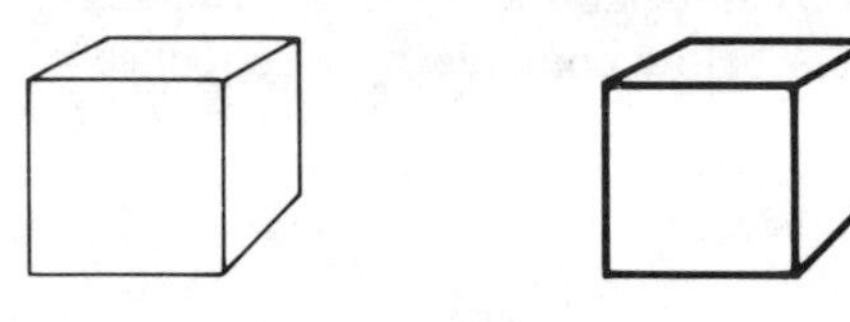

Here is one way. Since they are both cubes, they have the same shape. To conclude that they are congruent all that is left to do is to check that they have the same size as well. Probably the easiest way to do this is to push them together and see if an edge of one is congruent to an edge of the other.

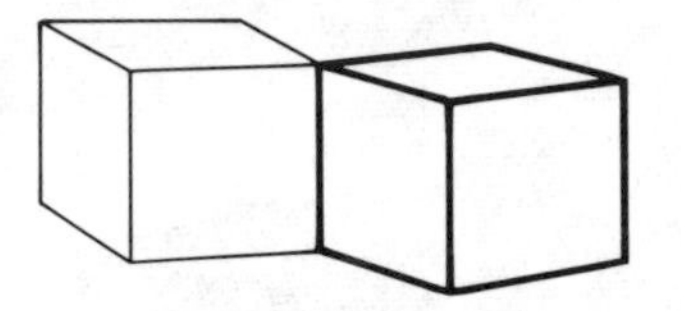

A more complicated way would be to make a plaster cast of one and see if the other fits into it exactly.

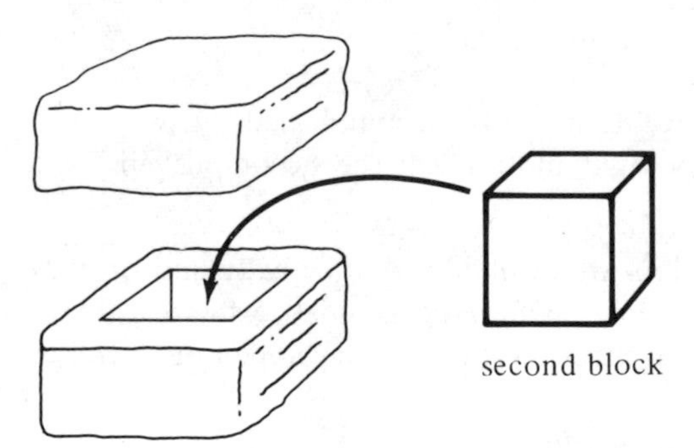

This amounts to checking that they both can be made to occupy the same set of points in space.

The second method reminds us that figures in space are really sets of points or locations in space. They are not made of wood. There is no reason why we cannot think of these figures as being moved through each other to coincide.

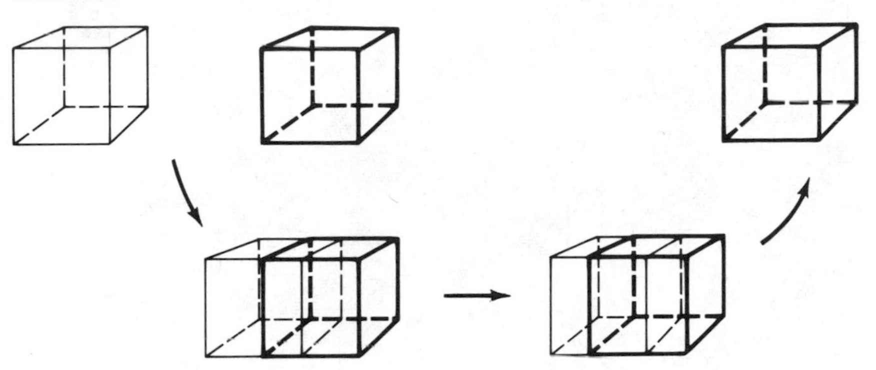

So it is reasonable to write a definition similar to the one for plane figures.

Two space figures are **congruent** if they can be made to coincide by a sequence of translations, rotations, and reflections.

There is one striking difference between the situations in the plane and in space, however. Pictures of congruent figures in a plane can actually be moved around by a sequence of slides, turns, and flips until they coincide. This is *not* the case in space. Consider this "tetrahedron" and its reflection. (Notice that in space we reflect in a *plane*, while in the plane we reflected in a line.)

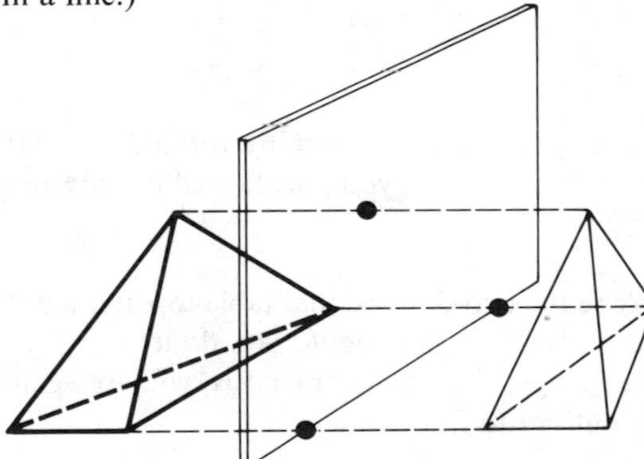

According to our definition these two figures are congruent. But if you think of picking them up and trying to make them coincide, you will soon discover that it cannot be done, even if you suppose that they can be slid through one another!

EXERCISES

1. How would you go about deciding if the geometric figures suggested by the following pairs of real objects are congruent?

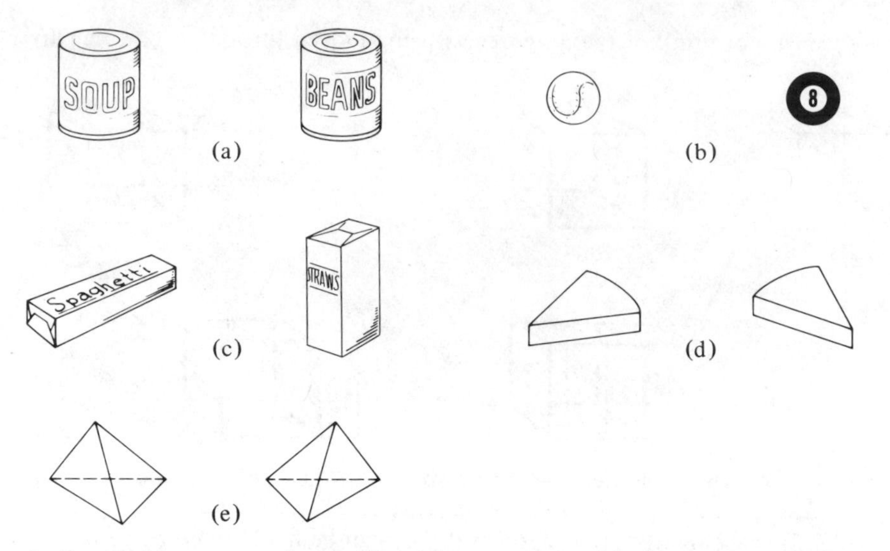

(e)

2. On a table are two sheets of translucent paper with these geometric figures drawn on them.

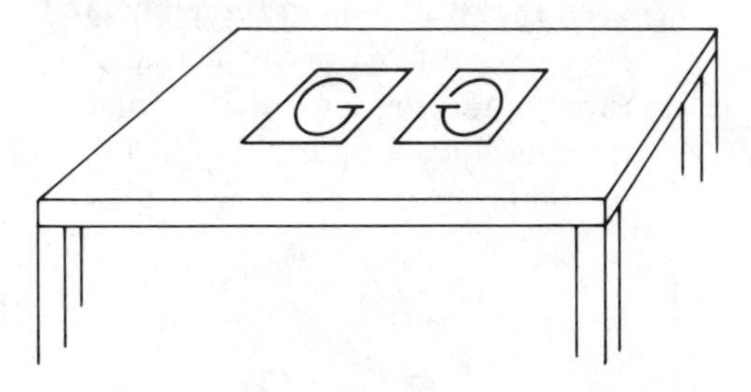

You are asked to make the figures coincide, but you are forbidden to lift either piece of paper off the table top. That is, you are allowed to slide and turn, but not to flip.

(a) Could you do it?

(b) If you were allowed to escape the two-dimensional table top and move the paper freely in three-dimensional space, could you do it?

(c) Can you make these two wire figures coincide? (That is, you are again allowed to slide and turn, but not to flip.)

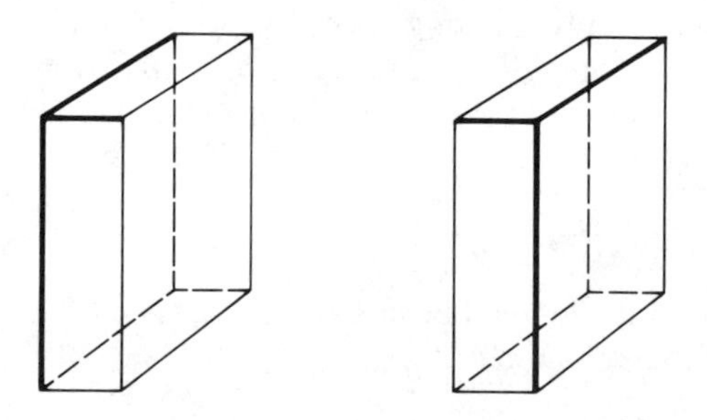

(d) Mathematicians study "spaces" of dimension greater than 3. Do you think that if you could somehow escape three-dimensional space and work in four-dimensional space that you could then make the wire figures coincide?

3. The figure on the left is the reflection in plane $\mathscr{P}$ of the figure on the right. What points in the left figure correspond to each of these points of the right figure under the reflection?

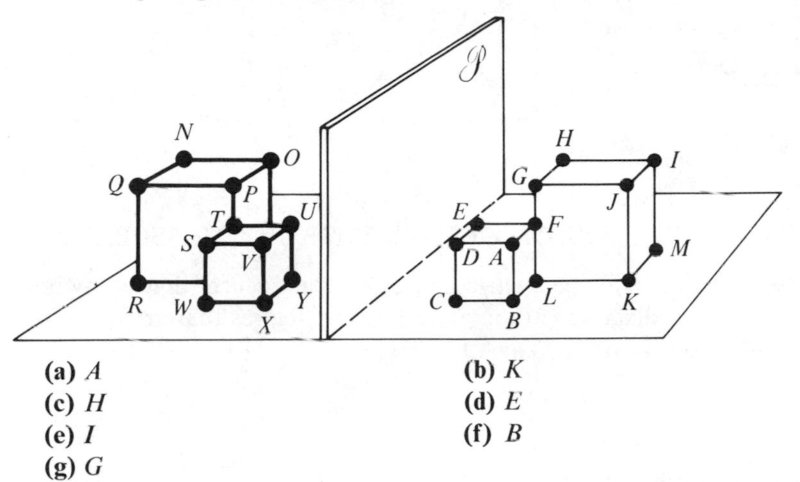

(a) A **(b)** K
(c) H **(d)** E
(e) I **(f)** B
(g) G

4. For each of the following pairs of figures, decide (i) whether they are congruent, and (ii) whether they could be physically slid and turned in (three-dimensional) space until they coincide.

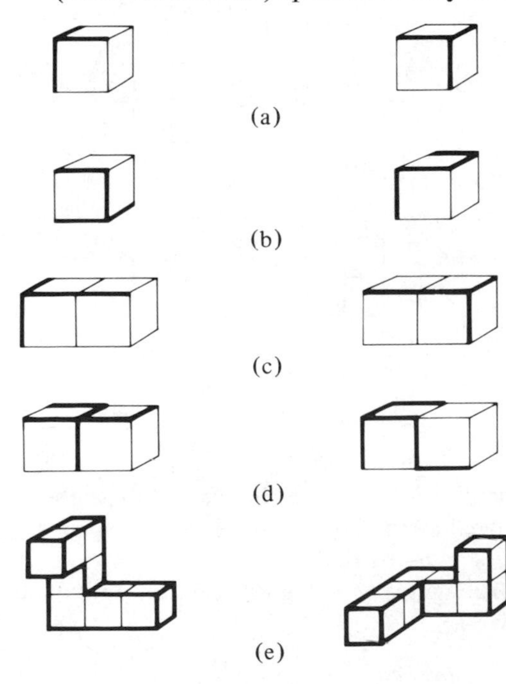

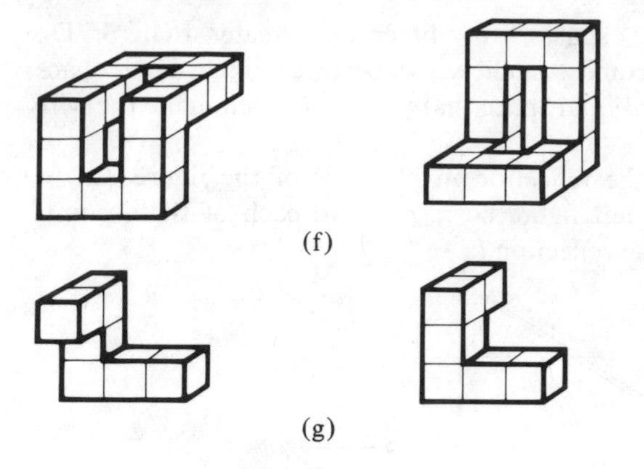

4.4 COMPARING SIZES OF FIGURES WITHOUT MEASURING

We have been classifying geometric figures by congruence, that is, by both size and shape. In diagram (2) there are just two figures that are congruent to the figure in diagram (1). Which two?

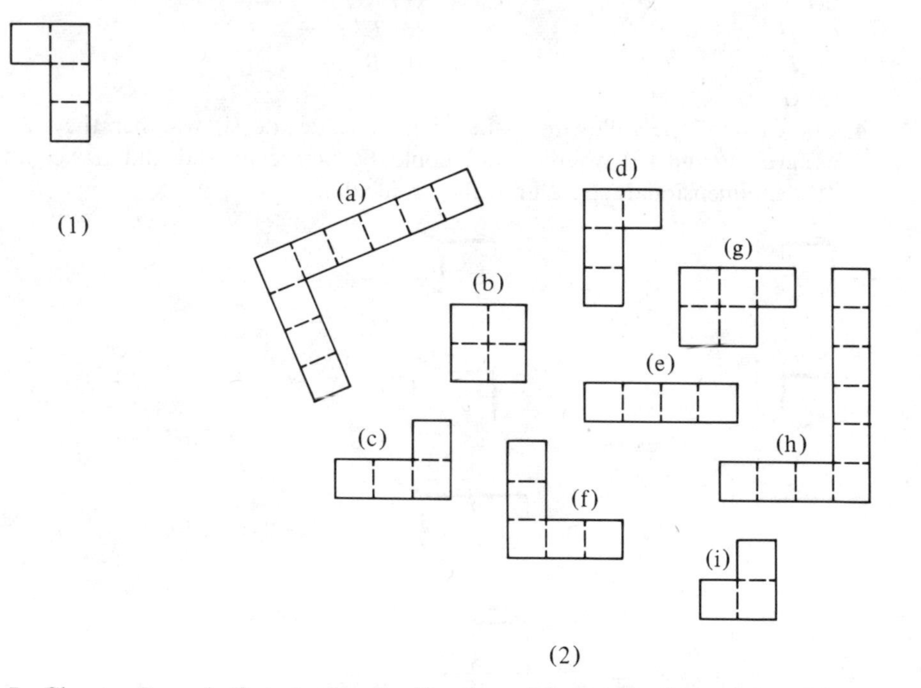

In Chapter 8 we shall study the classification of figures by shape only, without regard to size. Two figures that have the same shape will be called "similar." Which figures in diagram (2) are similar to the figure in diagram (1)? (It might be helpful to think of similar figures as being different photographic prints made from the same negative.)

In the rest of this chapter we shall concentrate on the idea of the "size" of a geometric figure. Which figures in diagram (2) have the same size (area) as the figure in diagram (1)? Which are larger? Which are smaller?

We all have good intuitive ideas about the relative sizes of objects. Which candy bar would you prefer?

Which snake is longer?

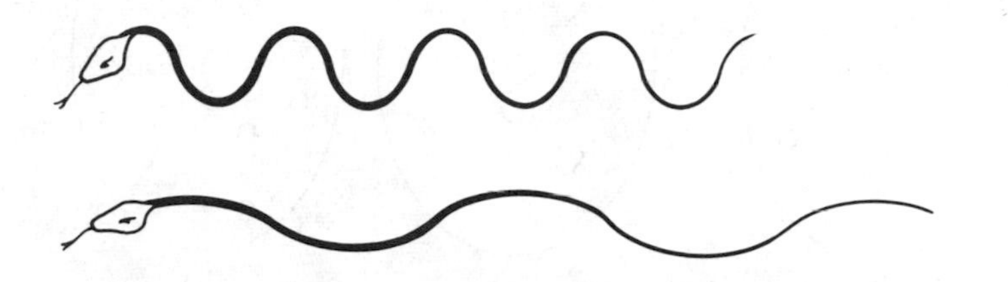

Can this can of soup be heated safely in the pan shown?

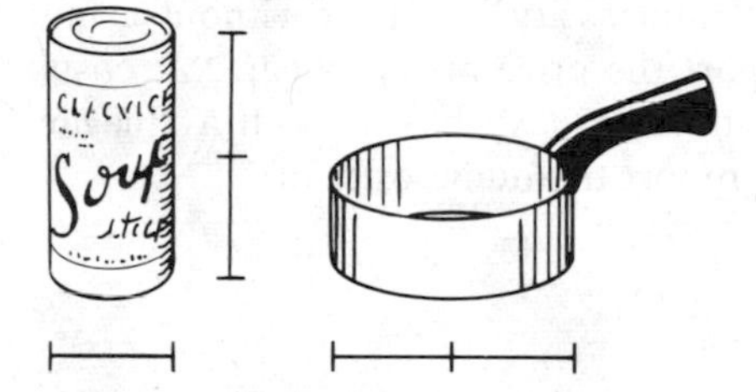

We do not have to know how to use a ruler or even know anything about numbers to answer these questions.

Comparing a *pair* of objects for size can often be done without numbers, but to describe the size of a *single* object without using numbers is usually clumsy. You could describe a small boy as being "as tall as" a putter. Or maybe you would say he is "taller" than a garbage can, but "shorter" than a parking meter. This is not a very exact description. Not everyone has a good idea of the heights of putters, garbage cans, and parking meters. Moreover, different putters have different lengths, as do different garbage cans and different parking meters. More people would understand you better if you said that the boy was 39 in. tall. Everyone knows about inches, and all inches are the same for everyone.

EXERCISES

1. *Without counting,* decide if set (i) has more, less, or the same number of elements as set (ii). Explain how you decided.

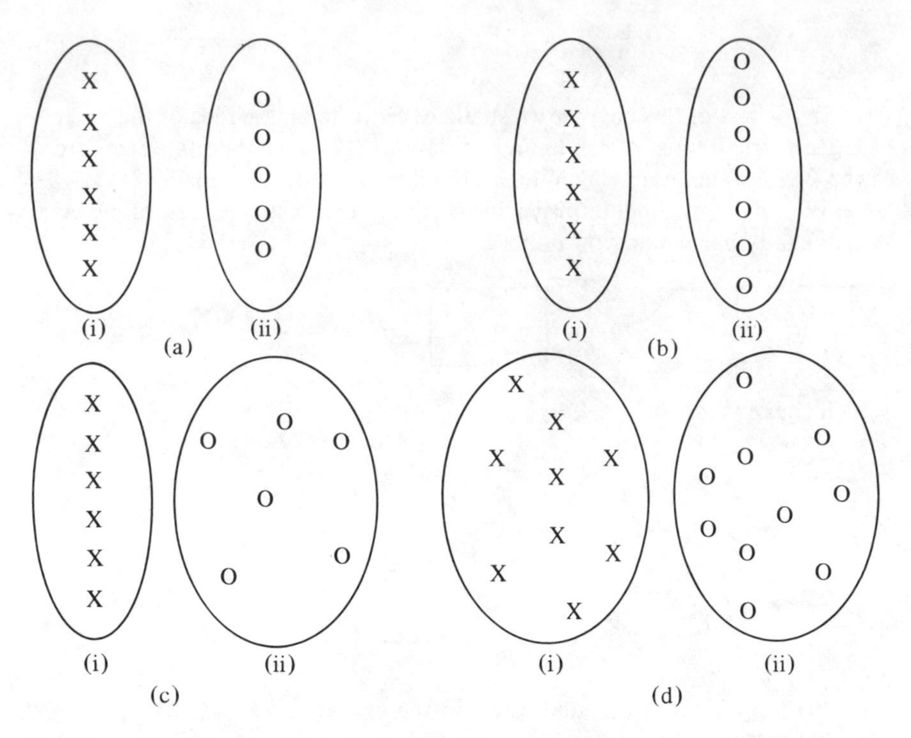

2. It is generally clumsy to report the "numerosity" of a set without using numbers. Here is how you could report the price of a new car. "It costs as many dollars for this new car as it takes navy beans to fill a vinegar jug." Without using numbers, try to report the numerosity of

(a) the set of students in your math class

(b) the set of people in your city

(c) the set of planets in the solar system

3. Try to report, without using numbers,

(a) the height of the ceiling in your bedroom

(b) the area of the ceiling in your bedroom

(c) the volume of the wastebasket in your bedroom

4. (a) When we say that a boy is as tall as a putter, are we describing the size of a single object or are we actually comparing two objects?

(b) When we say that a man is 6 ft tall, are we describing the size of a single object or are we actually comparing two objects?

(c) Is it possible to report the size of anything without somehow comparing it to something else?

5. It would not be a fair question to ask which figure below is "larger." We need to specify what aspect of the figures is being compared. We do this by using words such as "length," "area," and "volume."

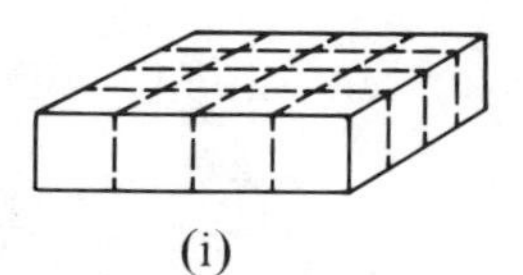

(i)

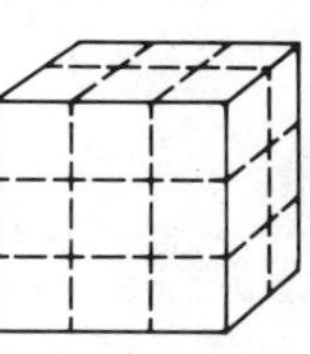

(ii)

(a) Which figure has more length?
(b) Which figure has more surface area?
(c) Which figure has more volume?

6. We would like to make more precise what we mean when we say that one plane figure is "smaller" than another.

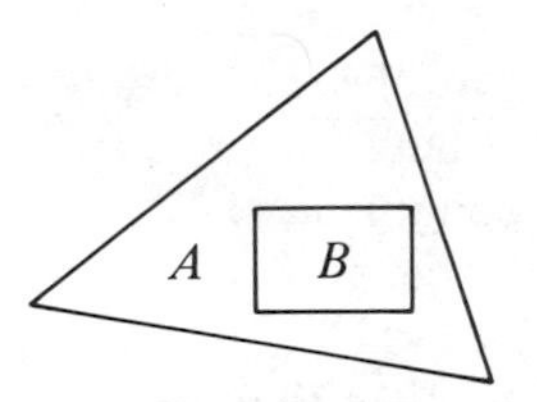

(a) In the diagram, which is smaller, the plane figure A or its subset B?
(b) If A is the entire square and B is the subset consisting of everything but the top edge, would you say that B has less area than A?

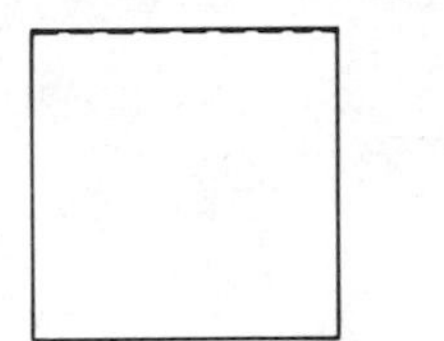

(c) If B is a subset of the plane figure A and if $A - B$ has dimension __________, then B is smaller than A.
(d) If C is congruent to a subset B of A and if $A - B$ has dimension 2, then . . . .
(e) If C is smaller (in area) than A, must C be congruent to a subset of A? Illustrate with a picture.
(f) Using the pictures you drew in (e), can C be cut up and reassembled so as to be congruent to a subset of A?

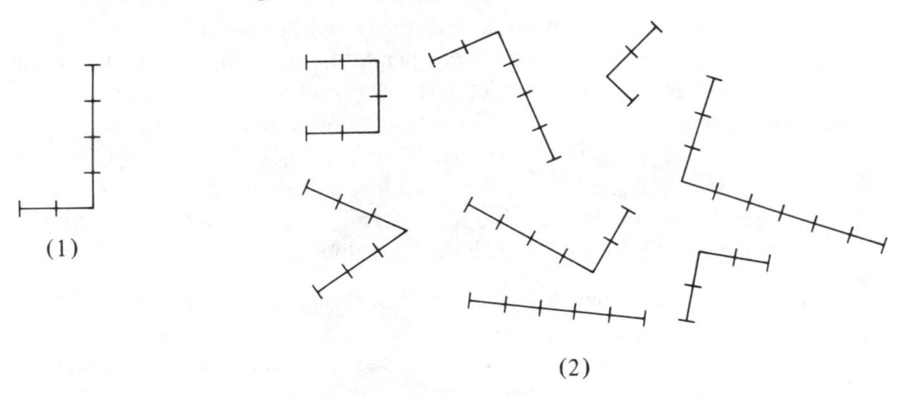

7. (a) Enclose with a simple closed curve those figures in diagram (2) that are similar to the figure in diagram (1). Shade the interior.
(b) Enclose with a simple closed curve those figures in diagram (2) that have the same size (total length) as the figure in diagram (1). Shade the interior.

(c) How are the figures in the doubly shaded region related to the figure in diagram (1)?

8. Mark these statements as true or false.

(a) If two figures are congruent, then they have the same size.

(b) If two figures are similar but not congruent, then they have different sizes.

(c) If two figures are not similar, then they do not have the same size.

(d) If two figures are not congruent, then they do not have the same size.

(e) If two figures are not similar, then they are not congruent.

(f) If two figures are similar, then they are either congruent or have different sizes.

(g) If two figures are similar and have the same size, then they are congruent.

(h) If two figures are not similar or do not have the same size, then they are not congruent.

4.5 THE PROCESS OF MEASUREMENT

If you were asked to measure the segment below,

---

you would probably lay your ruler down beside it and report that it is 4 in. long. If a European student were asked to do the same thing, he would probably lay his ruler down beside the segment and report that it is about 10 centimeters (cm) long. Both of your replies would involve two things: a number (4 or 10) and a unit of measure (inch or centimeter). These are the two essential ingredients in any problem involving measurement.

Many different kinds of measurement are possible. You might measure your age in years, your height in inches, your weight in pounds, your capacity in chocolate malts, your speed in miles per hour, or your intelligence in IQ points. In this chapter we restrict our attention to the measurement of geometric figures or of physical objects that suggest geometric figures. Thus, we shall always be measuring the "size" of an object. If the object is one-dimensional, we shall usually use the word **length** instead of size. If the object is two-dimensional, we shall use the word **area** instead of "size." If the object is three-dimensional, we shall use the word **volume** instead of "size."

To measure a geometric figure, the first thing required is a unit of measure.

> The **unit of measure** is a geometric figure having the same dimension as the figure to be measured.

A farmer might measure the *length* of a fence in *feet*, the *area* of a field in *acres*, and the *volume* of a corn crib in *bushels*. It would be absurd for him to try to measure the length of the fence in bushels, the area of the field in feet, or the volume of the corn crib in acres.

Having chosen a unit of measure, the next thing to do is to count how many copies of this unit are required to "fill up" the figure being measured. For example, if the triangle U is chosen as the unit of (area) measure, then the measure of the figure F with respect to U is 14, since 14 copies of U are

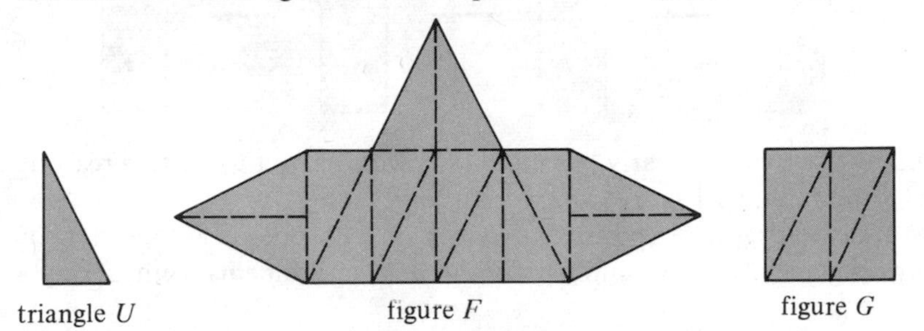

triangle U figure F figure G

required to "fill up" F. The measure of the figure G with respect to U is 4. Of course, the measure of U with respect to U is 1. The measure of figure H is not as easy to calculate.

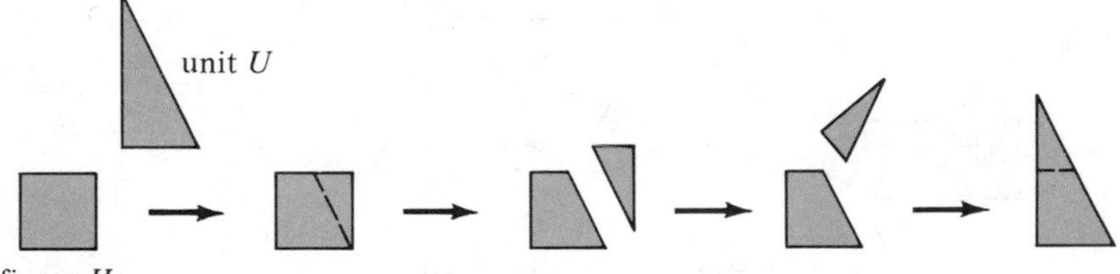

figure H

But if we cut and reassemble figure H as indicated, then obviously its measure with respect to U is 1. We have made the assumption that:

> The size of a geometric figure is not changed if the figure is cut up and reassembled.

(Mathematicians have devised pathological cutting-reassembling procedures that do in fact change size. But a description of these procedures is well beyond the scope of this text and the imagination of elementary school children.)

Measuring figure J presents still another problem. No matter how we cut up and reassemble figure J, it is not possible to fill the resulting figure with

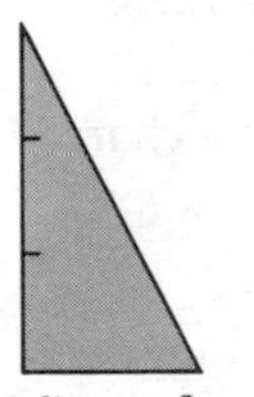

figure J

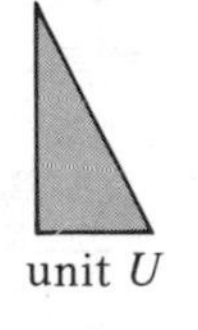

unit U

a whole number of copies of U. However, if we take two copies of J and cut and paste them as indicated,

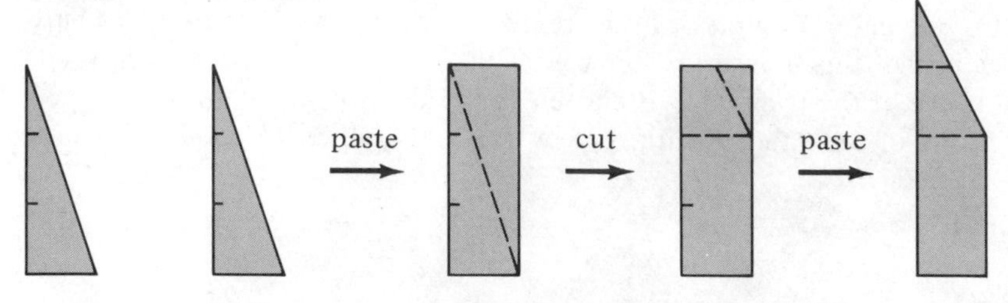

then the resulting figure clearly has measure 3 with respect to U. It is reasonable then to say that J has measure $\frac{3}{2}$ with respect to U.

We have assigned a (positive) number to each one of a collection of geometric figures. We can summarize our accomplishments with a single chart.

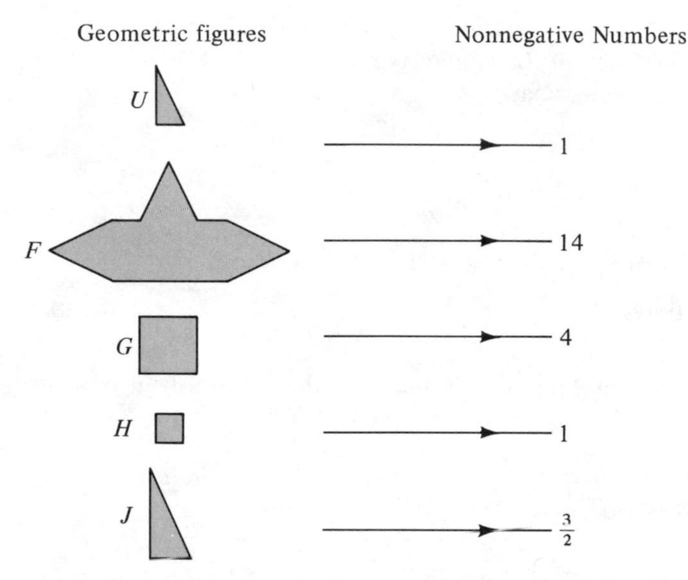

The chart suggests that there is a "measure function." (What is its "domain," that is, what are the inputs? What is its "range," that is, what are the outputs? Can you suggest a geometric figure whose measure with respect to U is zero?) We give this particular measure function the name m_U. The m suggests measure, the subscript U indicates what unit is being used. This symbolism allows us to replace the clumsy expression, "the measure of G with respect to U is 14," by the neat mathematical sentence $m_U(G) = 14$.

Length, area, and volume **measure functions** share the following characteristic properties, no matter what unit of measure is chosen.

---

1. Domain is a set of geometric figures.
2. Range is a set of nonnegative numbers.

3. If A is congruent to B, then $m(A) = m(B)$.
4. If A and B are disjoint,* then

$$m(A \cup B) = m(A) + m(B)$$

Properties 3 and 4 are mathematical statements of our assumption that the size of a figure remains unchanged when the figure is cut up and (perhaps) pasted back together.

Notice how we used these properties when we measured the figure H earlier:

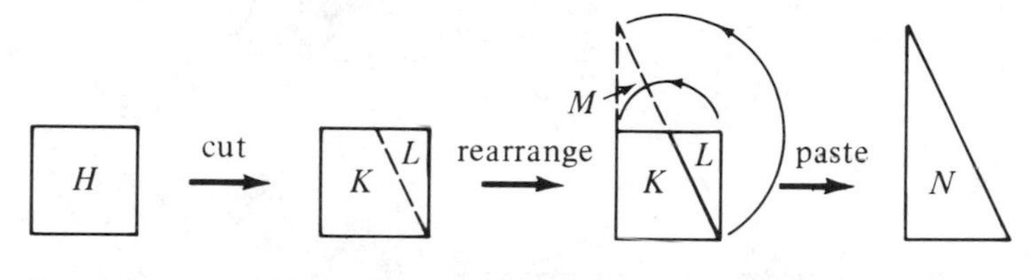

$$m(H) = m(K \cup L) \overset{(4)}{=} m(K) + m(L) \overset{(3)}{=} m(K) + m(M) \overset{(4)}{=} m(K \cup M) = m(N)$$

Property 4 is known as the **additivity property** of measure functions.

EXERCISES

1. In the diagram above, the reasoning at the "rearrange" step is this: L is congruent to M; therefore, $m(L) = m(M)$ by property 3; therefore, $m(K) + m(L) = m(K) + m(M)$ by the addition principle of equality. Explain why L is congruent to M.
2. We said that if a figure were two-dimensional, we would use the word "area" when referring to its "size." We agreed to reserve the word "length" for one-dimensional figures.
   (a) What sort of geometric figure have we really measured when we report that a rectangle has "length" 4 in.?
   (b) To what does the word "width" refer?
   (c) How is the word "thickness" or "breadth" used?
3. **(a)** If you tried to measure the area of a field in feet, how many copies of the unit would it take to "fill up" the field?
   (b) If you tried to measure the length of a fence in acres, how many copies of the fence would you have to glue together before you could fit a copy of the unit into their union?
4. Measure the length in inches of the figure below. Did you make use of property 4 of measure functions?

* It is enough to require only that A and B have no interior points in common.

5. Make use of property 3 to find, quickly, the length in inches of the figure below.

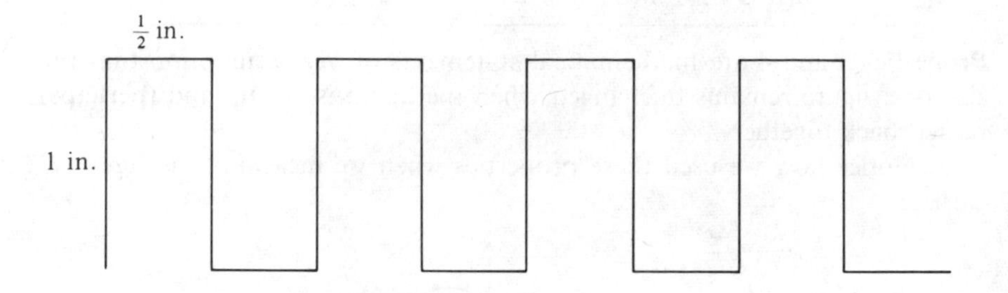

6. Using the figures below, compute

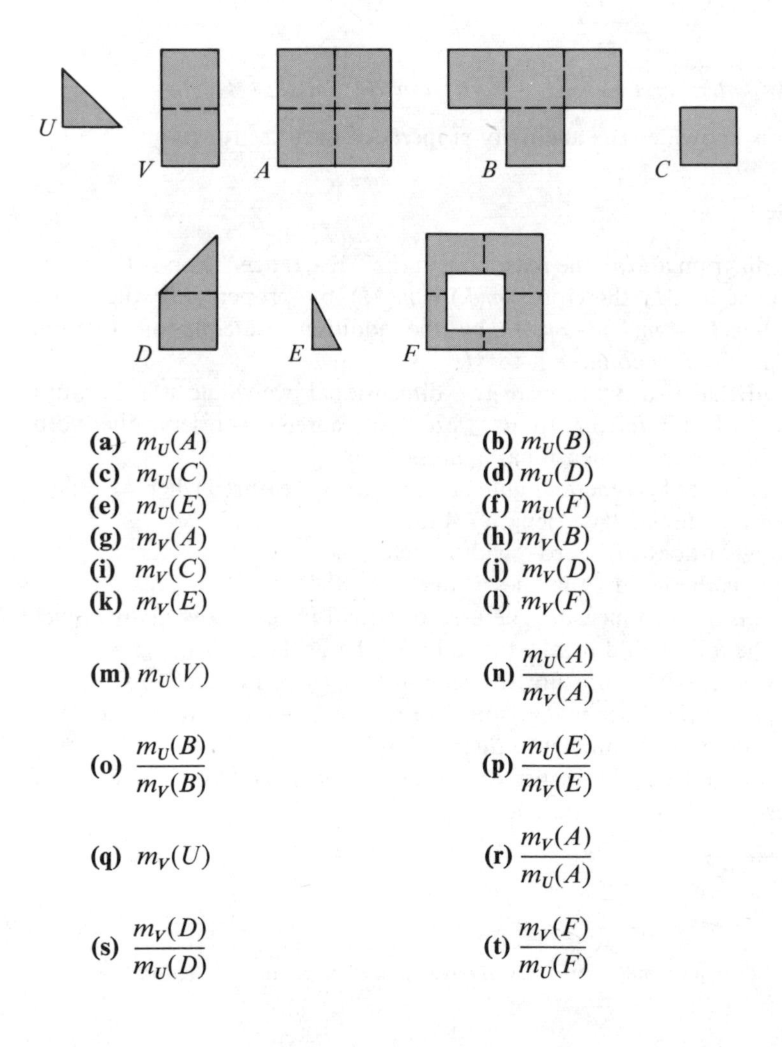

(a) $m_U(A)$ **(b)** $m_U(B)$
(c) $m_U(C)$ **(d)** $m_U(D)$
(e) $m_U(E)$ **(f)** $m_U(F)$
(g) $m_V(A)$ **(h)** $m_V(B)$
(i) $m_V(C)$ **(j)** $m_V(D)$
(k) $m_V(E)$ **(l)** $m_V(F)$

(m) $m_U(V)$ **(n)** $\dfrac{m_U(A)}{m_V(A)}$

(o) $\dfrac{m_U(B)}{m_V(B)}$ **(p)** $\dfrac{m_U(E)}{m_V(E)}$

(q) $m_V(U)$ **(r)** $\dfrac{m_V(A)}{m_U(A)}$

(s) $\dfrac{m_V(D)}{m_U(D)}$ **(t)** $\dfrac{m_V(F)}{m_U(F)}$

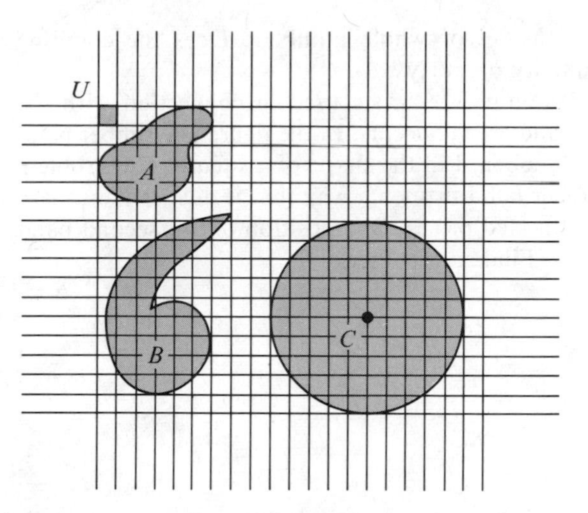

7. Estimate the following for the above figure:

(a) $m_U(A) \doteq$ ________

(b) $m_U(B) \doteq$ ________

(c) $m_U(C) \doteq$ ________

8. Supply reasons in this argument that if m is a measure function and $A \subset B$, then $m(A) \leq m(B)$.

(a) $m(B) = m[A \cup (B - A)]$ Why?

(b) $m[A \cup (B - A)] = m(A) + m(B - A)$ Why?

(c) $m(B - A) \geq 0$ Why?

(d) $m(A) + m(B - A) \geq m(A) + 0$ Why?

(e) $m(B) \geq m(A)$ Why?

9. When we measured the figure J (pp. 125–126), we drew a congruent

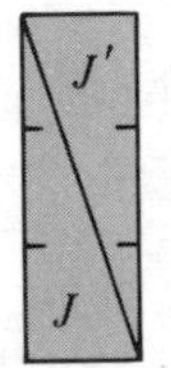

copy, J', of J and then argued this way:

i $2 \cdot m_U(J) = m_U(J) + m_U(J)$

ii $= m_U(J) + m_U(J')$

iii $= m_U(J \cup J')$

iv $= 3$

v Therefore $m_U(J) = \frac{3}{2}$.

(a) At which step in the argument was property 3 of measure functions used?

(b) At which step was property 4 of measure functions used?

10. Counting can be thought of as a function with domain a collection of (finite) sets and range the set of whole numbers. Does the counting function have the additivity property?

11. The geoboard (described on p. 99) can be used in connection with the teaching of congruence and area concepts. In the figure one rubber band has been placed on the geoboard in the shape of a square. The problem is to place another rubber band in such a way that it divides the square into two congruent, connected pieces. One position of this second band is shown by dotted lines. Find some others.

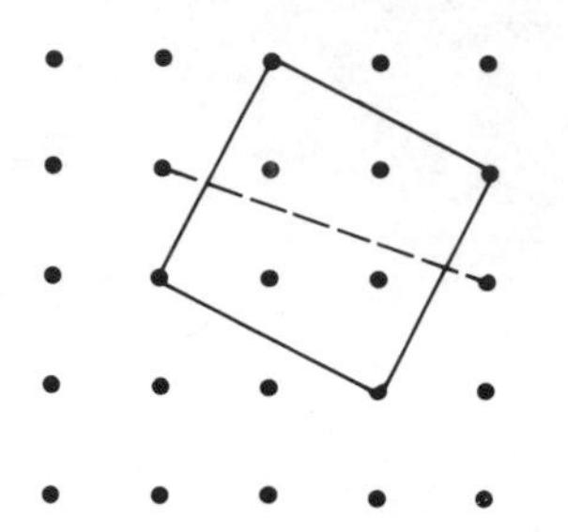

12. Now the problem is to place a second rubber band so that the square is divided into two connected pieces that have equal area, but are not congruent. One position is shown. Find some others.

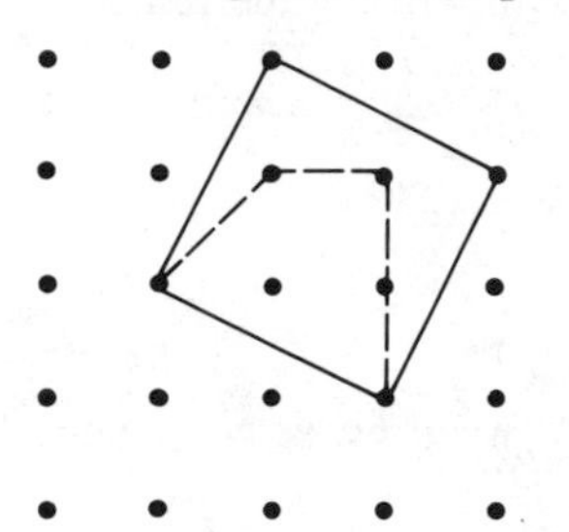

***13.** Two diagonally opposite corner squares have been removed from a checkerboard having 1 in. by 1 in. squares. You are given 31 1 in. by 2 in. dominoes.

(a) Try to cover the cut-down checkerboard completely with the dominoes.

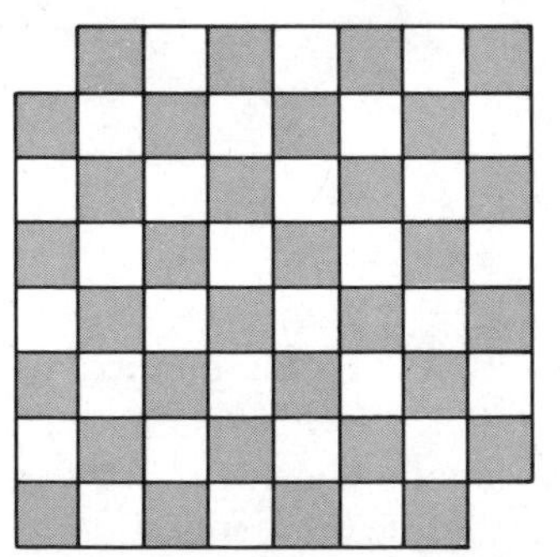

(b) Try to *prove* it is impossible to cover the cut-down checkerboard with the dominoes.

(c) If X denotes the checkerboard (cut-down) and U a domino, is $m_U(X)$ an integer?

(d) Does your answer to (c) tell you anything about (a) or (b)?

4.6 LENGTH AND PERIMETER

To measure one-dimensional figures a one-dimensional unit of measure is needed. The one in common use is a segment known as an **inch**.

an inch

In Chapter 5 we shall study other possible "units of linear measure."

The first figures to be measured for length in the elementary school are segments.

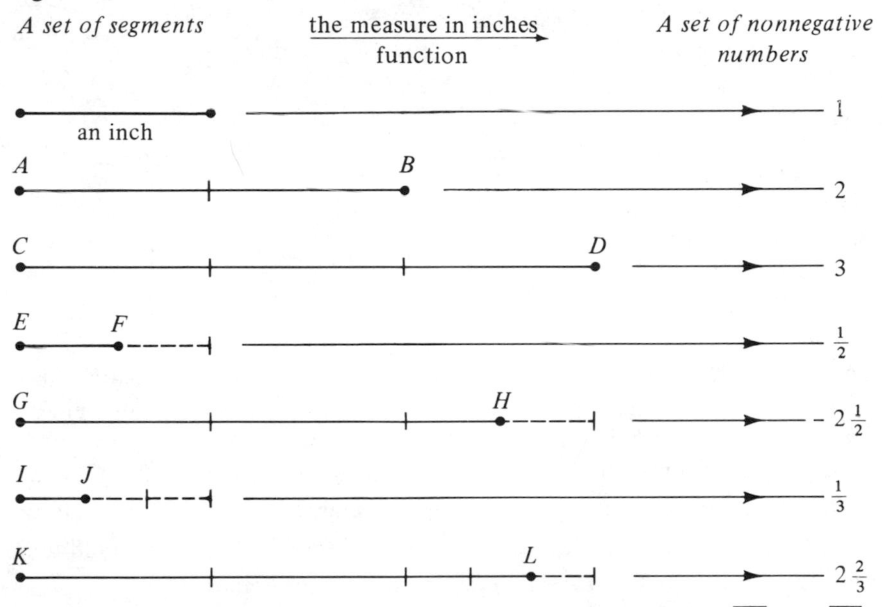

Among these segments the first ones to be measured would be $\overline{AB}$ and $\overline{CD}$. To measure them one needs only count how many inches, laid end to end, are needed to fill them up. (Initially, the child uses individual inches. Only later is he given a ruler on which the laying off and counting has been done for him.) A child who knows only the positive integers can assign an exact measure-in-inches to each of these two segments. For that reason $\overline{AB}$ and $\overline{CD}$ are called **integrally measurable** (in inches) segments. The other four segments in the diagram are **integrally nonmeasurable** (in inches). Given the segment $\overline{GH}$, a child who is not yet acquainted with fractions cannot assign it a number. All he can say is that its length is between 2 and 3 inches.

The introduction of rational numbers into arithmetic can be viewed as an attempt to make more segments measurable. All of the segments in the figure are **rationally measurable** (in inches), because each can be assigned one specific rational number as its measure-in-inches. In Chapter 8 (p. 356) we shall "construct" a segment that is **rationally nonmeasurable** (in inches). The existence of such segments can be used, in the upper elementary or junior high school grades, to motivate the introduction of irrational numbers. Once all of the real numbers, both rational and irrational, have been introduced, then every segment becomes measurable.

A special language and notation is in use for the measure-in-inches function. Instead of saying

the measure-in-inches of $\overline{AB}$ is 2

we say

$\overline{AB}$ is 2 in. long

and we write

$AB = 2$ in.

or, still more briefly,

$AB = 2''$

The symbol AB without a bar over it stands for the length of $\overline{AB}$. This symbol should be accompanied by a unit. The statement, $AB = 2$, is ambiguous. Unless a unit of measure has been agreed upon once and for all (as we have done in this section with the inch), there is no way of knowing if "$AB = 2$" means $\overline{AB}$ is 2 in. long, or 2 ft long, or 2 millimeters long, or 2 light years long.

In the figure, on p. 131, $\overline{EF}$ is assigned length $\frac{1}{2}$ because it would take two congruent copies of $\overline{EF}$, laid end to end, to have length 1.

$$2 \times \text{length of } \overline{EF} = 1 \Rightarrow$$
$$\text{length of } \overline{EF} = \tfrac{1}{2}$$

Length $2\frac{1}{2}$ is assigned to $\overline{GH}$ because $\overline{GH}$ is five times as long as $\overline{EF}$. (It would take five copies of $\overline{EF}$, laid end to end, to make a segment congruent to $\overline{GH}$.)

$$\text{length of } \overline{GH} = 5 \times \text{length of } \overline{EF}$$
$$= 5 \times \tfrac{1}{2} = \tfrac{5}{2} \quad \text{or} \quad 2\tfrac{1}{2}$$

Measuring $\overline{GH}$ by laying off copies of $\overline{EF}$ rather than the chosen unit, the inch, is known as "changing scale." $\overline{EF}$ is called a **subunit** of the inch.

The measure-in-inches function has all four of the characteristic properties of a measure function. In particular, the additivity property is used when, for example, the width of a 20-in. desk is measured with an ordinary 12-in. ruler.

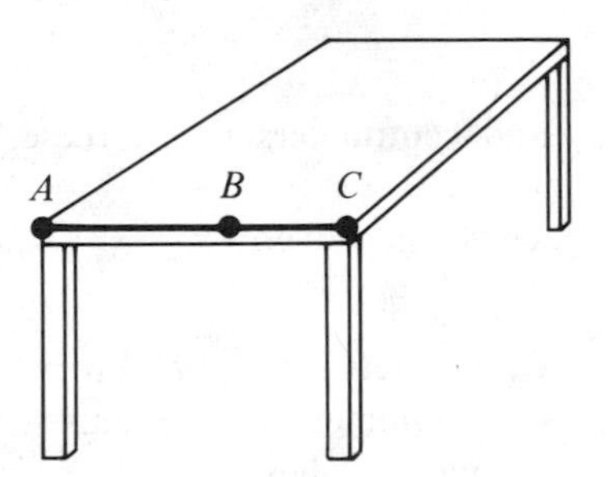

$$
\begin{aligned}
\text{width of desk} &= \text{measure of } \overline{AC} \\
&= \text{measure of } \overline{AB} \cup \overline{BC} \\
&\overset{!}{=} \text{measure of } \overline{AB} + \text{measure of } \overline{BC} \\
&= 12 + 8 \\
&= 20
\end{aligned}
$$

The additivity property suggests how one ought to assign lengths to a larger set of figures than just the set of segments; namely, the set of all polygonal curves. A polygonal curve such as the one below,

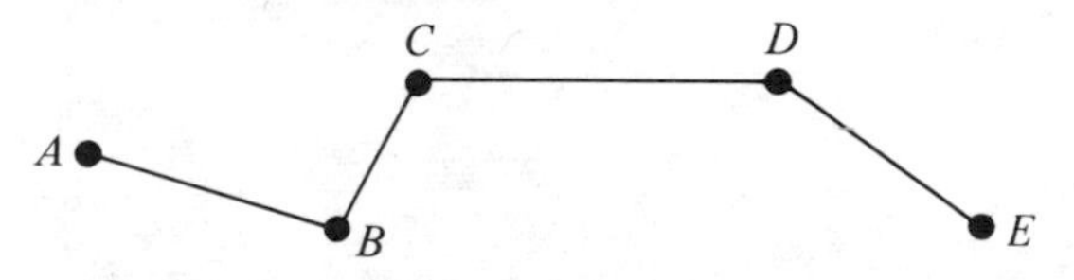

is assigned length $AB + BC + CD + DE$.

The domain of a length function such as the measure-in-inches function can be extended to include certain "nice," nonpolygonal curves as well.

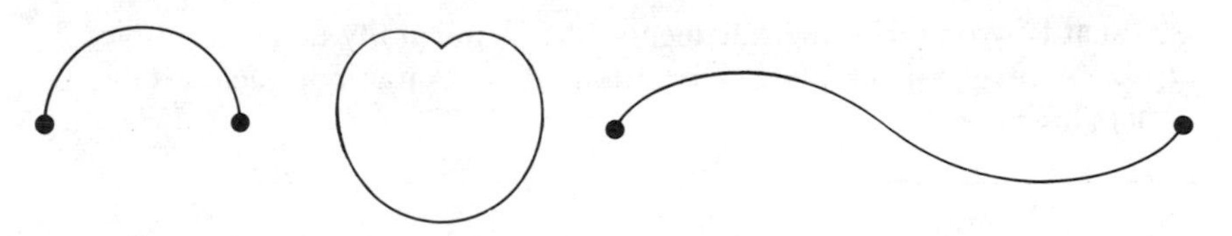

Mathematically, this extension is a very difficult problem involving concepts of calculus. To the elementary school child, however, it is not particularly tough. He simply uses a tape measure, or a piece of string along with his ruler.

These pliable measuring instruments are particularly useful for finding "perimeters."

The **perimeter** of a two-dimensional plane figure is the length of its boundary.

The length of a simple closed curve (one-dimensional) is also referred to, sometimes, as its perimeter.

EXERCISES

(You will need a ruler showing both inches and centimeters to do these exercises.)

1. Using the inch as unit, do two things for each segment below:

(i) Decide if it is or is not integrally measurable.
(ii) If it is, write down a statement of equality giving its measure; if it is not, write down a compound statement of inequality giving two integers between which its measure lies.

EXAMPLE

A *B*

(i) not integrally measurable
(ii) $1 < AB < 2$

(a)  *C* *D*

(b)  *E* *F*

(c)  *G* *H*

(d) *I* *J*

2. Repeat Exercise 1 using the centimeter as unit.

3. What is wrong with the statement, "$\overline{KL}$ is integrally nonmeasurable"?

4. Each polygonal curve below is rationally measurable in inches. Give the lengths of each.

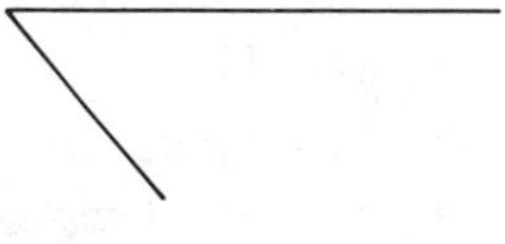

(a) 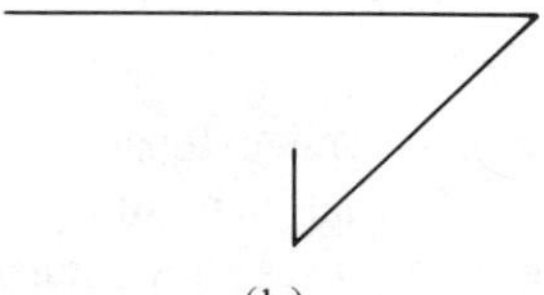 (b)

5. If two children were to measure the length of a classroom in chalkboard erasers, do you think they would agree on its length? Give some possible sources of disagreement.

6. What measuring instrument might be used to measure
(a) the length of a pencil
(b) the perimeter of the top rim of a wastebasket
(c) the diameter of a piece of wire
(d) the width of a city lot

7. In our description of the number line, we spoke glibly of the distance between points on it.
(a) What is meant by the distance between two points in space?
(b) What is the unit of measure on any number line?
With each real number is associated a point of the number line. It is also quite natural to associate with each real number a segment of the number line.
(c) What segment do you associate with the number $3\frac{1}{2}$? What is its length?
(d) What segment do you associate with the number -2.4? What is its length?

8. True or false:
(a) If two one-dimensional figures are congruent, then they have the same length.
(b) If two one-dimensional figures have the same length, then they are congruent.
(c) If two segments have the same length, then they are congruent.

TEACHING NOTE What seems to be the most natural sequence of concepts in the elementary grades, and the one we have outlined here, is that *congruence* of segments comes first. Very young children can manipulate Tinker Toy sticks and decide if they are congruent or not. No knowledge of numbers is required to do this. Later the idea of measuring *length* by laying off and counting congruent copies of a unit appears. Finally, the *distance* between two points is conceived of as the length of the segment joining them. In most contemporary high school geometry courses, for reasons of mathematical rigor and conciseness, the order of these concepts is reversed. The existence of a *distance* function is assumed, the *length* of a segment is defined to be the distance between its end points, and then two segments are defined to be *congruent* if they have the same length.

9. Ann has a brand new ruler and measures as below,

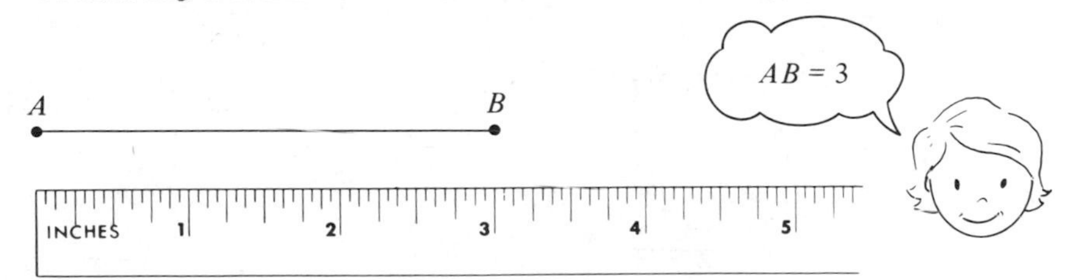

Joe has an old ruler that is badly battered and chewed. He measures as below,

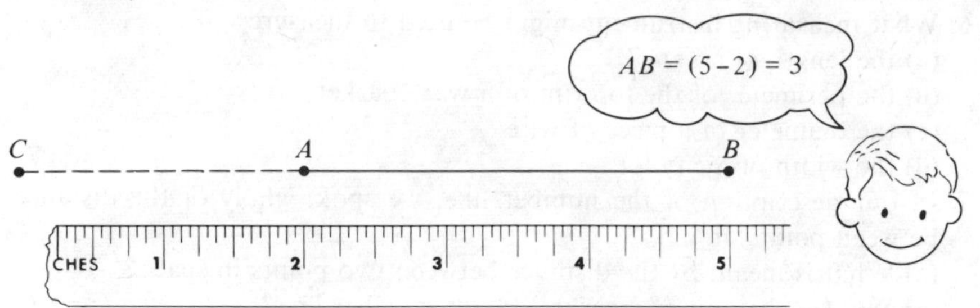

and he justifies his calculations this way. Let C be the point of $\overleftrightarrow{AB}$ that would correspond to the (missing) end of the ruler.

$$\overline{CB} = \overline{CA} \cup \overline{AB} \Rightarrow$$

$$m(\overline{CB}) = m(\overline{CA} \cup \overline{AB}) \Rightarrow$$

$$m(\overline{CB}) = m(\overline{CA}) + m(\overline{AB}) \Rightarrow$$

$$CB = CA + AB \Rightarrow$$

$$5 = 2 + AB \Rightarrow$$

$$AB = 5 - 2$$

Where does Joe make use of the "additive" property of measure functions?

10. Find the perimeters of these figures in inches.

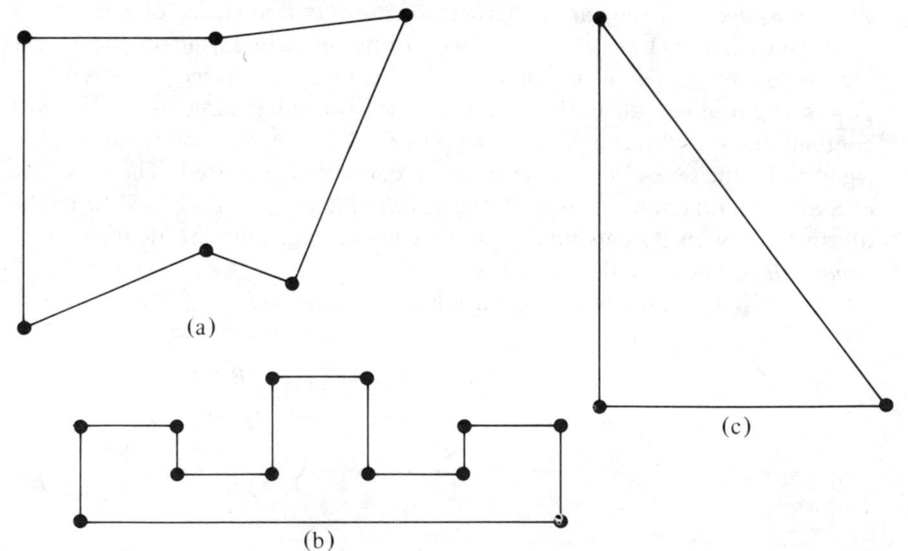

11. A forest of technical terms has grown up around the idea of "circle." First of all, a **circle** is a (one-dimensional) subset of a plane consisting of all points at a fixed distance from a given point of the plane, known as the **center** of the circle. The two-dimensional figure consisting of the circle and its interior is called a **disk**. Any segment having both of its end points on the circle is called a **chord**. A chord passing through the

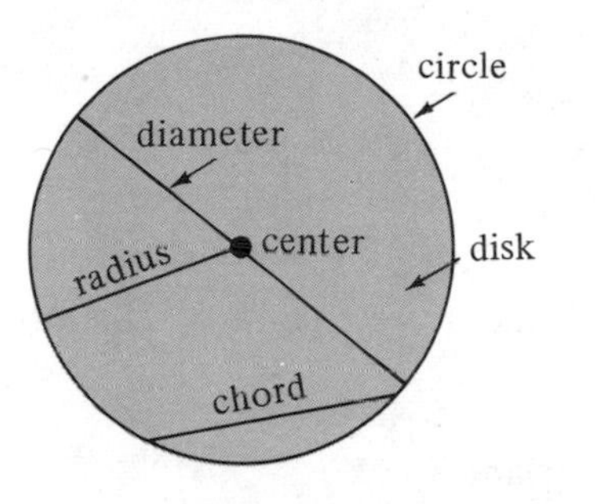

center of the circle is called **a diameter**. Any segment having the center of the circle as one end point and a point on the circle as the other is called **a radius**. The perimeter of the disk (or of the circle) is called the **circumference** of the circle. The length of any diameter is called **the diameter** of the circle. The length of any radius is called **the radius** of the circle.

Very often the word "circle" is used to mean disk, and sometimes the word "circumference" is used to mean circle. You will have to get used to this inconsistent usage and decide from context what is meant. Translate into technically correct language:

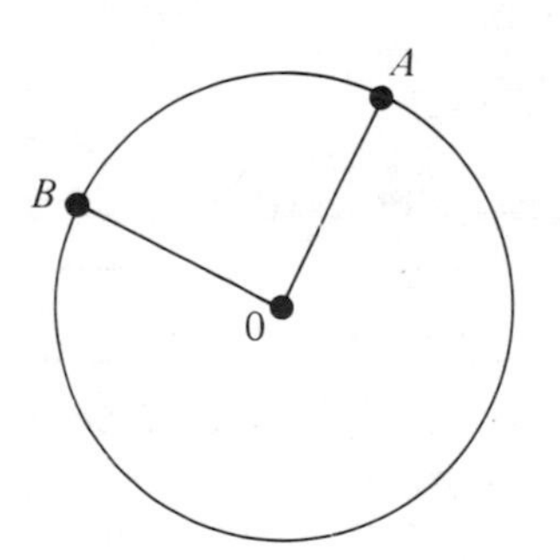

(a) "Point A lies on the circumference of the circle having center at 0 and passing through B."

(b) "The area of the circle is 2 square inches."

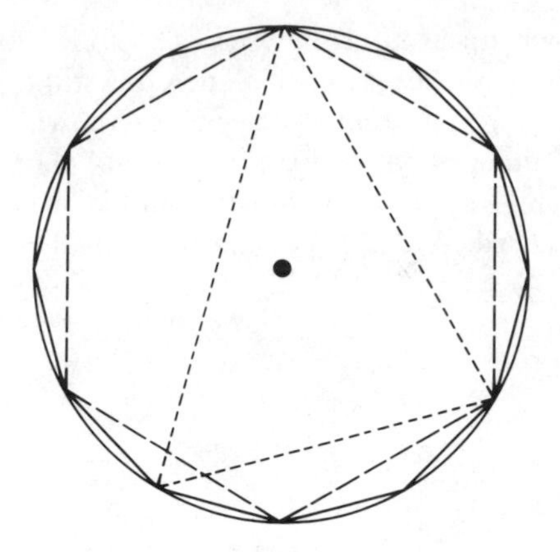

12. In this exercise write all measurements to the nearest tenth of a centimeter.

(a) Find the diameter, d, of the circle.
(b) Find the perimeter, p_1, of the (dotted) triangle.
(c) Find the perimeter, p_2, of the (dashed) hexagon.
(d) Why should it be true that $p_2 > p_1$?
(e) Find the perimeter, p_3, of the (solid) 12-gon.
(f) Why should it be true that $p_3 > p_2$?
(g) Express each ratio as a two-place decimal:

$$\frac{p_1}{d}, \quad \frac{p_2}{d}, \quad \frac{p_3}{d}$$

(h) Do you remember what number these successive ratios should be approaching?

13. For all circles the ratio,

$$\frac{\text{circumference}}{\text{diameter}}$$

is the same. This common number is known universally as π ("pi"). Thus

---

$$\text{circumference} = \pi \cdot \text{diameter}$$

The decimal expression for π begins $\pi = 3.14159....$ [Compare your answers to 12(g) with 3.14.] The number π cannot be expressed as a fraction. This was proved by Lambert in 1767. Many different rational numbers have been used to approximate the irrational number π. The Old Testament uses 3. Schools normally use 3.14 or $\frac{22}{7}$. The first 100,000 decimal places of π have recently been determined with the aid of an electronic computer. Arrange in order from smallest to largest: π, 3.14, $\frac{22}{7}$.

14. Using $\frac{22}{7}$ as an approximation for π, find the perimeter of each figure below.

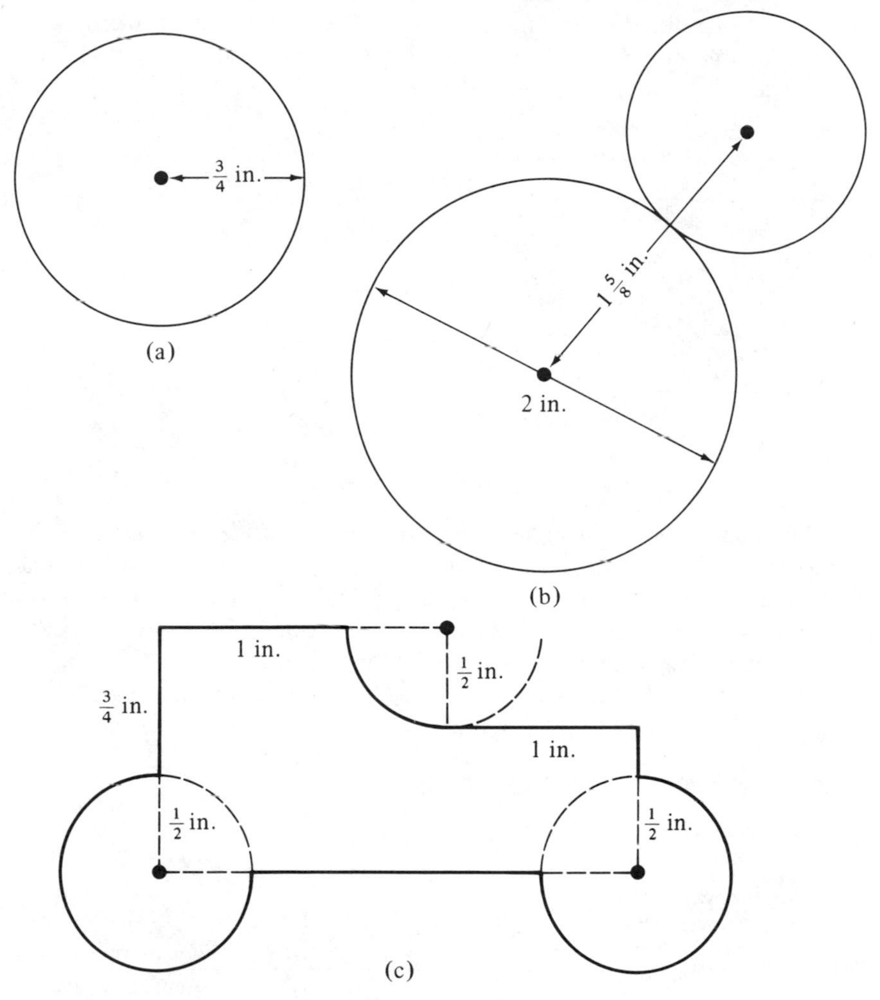

15. (Discussion) Relate this diagram to the child's mathematical progress through the elementary grades.

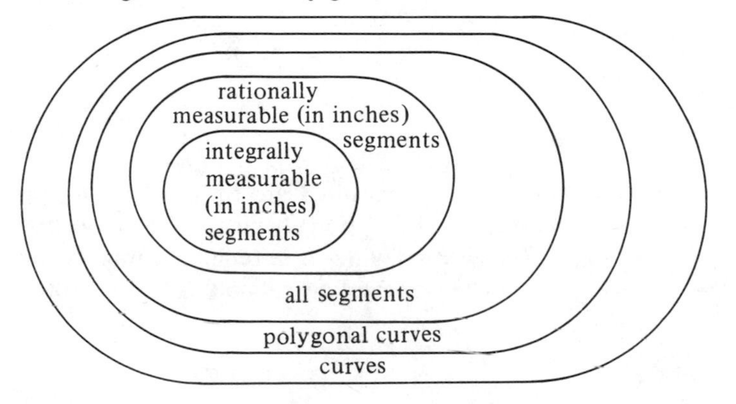

4.7 AREA AND AREA FORMULAS

To measure two-dimensional figures a two-dimensional unit of measure is needed. The one in common use is a square with one inch edges, called the **square inch.**

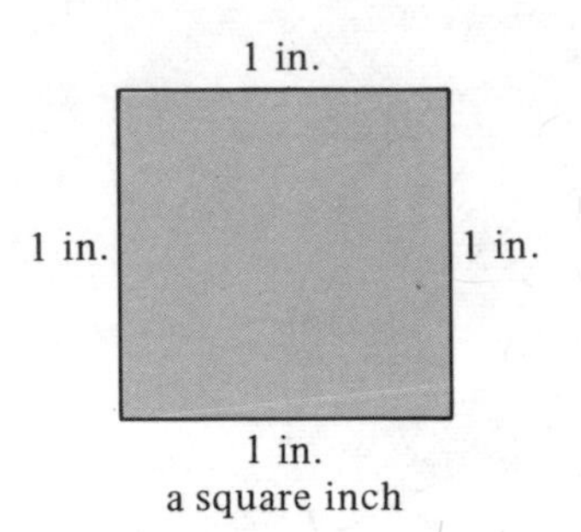

a square inch

As in the case of linear measurement, there is a definite sequence in the development of area measure. The first figures to be measured for area are rectangles, and among rectangles the first ones considered are those whose length and width are integrally measurable (in inches). The measuring process for such rectangles is simply a matter of counting how many square inches (shown shaded) need to be laid down, side by side, to cover the figure.

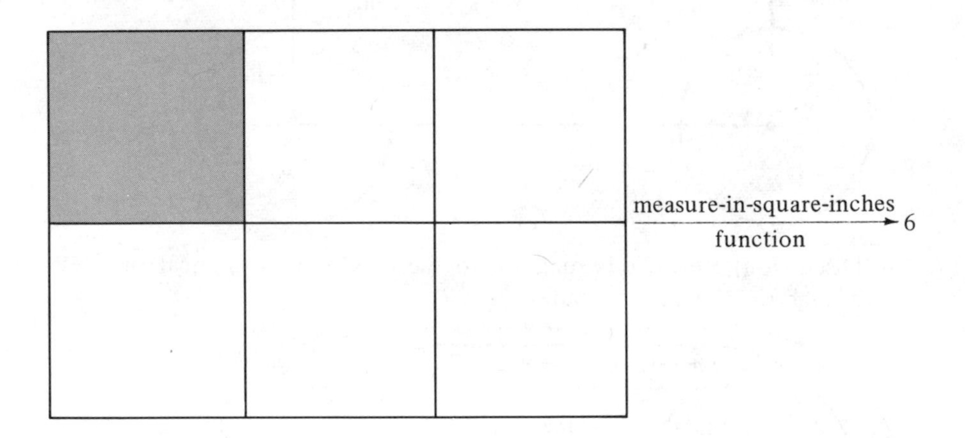

The next kind of rectangles to be measured are those whose length and width are integrally nonmeasurable, but rationally measurable, in inches. Measuring such rectangles can be viewed as laying off and counting appropriate subunits. In the following figure the appropriate subunit is a sixth of a square inch (shown shaded).

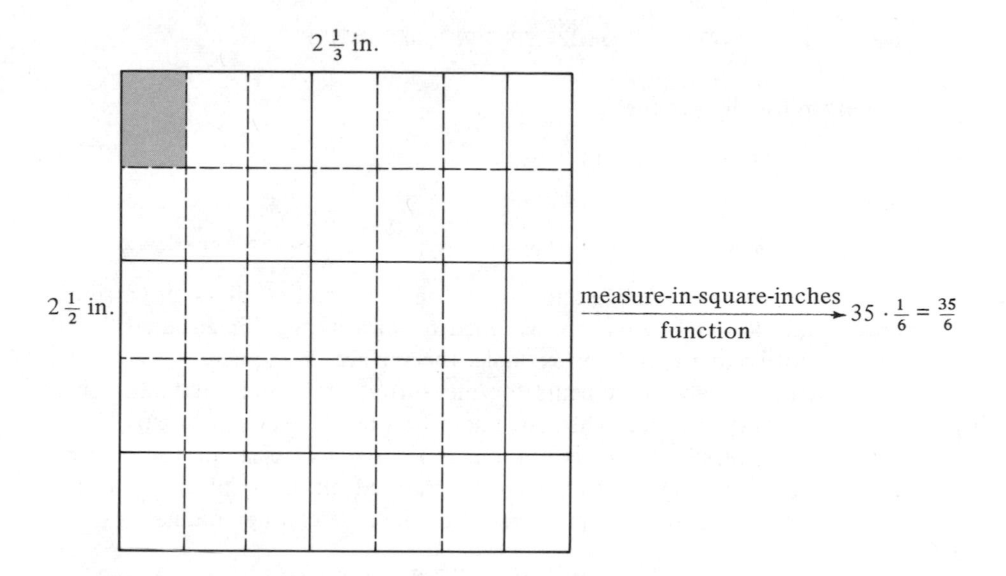

Making use of the fact that area is unchanged by cutting and pasting, we then enlarge the domain of the area-in-square-inches function to include a variety of simple polygonal regions.

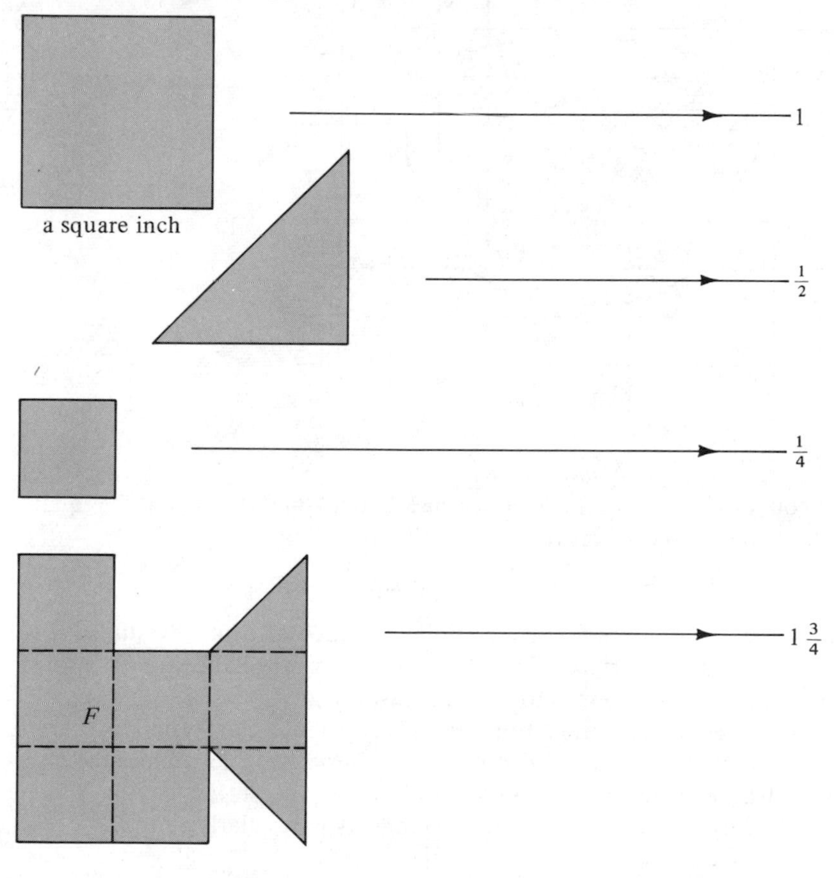

Again we usually say that

the area of F is $1\frac{3}{4}$ square inches

instead of

the measure in square inches of F is $1\frac{3}{4}$

Later certain "nice" nonpolygonal regions, such as disks, are assigned areas, and ultimately areas are assigned to various two-dimensional figures that do not lie in a plane, for example, the surface of a sphere.

There are few instruments for measuring area, and what ones there are are not very effective. The instrument that corresponds most closely to a ruler is the **grid sheet**. This is nothing but a sheet of clear plastic or other transparent material with square inches marked on it. By placing it over a two-dimensional plane figure, estimates of the area can be obtained.

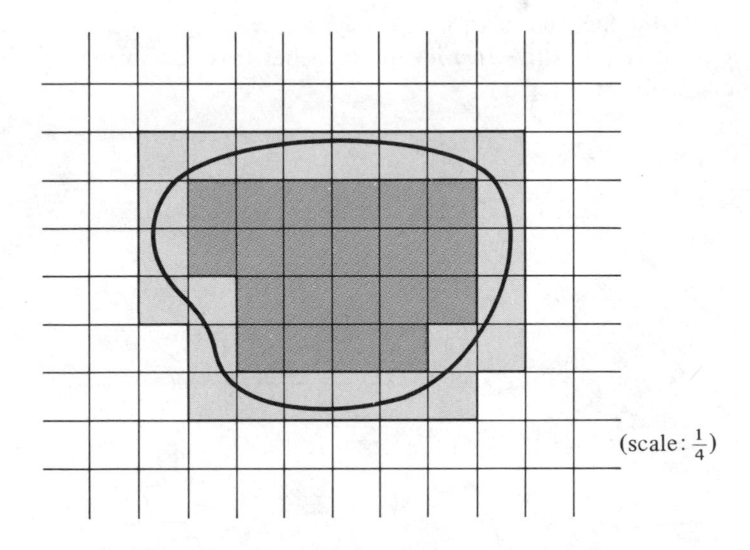

If we count the squares completely inside the figure and then count the squares touching the figure, we get the crude estimate

21 square inches < area figure < 45 square inches

Grid paper is not of much use for finding the area of bent, two-dimensional figures such as the surface of a sphere. There is no pliable area-measuring instrument that corresponds to a tape measure.

Since the physical process of measuring area is so clumsy and inaccurate, we look for other methods. We look for area *formulas* which reduce the problem of measuring area to the much easier task of measuring length.

The formula for the area of a rectangle is particularly simple.

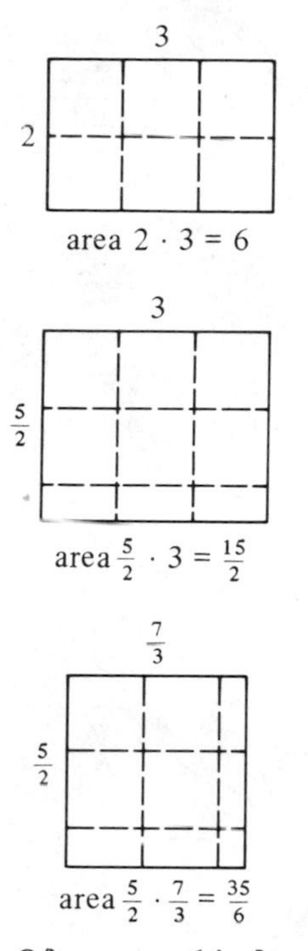

area of rectangle = length · width

Of course, this formula works only if the units of measure are consistently chosen. *Length and width must be measured in the same units* (inches or centimeters or feet...) *and then the area is given in the corresponding square units* (square inches or square centimeters or square feet...).

The formula for the area of a parallelogram is easily derived from the formula for a rectangle.

area of parallelogram = base · height

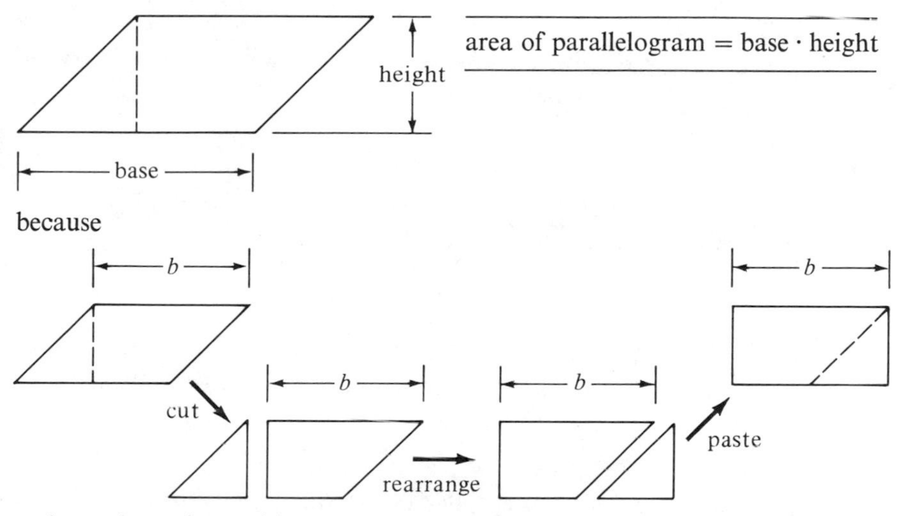

and area is unchanged by cutting and pasting.

The formula for the area of a triangle,

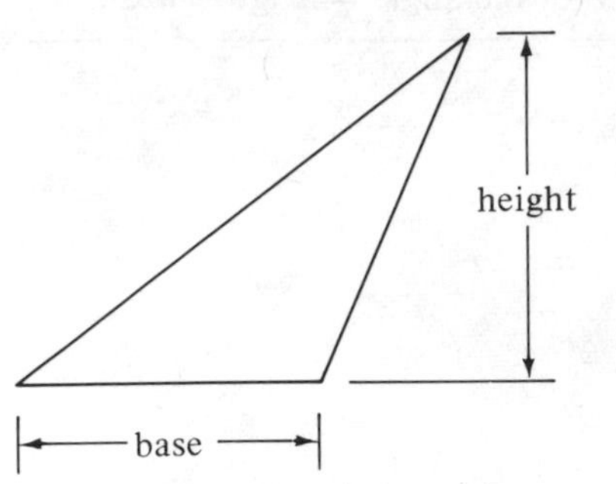

area of triangle $= \frac{1}{2} \cdot \text{base} \cdot \text{height}$

follows easily from the formula for parallelograms.

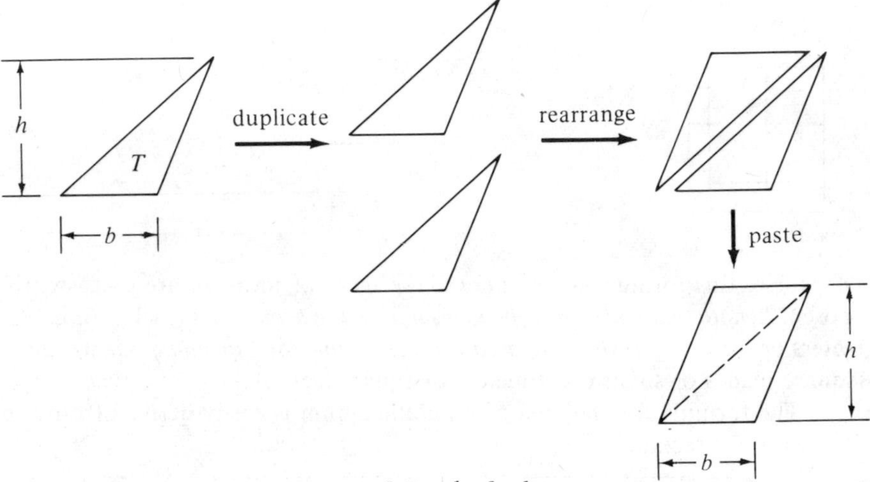

So $2 \cdot$ area of $T = b \cdot h$, or area of $T = \frac{1}{2} \cdot b \cdot h$.

Having the formula for the area of a triangle, we can find the area of any polygonal region using only a ruler.

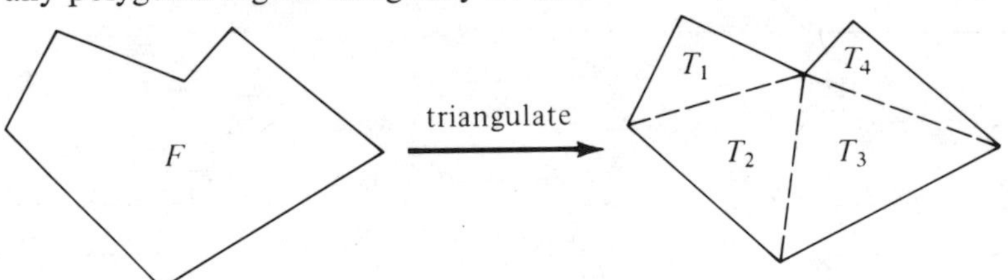

Now measure bases and heights of triangles T_1, T_2, T_3, T_4; compute their areas using the formula for triangles; and finally add up these areas to get the area of F.

EXERCISES

1. Using the longest edge of a dollar bill as the unit, measure the length and width of a sheet of your notebook paper (to the nearest half-dollar).

Using the face of the dollar bill as the unit, measure the area of the notebook paper. Does length × width = area? Why?

2. Suppose that your bedroom were 9 ft wide and 10 ft long, and you wanted the floor tiled with 1 ft by 1 ft tiles. How would you decide how many to order? How might a second grader decide?
3. Find the area.

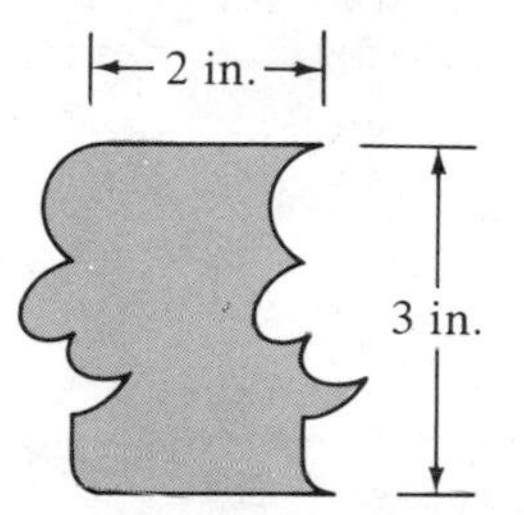

4. Find the outside surface area of this piece of pipe.

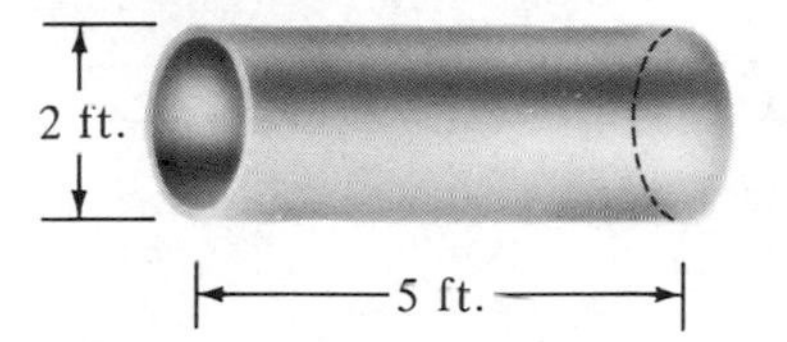

5. Describe an experiment for finding the surface area of a billiard ball.
6. Does a plane region with large perimeter have to have a large area? Draw some pictures to illustrate your answer. Could a region with a tiny perimeter have a huge area?
7. Which has more area, A or B? (The horizontal lines are parallel.)

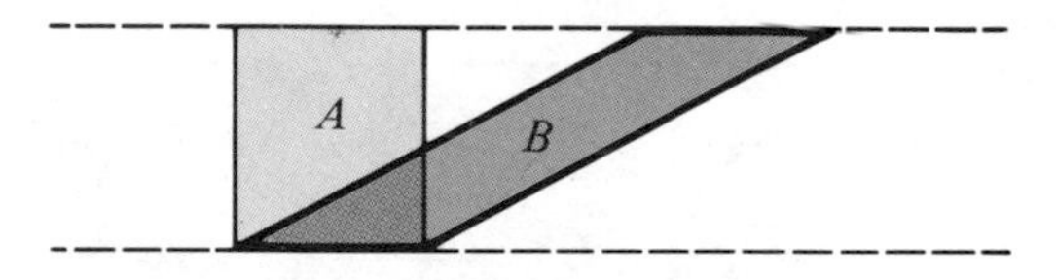

8. Study the diagram that follows and try to decide what the formula for the area of a circle of radius r must be. (How do the areas of the circles in the left column compare with the areas of the corresponding figures in the right column? The figure continues on p. 146.)

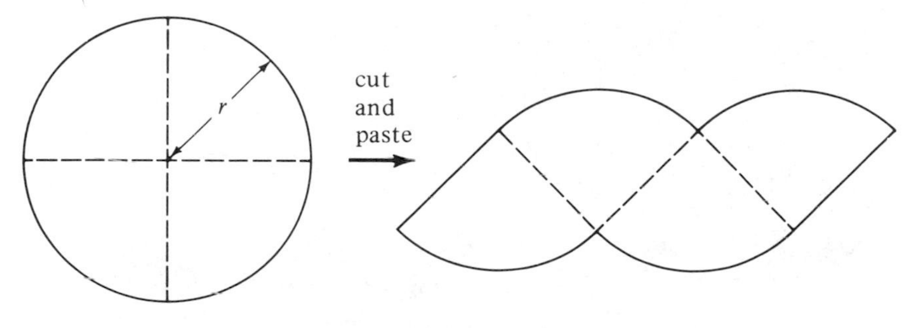

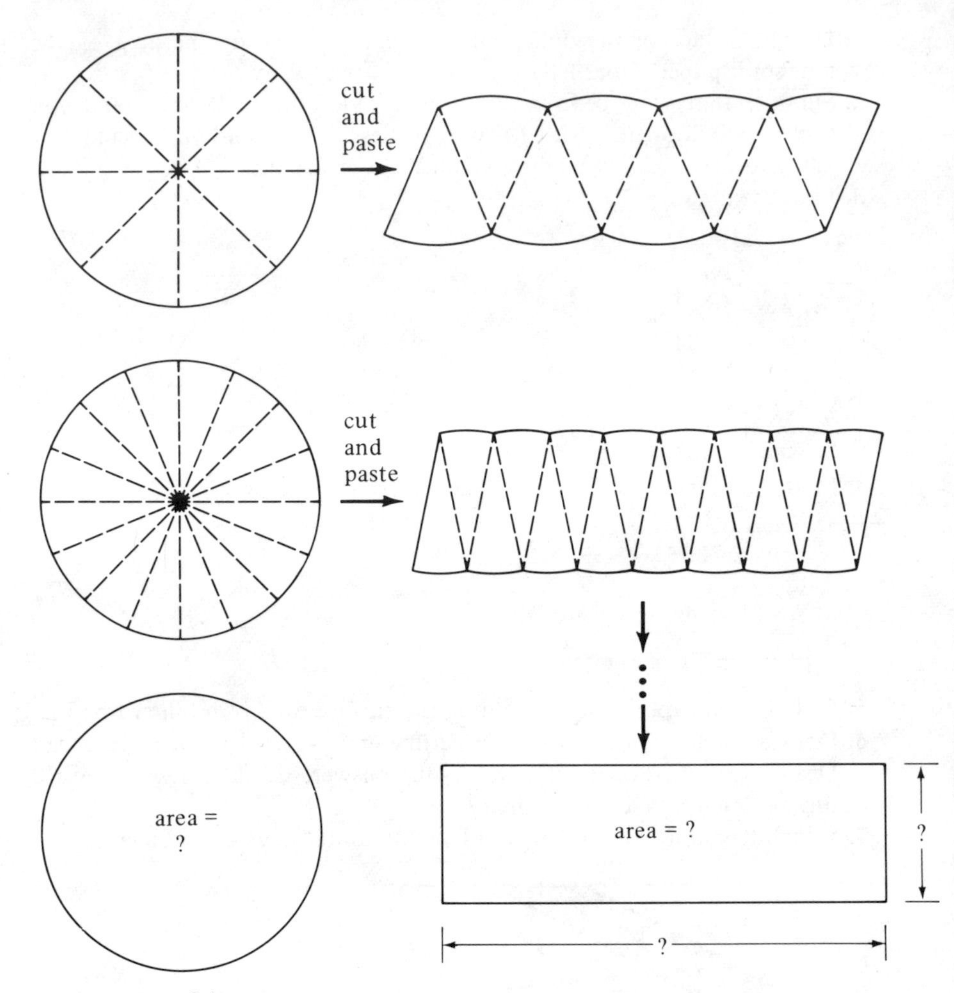

9. True or false:

(a) Congruent regions have the same area.

(b) If two regions have the same area, then they are congruent.

(c) Is (b) true if "regions" is replaced by "rectangles"?

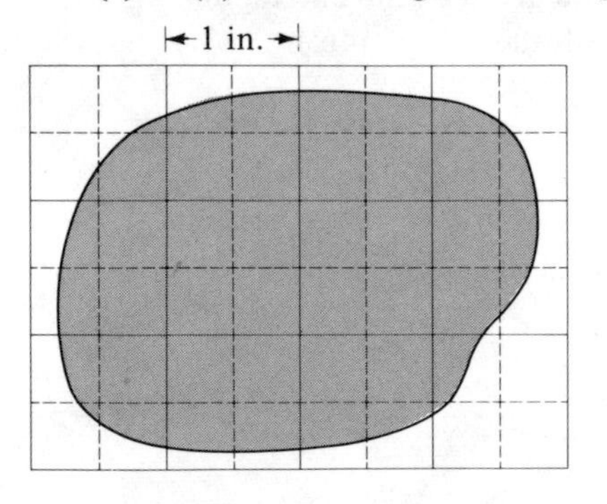

Figure for Exercise 10, p. 147

10. Count the large (solid) squares that lie inside the figure. Count the large squares that are inside or are cut by the figure. Now fill in the blanks.

______ square inches < area of figure < ______ square inches

Count again, this time using the small (dotted) squares, and fill in the blanks.

______ square inches < area of figure < ______ square inches

Repartition so each small square is cut up into four congruent smaller squares. Count again and fill in the blanks.

______ square inches < area of figure < ______ square inches

Estimate the area of the figure.

11. Find the areas of the figures below.

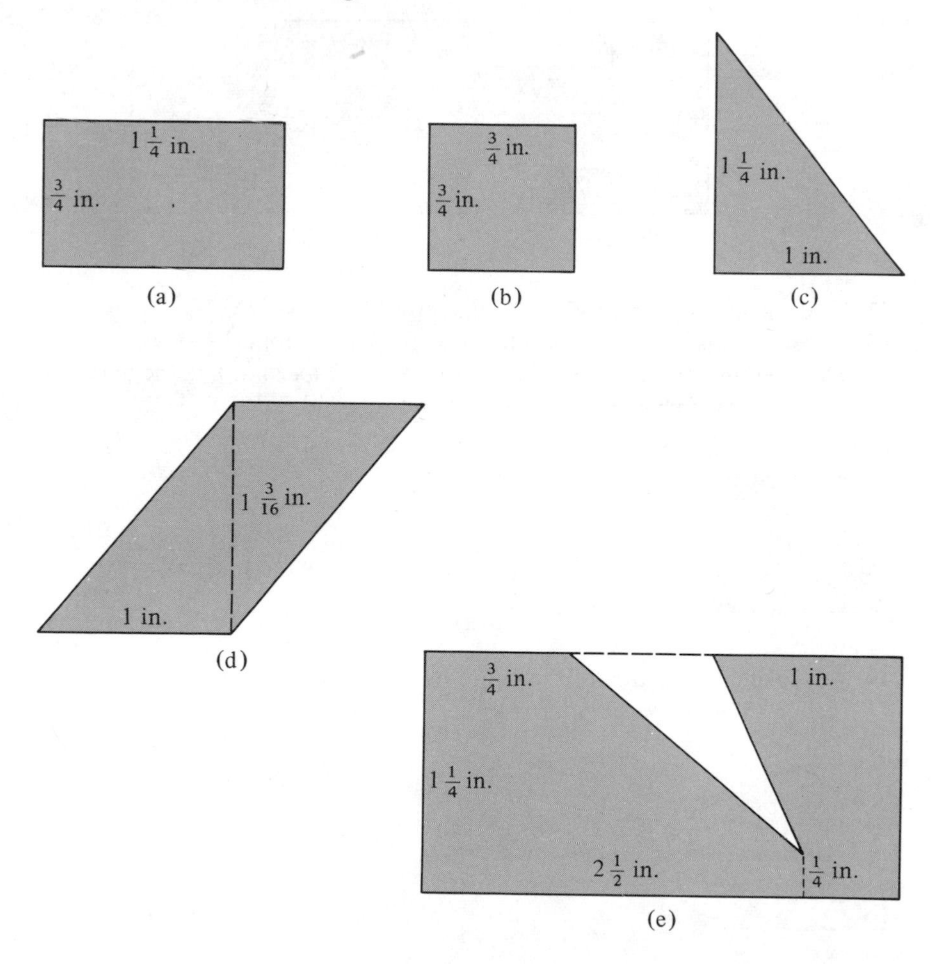

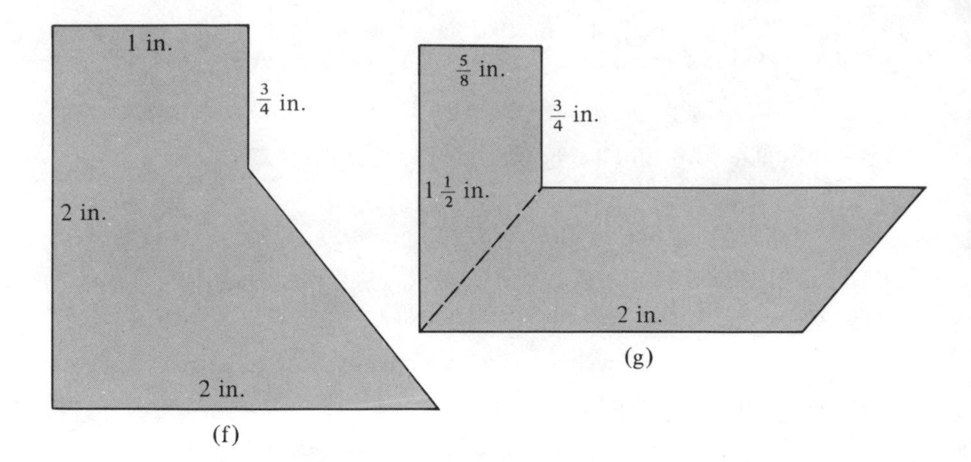

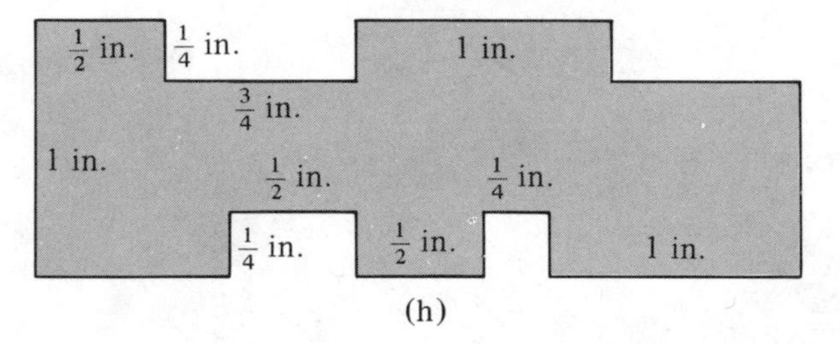

12. Find the areas of the regions in Exercise 14, p. 139.

13. A **trapezoid** is a quadrilateral having one pair of opposite sides parallel. These parallel sides are called the bases of the trapezoid. Use the diagram to discover a formula for the area of a trapezoid.

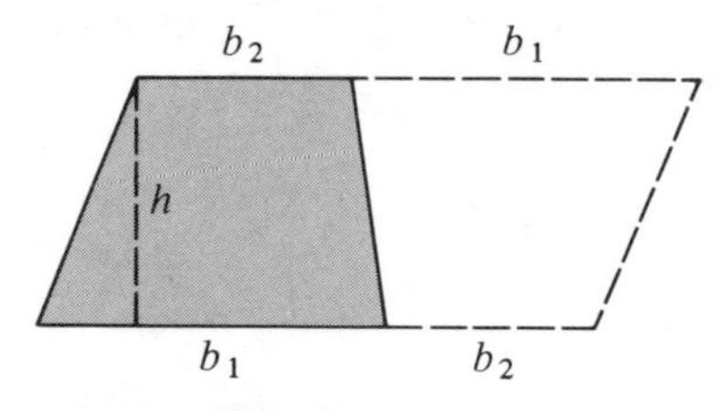

14. Approximate the area of the figure by finding the area of the polygonal region. (You will need your ruler.)

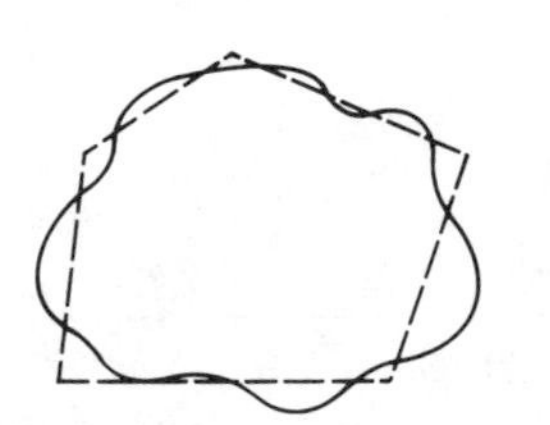

15. Approximate the area of the figure by first approximating it with a polygonal region.

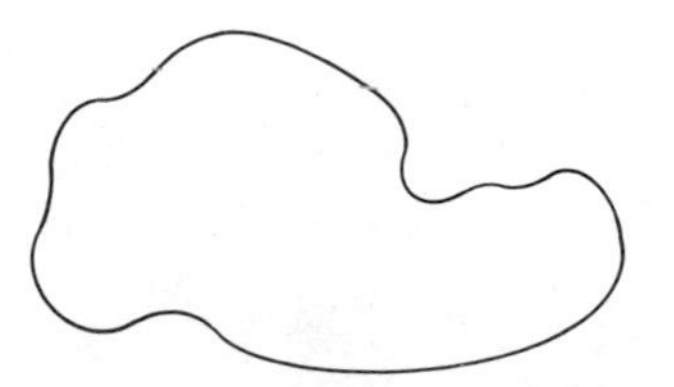

16. You just finished painting your garage and it took exactly 2 quarts of paint. How much paint should you buy to paint your house (ignore windows)?

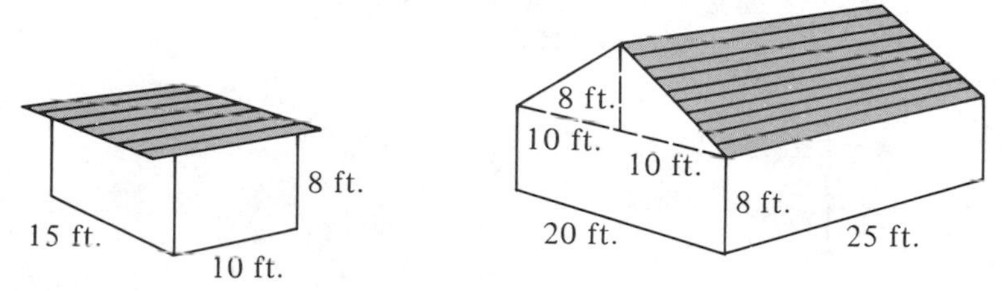

17. How many 2 in. by 3 in. by 8 in. bricks will you need to build a brick walk with the following shape. (Lay them so that the largest face is up.)

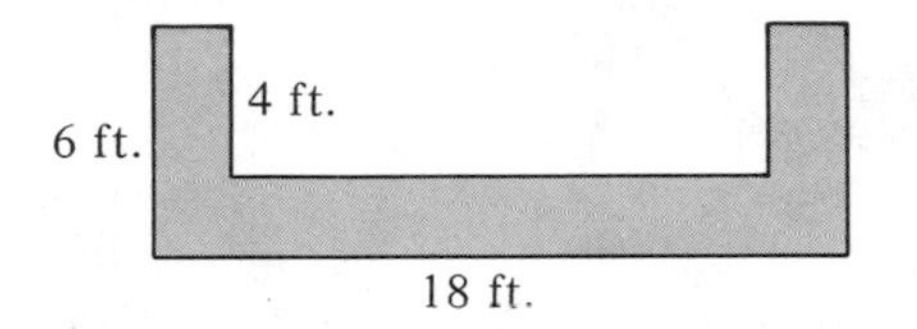

***18.** Explain why regions A (doubly shaded) and B (unshaded) have the same area.

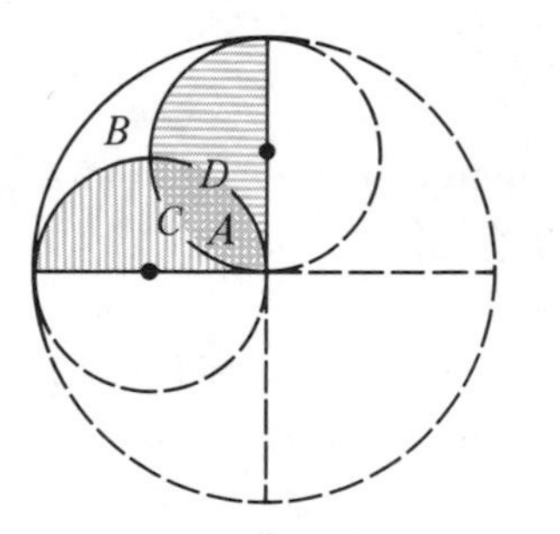

***19.** In the center of a large pasture stands a square shed, k ft on a side. A cow is tied to one corner of it by a rope $4k$ ft long. Find the area over which the cow can graze.

TEACHING NOTE The connection between measuring rectangles and multiplying numbers is often used to motivate the "numerator times numerator over denominator times denominator" rule for multiplying fractions. The development proceeds somewhat as follows.

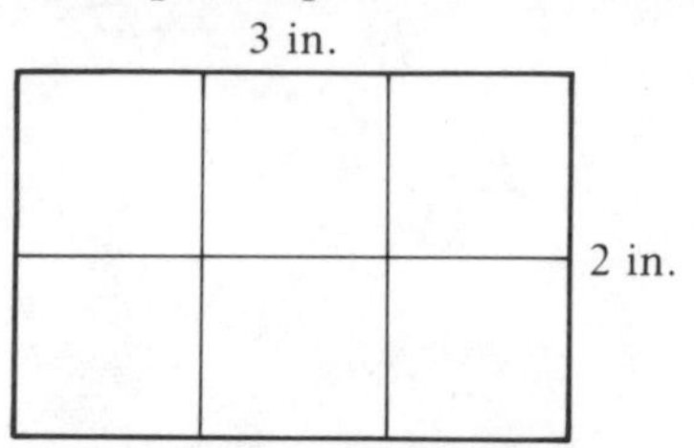

(i) For a rectangle with integral dimensions it is verifiable by counting that the (familiar whole-number) product of length times width gives area. This is a nice formula, which we want to preserve.

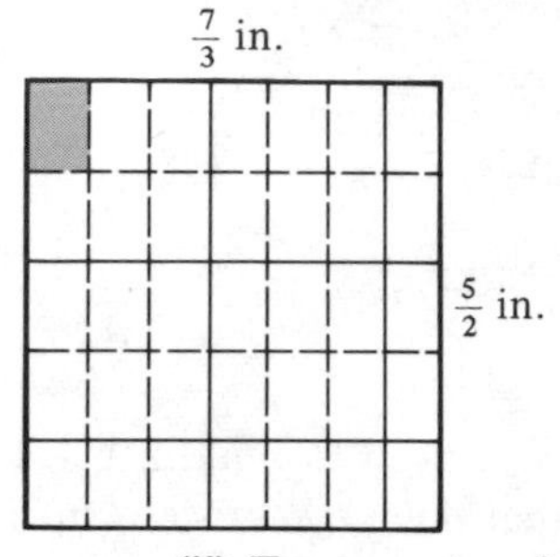

(ii) For a rectangle with rational dimensions, the area can be found by counting appropriate subunits (in this case, sixths of a square inch). There are 35 of these, so the area is $\frac{35}{6}$.

(iii) Let us define the product of rational numbers to preserve the nice formula, area = length × width. That is, define the product of two positive rational numbers a and b to be the area of an a by b rectangle. Thus $\frac{7}{3} \cdot \frac{5}{2} = \frac{35}{6}$. Lo and behold,

$$\frac{35}{6} = \frac{7 \cdot 5}{3 \cdot 2}$$

numerator times numerator over denominator times denominator.

4.8 VOLUME AND VOLUME FORMULAS

To measure three-dimensional figures, a three-dimensional unit of measure is needed. We shall usually use a cube each edge of which has length one inch. This figure is known as a **cubic inch**.

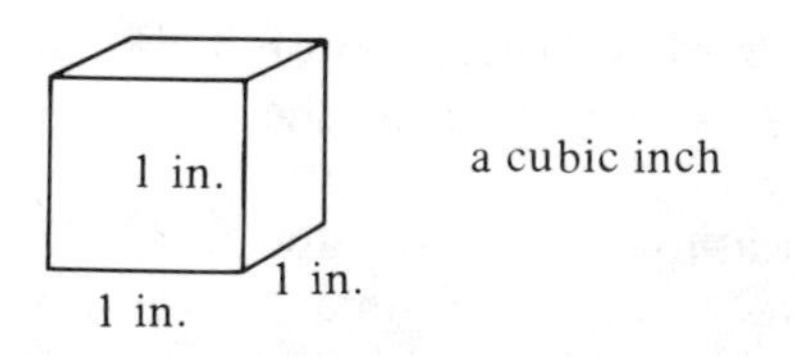

Other common "unit cubes" used in measuring volume are the cubic centimeter and the cubic foot.

The physical process of measuring area, you remember, is usually clumsy and inaccurate. But the physical process of measuring volume is even worse. How would you decide how many cubic inches would be required to fill up a bowling ball? How would you determine how many cubic feet would be required to fill up your refrigerator? Some indirect method of measurement would probably be used. The neatest indirect method is the use of an appropriate volume formula.

If the inside dimensions of your refrigerator are 2 ft deep, 2 ft wide, and $3\frac{1}{2}$ ft high, then, by counting cubes,

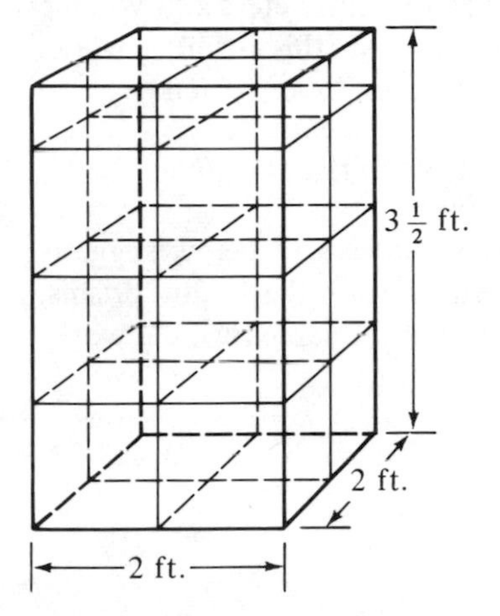

the volume is 14 cubic feet: 4 in the bottom layer, 4 in the next, 4 in the next, and 2 in the top layer. But also

$$14 = 2 \times 2 \times 3\tfrac{1}{2}$$

Two things can be concluded from this observation. One is that the volume of the refrigerator (in cubic feet) is the product of its depth, width, and height (in feet). The other is that the volume of the refrigerator (in cubic feet) is the product of the area of its base (in square feet) and its height (in feet).

The refrigerator is a special case of a very general type of figure called a **prism**.

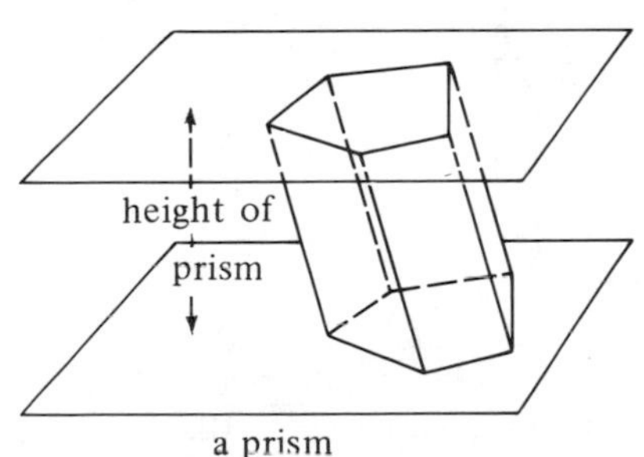

congruent polygonal "bases" (two) lie in parallel planes

"lateral faces" (five here) are all parallelograms

a prism

The refrigerator is a **right rectangular prism**, "rectangular" because that is the shape of the bases, "right" because its lateral edges form right angles with the base edges. The prism above is an ordinary (not "right") pentagonal prism.

The formula for the volume of any prism is

---

volume of prism = area of base × height

---

and, of course, this formula works only if compatible units are used. (If height is measured in inches, then area of base must be measured in square inches and volume will be given in cubic inches. If height is measured in centimeters, then area of base must be measured in square centimeters and volume will be given in cubic centimeters.) We have seen that this formula works for a right rectangular prism, in which case it can even be rewritten as

---

volume of right rectangular prism = length × width × height

(Again, the units must be chosen consistently.) Exercise 17 below suggests why the base-times-height volume formula works for other right prisms. The following diagram might help convince you that the same formula works for nonright prisms as well.

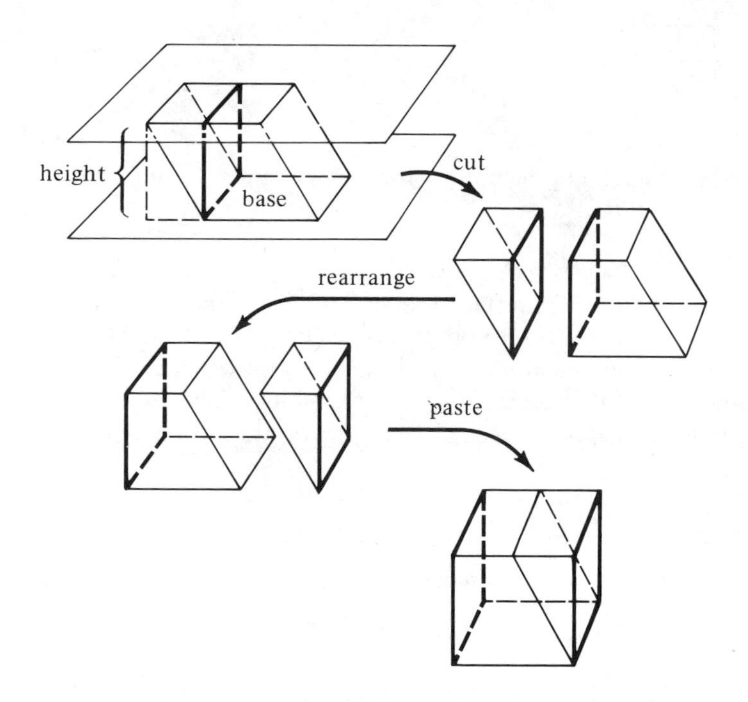

NOTE A much more extensive investigation of volume (and surface area) formulas for a variety of space figures is carried out in Chapter 10.

EXERCISES

1. Refer to the preceding diagram and give reasons for (a) and (b) below:

$$\text{volume of original prism} \overset{(a)}{=} \text{volume of final rectangular prism}$$
$$= \text{area of base of rectangular prism} \times \text{height of rectangular prism}$$
$$\overset{(b)}{=} \text{area of base of original prism} \times \text{height of original prism}$$

2. You have one cubic inch of bowling ball material and it weighs .3 oz. A bowling ball weighs 16 lb. What is the volume of the bowling ball in cubic inches?

3. Using a glass container calibrated in inches as shown,

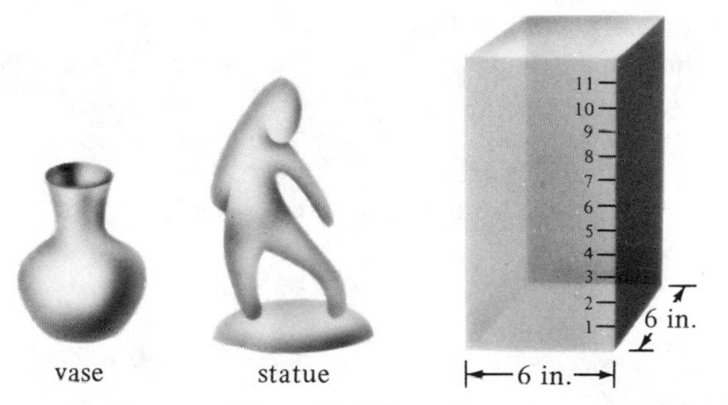

(a) How would you find the volume (capacity) of the vase in cubic inches?
(b) How would you find the volume (amount) of clay, in cubic inches, used in the statue (without ruining it)?

4. You decide the statue is not worth keeping, but you are still curious how much clay it required, so you roll it out with a rolling pin into a disk 1 in. thick. If the disk has a diameter of 10 in., what was the volume of the statue? To what sort of figure did you extend the base × height formula to answer the question?

5. You take another statue and roll it out with your hands into a clay "worm" of length 4 ft. When you roll out a cubic inch of clay into a worm of the same diameter, it is only $1\frac{1}{2}$ in. long. What was the volume of the statue?

6. Which prism has greater volume, *A* or *B*?

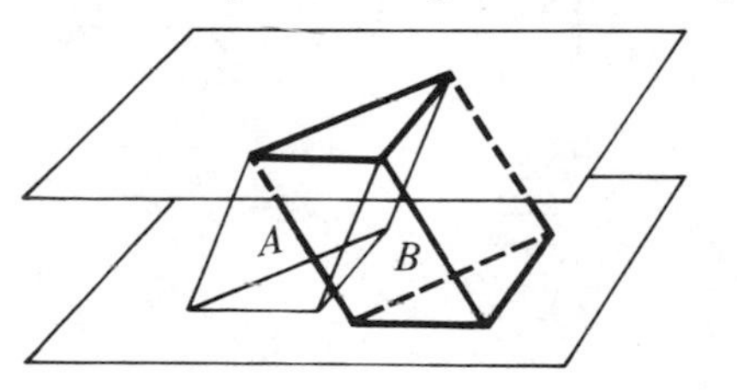

7. This experiment requires 16 1 in. × 2 in. × 4 in. chalkboard erasers and a 4 in. × 4 in. × 8 in. cardboard carton.

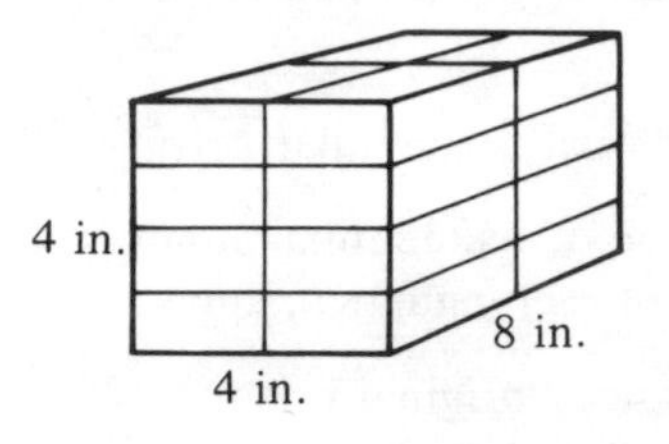

(a) Using the eraser as the unit of volume, find the volume of the carton.
(b) Using the largest face of the eraser as the unit of area, find the area of the bottom of the carton.
(c) Using the longest edge of the eraser as the unit of length, find the length, width, and height of the carton.
(d) Does volume = base × height? Why?
(e) Does volume = length × width × height? Why?

8. Cereal box A has dimensions 4 in. by 4 in. by 7 in. Cereal box B has dimensions 5 in. by 2 in. by 11 in.
(a) Which box has greater volume?
(b) Which box has greater surface area?

9. Is there some relation between surface area and volume? (Compare with Exercise 6, p. 153)

10. A deck of cards is deformed as shown:

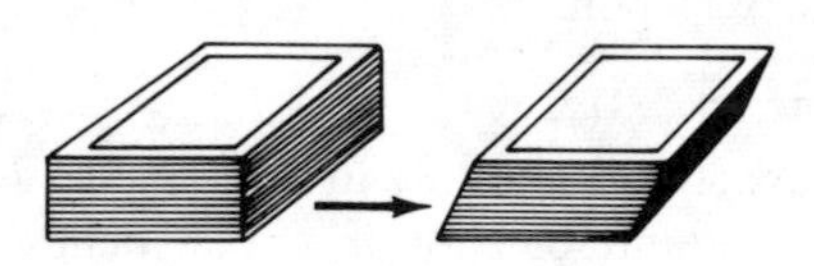

(a) Does area of base change?
(b) Does height change?
(c) Does volume change?
(d) What formula does this illustrate?

11. Find the volumes of these figures. (Segments that look parallel are intended to be parallel. Segments that look perpendicular are intended to be perpendicular.)

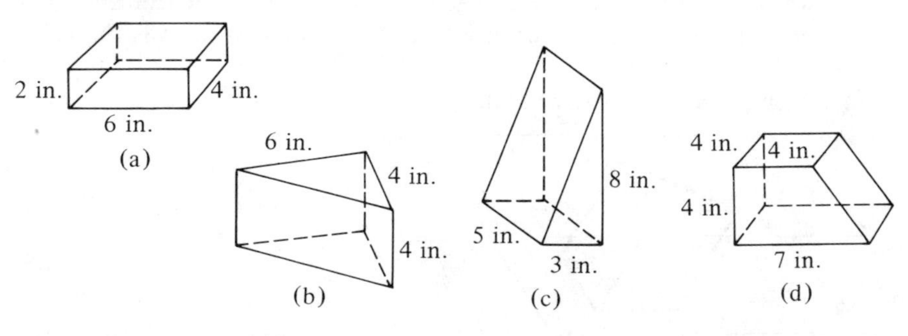

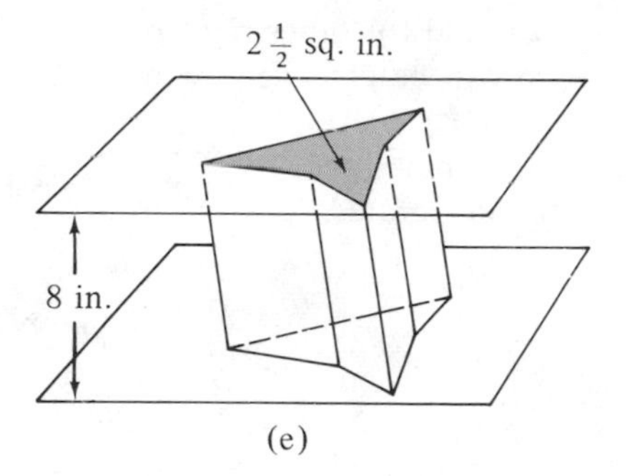

(e)

12. You want to buy an air conditioner for your house and need to know how many cubic feet of space there are to be cooled.

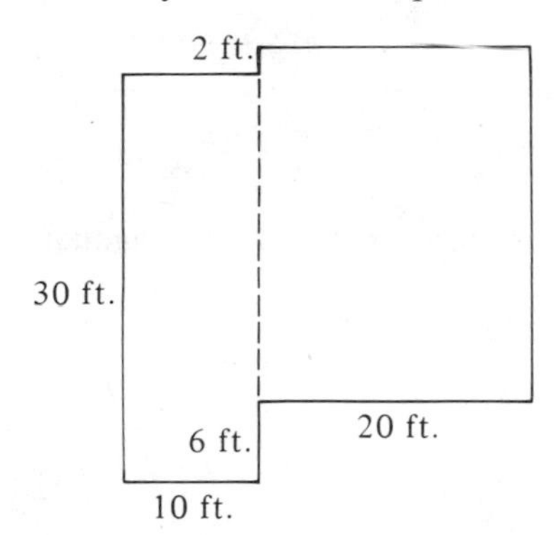

If your house is one story high, has 8-ft ceilings, and has this floor plan, what is its volume?

13. A swimming pool has the side view shown below. If it is 15 ft wide, how much water can it hold?

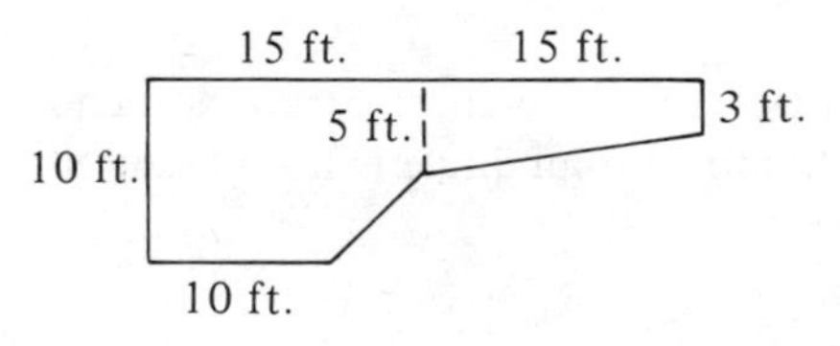

14. During a heavy rainfall $1\frac{1}{2}$ in. of water collects in the bottom of a can with a 10-in. diameter. How many cubic inches of water are in the can?

15. If you want to fill a 6 ft by 6 ft sandbox to a depth of 18 in., how many cubic yards of sand should you order?

16. True or false:

(a) Congruent figures have the same volume.

(b) If two figures have the same volume, then they are congruent.

17. For right rectangular prisms the formula $V = B \cdot h$ can be derived experimentally by filling various cardboard boxes with wooden blocks. Cutting and reassembling experiments, possibly using pieces of cheese, can you suggest why the formula should continue to hold for other right prisms. (Exercise 1 has suggested how the formula can be extended from right prisms to nonright prisms.)

(a) Describe a cut-and-reassemble experiment that shows that $V = B \cdot h$ for a right prism with bases that are parallelograms.

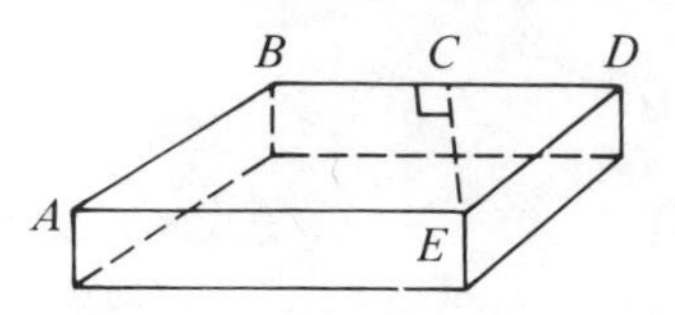

(b) Describe a derivation of the same formula for right prisms with triangular bases.

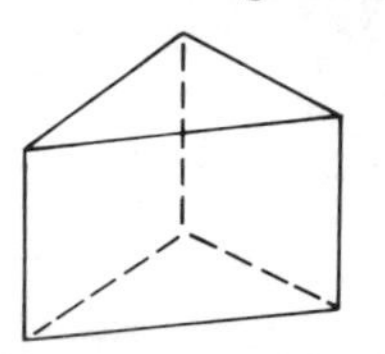

(c) Describe a derivation of the same formula for a right pentagonal prism as shown.

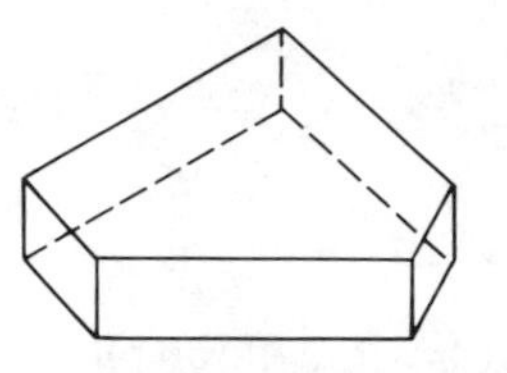

(d) How might you justify the validity of the same formula for right circular cylinders?

4.9 ANGLE MEASURE

An angle, you remember, is the union of two rays that share a common end point. So an angle is a one-dimensional figure. But an angle is not measured

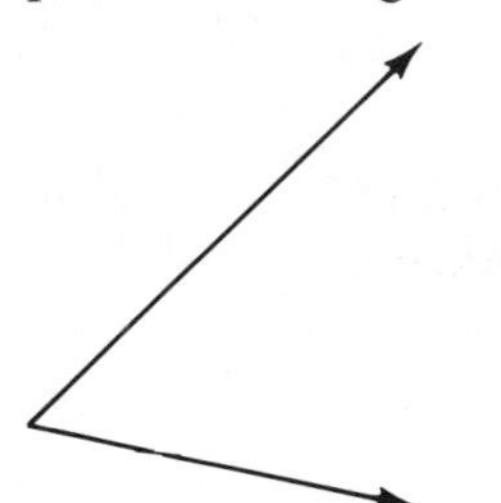

as other one-dimensional figures are. It would be foolish to try to measure the length of an angle, since all angles have infinite length. It would also be pointless to measure the area of the interior of an angle, since all angles have interiors with infinite area.

The unit of measure for measuring angles is another angle. The one we commonly use is called a **degree**.

a degree

To measure an angle is to determine how many degrees, laid side by side (adjacent to each other), are needed to fill up the interior of the angle being measured. The degree was chosen of such a size that 360 of them, laid side by side, fill up a plane. As before, the assignment of numbers to angles can be described as a function. Its domain is a set of angles; its range can be described in set builder notation as $\{x \in R \mid 0 < x \leq 180\}$.

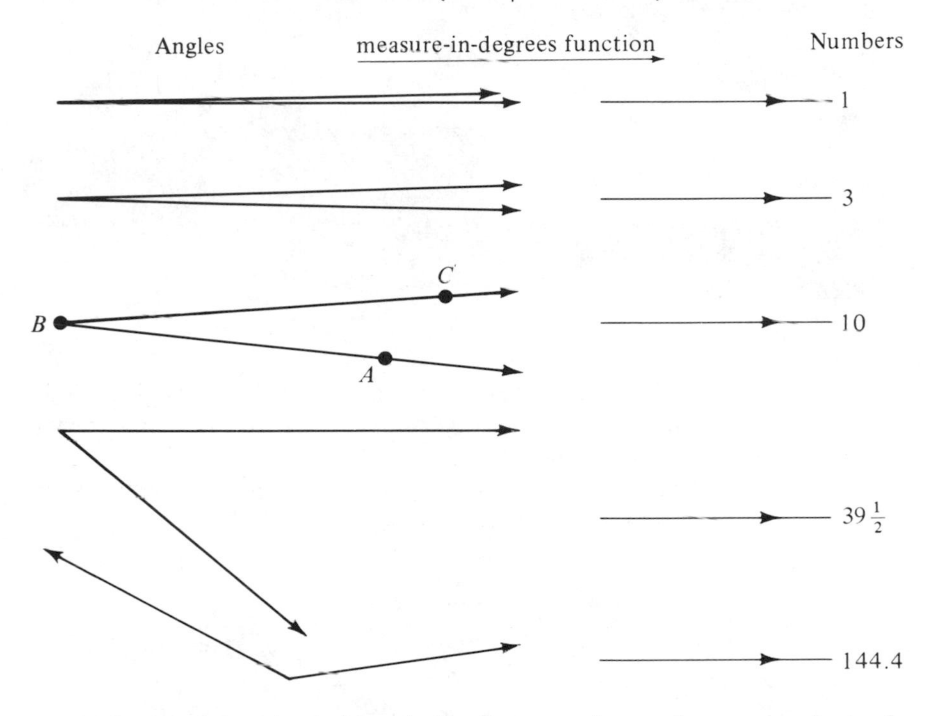

There is a standard abbreviation for reporting angle measure. Instead of writing

the measure-in-degrees of $\angle ABC$ is 10

we write

$m \angle ABC = 10°$

and read it

The measure of angle A, B, C is 10 degrees.

Occasionally, we simply write

$$m\,\angle ABC = 10$$

in which case the degree is understood to be the unit of measure.

The instrument for measuring angles is the **protractor**. It is quite similar to a ruler. The protractor has many copies (180) of the unit of angle measure (the degree) already arranged side by side, just as the ruler has many copies (usually 12) of the unit of linear measure (usually the inch) already arranged end to end.

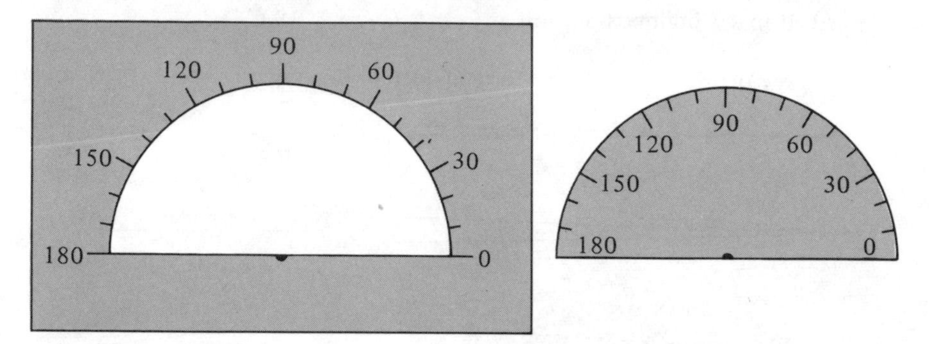

two kinds of protractors

One difficulty you might face in teaching angle measurement to children is that you know too much about angles. In high school trigonometry you may have worked with "ordered angles," angles having an initial side and a terminal side. When working with such angles in a "counterclockwise oriented plane," it makes good sense to write

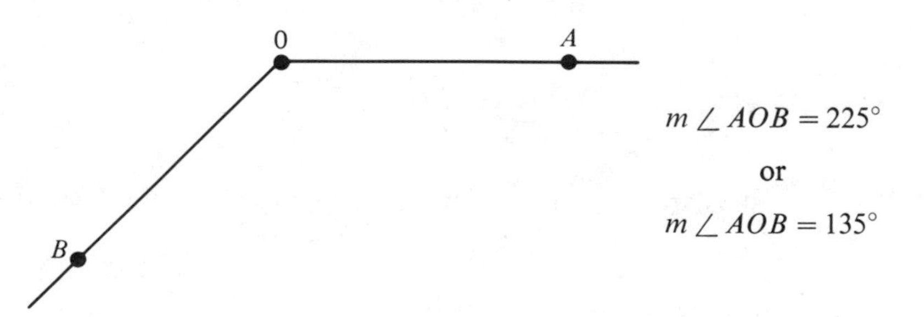

$$m\,\angle AOB = 225°$$

or

$$m\,\angle AOB = 135°$$

But in the elementary school an angle is simply a set of points. There is no preferred initial side. Also, no orientation is given to the plane. In the elementary school every angle has positive degree measure less than or equal to 180. For the figure

$$m\,\angle AOB = 135° = m\,\angle BOA$$

A consequence of this fact is that the measure-in-degrees function falls a little short of being a true measure function. It is still true that its domain is a set of geometric figures, its range is a set of positive real numbers, and congruent angles are assigned the same number, but the additivity property occasionally fails. Angle measures do not always "add up."

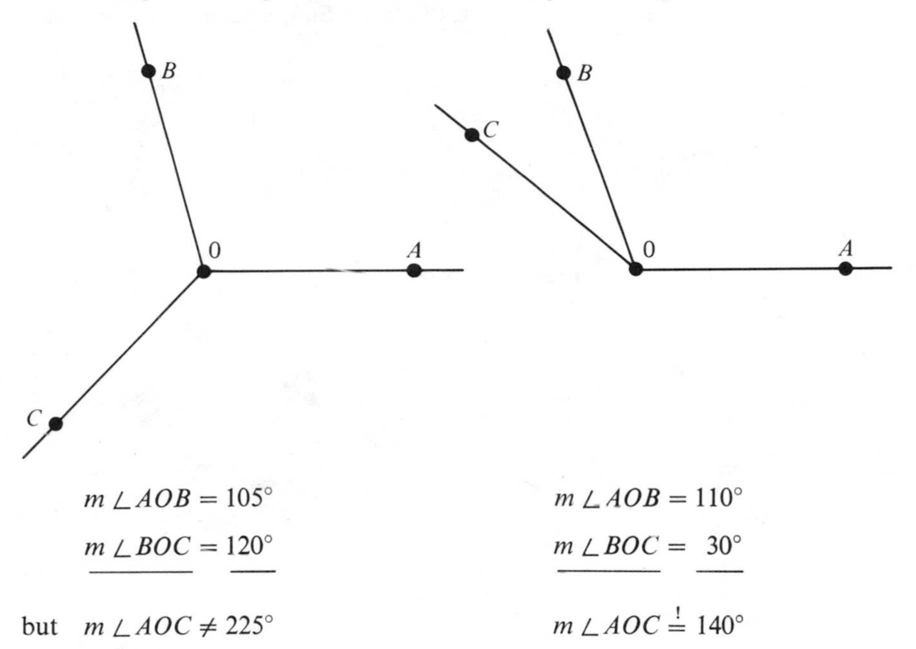

$m\angle AOB = 105°$	$m\angle AOB = 110°$
$m\angle BOC = 120°$	$m\angle BOC = 30°$
but $m\angle AOC \neq 225°$	$m\angle AOC \overset{!}{=} 140°$

in fact

$$m\angle AOC = 135°$$

The weakened form of the additivity property can be stated in two cases:

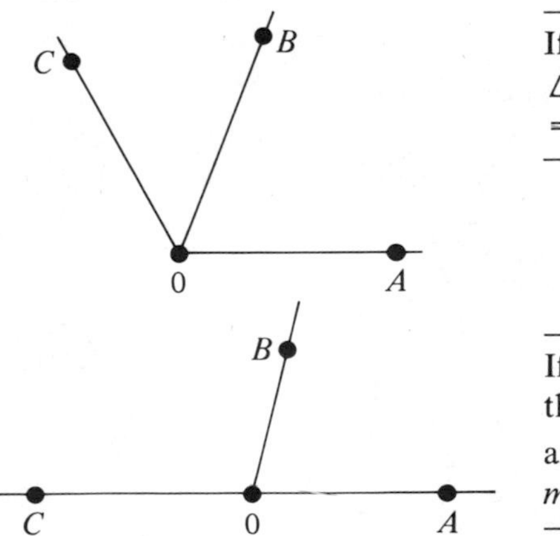

If point B lies in the interior of $\angle AOC$, then $m\angle AOB + m\angle BOC = m\angle AOC$.

If $\angle AOC$ is a straight angle (and thus has no interior) and if B is any point not on $\overleftrightarrow{AC}$, then $m\angle AOB + m\angle BOC = 180$.

Additivity property for angle measure.

NOTE In most contemporary high school geometry texts, the first half of this property is referred to as the *angle addition postulate*, and the second half is referred to as the *angle supplement postulate*. Any two angles whose degree measures add up to 180 are called **supplementary angles**. The angle supplement postulate states that any two angles that form a linear pair are supplementary. Of course the converse of this is false: Supplementary angles need not be adjacent.

EXERCISES

1. Ann has a brand new protractor and measures angles as shown.

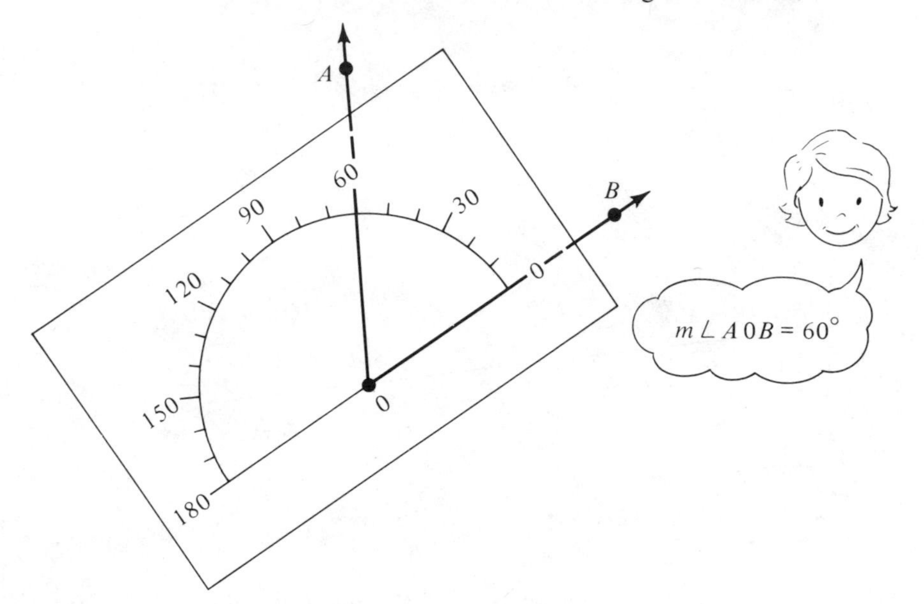

Joe has been chewing on his protractor. He measures angles as below.

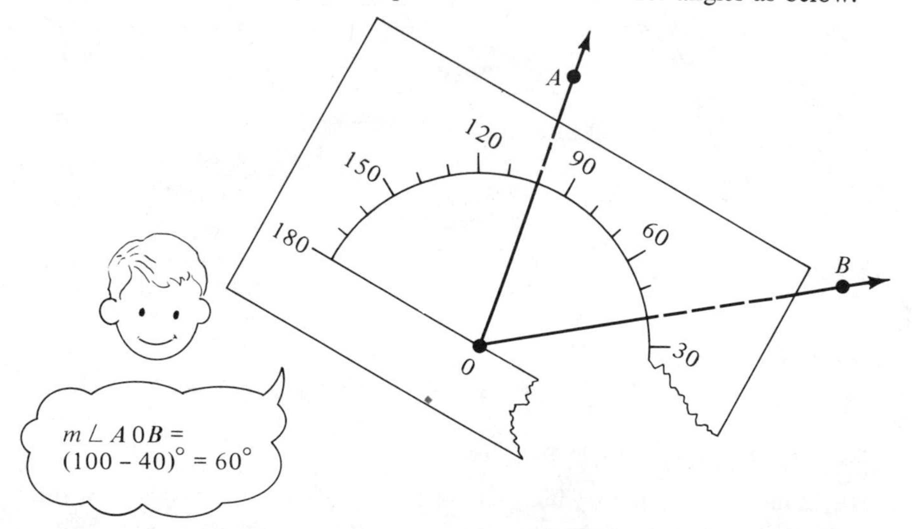

How might Joe measure a 170° angle?

2. **(a)** Measure $\angle ABC$ with your protractor.

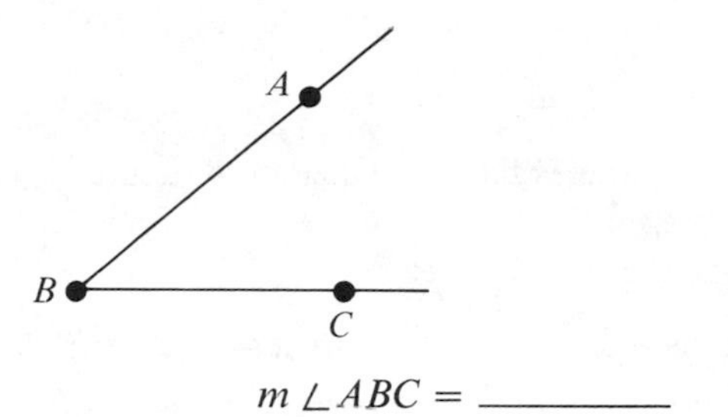

$m\angle ABC =$ ________

(b) Use your protractor to locate a point D on the side of $\overleftrightarrow{BC}$ opposite A so that $m\angle CBD = 42°$.

(c) Do $\angle ABC$ and $\angle CBD$ appear to be congruent?

(d) Describe a rigid motion of $\angle ABC$ that would make it coincide with $\angle CBD$.

3. A basic fact about angle measure is that two angles are congruent if and only if they have the same measure.

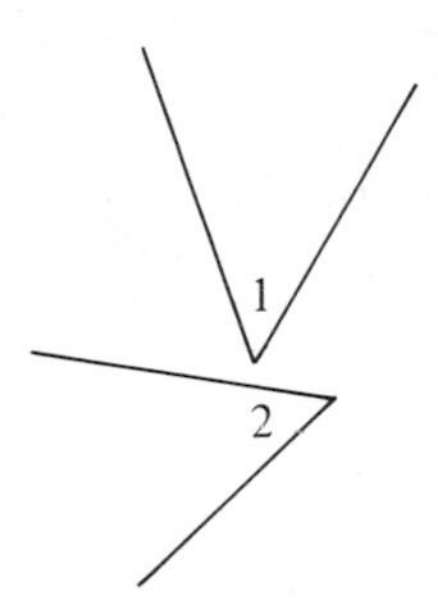

$\angle 1 \cong \angle 2 \Longleftrightarrow m\angle 1 = m\angle 2$ Equal measures congruence criterion for angles.

This property can be verified experimentally. For the left-to-right implication:

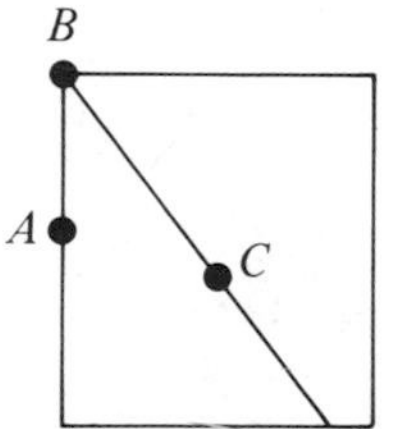

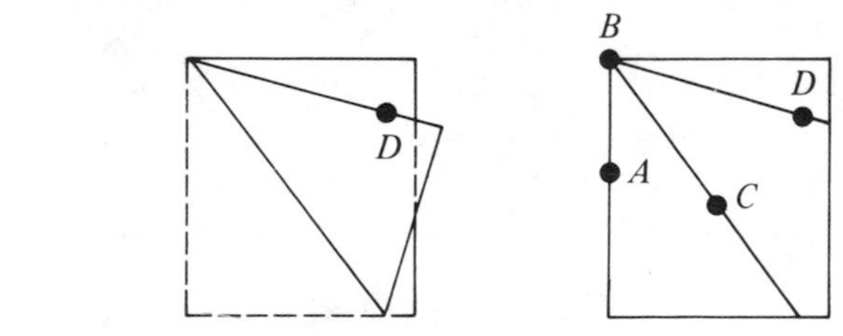

1. Draw three points on a sheet of paper as shown.

2. Draw $\overleftrightarrow{BC}$.

3. Fold the paper along $\overleftrightarrow{BC}$ and draw a point D anywhere along the upper diagonal.

4. Unfold the paper and draw $\overleftrightarrow{BD}$. Why is it true that $\angle ABC \cong \angle CBD$?
5. Measure $\angle ABC$ and $\angle CBD$ and see if $m \angle ABC = m \angle CBD$

For the right-to-left implication:

1. Use a protractor to draw two angles having the same measure.
2. Cut out the angles with a scissors.
3. See if the cut-outs can be made to coincide.

4. The additivity property for angle measure was stated in terms of *pairs* of angles. Give arguments (based on the additivity property for pairs of angles) to show that

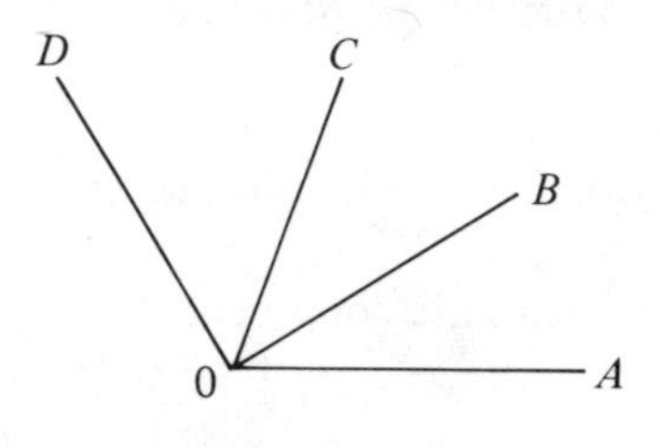

(a) for this figure

$$m \angle AOD = m \angle AOB + m \angle BOC + m \angle COD$$

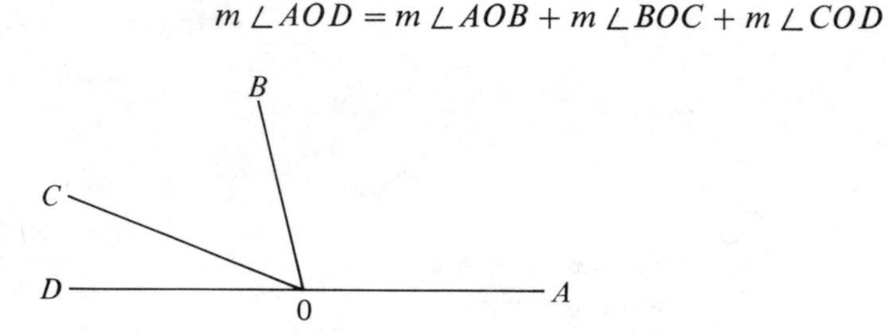

(b) for this figure

$$m \angle AOB + m \angle BOC + m \angle COD = 180°$$

Is it clear that the additivity properties of angle measure apply to suitable *triples*, *quadruples*, ... of angles?

5. Many basic properties of angles can be discovered by the use of a protractor and the equal measures congruence criterion. One of them is this:

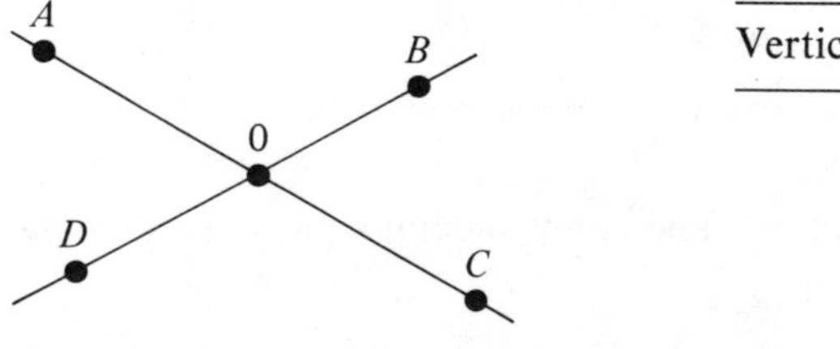

Vertical angles are congruent.

Simple measuring shows that

$$m\angle AOB = m\angle COD$$

and hence

$$\angle AOB \cong \angle COD$$

(a) Can you describe a rigid motion of $\angle AOB$ that will make it coincide with $\angle COD$?

Another basic property of angles is this:

The sum of the (degree) measures of the three angles of a triangle is 180.

Besides the obvious measure-and-add activity, there is a manipulative experiment for establishing this property. It involves tearing apart a triangle and rearranging the pieces as shown.

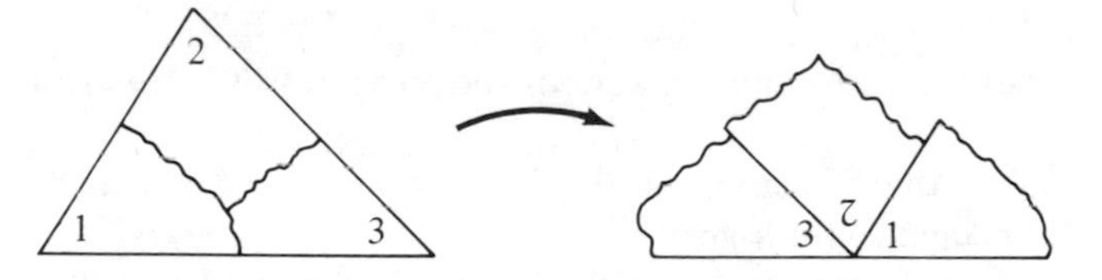

One observes the straight line in the right-hand figure and concludes that $m\angle 1 + m\angle 2 + m\angle 3 = 180$.

(b) What property of angle measure is used in drawing this conclusion?

(c) What property of angle measure assures us that moving the angles about does not change their measure? (For example, how do we know that $m\angle 2 = m\angle 2$?)

6. Without measuring, find the measure of $\angle AOB$. (Everything that looks like a straight line is intended to be one.)

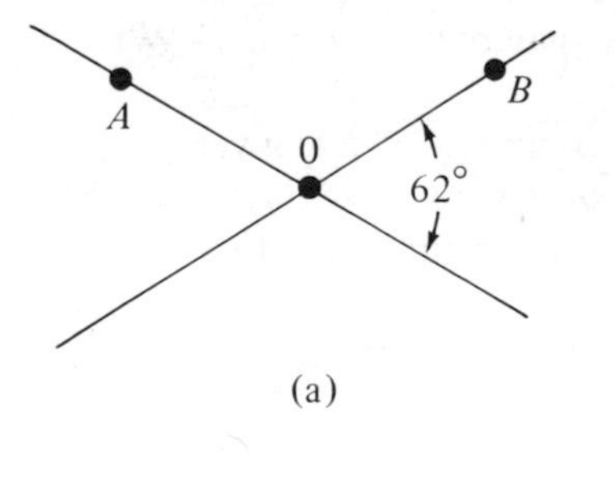

(a)

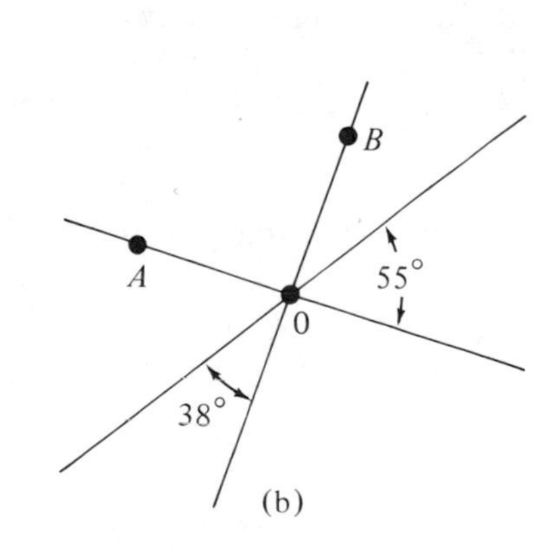

(b)

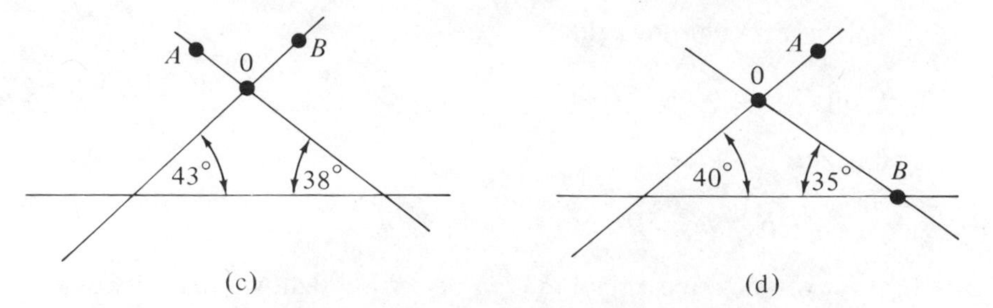

7. The convex pentagon $ABCDE$ has five **interior angles**: $\angle ABC$, $\angle BCD$, $\angle CDE$, $\angle DEA$, $\angle EAB$. Determine the sum of their measures.

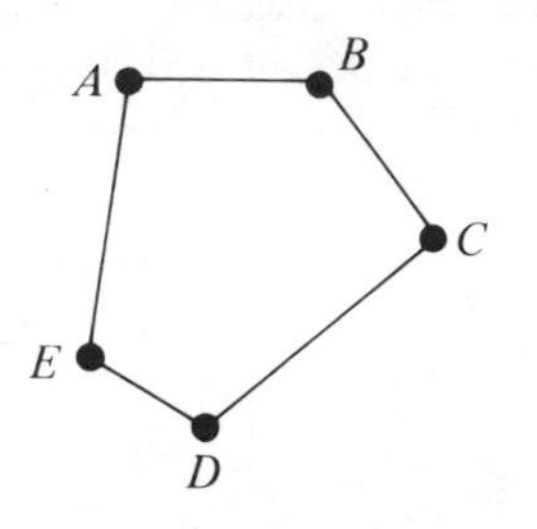

8. Sketch a convex hexagon and find the sum of the measures of its interior angles.
9. What is the sum of the measures of the interior angles of a convex **(a)** heptagon? **(b)** octagon? **(c)** n-gon?
10. A polygon is called **equilateral** if all of its sides are congruent. It is called **equiangular** if all of its interior angles are congruent. It is called **regular** if it is both equilateral and equiangular.
    - **(a)** Using a compass, sketch an equilateral triangle. Check with a protractor to see if it is also equiangular.
    - **(b)** Using a protractor, sketch an equiangular triangle. Check with a ruler to see if it is also equilateral.
    - **(c)** Sketch a quadrilateral that is equilateral but not regular. Do you know a special name for such a figure?
    - **(d)** Sketch a quadrilateral that is equiangular but not regular. What do you call such a figure?
11. The sum of the measures of the interior angles of a convex n-gon is __________. Thus, each interior angle of a regular n-gon has measure __________. Use this fact to draw a regular pentagon with 1-in. sides. (Use ruler and protractor.)
12. Equilateral triangles can be arranged to "fill up" a plane.

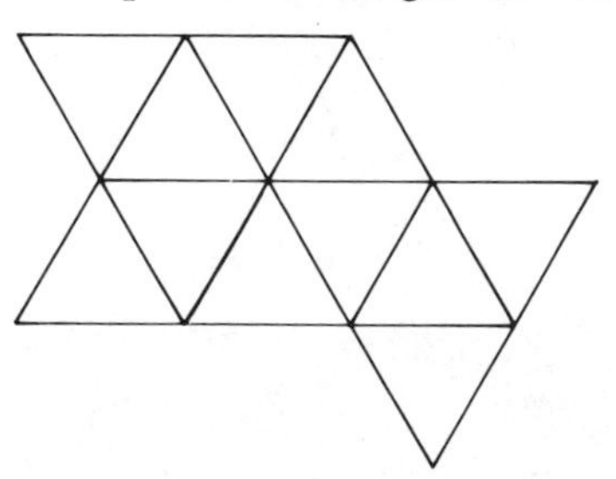

(a) Can squares be arranged to fill up a plane? (Relate your answer to the choice of squares as units of area measure.)
(b) Can regular pentagons be arranged to fill up a plane?
(c) Can regular hexagons be arranged to fill up a plane?
(d) Can regular heptagons be arranged to fill up a plane?
(e) Can regular n-gons be arranged to fill up a plane if $n > 6$?
(f) What regular polygons do bees use in building their hives?

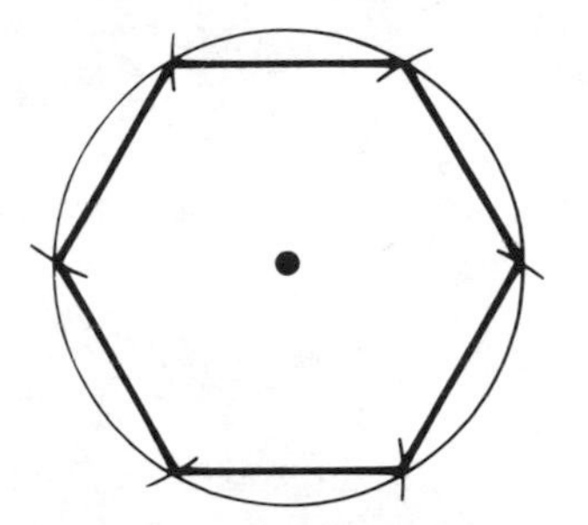

13. The diagram suggests how a regular hexagon can be constructed using only a compass and a straightedge. Why is it that after swinging the third arc you are exactly half way around the circle?

4.10 PARALLELS, TRANSVERSALS, AND ANGLES

When two (coplanar) lines, such as ℓ and n, are cut by a third line, such as p, that third line is referred to as a **transversal** of the first two.

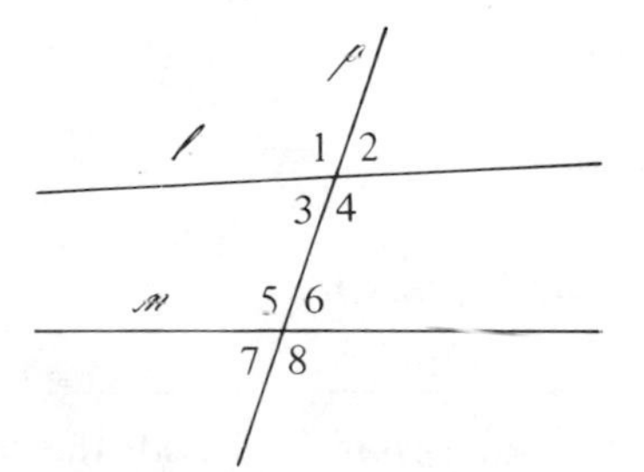

Figure 1.

The angles that appear in such a figure are given special names. Angles 3, 4, 5, 6 are referred to as **interior** angles; angles 1, 2, 7, 8 as **exterior** angles. Various pairs of these angles are also given special names:

Alternate interior angles: $\angle 3$ and $\angle 6$, $\angle 4$ and $\angle 5$
Corresponding angles: $\angle 1$ and $\angle 5$, $\angle 2$ and $\angle 6$, $\angle 3$ and $\angle 7$, $\angle 4$ and $\angle 8$

The question of whether lines ℓ and n are parallel or not can be settled by measuring a pair of these angles. Use your protractor to measure angles

3 and 6 in Figure 1. Are these angles congruent? Do lines ℓ and n appear to be parallel? Would you agree with this statement:

If $\ell \nparallel n$, then $\angle 3 \ncong \angle 6$.

Would you agree with this one:

If $\angle 3 \cong \angle 6$, then $\ell \parallel n$.

Now measure angles 3 and 6 in Figure 2.

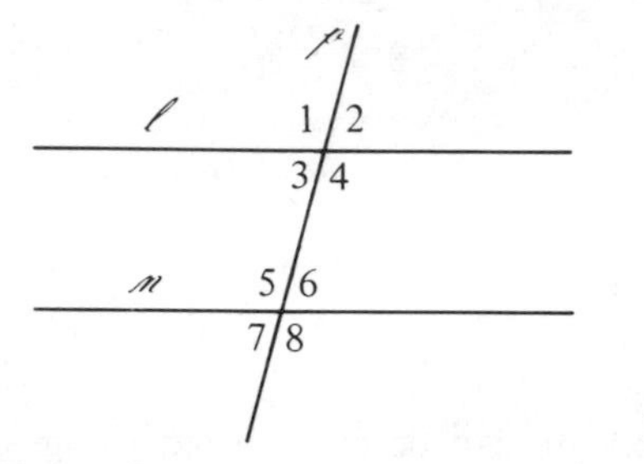

Figure 2.

Are they congruent? Do lines ℓ and n appear to be parallel? Would you agree with the statement:

If $\ell \parallel n$, then $\angle 3 \cong \angle 6$?

Is this statement any different from the ones suggested by Figure 1?

In words, the two basic facts about parallels, transversals, and angles are these:

If alternate interior angles are congruent when two lines are cut by a transversal, then the two lines are parallel.

When two parallel lines are cut by a transversal, alternate interior angles are congruent.

We have encountered several properties of angles in this and the previous section. We list them here and look for some of the logical connections among them in the exercises that follow.

Property 1

$\angle 1 \cong \angle 2 \Leftrightarrow m\angle 1 = m\angle 2$

Property 2 (Additivity Property of Angle Measure)

$m\angle AOB + m\angle BOC = m\angle AOC$, which is 180 in the case of collinear points A, O, C.

Property 3

Vertical angles are congruent.

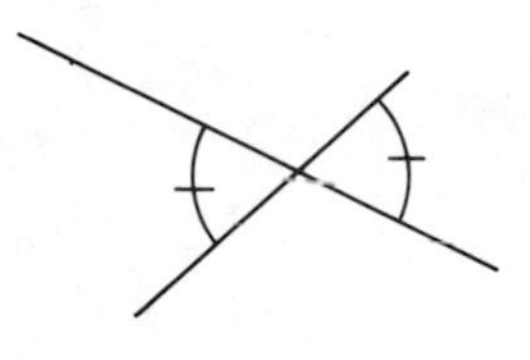

Property 4

The sum of the measures of the interior angles of a triangle is 180.

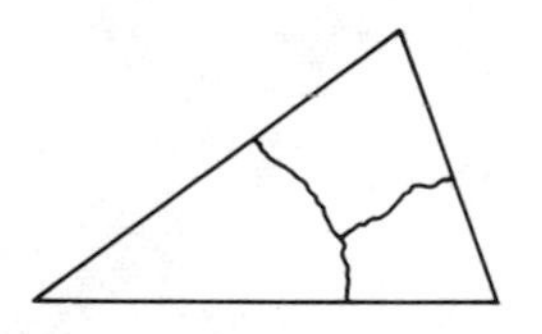

Property 5

If $\angle 1 \cong \angle 2$, then $n \parallel \ell$.

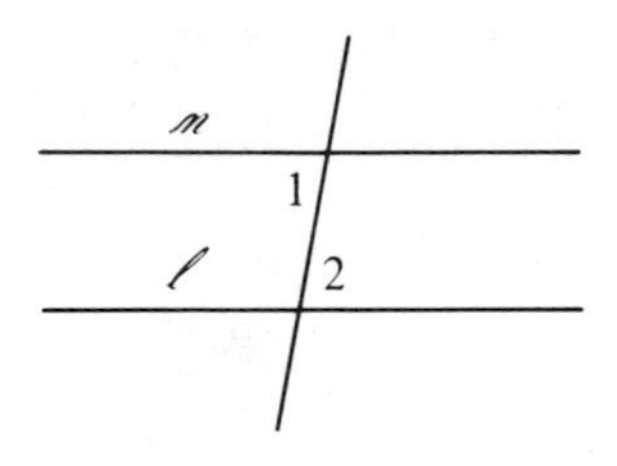

Property 6

If $n \parallel \ell$, then $\angle 1 \cong \angle 2$.

EXERCISES

1. Use properties 1, 2 and the figure to prove property 3: $\angle 1 \cong \angle 3$. (Common properties of equality and inequality such as the additive properties, the transitive properties, and substitution can also be used as reasons.)

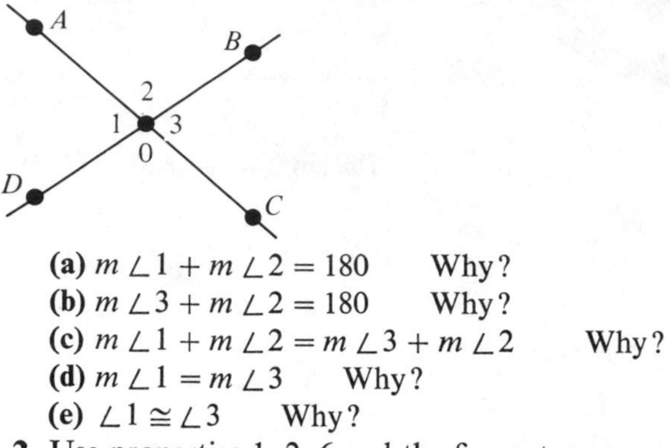

(a) $m\angle 1 + m\angle 2 = 180$ Why?
(b) $m\angle 3 + m\angle 2 = 180$ Why?
(c) $m\angle 1 + m\angle 2 = m\angle 3 + m\angle 2$ Why?
(d) $m\angle 1 = m\angle 3$ Why?
(e) $\angle 1 \cong \angle 3$ Why?

2. Use properties 1, 2, 6 and the figure to prove property 4.

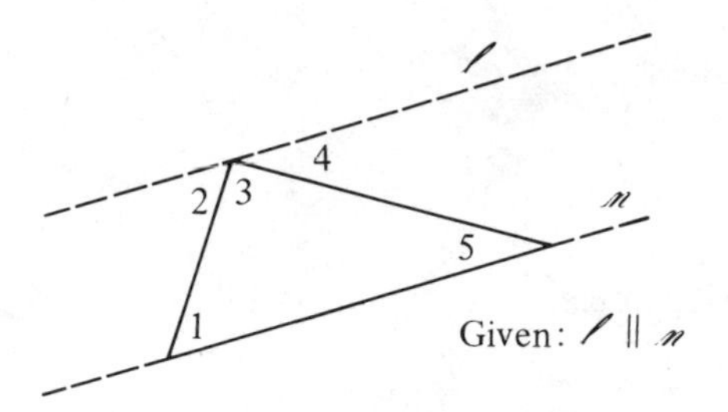

(a) $\angle 4 \cong \angle 5$ Why?
(b) $m\angle 4 = m\angle 5$ Why?
(c) $\angle 1 \cong \angle 2$ Why?
(d) $m\angle 1 = m\angle 2$ Why?
(e) $m\angle 1 + m\angle 3 + m\angle 5 = m\angle 2 + m\angle 3 + m\angle 4$ Why?
(f) $m\angle 2 + m\angle 3 + m\angle 4 = 180$ Why?
(g) $m\angle 1 + m\angle 3 + m\angle 5 = 180$ Why?

3. We shall use properties 1, 2, 4 to prove property 5, "if $\angle 1 \cong \angle 2$, then $n \parallel \ell$." Our plan will be to suppose $n \nparallel \ell$ and show that then $\angle 1 \ncong \angle 2$. (Does this sound like a logical plan?)

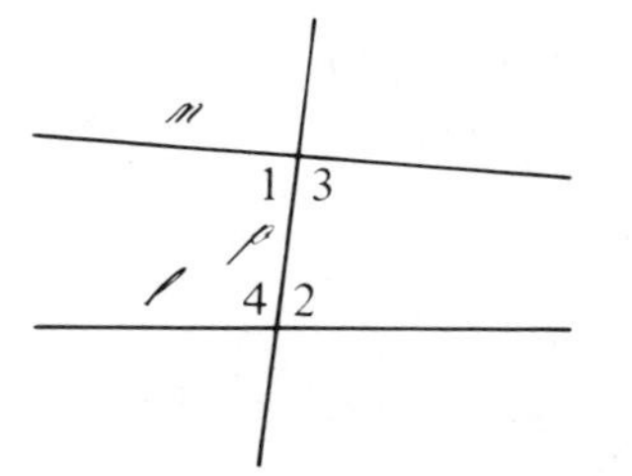

(a) We suppose that $n \nparallel \ell$.
(b) n and ℓ intersect Why?
(c) Let us say they intersect to the right of p.
(d) $m\angle 2 + m\angle 3 < 180$ Why?
(e) $m\angle 1 + m\angle 3 = 180$ Why?
(f) $m\angle 2 + m\angle 3 < m\angle 1 + m\angle 3$ Why?
(g) $m\angle 2 < m\angle 1$ Why?
(h) $\angle 2 \ncong \angle 1$ Why?

4. Replace step (c) above by "Let us say they intersect to the left of p." Rewrite steps (d), (e), (f), (g), and (h).

5. Write a statement-reason proof of this property:

> If parallel lines are cut by a transversal, then corresponding angles are congruent.

In your proof use any of the six properties as reasons. Refer to the figure below, assume that $\ell \parallel n$, and first prove that $\angle 1 \cong \angle 5$. Then prove that $\angle 4 \cong \angle 8$, $\angle 2 \cong \angle 6$, and $\angle 3 \cong \angle 7$.

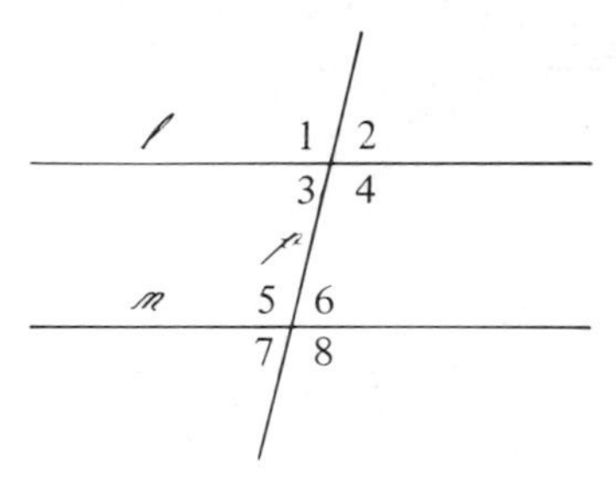

6. Write a statement-reason proof of this property:

> If corresponding angles are congruent when two lines are cut by a transversal, then the two lines are parallel.

In your proof refer to the figure above, assume that $\angle 1 \cong \angle 5$, and deduce that $\ell \parallel n$. Use any of the first six properties as reasons.

7. Write a statement-reason proof of this property, using any previous properties as reasons. Refer to the figure of Exercise 5 and show that $m\angle 4 + m\angle 6 = 180$.

> If parallel lines are cut by a transversal, then interior angles on the same side of the transversal have measure sum 180.

8. Write a statement-reason proof of this property, using any previous properties as reasons. Refer to the figure in Exercise 5 and assume that $m\angle 4 + m\angle 6 = 180$.

If, when two lines are cut by a transversal, the interior angles on one side of the transversal have measure sum 180, then the two lines are parallel.

9. Describe a procedure for drawing, with straightedge and protractor, a line through P parallel to ℓ.

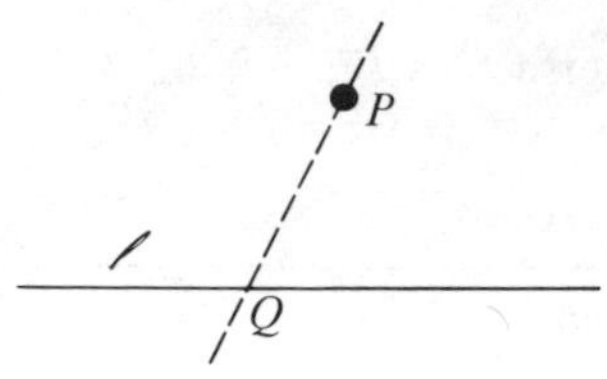

Step 1 Draw a line through P intersecting ℓ at, say, Q.

Step 2 . . .

10. Assuming that lines ℓ and n are parallel, $m\angle 2 = 55°$, and $m\angle 10 = 50°$, fill in the blanks and give, in a few words, a reason for your answer.

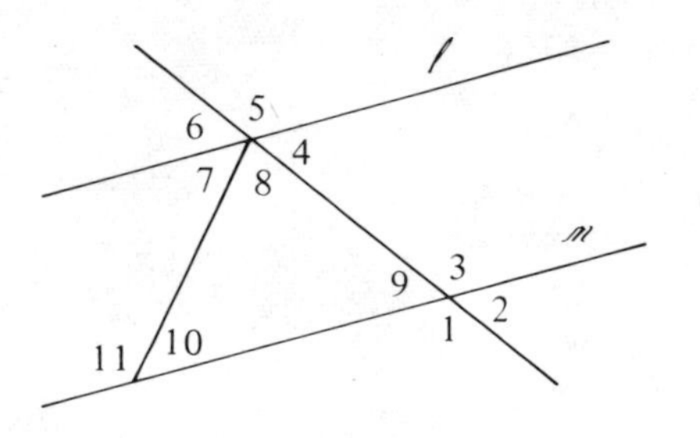

(a) $m\angle 9 =$ ________ because ________
(b) $m\angle 8 =$ ________ because ________
(c) $m\angle 7 =$ ________ because ________
(d) $m\angle 11 =$ ________ because ________

11. (a) Which two rays are collinear: $\overrightarrow{OA}$ and $\overrightarrow{OC}$, or $\overrightarrow{OB}$ and $\overrightarrow{OD}$?
(b) $m\angle AOD = ?$

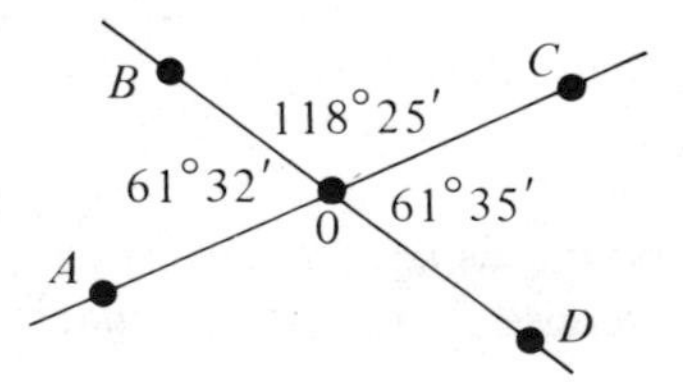

VOCABULARY

4.1 Introduction

metric

4.2 Congruence of Plane Figures

congruent (plane figures)
rotation (turn)
rigid motion
right angles
translation (slide)
reflection (flip)
perpendicular lines
line of reflection

4.3 Congruence of Figures in Space

congruent (space figures)

4.4 Comparing Sizes of Figures Without Measuring

similar

4.5 The Process of Measurement

unit of measure
additivity property (of measure functions)
volume
measure function, $m_U(A)$
length
area

4.6 Length and Perimeter

inch
integrally nonmeasurable
rationally nonmeasurable
subunit
circle
disk
diameter (a, the)
circumference
integrally measurable
rationally measurable
$AB = 2$ in. notation
perimeter
center
chord
radius (a, the)
π

4.7 Area and Area Formulas

square inch
grid sheet
trapezoid

4.8 Volume and Volume Formulas

cubic inch
prism
right rectangular prism

4.9 Angle Measure

degree
protractor
additivity property for angle measure
supplementary angles
equal measures criterion for congruence of angles
interior angles (of a polygon)
equilateral (polygon)
equiangular (polygon)
regular (polygon)

4.10 Parallels, Transversals, and Angles

transversal
exterior angles
corresponding angles
interior angles
alternate interior angles

USING MEASUREMENT

OVERVIEW

In Chapter 4 we developed the *theory* of measurement. We investigated the basic *ideas* of congruence, unit of measure, and measure function. We dissected *abstract* geometric figures in our mathematical laboratory and derived formulas for area and volume. In this chapter we consider some of the *practical* matters connected with the measurement of *real* things.

Since a good deal of school time is spent in studying various units of measure, we devote the first three sections to that topic. This provides a natural context within which to do work with applications of mathematics.

The next four sections are devoted to concepts associated with approximation. Any time that something in the real world is measured, an approximate measurement results. To work with these approximations it is crucial to know how "good" they are. It is difficult to argue with the statement,

Everyone weighs approximately one hundred pounds

because no claim is made as to how "good" that approximation is. A goal of this chapter is to attach some quantitative meanings to the word "good." Two main meanings are in general use. An approximation can be "good" because it is "precise," or it can be "good" in the sense that it is "accurate."

Careful, and distinct, meanings are assigned to "precise" and "accurate" in Section 5.4. In the next three sections common notational conventions for reporting how good an approximation is are described. The material of Sections 5.6 and 5.7 is rather technical, and the notational conventions described are used and understood primarily by the community of scientists and engineers. Although some of this material appears in the junior high school curriculum, very little is found in the elementary school. We consider these sections optional.

The final section, also optional, is devoted to some of the uses of measurement in science.

5.1 BRITISH AND METRIC UNITS OF LENGTH

The system of units we have been using is known as the **British** or British-American system. The common units of length in this system and their relations to each other are familiar to us.

BRITISH UNITS OF LENGTH

12 inches (in.) = 1 foot (ft)
3 feet = 1 yard (yd)
5280 feet = 1 mile (mi)

During the course of history many other units of length have been used. Most of them had colorful origins, but lacked uniformity, the *rod* originally was the length of the left feet of 16 men chosen at random. It has been now standardized as $16\frac{1}{2}$ ft. The *barleycorn* was the length of a grain of barley. It has now been standardized as $\frac{1}{3}$ in. Legend has it that the *yard* itself originated as the distance from the tip of King Henry I's nose to the end of his thumb. It has now been standardized as .91440183 meter (m).*

Modern society requires much more careful definition of units. The system of units in use in most of the civilized world is the **metric system**. The basic unit of length in this system is the **meter**. (You probably noticed that the yard is now defined in terms of the meter. This should suggest how the competition between the British and the metric systems is progressing.) A meter was defined originally, by a commission of French scientists near the end of the eighteenth century, to be one ten-millionth of the distance from the North Pole to the equator measured along the meridian through Paris. Nowadays the meter is defined in terms of the wavelength of the orange-red light from the isotope krypton-86 measured in a vacuum. The other units of length in the metric system are defined in terms of the meter using only powers of ten.

* This definition of the yard in terms of the meter is not universally accepted. Two other common definitions are: 1 yard = 3600/3937 m, 1 yard = .9144 m.

METRIC UNITS OF LENGTH

1 millimeter (mm)	= 1/1000 meter (m)	= $1/10^3$ m
1 centimeter (cm)	= 1/100 m	= $1/10^2$ m
1 decimeter (dm)	= 1/10 m	= $1/10^1$ m
1 meter		
1 dekameter (dkm)	= 10 m	= 10^1 m
1 hectometer (hm)	= 100 m	= 10^2 m
1 kilometer (km)	= 1000 m	= 10^3 m

It is frequently necessary to convert a measurement made in one unit to a measurement in another unit of the *same system*. The examples that follow show how this is done and show why most people prefer the metric system to the British system.

EXAMPLE How many inches are there in 197 miles?

$$
\begin{aligned}
197 \text{ miles} &= 197 \times 5280 \text{ ft} \\
&= 197 \times 5280 \times 12 \text{ in.} \\
&= 12{,}481{,}920 \text{ in.} \qquad \text{(Check the multiplication.)}
\end{aligned}
$$

EXAMPLE How many rods are there in a mile?

$$
\begin{aligned}
1 \text{ mile} &= 5280 \text{ ft} \\
&= \frac{5280}{16\frac{1}{2}} \text{ rods} \\
&= 320 \text{ rods} \qquad \text{(Check the calculation.)}
\end{aligned}
$$

EXAMPLE How many centimers are there in 197 m?

$$
\begin{aligned}
197 \text{ m} &= 197 \times 100 \text{ cm} \\
&= 19{,}700 \text{ cm} \qquad \text{(Was this product difficult to compute?)}
\end{aligned}
$$

EXAMPLE How many millimeters are there in a hectometer?

$$
\begin{aligned}
1 \text{ hm} &= 100 \text{ m} \\
&= 100 \times 1000 \text{ mm} \\
&= 100{,}000 \text{ mm} \qquad \text{(Was this product difficult to compute?)}
\end{aligned}
$$

Occasionally it is necessary to convert a measurement made in the units of one system to a measurement in units of the *other system*. In order to do this **conversion factors** must be used. By definition,

$$1 \text{ yd} = .91440183 \text{ m}$$

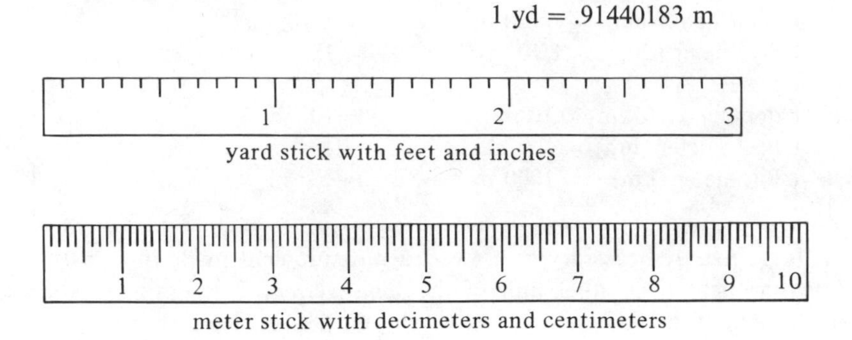

yard stick with feet and inches

meter stick with decimeters and centimeters

From this single relation all of the other conversion factors can be computed. For example,

$$1 \text{ in.} = \tfrac{1}{36} \text{ yd} = \frac{.91440183}{36} \text{ m}$$

$$\doteq .02540005 \text{ m}$$

$$\doteq 2.540005 \text{ cm}$$

(Read the symbol " $\doteq$ " as "approximately equals.")

The table that follows lists the most commonly used conversion factors rounded to two decimal places.

CONVERSION FACTORS FOR LENGTH MEASURE

1 in.	$\doteq$ 2.54 cm
1 cm	$\doteq$ .39 in.
1 yd	$\doteq$ .91 m
1 m	$\doteq$ 39.37 in. $\doteq$ 1.09 yd
1 mi	$\doteq$ 1.61 km
1 km	$\doteq$ .62 mi

The conversion factors can be used to convert measurements in the metric system to measurements in the British system and vice versa.

EXAMPLE Approximately how many centimeters are there in 1 ft?

$$1 \text{ ft} = 12 \text{ in.} \doteq 12 \times 2.54 \text{ cm}$$

$$\doteq 30.48 \text{ cm}$$

E X A M P L E Approximately how many feet are there in 1 km?

$$
\begin{aligned}
1 \text{ km} = 1000 \text{ m} &\doteq 1000 \times 1.09 \text{ yd} \\
&\doteq 1090 \text{ yd} \\
&\doteq 1090 \times 3 \text{ ft} \\
&\doteq 3270 \text{ ft}
\end{aligned}
$$

EXERCISES

1. What is unsatisfactory about this recipe for a chocolate malt: 4 lumps of chocolate ice cream, 1 glass of milk, 1 spoon of malt powder?

2. The metric system is the official one in most countries of the world except the United States and Canada. Scientists everywhere use it. There is good reason to believe that its use in this country will increase steadily and ultimately supplant the British system. Already many elementary textbooks include work with metric units. If you are not yet familiar with the names of these units, the questions that follow might suggest ways of keeping straight the meanings of the prefixes.

(a) What do you call a hundredth of a dollar?
(b) What do you call a thousandth of a dollar?
(c) How many years in a century?
(d) How many years in a millenium?
(e) What vowel precedes "-meter" in the three smallest units?
(f) For the metric units we listed, *L*atin prefixes go with units *l*ess than a meter while *G*reek prefixes go with units *g*reater than a meter. Have you noticed any other patterns that will help you to remember the order of these metric units?

3. Fill in the blanks without looking back at the table of metric units.

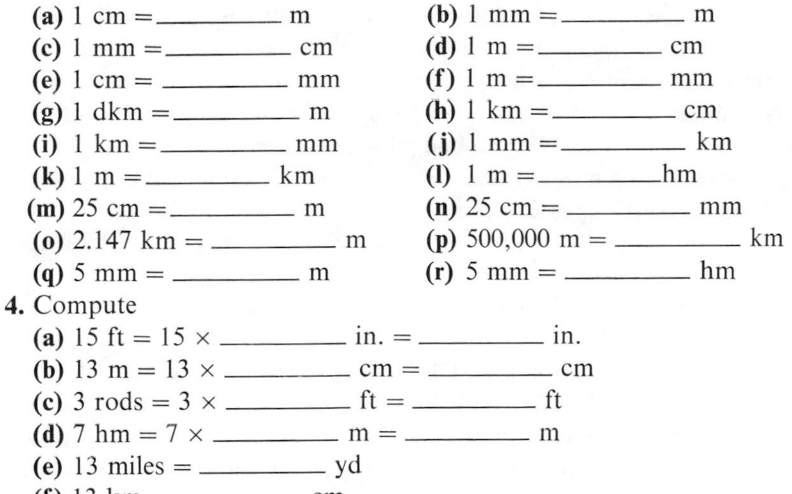

(a) 1 cm = ________ m **(b)** 1 mm = ________ m
(c) 1 mm = ________ cm **(d)** 1 m = ________ cm
(e) 1 cm = ________ mm **(f)** 1 m = ________ mm
(g) 1 dkm = ________ m **(h)** 1 km = ________ cm
(i) 1 km = ________ mm **(j)** 1 mm = ________ km
(k) 1 m = ________ km **(l)** 1 m = ________ hm
(m) 25 cm = ________ m **(n)** 25 cm = ________ mm
(o) 2.147 km = ________ m **(p)** 500,000 m = ________ km
(q) 5 mm = ________ m **(r)** 5 mm = ________ hm

4. Compute

(a) 15 ft = 15 × ________ in. = ________ in.
(b) 13 m = 13 × ________ cm = ________ cm
(c) 3 rods = 3 × ________ ft = ________ ft
(d) 7 hm = 7 × ________ m = ________ m
(e) 13 miles = ________ yd
(f) 13 km = ________ cm

5. Use the conversion tables and fill in the blanks, rounding the final answer to the nearest whole number.

(a) 10 cm = 10 × 1 cm ≐ 10 × ________ in. ≐ ________ in.

(b) 100 yd = 100 × 1 yd ≐ 100 × ________ m ≐ ________ m

(c) 72 in. = 72 × 1 in. ≐ 72 × ________ cm ≐ ________ cm

(d) 65 miles = 65 × 1 mile ≐ 65 × ________ km ≐ ________ km

(e) $3\frac{1}{2}$ miles ≐ ________ km

(f) 2 ft 5 in. ≐ ________ cm

6. Assuming that 1 yard is *exactly* .9144 m, find exactly how many centimeters there are in an inch.

7. What is wrong with this proof that 18 = 0?

1 mile = 1.61 km, but 1 km = .62 mile,

so

1 mile = 1.61 × .62 mile = .9982 mile

Thus,

1 = .9982, so 10,000 = 9982, so 18 = 0

8. Which is faster, a car that goes 100 miles/hr or one that goes 100 km/hr?

9. A "youth-size 7" shoe fits a foot of length (12 + 7) barleycorns. A "youth-size 9" shoe fits a foot of length (12 + 9) barleycorns. For adults add 25 instead of 12 barleycorns to the size. For example, an "adult-size 8" shoe fits a foot of length (25 + 8) barleycorns.

(a) If a boy wears an adult size 8 shoe, how long is his foot in barleycorns? In inches?

(b) If an infant wears a size 2 shoe, how long is his foot (in inches)?

(c) What size shoe does a man with a 12-in. foot wear?

(d) What size shoe does a child with an 8-in. foot wear?

10. Jack can run a mile in 6 min. Pierre can run 1500 m in 5 min and 50 sec. Which boy has greater average speed?

11. If the diameter of an American golf ball is 1.68 in. and the diameter of a British golf ball is 4.11 cm, which ball is larger?

12. Most materials expand when heated. The *coefficient of thermal expansion* for cast brass is .00001875. This means that if the temperature of a brass rod 1 m long is raised 1 degree Centigrade, the rod's length will increase by .00001875 m. The temperature of a brass rod 4 m long is raised 80 degrees Centigrade. By how much does the length increase?

***13.** Joe has a single-speed bicycle with a front sprocket with 30 teeth and a rear sprocket with 18 teeth. The teeth on each sprocket are $\frac{3}{4}$ in. apart.

The outside diameter of his rear tire is 28 in. The outside diameter of his front tire is 20 in.

(a) If he never coasts, how many times will Joe have to push the pedals around to go 1 mile? (Use $\pi \doteq \frac{22}{7}$.)

(b) If he can make his pedals go around 100 times per minute, how many miles per hour can he go?

***14.** The distance from the center to the outermost groove on a certain long-playing record is $5\frac{3}{4}$ in. The distance from the center to the innermost groove is $2\frac{1}{8}$ in. The playing time, at $33\frac{1}{3}$ rpm, is 24 min.

(a) How far apart are the grooves on the record?

(b) What (approximately) is the total length of the groove? (Use $\pi \doteq \frac{22}{7}$.)

(*Hint*: Make the simplifying assumption that there is not one groove, but many concentric circular grooves all the same distance apart).

5.2 BRITISH AND METRIC UNITS OF AREA AND VOLUME

The common units of area and volume are based on the common units of length.

BRITISH UNITS OF AREA AND VOLUME

Area	*Volume*
square inch	cubic inch
1 sq ft = 12^2 sq in. = 144 sq in.	1 cu ft = 12^3 cu in. = 1728 cu in.
1 sq yd = 3^2 sq ft = 9 sq ft	1 cu yd = 3^3 cu ft = 27 cu ft

METRIC UNITS OF AREA AND VOLUME

Area	*Volume*
square cm	cubic cm (cc)
1 sq m = 100^2 sq cm = 10,000 sq cm	1 cu m = 100^3 cc = 1,000,000 cc
1 sq km = 1000^2 sq m = 1,000,000 sq m	

A notational scheme has been devised that makes converting units, within a system, a simple mechanical task. Instead of writing

area of desk top = 4 sq ft

we write

$$\text{area of desk top} = 4\ \text{ft}^2 \text{ or } 4 \cdot \text{ft}^2$$

If we want to find the area in square inches, we substitute "12 in." for "ft":

$$\begin{aligned}\text{area of desk top} &= 4 \cdot (12 \cdot \text{in.})^2\\ &= 4 \cdot 12^2 \cdot \text{in.}^2\\ &= 576 \cdot \text{in}^2 \text{ or } 576\ \text{in.}^2\end{aligned}$$

Similarly, if we wish to convert a measurement of 100,000 cubic inches to cubic yards, we calculate

$$\begin{aligned}100{,}000 \cdot \text{in.}^3 &= 100{,}000 \cdot (\tfrac{1}{36} \cdot \text{yd})^3\\ &= 100{,}000 \cdot (\tfrac{1}{36})^3 \cdot \text{yd}^3\\ &= \frac{100{,}000}{46{,}656} \cdot \text{yd}^3 \doteq 2.14\ \text{yd}^3\end{aligned}$$

In the metric system conversion problems are much easier. To find the number of square meters in a square field 7 hm on a side:

$$\begin{aligned}(7 \cdot \text{hm})^2 &= 49 \cdot \text{hm}^2\\ &= 49 \cdot (100 \cdot \text{m})^2\\ &= 49 \cdot 100^2 \cdot \text{m}^2\\ &= 490{,}000\ \text{m}^2\end{aligned}$$

To find the number of cubic centimeters in 3562 cubic millimeters:

$$\begin{aligned}3562 \cdot \text{mm}^3 &= 3562 \cdot (\tfrac{1}{10} \cdot \text{cm})^3\\ &= 3562 \cdot (\tfrac{1}{10})^3 \cdot \text{cm}^3\\ &= 3.562\ \text{cc}\end{aligned}$$

In practical problems the safest procedure is to convert all units to a common one as soon as possible.

EXAMPLE How many cubic feet of wood are in a plank 2 in. thick, 10 in. wide, and 5 yd long?

$$\begin{aligned}\text{volume} &= \text{length} \times \text{width} \times \text{thickness}\\ &= 15\ \text{ft} \times \tfrac{10}{12}\ \text{ft} \times \tfrac{2}{12}\ \text{ft}\\ &= 15 \times \tfrac{10}{12} \times \tfrac{2}{12}\ \text{ft}^3 = \tfrac{300}{144}\ \text{ft}^3 = 2\tfrac{1}{12}\ \text{ft}^3\end{aligned}$$

Conversion factors have been computed for translating area and volume measurements from British to metric units and vice versa.

CONVERSION FACTORS FOR AREA MEASURE

$1 \text{ in.}^2 \doteq 6.45 \text{ cm}^2$
$1 \text{ cm}^2 \doteq .16 \text{ in.}^2$
$1 \text{ yd}^2 \doteq .84 \text{ m}^2$
$1 \text{ m}^2 \doteq 1.20 \text{ yd}^2$
$1 \text{ mi}^2 \doteq 2.59 \text{ km}^2$
$1 \text{ km}^2 \doteq .39 \text{ mi}^2$

CONVERSION FACTORS FOR VOLUME MEASURE

$1 \text{ in.}^3 \doteq 16.39 \text{ cc}$
$1 \text{ cc} \doteq .06 \text{ in.}^3$
$1 \text{ ft}^3 \doteq .03 \text{ m}^3$
$1 \text{ m}^3 \doteq 35.31 \text{ ft}^3$

These conversion factors are calculated from the corresponding factors for units of length. For example,

$$\begin{aligned} 1 \text{ square inch} &= 1 \cdot \text{in.}^2 \\ &\doteq 1 \cdot (2.54 \text{ cm})^2 \\ &\doteq 2.54^2 \text{ cm}^2 \\ &\doteq 6.45 \text{ cm}^2 \end{aligned}$$

The conversion factors are used as before.

EXAMPLE How many square centimeters are there in 16 square feet?

$$\begin{aligned} 16 \text{ ft}^2 &= 16 \times (12 \times \text{in.})^2 \\ &= 16 \times 144 \text{ in.}^2 \\ &\doteq 16 \times 144 \times 6.45 \text{ cm}^2 \\ &\doteq 14{,}860 \text{ cm}^2 \end{aligned}$$

EXERCISES

1. How many cubic yards of cement (to the nearest tenth of a cubic yard) are needed to make a sidewalk 5 in. thick, 44 in. wide, and 70 ft long?

2. How many cubic meters of cement (to the nearest tenth of a cubic meter) are needed to make a sidewalk 8 cm thick, 11 dm wide, and 26 m long? Was this calculation easier than the one in Exercise 1?

3. Show how the common factor, .16, in the relation, 1 $\text{cm}^2 \doteq .16$ in.^2, could be derived.

4. Fill in the blanks. [Use whole-number approximations in parts (f), (g), (h).]

(a) 1 km = __________ m
(b) 1 km^2 = __________ m^2
(c) 1 km^3 = __________ m^3
(d) 3 ft in 1 yd $\Rightarrow$ __________ ft^3 in 1 yd^3
(e) 2.54 cm $\doteq$ 1 in. $\Rightarrow$ __________ $\text{cm}^2 \doteq 1$ in.^2
(f) 26 $\text{mi}^2 \doteq$ __________ km^2
(g) 84 $\text{m}^3 \doteq$ __________ ft^3
(h) 16 $\text{in.}^2 \doteq$ __________ cm^2

5. In the figure the dimensions of a box are given in linear units known as Hufas (H) and Pufas (P).

(a) Find the conversion factor relating Hufas and Pufas:

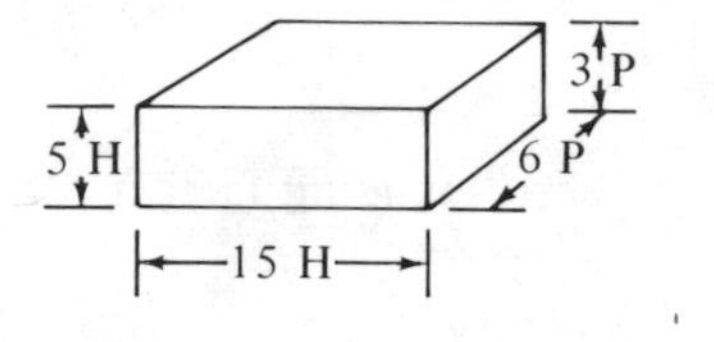

1 H = __________ P

(b) 1 H^2 = __________ P^2
(c) What is the volume of the box in cubic Hufas?

6. Find the area in square meters of a rectangle 275 cm wide and 1200 cm long.

7. A trench for a natural gas pipeline is to be 12 km long, 2 m deep, and 60 cm wide. What volume of earth, in cubic meters, will be excavated?

8. A sheet of steel is 3.6 m long, 1.5 m wide, and 4 cm thick. How much does this sheet weigh in grams? (One cc of steel weighs 7.7 grams.)

9. A football stadium has a grass area of about 140 yd by 90 yd. A sack of grass fertilizer is advertised to cover 10,000 square feet. How many sacks will be required to fertilize the stadium grass?

10. An acre originally was the amount of land that a man could plow in one day with a yoke of oxen. It has now been standardized as $1/640$ mi^2. How many square feet are there in an acre? What fraction of an acre is a city lot which is 60 ft wide and 110 ft deep?

11. Water weighs about 62.4 lb per cubic foot. How much weight is a flat supermarket roof, with dimensions 80 ft by 60 ft, supporting after a 16-in. snow fall? (You may assume that 10 in. of snow is equivalent to 1 in. of rain. Round your answer to the nearest 1000 lb.)

12. A floating object is known to displace its own weight of water. A 6 in. by

8 in. by 6 ft railroad tie is floating in a pond with its largest face up so that only 2 in. of it shows above the surface of the water. How heavy is the railroad tie (to the nearest pound)?

13. If a car tire has outside diameter 2 ft and is 6 in. wide, how many square inches of treaded surface does it have? (Use $\pi \doteq \frac{22}{7}$.)

14. A race track is 1500 m around and has the shape below. How many square yards of sod (to the nearest thousand) are required to sod the infield?

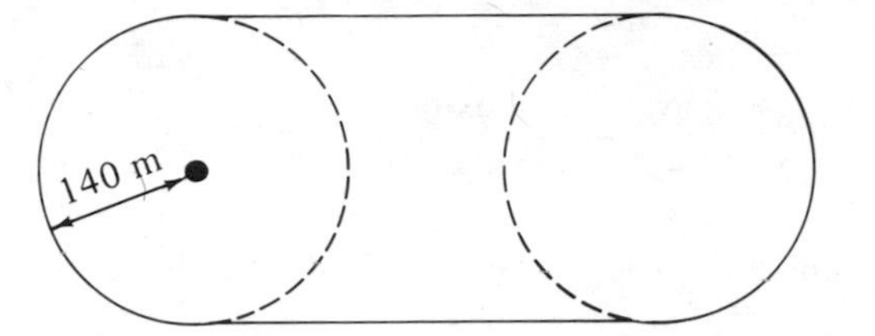

15. A nurse is given a cylindrical glass container with inside diameter 1.4 cm and is asked to calibrate it to show its capacity in cubic centimeters. How far apart should the calibration marks be? (Use $\pi = \doteq \frac{22}{7}$.)

5.3 UNITS OF ANGLE MEASURE

There are three systems of angle measure in general use. The most familiar system uses the **degree** as the basic unit. A degree, you recall, is an angle of such a size that 360 of them fill up a plane (when placed adjacent to each other). Equivalently, there are 180 degrees in a straight angle or 90 degrees in a right angle. The degree itself can be thought of as being composed of 60 smaller (congruent) angles called **minutes**. Each minute in turn can be thought of as being composed of 60 still smaller (congruent) angles called **seconds**.

1 right angle	=	90 degrees (90°)
1 degree	=	60 minutes (60′)
1 minute	=	60 seconds (60″)

EXAMPLE If $m \angle BOA = 47°28'53''$ and $m \angle COB = 35°41'10''$, what is $m \angle COA$?

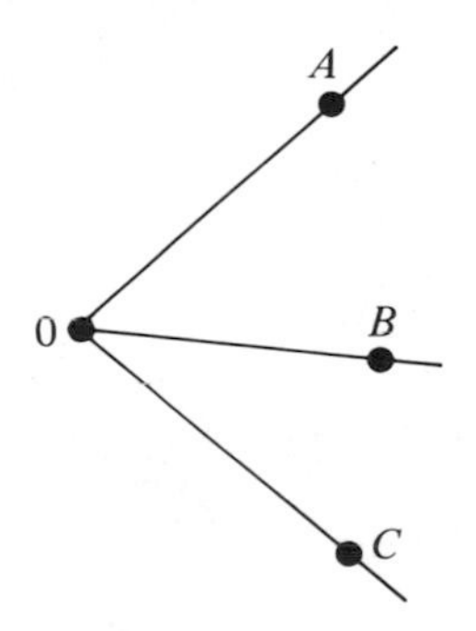

$$\begin{aligned} m\angle COA &= 47°28'53'' + 35°41'10'' \\ &= 82°69'63'' = 82°69'(60+3)'' \\ &= 82°70'3'' = 82°(60+10)'3'' \\ &= 83°10'3'' \end{aligned}$$

EXAMPLE If, in the above figure, $m\angle BOA = 51°43'45''$ and $m\angle COA = 82°37'30''$, what is $m\angle COB$?

$$m\angle COB = \begin{array}{r} 82°37'30'' \\ -51°43'45'' \end{array} = \begin{array}{r} 82°36'90'' \\ -51°43'45'' \end{array} = \begin{array}{r} 81°96'90'' \\ -51°43'45'' \end{array} = 30°53'45''$$

EXAMPLE What size angle has the property that 1000 of them fill up a plane?

Solution If 1000 are to fill up a plane, then the angle must be 1/250th of a right angle.

$$\begin{aligned} \tfrac{1}{250} \times 90 \text{ degrees} &= \tfrac{9}{25} \text{ degree} = \tfrac{9}{25} \times 60 \text{ minutes} \\ &= 21\tfrac{3}{5} \text{ minutes} = 21' + \tfrac{3}{5} \times 60 \text{ seconds} \\ &= 21'36'' \end{aligned}$$

Another system of angle measure that is used frequently in mathematics and physics employs the **radian** as unit of measure. A radian is an angle of such a size that there are π of them in a straight angle. Since $\pi \doteq 3.1416$ and since there are 180 degrees in a straight angle,

$$\begin{aligned} 1 \text{ radian} &\doteq \frac{180}{3.1416} \text{ degrees} \\ &\doteq 57.30 \text{ degrees} \\ &\doteq 57°18' \end{aligned}$$

Similarly,

$$\begin{aligned} 1 \text{ degree} &\doteq \frac{3.1416}{180} \text{ radian} \\ &\doteq .017 \text{ radian} \end{aligned}$$

EXAMPLE What is the radian measure of a 30° angle?

First Solution

$$\begin{aligned} 30° &\doteq 30 \times .017 \text{ radian} \\ &\doteq .51 \text{ radian} \end{aligned}$$

Second Solution

$$\begin{aligned} 30^\circ &= \frac{30}{180} \text{ straight angle} \\ &= \frac{1}{6} \text{ straight angle} \\ &= \frac{\pi}{6} \text{ radian} \\ &\doteq \frac{3.1416}{6} \text{ radian} \\ &\doteq .5236 \text{ radian} \end{aligned}$$

EXAMPLE What is the degree measure of an angle of radian measure 2.6?

$$\begin{aligned} 2.6 \text{ radians} &\doteq 2.6 \times 57.30 \text{ degrees} \\ &\doteq 149^\circ \end{aligned}$$

A third system of angle measure is used by the military. The unit of measure is the **mil**. A mil is an angle of such size that 6400 mils = 360°. Thus

$$\begin{aligned} 1 \text{ mil} &= \frac{360}{6400} \text{ degrees} \\ &= \frac{360 \times 60}{6400} \text{ minutes} \\ &= 3\tfrac{3}{8} \text{ minutes} = 3'22\tfrac{1}{2}'' \end{aligned}$$

and

$$1 \text{ degree} = \frac{6400}{360} \text{ mils} \doteq 17.78 \text{ mils}$$

CONVERSION FACTORS FOR ANGLE MEASURE

1 degree $\doteq$ .017 radian
1 radian $\doteq 57.30^\circ \doteq 57^\circ 18'$
1 degree $\doteq$ 17.78 mils
1 mil $\doteq .056^\circ \doteq 3'22\tfrac{1}{2}''$

EXAMPLE How many mils in a 45° angle?

First Solution

$$45° \doteq 45 \times 17.78 \text{ mils}$$

$$\doteq 800.1 \text{ mils}$$

Second Solution

$$45° = \tfrac{1}{4} \text{ straight angle}$$

$$= \tfrac{1}{4} \times 3200 \text{ mils}$$

$$= 800 \text{ mils}$$

EXERCISES

1. Fill in the blanks:
(a) 1 degree = _________ seconds
(b) $\frac{1}{4}$ of a right angle = _________ ° _________ ′
(c) $\frac{1}{8}$ of a right angle = _________ ° _________ ′
(d) $\frac{1}{16}$ of a right angle = _________ ° _________ ′ _________ ″
2. Find $m \angle AOC$ if

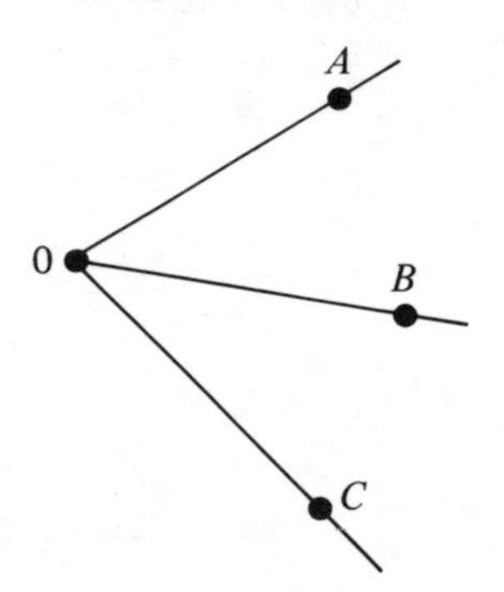

(a) $m \angle AOB = 42°55'18''$ and $m \angle BOC = 35°8'35''$
(b) $m \angle AOB = 40°27'5''$ and $m \angle BOC = 36°38'57''$
(c) $m \angle AOB = 41°27''$ and $m \angle BOC = 37°10'40''$
(d) $m \angle AOB = 40°28'17''$ and $m \angle BOC = 39°31'43''$
3. In the figure above, find $m \angle AOB$ if $\angle AOC$ is a right angle and
(a) $m \angle BOC = 35°19'47''$
(b) $m \angle BOC = 37°8'25''$
(c) $m \angle BOC = 40°29''$
(d) $m \angle BOC = 42°15'$
4. Find $m \angle A$ in the following figures. Do not try to use a protractor.

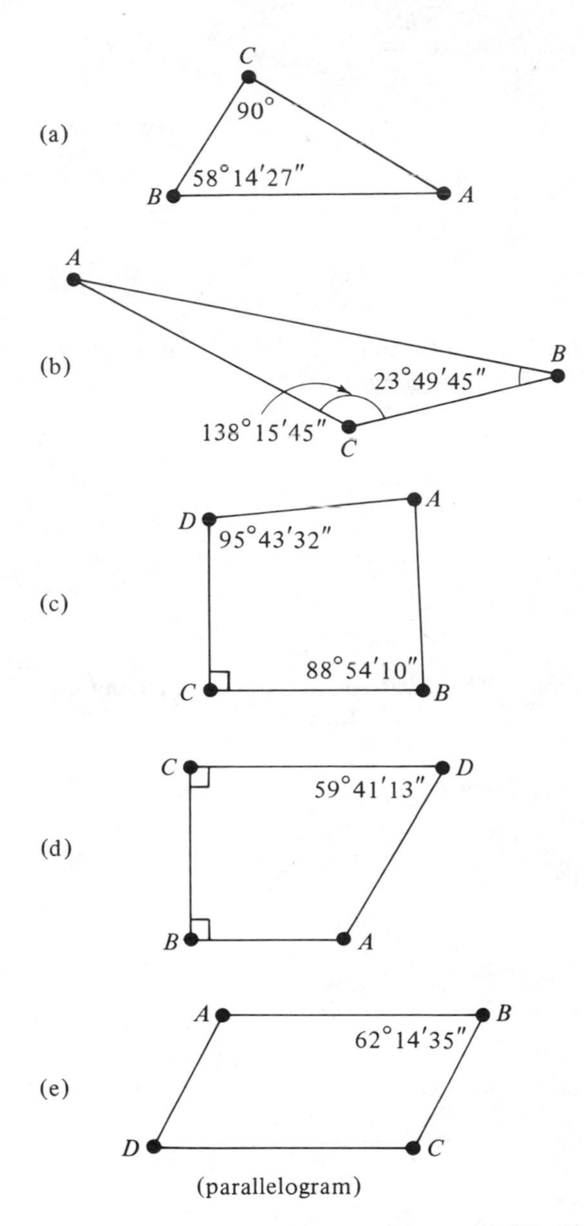

(parallelogram)

5. If a distant radio tower appears to be 27°15′ East of North and a mountain peak appears to be 19°50′ North of West, what size angle does the observer perceive between the radio tower and the mountain peak?

6. One pie is to be shared equally by seven people. What size "central angle" should each piece have (to the nearest minute)?

7. The following diagram shows a portion of a regular polygon "inscribed in a circle."
How many sides has the polygon?

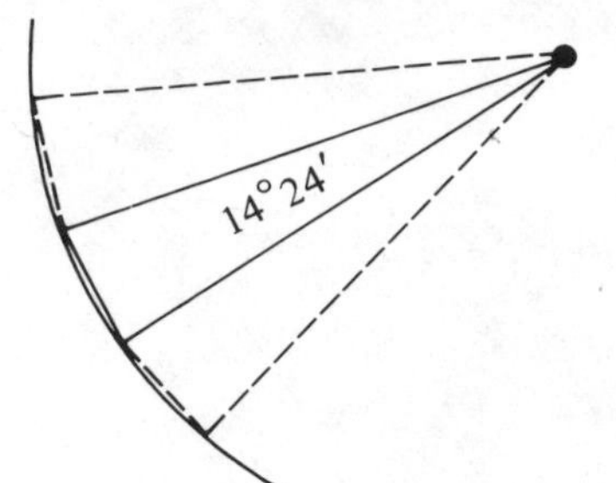

8. Find $m\angle AOF$ if $m\angle AOB = 16°39'47''$ and all of the angles, $\angle AOB$, $\angle BOC$, $\angle COD$, $\angle DOE$, $\angle EOF$, are congruent.

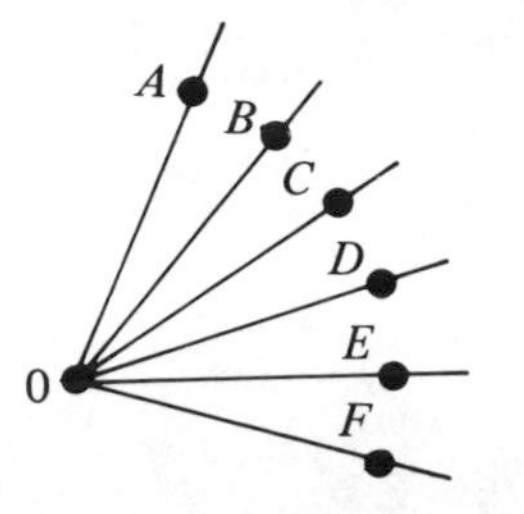

9. Fill in the blanks. Do not replace π by a fraction or a decimal. Do not use conversion factors.

(a) 1 right angle = ________ radians
(b) $\frac{1}{2}$ right angle = ________ radians
(c) 45° = ________ radians
(d) 30° = ________ radians
(e) 120° = ________ radians
(f) $\pi/10$ radian = ________°
(g) $\pi/5$ radian = ________°
(h) 1 right angle = ________ mils
(i) $\frac{1}{2}$ right angle = ________ mils
(j) 45° = ________ mils
(k) 22°30′ = ________mils
(l) 200 mils = ________° ________′
(m) 100 mils = ________° ________′ ________″
(n) $\frac{1}{64}$ right angle = ________ mils
(o) 800 mils = ________ radians

10. Fill in the blanks. Use the table of conversion factors.

(a) 7° $\doteq$ ________ radians
(b) 1.5 radians $\doteq$ ________°
(c) 135 mils $\doteq$ ________°
(d) 30° $\doteq$ ________ mils

11. The sum of the measures of the angles of any triangle in degrees is ________, in mils is ________, and in radians is ________.

When an angle is drawn with its vertex at the center of a circle, the length of the **arc** of the circle cut off by this **central angle** is proportional to the size of the angle.

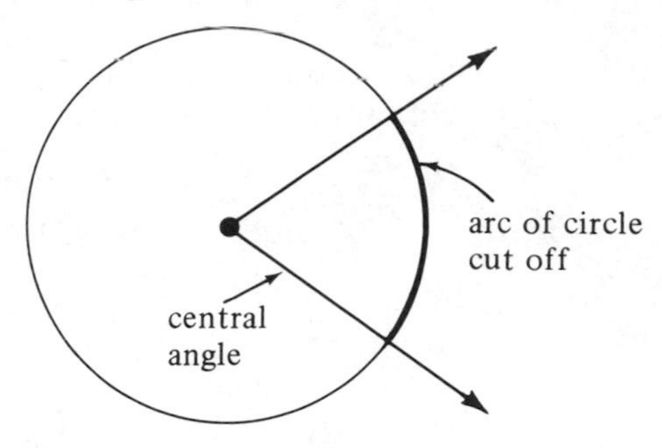

EXAMPLES A 90° central angle cuts off an arc whose length is $\frac{90}{360}(=\frac{1}{4})$ times the circumference of the circle. A 30° central angle cuts off an arc whose length is $\frac{30}{360}$ $(=\frac{1}{12})$ times the circumference of the circle. A 5° central angle cuts off an arc whose length is $\frac{5}{360}$ $(=\frac{1}{72})$ times the circumference of the circle.

12. A circle has circumference 60 ft. What length of arc is cut off by a central angle of measure

(a) 36° **(b)** 30° **(c)** 54° **(d)** 64 mils

13. On a circle of radius 4 in., what length of arc (to the nearest tenth of an inch) is cut off by a central angle having the following measure? (Use $3.14 \doteq \pi$.)

(a) 36° **(b)** 14° **(c)** 1.3 radians **(d)** 2.63 radians

14. Use your protractor and ruler to find the length of each arc to the nearest tenth of an inch.

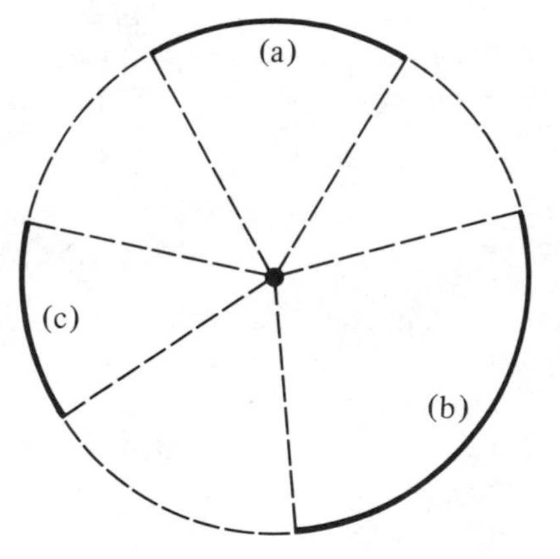

15. (a) If a circle has radius r (feet), what is its circumference (in feet)?

(b) If a central angle in this circle has radian measure 1.34, what length of arc does it cut off?

(c) If a central angle in this circle has degree measure 32, what length of arc does it cut off?

(d) Compare the computations you performed in parts (b) and (c). Is there any reason for preferring the radian over the degree as unit of angle measure?

16. Suppose that a circle has radius r and a central angle of that circle has radian measure m. Then the length ℓ of arc cut off by that central angle is related to r and m by the formula _________ = _________.

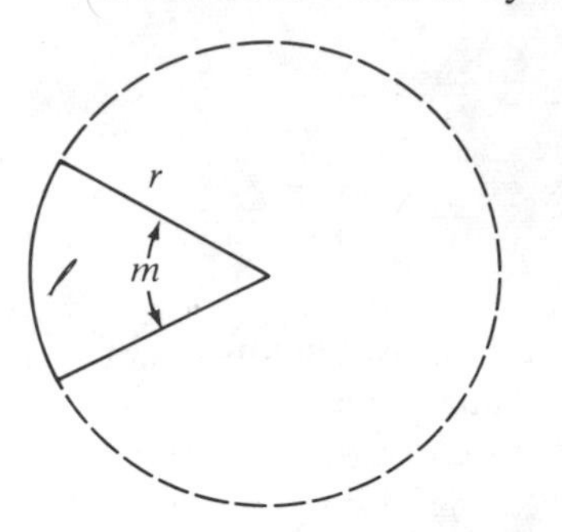

17. (a) If the minute hand on a clock moves steadily rather than in jerks, how long does it take it to rotate through an angle of $1°$?

(b) Through an angle of $1'$?

18. If the clock has a sweep second hand, how long does it take this hand to rotate through an angle of $1''$?

19. (a) Through what size angle does the hour hand rotate in 1 hour?

(b) 1 minute?

(c) 1 second?

20. (a) If a circle has radius 1000 yd, show that the length of arc cut off by a central angle of 1 mil is just about 1 yd.

(b) An artillery piece is firing at a target 5000 yd away. A forward observer reports that the range is fine, but that the shells are landing 100 yd to the left (relative to the gunner) of the target. How many mils to the right should the muzzle be rotated?

5.4 APPROXIMATE MEASUREMENT

When we studied the *theory* of measurement in Chapter 4, we worked in the *world of mathematics*. In that world we chose a unit segment and then considered other segments of length *exactly* 2 units, exactly 4 units, exactly $\frac{1}{3}$ unit, and so on. In this chapter we are studying the *use* of measurement in the *real world*. In the real world nothing ever seems to turn out exactly as advertised. The "twelve-inch" ruler you trustingly bought at the drugstore is undoubtedly *not exactly* 12 in. long. It is probably closer to $11\frac{255}{256}$ or $12\frac{1}{256}$ in. long. The "$8\frac{1}{2} \times 11$" paper you write on is not exactly $8\frac{1}{2}$ in. wide and 11 in. long. The "two by fours" you buy at the lumber yard are certainly not exactly 2 in. thick and 4 in. wide. They are more nearly $1\frac{5}{8}$ in. thick and $3\frac{5}{8}$ in. wide. The dimensions 2 in. and 4 in. are *approximations* to the nearest inch.

The dimensions $1\frac{5}{8}$ in. and $3\frac{5}{8}$ in. are approximations to the nearest eighth inch.

Measuring any real thing (including a drawing on a piece of paper) produces only an approximation.

Since we have to live in the real world, we need to work out some means of coexisting with errors and approximations. The way we do this is to accompany each approximate measurement with a note telling how "good" the approximation is. For example, we may report the length of the segment below as

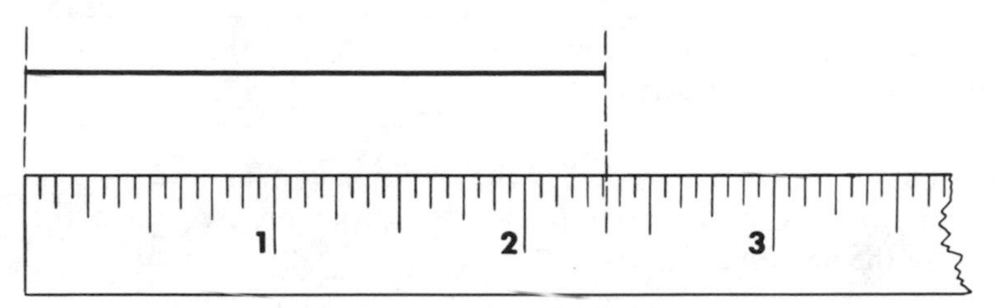

2 in., to the nearest inch
$2\frac{1}{2}$ in., to the nearest half inch
$2\frac{1}{4}$ in., to the nearest quarter inch
$2\frac{3}{8}$ in., to the nearest eighth inch
$2\frac{5}{16}$ in., to the nearest sixteenth inch

When we report its length as "2 in., to the nearest inch" we are choosing the inch as our unit of measure and asserting that the length of the segment is nearer to 2 in. than to either 1 in. or 3 in. This means that the actual length of the segment is between $1\frac{1}{2}$ in. and $2\frac{1}{2}$ in. That is,

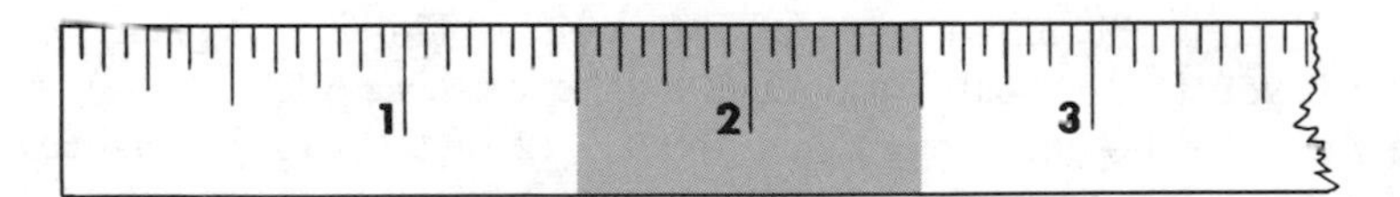

the actual length of the segment (somewhere between $1\frac{1}{2}$ in. and $2\frac{1}{2}$ in.) differs from the measurement we report (2 in.) by no more than *half* of the unit we chose to measure in (the inch).

The greatest possible difference between actual measurement and reported measurement is called the **greatest possible error** (**g.p.e.**) of the (reported) measurement.

In this case the greatest possible error is $\frac{1}{2}$ in.

When we report its length as "$2\frac{3}{8}$ in., to the nearest eighth inch" we are choosing the eighth inch as our unit of measure and asserting that the length of the segment is nearer to $2\frac{3}{8}$ in. than to either $2\frac{2}{8}$ in. or $2\frac{4}{8}$ in. This means that the actual length of the segment is between $2\frac{3}{8} - \frac{1}{16} = 2\frac{5}{16}$ in. and $2\frac{3}{8} + \frac{1}{16} = 2\frac{7}{16}$ in.

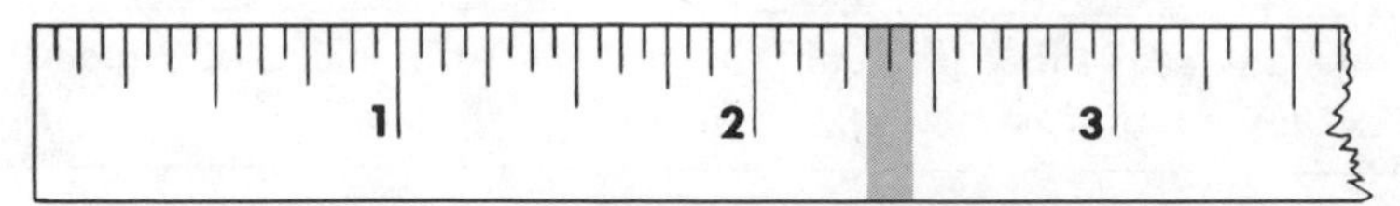

That is, the actual length of the segment (somewhere between $2\frac{5}{16}$ in. and $2\frac{7}{16}$ in.) differs from the measurement we report ($2\frac{3}{8}$ in.) by no more than *half* of the unit we chose to measure in (the eighth inch). The greatest possible error in this measurement is $\frac{1}{16}$ in.

The greatest possible error of a measurement is half of the (smallest) unit of measure used.

This smallest unit used is sometimes called the **precision unit**, and we say that the measurement to the nearest eighth inch is "more precise" than the measurement to the nearest inch.

The smaller the unit of measure used, the **more precise** the measurement.

When we measure small things, it is more important that we be precise than when we measure large things. You probably report your weight to the nearest pound and no one considers you sloppy or dishonest for doing so. The weight of a newborn baby, however, is usually reported to the nearest ounce. A proud mother with a 6 lb 7 oz baby might be upset if you said that it weighed 6 lb.

If you measure your weight in pounds and report it as 92 lb, then the g.p.e. is $\frac{1}{2}$ lb, which is only about 1/184 of your weight. This is a negligible fraction. But if you weigh a baby in pounds and report it as 6 lb, then, although the g.p.e. is still $\frac{1}{2}$ lb, it represents about 1/12 of the baby's weight (1/11 if the baby weighs $5\frac{1}{2}$ lb, 1/13 if the baby weighs $6\frac{1}{2}$ lb). The error in approximating the baby's weight to the nearest pound is relatively more serious than the error in approximating your weight to the same unit.

The **relative error** of a measurement is the fraction,

$$\frac{\text{g.p.e.}}{\text{measurement}}$$

The relative error in the measurement of your weight in pounds is

$$\frac{\frac{1}{2}}{92} = \frac{1}{184}$$

The relative error in the measurement of the baby's weight in pounds is

$$\frac{\frac{1}{2}}{6} = \frac{1}{12}$$

Sometimes the fraction, g.p.e./measurement, is converted to a percent. Then it is called the **percent of error**. The percent of error in the measurement of your weight is about 0.54%. The percent of error in the measurement of the baby's weight is about 8.33%. We say that the measurement of your weight is "more accurate" than the measurement of the baby's weight because the relative error (or percent of error) is smaller.

The smaller the relative error, the **more accurate** the measurement.

EXAMPLE If the length of a sword is reported as "$32\frac{1}{2}$ in. to the nearest quarter inch," then:

the unit of measure is the quarter inch

the g.p.e. is $\frac{1}{2} \times \frac{1}{4}$ in. $= \frac{1}{8}$ in.

the relative error is $\frac{\text{g.p.e.}}{\text{measurement}} = \frac{\frac{1}{8}}{32\frac{1}{2}\text{ in}} = \frac{\frac{1}{8}}{\frac{65}{2}}$

$= \frac{1}{8} \times \frac{2}{65} = \frac{1}{260}$

the percent of error is $\frac{100}{260}\% \doteq 0.4\%$

EXERCISES

1. Measure this segment to the nearest inch, $\frac{1}{2}$ in., $\frac{1}{4}$ in., $\frac{1}{8}$ in., $\frac{1}{16}$ in.

2. Fill in this chart.

REPORTED MEASUREMENT	g.p.e.	RELATIVE ERROR	PERCENT OF ERROR TO THE NEAREST TENTH OF A PERCENT
(a) "10 ft, to the nearest foot"			
(b) "32 in., to the nearest inch"			
(c) "40 in., to the nearest quarter inch"			

REPORTED MEASUREMENT	g.p.e.	RELATIVE ERROR	PERCENT OF ERROR TO THE NEAREST TENTH OF A PERCENT
(d) "16 in.^{2}, to the nearest square inch"			
(e) "6 gallons, to the nearest quart"			
(f) "10 yd, to the nearest inch"			
(g) "$16\frac{3}{4}$ kazooties, to the nearest glimp" (There are 64 glimps in 1 kazootie.)			
(h) "275 km to the nearest kilometer"			
(i) "6 tons to the nearest hundreds pounds"			
(j) "200 cc to the nearest cubic centimeter"			

3. Arrange the measurements (a)–(j) of Exercise 2 in order of increasing accuracy.

4. Can the measurements (a)–(j) be arranged in order of increasing precision?

5. Arrange the linear measurements of Exercise 2 in order of increasing precision. (The kazootie is not a unit of linear measure.)

6. Joe reports his height as 5′2″ to the nearest inch.

(a) Find the g.p.e. in his measurement.

(b) Find the relative error.

(c) Find the approximate percent of error.

Joe reports the length of his foot as $8\frac{3}{4}$ in. to the nearest quarter inch.

(d) Find the g.p.e. in this measurement.

(e) Find the relative error.

(f) Find the approximate percent of error.

(g) Which of Joe's two measurements is more precise?

(h) Which is more accurate?

7. The length and width of a porch are reported, to the nearest foot, as 12 ft and 7 ft respectively. Assume that the person who did the measuring made no mistakes.

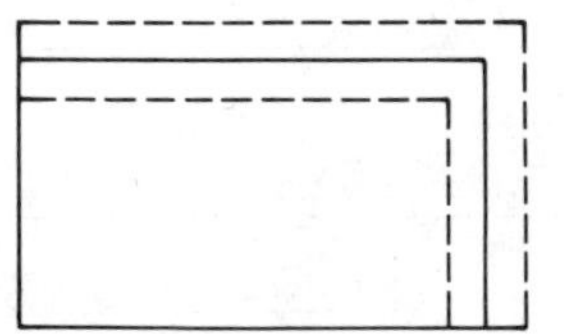

(a) How short could the porch be?
(b) How long?
(c) How narrow?
(d) How wide?
(e) How small, in square feet, could the porch be?
(f) How large?
(g) What is the greatest possible error in reporting the area to be $12 \times 7 = 84$ ft^2?

8. A reliable man reports that a box has dimensions 20 in. by 14 in. by 10 in., all dimensions to the nearest inch.
(a) What is the largest volume the box might have?
(b) What is the smallest?
(c) What is the greatest possible error in reporting the volume to be $20 \times 14 \times 10 = 2800$ in.^{3}?

5.5 REPORTING APPROXIMATE MEASUREMENT: THE BASIC AGREEMENT

In the previous section we remarked that we really should "accompany each approximate measurement we report with a note telling how 'good' the approximation is." The safest, most easily understood way to do this is to actually name the (smallest) unit in which the measurement was made (rounded off).

5 ft $8\frac{3}{4}$ in. to the nearest quarter inch
$6\frac{3}{4}$ in. to the nearest sixteenth inch
100 m to the nearest tenth of a meter

This safe, sure way of reporting approximations involves quite a bit of writing. People who do a lot of work with measurement have come to an agreement that makes this much writing unnecessary. We illustrate the agreement with an example.

EXAMPLE When we measure the length of an object to the nearest *sixteenth inch*, and it turns out to be $6\frac{12}{16}$ in., we should report it as $6\frac{12}{16}$ in. We should *not* report it as $6\frac{3}{4}$ in. or 6.75 in. The agreement is this.

The measurement of an object is reported in the same (smallest) unit in which it was made.

As long as everyone follows this rule, it is not necessary to state in words what smallest unit of measurement (precision unit) was used. It appears right in the reported measurement!

EXAMPLE If the length of a pencil is reported as $6\frac{3}{4}$ in., we assume that the measurement was made to the nearest *quarter inch*. We know then that the g.p.e. is $\frac{1}{2} \times \frac{1}{4} = \frac{1}{8}$ in., that the actual length of the pencil is between $6\frac{5}{8}$ in. and $6\frac{7}{8}$ in., and that the relative error in the measurement, $6\frac{3}{4}$ in., is

$$\frac{\frac{1}{8}}{6\frac{3}{4}} = \frac{1}{54}$$

EXAMPLE If the length of a pencil is reported as 6.75 in. (that is, $6\frac{75}{100}$ in.), we assume that the measurement was made to the nearest *hundredth* of an inch. Thus the g.p.e. is $\frac{1}{200} = \frac{5}{1000}$ in., the actual length of the pencil is between 6.745 in. and 6.755 in., and the relative error in the measurement, 6.75 in., is

$$\frac{\frac{1}{200}}{6.75} = \frac{1}{1350}$$

EXAMPLE How do you report that a pole is 10 ft long to the nearest tenth of a foot?

Solution 10.0 ft

EXAMPLE How do you report that a pole is 10 feet long to the nearest inch?

Solution 10 ft 0 in.

EXAMPLE What can you conclude from the reported measurement, 6.3400 m?

Solution

(1) Since $6.3400 = 6\frac{3400}{10{,}000}$, the measurement was made in *ten-thousandths* of a meter,
(2) the g.p.e. is $\frac{1}{20{,}000} = \frac{5}{100{,}000}$ m,
(3) the actual measurement is between 6.33995 m and 6.34005 m,
(4) the relative error in the measurement, 6.3400 m is

$$\frac{1/20{,}000}{6.3400} = \frac{1}{2 \times 63{,}400} = \frac{1}{126{,}800} \doteq .000008$$

We sometimes say that the measurement 6.3400 m has five **significant digits**. Each of the digits 6, 3, 4, 0, 0 represents a certain (whole) number of units (precision units) in which the measurement was made. Each of the digits is important. None can be left out. The reported measurement 6.340 m would carry with it very different information. In what (smallest) unit was this measurement made? What is the g.p.e.? Between what two numbers does the actual measurement fall? What is the relative error?

EXERCISES

1. Measure this segment to the nearest
   (a) inch
   (b) $\frac{1}{2}$ in.
   (c) $\frac{1}{4}$ in.
   (d) $\frac{1}{8}$ in.
   (e) $\frac{1}{16}$ in.

and report each of these measurements in the manner agreed upon.

2. Using an ordinary ruler, would it be honest to report the length of a pencil as $5\frac{64}{128}$ in.?

3. Rewrite each of these measurements using the agreement and no words.
   (a) $6\frac{3}{4}$ in. to the nearest sixteenth inch
   (b) 100 m to the nearest tenth of a meter
   (c) 32 in. to the nearest quarter inch
   (d) 8.6 m to the nearest centimeter
   (e) $5\frac{1}{2}$ gallons to the nearest pint
   (f) $16\frac{3}{4}$ kazooties to the nearest glimp
   (Recall that there are 64 glimps in 1 kazootie.)

4. The reported measurement is 6.3 m.
   (a) In what smallest unit was this measurement made?
   (That is, what is the precision unit?)
   (b) What is the g.p.e.?
   (c) Between what two values does the actual measurement fall?
   (d) What is the relative error?

5. Answer the same four questions for the reported measurement 6.34 m.

6. Answer the same four questions for the reported measurement 6.34000 m.

7. When a single object is measured several times (in meters), which measurement is more precise, one with many significant digits or one with few? Which is more accurate?

8. Fill in the chart.

REPORTED MEASUREMENT	PRECISION UNIT	g.p.e.	RELATIVE ERROR
(a) 14 in.			
(b) 65 cc			
(c) $3\frac{16}{32}$ in.			
(d) 7.4 yd^2			
(e) 5 ft $2\frac{3}{4}$ in.			

REPORTED MEASUREMENT	PRECISION UNIT	g.p.e.	RELATIVE ERROR
(f) 15.7°			
(g) 24°8′			
(h) 6.045 gallons			
(i) 9 days, 7 hr, 30 min, and 15 sec			
(j) 1.5 volts			

9. Use your protractor to measure each angle below to the nearest degree. Report the measurements using the agreement on how to report. What is the g.p.e. in each case? What is the relative error in each case?

10. By means of a reflected laser beam the average distance from Earth to Moon has now been determined as 226,970.9 miles. Find the relative error in this measurement.

5.6 REPORTING APPROXIMATE MEASUREMENT: PLUS-OR-MINUS AND INTERVAL NOTATION (OPTIONAL)

Plus-or-minus notation is another way of reporting how good an approximate measurement is. You may have heard expressions of the sort, "my car is 12 ft long, give or take a foot." Plus-or-minus notation is a symbolic way of reporting such a "give-or-take" measurement.

length of car $= 12 \pm 1$ ft (twelve plus or minus one foot)

Actually, plus-or-minus notation is a kind of mathematical slang and is not to be translated literally. The above expression does *not* mean "length of car $= 12 + 1 = 13$ ft or length of car $= 12 - 1 = 11$ ft." What it does mean is that the length of the car is *between* 11 and 13 ft. A less barbarous way of symbolizing this information is in **interval-of-measure** notation:

11 ft $\leq$ length $\leq$ 13 ft

Notice that the length of the car, to the nearest foot, is *not* necessarily 12. The length to the nearest foot could be 11, 12, or 13 ft.

1 2 3 4 5 6 7 8 9 10 11 12 13 14

The greatest possible error you could make by using 12 ft for the length of the car would be 1 ft, *not* $\frac{1}{2}$ ft. The relative error would be $\frac{1}{12}$.

EXAMPLE If the diameter of a pipe is reported as $63.542 \pm .001$ cm, then the interval of measure is

$$63.541 \text{ cm} \leq \text{diameter} \leq 63.543 \text{ cm}$$

The greatest possible error in using 63.542 as diameter is .001, and the relative error is

$$\frac{.001}{63.542} = \frac{1}{63{,}542} \doteq .00002$$

EXAMPLE

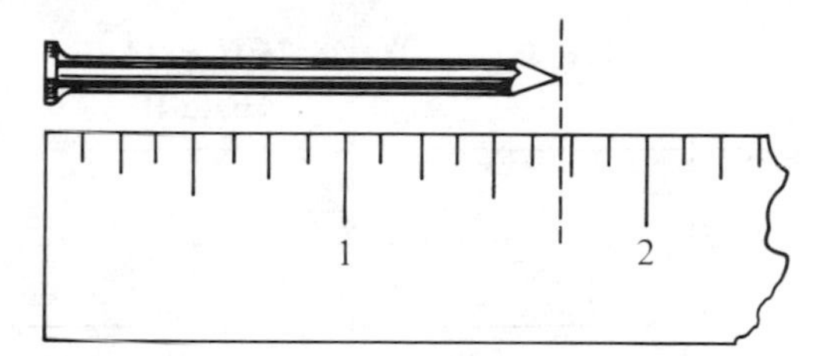

What is the interval of measure of the nail? What is its length in "plus-or-minus" notation? What is the g.p.e. and the relative error in this latter reported measurement?

Solution

$$1\tfrac{5}{8} \text{ in.} < \text{length of nail} < 1\tfrac{6}{8} \text{ in.}$$

$$\text{length} = (1\tfrac{11}{16} \pm \tfrac{1}{16}) \text{ in.}$$

$$\text{g.p.e.} = \tfrac{1}{16} \text{ in.}$$

$$\text{rel. error} = \frac{\frac{1}{16}}{1\frac{11}{16}} = \frac{1}{27}$$

EXERCISES

1. You have a 12-in. shoe and you step off the length of a car. Eleven steps are not enough; twelve are too many.
(a) What interval of measure would you report for the car?

(b) How would you report its length in plus-or-minus notation?

2. Rewrite each measurement M in plus-or-minus notation.

(a) $12 \text{ ft} \le M \le 16 \text{ ft}$
(b) $10.3 \text{ in.} \le M \le 10.9 \text{ in.}$
(c) $44° \le M \le 45°$
(d) $3.5 \text{ yd} \le M \le 3.6 \text{ yd}$
(e) $8\frac{1}{2} \text{ in.} \le M \le 9\frac{1}{4} \text{ in.}$

3. Rewrite each measurement M in interval notation.

(a) $M = 17 \pm .1 \text{ m}$
(b) $M = 240 \pm 5 \text{ mi}$
(c) $M = 25.3 \pm .07 \text{ cc}$
(d) $M = 180 \pm 2 \text{ acres}$
(e) $M = 25° \pm 10'$

4. Fill in the chart.

REPORTED MEASUREMENT	PRECISION UNIT	g.p.e.	ACTUAL MEASURE M IN ± NOTATION	ACTUAL MEASURE IN INTERVAL NOTATION
Example: 6.4 ft	$\frac{1}{10} = .1$ ft	$\frac{1}{20} = .05$ ft	$M = 6.4 \pm .05$ ft	$6.35 \text{ ft} \le M \le 6.45 \text{ ft}$
(a) 16 mm				
(b) $9\frac{8}{16}$ in.				
(c) 2.7 cm^2				
(d) 24.7°				
(e) 4 lb 5 oz				
(f) 2 hr 25 min				

5. Give the g.p.e. and the relative error for each measurement.

(a) $M = 24 \pm 1$ in.
(b) $M = 1.43 \pm .08$ m
(c) $M = 100 \pm 3$ yd

6. Suppose that you know your shoe is $10 \pm \frac{1}{2}$ in. long. Suppose that you step off the length of a building and get 80 steps.

(a) What is the length of the building, in ± notation?

Suppose that you step off its width and get 48 steps.

(b) What is its width, in ± notation?
(c) How many square feet does it occupy, in interval notation?

7. You are told that a field is 118 ft long and 26 ft wide.

(a) What (smallest) unit of measure was used in the two measurements?
(b) What is the g.p.e. in each measurement?
(c) Express the length, ℓ, in interval notation.
(d) Express the width, w, in interval notation.
(e) How large could the area of the field be?
(f) How small could the area of the field be?
(g) Express the area in interval notation.
(h) Express the area in $\pm$ notation.
(i) Could the area differ from $26 \times 118 = 3068$ ft^2 by more than $\frac{1}{2}$ ft^2?
(j) Is it technically correct to say that the area is 26×118 ft^2?

8. The dimensions of a cardboard box are reported to be 6 in. by 8 in. by 12 in.
(a) What (smallest) unit of measure was used in the three measurements?
(b) Express the volume of the box in interval notation.
(c) Express the volume of the box in $\pm$ notation.
(d) Is it technically correct to say that the volume is $6 \times 8 \times 12$ in.^{3}?

9. Absolute value notation is closely related to plus-or-minus and interval notation, and is also used sometimes in reporting measurement. Fill in the table.

$\pm$ NOTATION	INTERVAL NOTATION	ABSOLUTE VALUE NOTATION
Example: $M = 8 \pm 1$ m	$7 \text{ m} \leq M \leq 9 \text{ m}$	$\lvert M - 8 \rvert \leq 1$ (m)
(a) $M = 5.4 \pm 1$ ft		
(b)	$40° \leq M \leq 45°$	
(c)		$\lvert M - 2.6 \rvert \leq .1$ (cm)
(d)	$\$900 \leq M \leq \$1000$	

5.7 SCIENTIFIC NOTATION (OPTIONAL)

Another convenient way of reporting approximate measurement, particularly when the measurement is very small or very large, is by means of scientific notation. To understand scientific notation it is necessary first to recall the meanings of such symbols as

$$10^4, \quad 10^0, \quad 10^{-3}$$

The entire symbol "10^4" is referred to as the fourth **power** of 10, the "4" by itself is referred to as the **exponent** and the "10" by itself is referred to as the **base**.

Positive integer exponents signify repeated multiplication:

$$10^4 = 10 \times 10 \times 10 \times 10 \text{ (4 factors of 10)} = 10{,}000$$

$$10^{12} = 10 \times 10 \times \cdots \times 10 \text{ (12 factors of 10)} = 1{,}000{,}000{,}000{,}000$$

By agreement, the *zeroth* power of any nonzero number is one:

$$10^0 = 1$$

The examples below suggest how *negative* integer exponents are interpreted.

$$10^{-4} = \frac{1}{10^4} = \frac{1}{10{,}000} = .0001$$

$$10^{-12} = \frac{1}{10^{12}} = \frac{1}{1{,}000{,}000{,}000{,}000} = .000\ 000\ 000\ 001$$

The general definitions of integral powers (of the base 10) are these:

---

1. If n is a positive integer,

$$10^n = 10 \times 10 \times \cdots \times 10 \text{ } (n \text{ factors}) = \underbrace{100 \cdots 0}_{n \text{ zeros}}$$

2. $10^0 = 1$

3. If n is a positive integer,

$$10^{-n} = \frac{1}{10^n} = \underbrace{.00 \cdots 01}_{n - 1 \text{ zeros}}$$

One basic property that makes working with powers a pleasure is this: To *multiply* two powers of the same base you need only *add* the exponents (and retain the base). In symbols (using the base 10),

---

$10^m \cdot 10^n = 10^{m+n}$ for any integers m and n

For example,

$$10^4 \cdot 10^3 = 10^{4+3} = 10^7$$

$$10^2 \cdot 10^{-5} = 10^{2-5} = 10^{-3}$$

$$10^{-4} \cdot 10^4 = 10^{-4+4} = 10^0$$

A second property that gives practical importance to working with powers of 10 is that:

Every positive (terminating) decimal is expressible as a decimal between 1 and 10 times an integer power of 10.

For example,

$$275.43 = 2.7543 \times 10^2$$
$$5{,}463{,}288 = 5.463288 \times 10^6$$
$$2.75 = 2.75 \times 10^0$$
$$.044 = 4.4 \times 10^{-2}$$
$$.00000096 = 9.6 \times 10^{-7}$$

When a number is expressed as the product of a decimal between 1 and 10 and a power of 10, we say that it is expressed in **scientific notation.**

Irrational numbers and repeating decimal rationals cannot be expressed exactly in scientific notation, but approximations to them can. For example,

$$\tfrac{1}{3} = .333\ldots \doteq 3.33 \times 10^{-1}$$
$$\pi \doteq 3.1416 = 3.1416 \times 10^0$$
$$\tfrac{1}{11} = .0909\ldots \doteq 9.09 \times 10^{-2}$$

One of the major reasons for using scientific notation is exactly this usefulness as a means of reporting approximate measurement. Another reason is, of course, that it is a concise way of expressing very large and very small numbers.

EXAMPLE Compute $30{,}000{,}000{,}000 \times .000002$.

Solution

$$30{,}000{,}000{,}000 = 3 \times 10^{10}$$
$$.000002 = 2 \times 10^{-6}$$
$$3 \times 10^{10} \times 2 \times 10^{-6} = 3 \times 2 \times 10^4 = 60{,}000$$

Whenever a measurement is reported in scientific notation, the precision of that measurement is conveyed automatically by the scientific notation. We illustrate with examples how this is done.

EXAMPLE When the diameter of the earth is reported to be 4.2×10^7 ft, it is to be understood that the "4" and the "2" are the significant digits;

that is,

the diameter is closer to 4.2×10^7 ft than to 4.1×10^7 ft or 4.3×10^7 ft.

In interval notation,

$$4.15 \times 10^7 \text{ ft} \leq \text{diameter of earth} \leq 4.25 \times 10^7 \text{ ft, or}$$

$$41{,}500{,}000 \text{ ft} \leq \text{diameter of earth} \leq 42{,}500{,}000 \text{ ft}$$

In plus-or-minus notation,

$$\text{diameter of earth} = (4.2 \pm .05) \times 10^7 \text{ ft}$$
$$= 42{,}000{,}000 \pm 500{,}000 \text{ ft}$$

But no matter what notation is used,

$$\text{g.p.e.} = .05 \times 10^7 = 500{,}000 \text{ ft}$$

$$\text{rel. error} = \frac{.05 \times 10^7}{4.2 \times 10^7} = \frac{.05}{4.2} = \frac{1}{84}$$

$$\text{percent of error} \doteq 1.2\%$$

EXAMPLE When the diameter of a hydrogen atom is reported to be 3.48×10^{-10} ft, it is to be understood that

the "3," "4," and "8" are significant digits;

that is,

the diameter is closer to 3.48×10^{-10} ft than to 3.47×10^{-10} ft or 3.49×10^{-10} ft.

In interval notation,

$$3.475 \times 10^{-10} \text{ f} \leq \text{diameter of hydrogen atom} \leq 3.845 \times 10^{-10} \text{ ft}$$

In plus-or-minus notation,

$$\text{diameter of hydrogen atom} = (3.48 \pm .005) \times 10^{-10} \text{ ft}$$

For this measurement,

$$\text{g.p.e.} = .005 \times 10^{-10} = .000\,000\,000\,000\,5 \text{ ft}$$

$$\text{rel. error} = \frac{.005 \times 10^{-10}}{3.48 \times 10^{-10}} = \frac{5}{3480} = \frac{1}{696}$$

percent of error $\doteq 0.14\%$

The two previous examples can be used to illustrate the value of scientific notation in computations. Suppose that we want to compare the diameter of the earth to the diameter of a hydrogen atom.

$$\frac{\text{diam. of earth}}{\text{diam. of hydrogen atom}} \doteq \frac{4.2 \times 10^7}{3.48 \times 10^{-10}} = \frac{4.2}{3.48} \times 10^7 \times 10^{10}$$

$$= \frac{4.2}{3.48} \times 10^{17} \doteq 1.2 \times 10^{17}$$

INTERPRETATION If hydrogen atoms could be laid side by side, a chain made up of

120,000,000,000,000,000

of them would be about as long as a diameter of the earth.

Notice that the answer above was written with just two significant digits. The reason why is this. Even though the division of 4.2 by 3.48 could be carried out to give approximations 1.21, 1.207, . . . , which seem to be better than 1.2, these values cannot be relied on. The reason is that the quotient 4.2/3.48 is itself derived from approximations. The divisor could be as large as 3.485 or as small as 3.475; the dividend could be as small as 4.15 or as large as 4.25. Thus the quotient could be as small as

$$\frac{4.15}{3.485} = 1.19^{+}$$

or as large as

$$\frac{4.25}{3.485} = 1.23^{-}$$

Thus it *is* safe to say that the quotient is 1.2 to the nearest tenth, since it must lie between 1.15 and 1.25; but it is *not* safe to say that the quotient is 1.21 to the nearest hundredth, since it need *not* lie between 1.205 and 1.215.

The most careful sort of statement that could be made is that

$$\frac{4.15}{3.485} \times 10^{17} \le \frac{\text{diam. of earth}}{\text{diam. of hydrogen atom}} \le \frac{4.25}{3.475} \times 10^{17}$$

or

$$1.19 \times 10^{17} \le \frac{\text{diam. of earth}}{\text{diam. of hydrogen atom}} \le 1.23 \times 10^{17}$$

This careful kind of error analysis can be carried out in any specific problem, but it is tedious. A fairly risky rule of thumb is to keep only as many digits in your answer as the least number of significant digits in any of the pieces of data used. That is what we did in our problem. In our problem the worst piece of data (the measurement of the earth's diameter) had just two significant digits, so we wrote our answer using two digits. We were lucky that the rule of thumb worked: As the subsequent careful error analysis showed, both significant digits were justified. Exercise 10 below provides an example where the rule of thumb fails.

EXERCISES

1. Write as a decimal:
(a) 10^6 **(b)** 10^{-6} **(c)** 10^0

2. Write as a power of 10:
(a) .00001 **(b)** 1 **(c)** 1,000,000,000

3. Express in scientific notation:
(a) 315.046 **(b)** .00287 **(c)** .0000002

4. Express as a decimal:
(a) 1.75×10^4 **(b)** 3.1×10^{-7} **(c)** 9×10^{-9}

5. Verify the rule for adding exponents as in the example. (An arbitrary nonzero base b has been used in place of 10.)

EXAMPLE

$$b^5 \cdot b^{-3} = b^5 \cdot \frac{1}{b^3} \quad \text{(interpretation of negative exponents)}$$

$$= (b \cdot b \cdot b \cdot b \cdot b) \cdot \frac{1}{(b \cdot b \cdot b)} \quad \text{(interpretation of positive exponents)}$$

$$= b \cdot b \quad \text{(property of fractions)}$$

$$= b^2 \quad \text{(interpretation of positive exponents)}$$

$$= b^{5-3} \quad \text{(property of integers)}$$

(a) $b^3 \cdot b^2$
(b) $b^3 \cdot b^{-7}$
(c) $b^2 \cdot b^{-2}$
(d) $b^{-2} \cdot b^{-4}$
(e) $b^0 \cdot b^{-3}$

6. Explain why $1/10^{-7} = 10^7$

7. Write as a power of 10:

(a) $\frac{10^{12}}{10^7}$ **(b)** $\frac{10^{-8}}{10^5}$ **(c)** $\frac{10^{-2}}{10^4}$ **(d)** $\frac{10^7}{10^{-3}}$

(e) $10^5 \cdot 10^{-2}$ **(f)** $10^{10} \cdot 10^0$ **(g)** $10^{-5} \cdot 10^{-2}$ **(h)** $10^{-10} \cdot 10^3$

8. Fill in the chart.

	A	B
REPORTED MEASUREMENT	1.25×10^6 cm	4.9×10^{-19} cm
greatest possible error		
relative error		
percent of error		
plus-or-minus notation		
interval notation		

9. Which of the measurements, A or B, in Exercise 8 is more precise? Which is more accurate?

10. A rectangular field is reported to be 1.2×10^2 ft wide and 1.4×10^3 ft long.
(a) Write an interval for its actual width, w.
(b) Write an interval for its actual length, ℓ.
(c) Compute the minimum area the field might have.
(d) Compute the maximum area the field might have.
(e) Verify that $1.2 \times 1.4 = 1.68$.
(f) Explain why "$A = 1.68 \times 10^5$ ft" is a misleading report of the area of the field.
(g) Explain why "$A = 1.7 \times 10^5$ ft" is also a misleading report.
(h) Explain why "$A = 2 \times 10^5$ ft" is not a misleading report.

11. The presence or absence of zeros in scientific notation is extremely important. If the population of a city is reported to be 5.3×10^6 people, then only the "5" and the "3" are significant. We can only be certain that

$$5.25 \times 10^6 \leq \text{population} \leq 5.35 \times 10^6$$

or, in plus-or-minus notation,

$$\text{population} = (5.3 \pm .05) \times 10^6 = 5{,}300{,}000 \pm 50{,}000 \text{ people}$$

If, however, the population is reported to be 5.300×10^6 people, then the "5", "3", and two "0"s are significant. We can infer that

$$5.2995 \times 10^6 \leq \text{population} \leq 5.3005 \times 10^6$$

or, in plus-or-minus notation,

$$\text{population} = (5.300 \pm .0005) \times 10^6 = 5{,}300{,}000 \pm 500 \text{ people}$$

Complete the chart:

POPULATION REPORTED AS	⇒POPULATION BETWEEN	⇒UNIT OF MEASURE (PRECISION UNIT)	⇒g.p.e.	⇒PERCENT OF ERROR
5.3×10^6	5,250,000 and 5,350,000	100,000	50,000	
5.30×10^6				
5.300×10^6	5,299,500 and 5,300,500	1000	500	
5.3000×10^6				
5.30004×10^6				
5.300037×10^6				

5.8 MASS, WEIGHT, AND CAPACITY (OPTIONAL)

Until now, we have concentrated on the measurement of geometric figures or of objects that suggest geometric figures. We have worked mainly with length, area, volume, and angle measure. Occasionally measurements of weight and time were mentioned. Many other things could also have been mentioned. Temperature can be measured in degrees *Fahrenheit* or degrees *Centigrade*. Intensity of light can be measured in *candlepowers*. Intensity of sound can be measured in *decibels*. Work can be measured in *foot-pounds*, power in *horse-powers*. The number of measurable quantities and units of measure is astonishing. A recent edition of the C.R.C. *Handbook of Chemistry and Physics* devotes 52 pages to measurement and units of measure. In this section we shall look briefly at just three closely related measurable quantities: mass, weight, and capacity.

TEACHING NOTE Our reason for dabbling in these areas, which are as much science as they are mathematics, is this. One of the more persistent dreams of educators has been to break down the "artificial barriers" that exist between such closely related school subjects as math and science. Several recent experimental projects have been directed toward producing an integrated math-science curriculum. This movement seems destined to influence the elementary school mathematics curriculum for some time. The future elementary school teacher can expect to find science in the math texts and math in the science texts.

Mass

Roughly speaking, the **mass** of an object is the "amount" of matter in it. One object may have more mass than another because it is larger (a railroad tie has more mass than a baseball bat) or because the matter of which it is composed is more tightly packed (a marble has more mass than a marshmallow). It is easy to decide, using a balance, which of two objects has more mass.

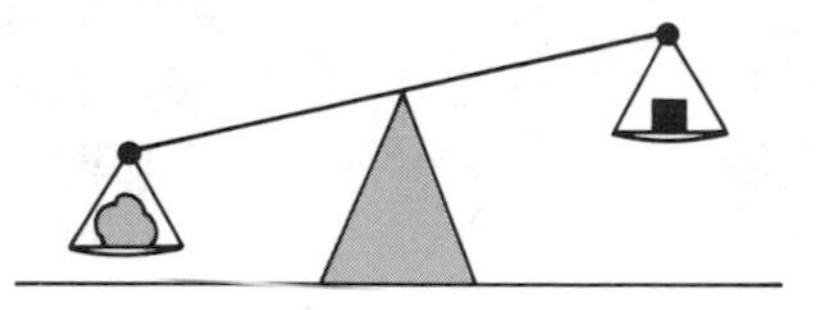

object with more mass | object with less mass

This experiment can be carried out wherever there is gravity. It will work as well on the moon as on earth. It would not work in an orbiting satellite.

The metric unit of mass is the gram. By definition:

One **gram** is the mass of one cubic centimeter of water under standard conditions of temperature (39.2°F) and pressure (sea level).

The other metric units of mass are defined in terms of the gram. For example,

1 milligram (mg) = 1/1000 gram (g) 1 kilogram (kg) = 1000 g 1 metric ton = 1000 kg = 1,000,000 g

Weight

The balance experiment might suggest to you that mass is the same as weight. This is *not* so. Roughly speaking, the **weight** of an object is the force exerted on that object by gravity. Weight is a function of both mass and gravity. When gravity is less, as on the moon, weights will also be less. When gravity is greater, as on Jupiter, weights will also be greater.

EXAMPLE The gravity of the moon is about one-sixth that of the earth. If a 180-lb astronaut were to step on an ordinary spring-operated bathroom scale on the moon, he would weigh about $\frac{1}{6} \times 180 = 30$ lb. His mass, however, would be the same as on earth. The gravity of Jupiter is about three times that of earth. On Jupiter the same astronaut would weigh about $3 \times 180 = 540$ lb. His mass would still be the same as it was on earth, on the moon, or in flight.

The metric units of weight were chosen in a particularly convenient way. The basic metric unit of weight is called the **gram**, the same as the basic unit of mass, and is defined to be the weight of 1 cc of water under standard conditions of temperature and pressure. That is, under those conditions,

1 cc of water has mass 1 gram and weight 1 gram.

The other metric units of weight are defined in terms of the gram exactly as the other units of mass were.

The British units of weight are familiar to most of us already. The pound is defined in terms of the gram. 1 lb ≐ 453.60 g. The other British units of weight are defined in terms of the pound.

BRITISH UNITS OF WEIGHT

1 pound (lb) = 16 ounces (oz)
1 hundredweight (cwt) = 100 lb
1 ton = 2000 lb

Again, we have conversion factors between the two systems:

CONVERSION FACTORS FOR WEIGHT MEASURE

Metric Weights		*British (Avoirdupois) Weights*
1 g	≐	.035 oz
28.35 g	≐	1 oz
.46 kg	≐	1 lb
1 kg	≐	2.20 lb

Capacity

As you have probably come to expect, there is a very sensible system of metric units for liquid capacity, and a colorful but clumsy British system. The basic metric unit is the liter.

A **liter** is defined to be the amount of liquid that would fill a volume of 1000 cubic centimeters.

Since a gram was defined to be the weight (and mass) of 1 cc of water,

1 liter of *water* weighs 1000 grams

The other metric units of liquid measure are defined in terms of the liter using the familiar prefixes. For example, 1 *milliliter* (ml) is 1/1000 liter, or 1 cc of liquid. The British units for liquid measure are these:

BRITISH UNITS OF LIQUID MEASURE

1 gallon = 4 quarts
1 quart = 2 pints
1 pint = 4 gills
1 gill = 7.22 in.^{3}

The conversion factors relating the two systems are as follows:

CONVERSION FACTORS FOR LIQUID MEASURE

1 liter ≐ 1.06 quarts
1 quart ≐ .95 liter

EXERCISES

1. Two identical appearing balls are floating weightlessly in a rocket ship. One is made of lead, the other of aluminum. How might an astronaut decide which had more mass?

2. The standard temperature of 39.2°F was chosen because at that temperature water is most "dense." That means that if 39.2°F water is either warmed or cooled it will expand slightly.
- **(a)** Does 1 cc of 36° water weigh more or less than 1 cc of 39.2° water?
- **(b)** In winter, why does ice first appear at the top of a pond instead of at the bottom?
- **(c)** In summer, where is the coolest water in a pond?

3. How many:
- **(a)** milligrams in a gram
- **(b)** grams in a kilogram
- **(c)** milligram in a kilogram
- **(d)** milliliters in a liter
- **(e)** liters in a kiloliter

4. Fill in the blanks using the tables of conversion factors.
- **(a)** 10 oz ≐ ________ g
- **(b)** 10 kg ≐ ________ lb
- **(c)** 1 metric ton ≐ ________ tons
- **(d)** 110 lb ≐ ________ kg
- **(e)** 650 g ≐ ________ lb ________ oz
- **(f)** 1 gallon ≐ ________ liters

(g) 1 liter ≐ ________ pints
(h) 500 cc ≐ ________ pints

5. Balance scales come equipped with brass weights of various sizes. The weights go in one pan, the object to be weighed in the other.

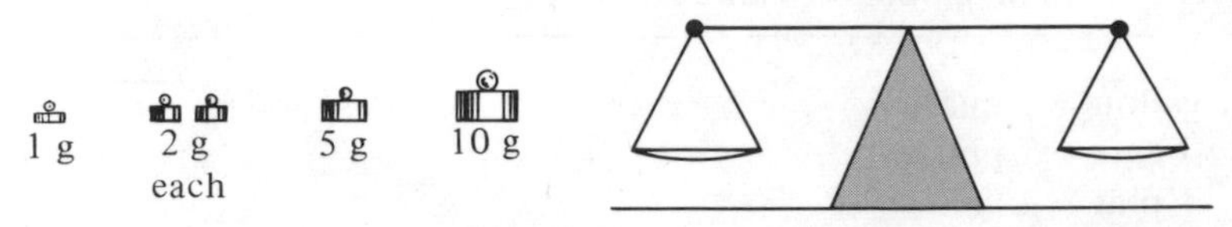

(a) Using only the five weights pictured, can objects of all weights from 1 to 20 g be weighed?
(b) An astronaut takes the balance scale and weights to the moon and weighs a moon rock. Eighteen grams are required to balance the rock. How much will it weigh on earth?
(c) He also takes along a spring scale calibrated in grams. When the same rock is placed on this scale, on the moon, what will the reading be? What will it be on earth?

6. An astronaut weighs 170 lb on earth.
(a) What is his earth weight in kilograms?
(b) What is his mass in kilograms?
(c) What is his moon weight in kilograms?
(d) If the astronaut can carry comfortably a load weighing 15 kg on earth, what earth weight of clothing and equipment can he be expected to carry on the moon? [Assume that his muscles, like the springs in Exercise 5(c), are unaffected by changes in gravity.]

7. Where would you prefer to do the following, on the earth or the moon?
(a) do 100 pushups
(b) diet to lose 5 lb
(c) pole vault
(d) run a low hurdles race

8. How much does a gallon of water weigh in kilograms? in pounds?

9. How many pounds does a liter of water weigh?

10. The **density** of a material is usually given by telling how many grams of it there are in one cubic centimeter. For example,

density of lead ≐ 11 g per cc
density of gold ≐ 19.3 g per cc
density of magnesium ≐ 1.7 g per cc

(a) What is the density of water?
(b) How much does a block of lead with dimensions 5 cm × 8 cm × 2 cm weigh?
(c) If gold costs $105 per ounce, how much does 1 g cost?
(d) How many cubic centimeters of gold will $1.00 buy?
(e) Which would be easier to lift, a 5-in. cube of magnesium on earth or a 5-in. cube of lead on the moon?

(f) It takes about ________ cc of water to weigh 1 g on the moon.

(g) How much does 1 m^3 of gold weigh on earth?

(h) Ebony wood has a density between 1.11 and 1.33 g per cc. Does ebony float? Would ebony float on the moon?

11. If a man lifts a 60-lb weight to a height of 4 ft above the ground, we say he has done $60 \times 4 = 240$ **foot-pounds** (ft-lb) of work. How many foot-pounds of work are done in

(a) raising a 16-lb bowling ball $3\frac{1}{2}$ ft?

(b) throwing a 5-oz baseball 80 ft into the air?

(c) raising a $2\frac{1}{2}$-ton vehicle 7 ft off the ground on a greaserack?

12. Power can be thought of as the rate of doing work. A **horsepower** is the amount of power that will do 550 ft-lb of work per second. Thus, if it takes you 1 sec. to lift the bowling ball of Exercise 11(a) (that is, to do 56 ft-lb of work), then you produce an average of $\frac{56}{550}$, or about 0.1, horsepower during that second.

(a) If it takes you $\frac{1}{4}$ sec to launch the ball of Exercise 11(b), then during that $\frac{1}{4}$ sec you produce an average of ________ horsepower.

(b) If it takes the greaserack of Exercise 11(c) 30 sec to do its lifting, what average power was it expending during that time?

VOCABULARY

5.1 British and Metric Units of Length

British system
metric system
conversion factors
milli-, centi-, deci-, deka-, hecto-, kilo-

5.2 British and Metric Units of Area and Volume

ft^2, cm^3, etc.

5.3 Units of Angle Measure

degree, minute, second
radian
mil
arc
central angle

5.4 Approximate Measurement

greatest possible error, g.p.e.
more precise
percent of error
precision unit
relative error
more accurate

5.5 Reporting Approximate Measurement: The Basic Agreement

significant digits

5.6 Reporting Approximate Measurement: Plus-or-Minus and Interval Notation

plus-or-minus notation
interval of measure

5.7 Scientific Notation

power
exponent
base
scientific notation

5.8 Mass, Weight, and Capacity

mass
weight
liter
foot-pound
gram
capacity
density
horsepower

PROBABILITY

OVERVIEW

In this chapter the theory of measurement that was developed in Chapter 4 and applied in Chapter 5 is extended to include elementary probability. Although not yet in commercial textbooks for the elementary school, the heavy emphasis on probability in all of the present experimental curriculum projects guarantees that elementary probability will soon appear in the K-6 curriculum.

Section 6.1 is a transitional section that does three things. It presents the view of the rational numbers as operators which is so useful later on in working with probability trees. It introduces the technique of "measuring" subsets of a finite set by comparing them to the entire set, which justifies viewing probability as a branch of measure theory. It provides an initial experience in observing the long-range stability of relative frequencies.

The remainder of the chapter, with the exception of Section 6.5, is devoted to the basic concepts of elementary probability theory. These concepts are the ones that one must understand to make sense of the following, over-simplified version of what constitutes a probability problem.

Every probability problem involves an outcome set. The probabilities of these outcomes are determined from either statistical data on past performances of the experiment or by a priori judgements of their likelihood. The probability of an event is equal to the sum of the probabilities of the outcomes that make up that event. To find the probability of an event, then, it is enough to determine what outcomes make it up and what the probability of each such outcome is. In the case of an experiment with equally likely outcomes, it is necessary to know only how many outcomes are in the outcome set and how many are in the event in question.

In Section 6.2 the ideas of outcome set and the probability of an outcome are introduced. Situations are presented that illustrate two ways in which probabilities of outcomes can be decided upon: (1) by a priori judgement based on faith in "honest" coins, "well-shuffled" decks, the shape of spinners, the constituency of urns; (2) by observation of the relative frequencies with which various outcomes occur in many performances of an experiment.

In Section 6.3 the extension to probabilities of events is made. The events are described in set notation, by ordinary sentences (both simple and compound), and by mathematical open sentences.

The use of tree diagrams to represent multistage experiments is illustrated in Section 6.4 and extended in Section 6.5 to a variety of real problems, most of which are not probability problems.

Sections 6.6–6.8 are concerned with the problem of counting large outcome sets and large events, and using the numbers obtained to find probabilities in experiments having equally likely outcomes. Techniques and formulas for counting permutations and combinations are developed and applied.

In Section 6.9 (optional) some simple identities, involving combination numbers, are drawn from Pascal's triangle and then used to give a complete solution to the counting problem of Polya which was posed in Chapter 1 (pp. 30–31).

6.1 COMPARISON, MEASUREMENT, AND POSITIVE RATIONAL NUMBERS

Basically, to measure an object is to compare its size to that of some standard object called the unit of measure. Such comparisons are usually describable by positive rational numbers (represented by fractions).

EXAMPLE The statement "$\overline{AB}$ is $\frac{5}{2}$ in. long" might bring to mind a picture of an inch, partitioned into 2 congruent parts and the segment, $\overline{AB}$, partitioned into 5 such parts.

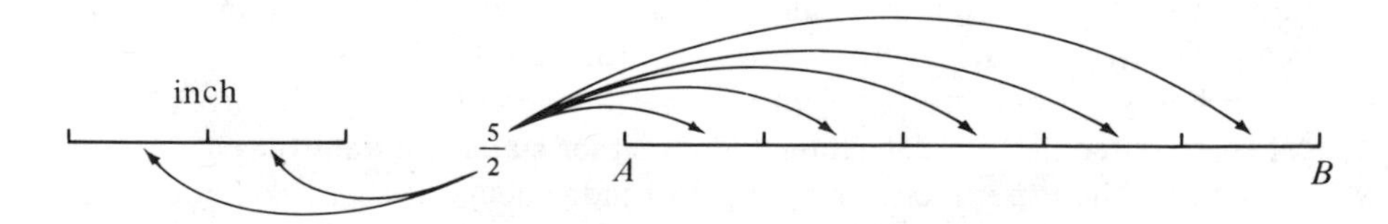

Instead of viewing $\frac{5}{2}$ as above, as passively describing a size comparison between two sets, some people prefer to view $\frac{5}{2}$ as an **operator** actively shrinking and stretching the inch into the segment, $\overline{AB}$.

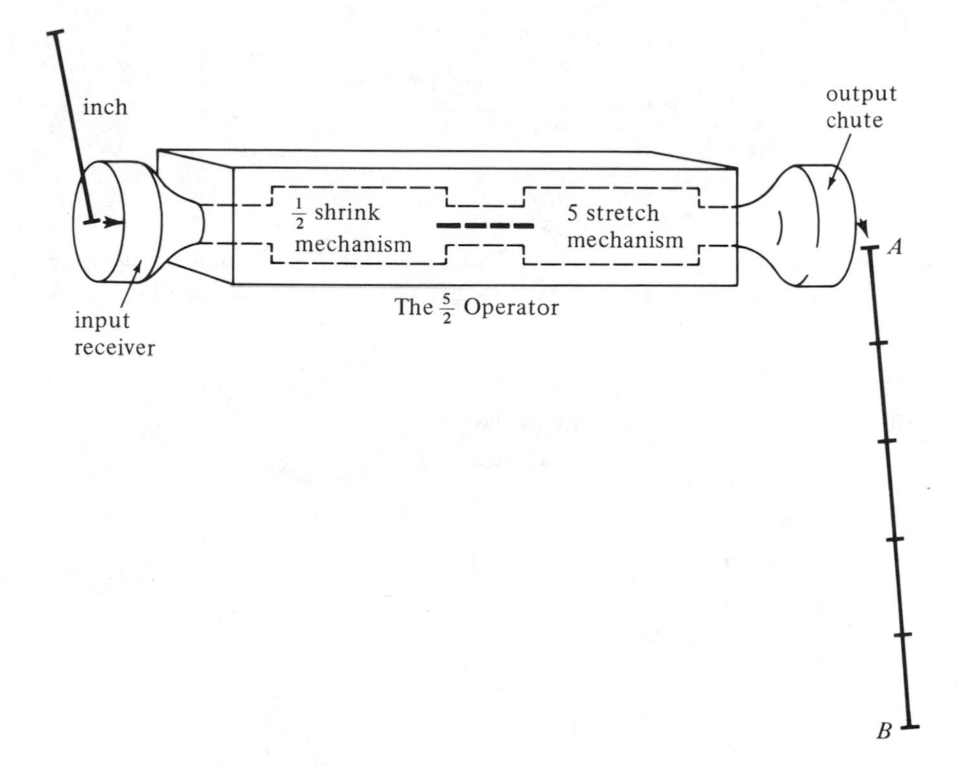

The $\frac{5}{2}$ Operator

EXAMPLE The statement "area of R is $\frac{3}{4}$ in.^{2}" can again bring various pictures to mind:

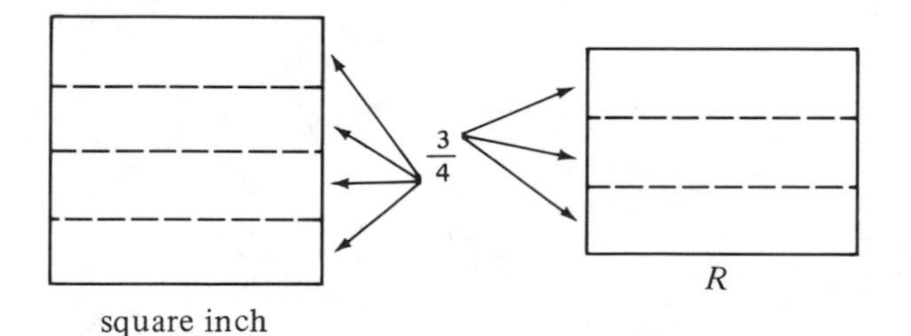

$\frac{3}{4}$ in a static comparison role

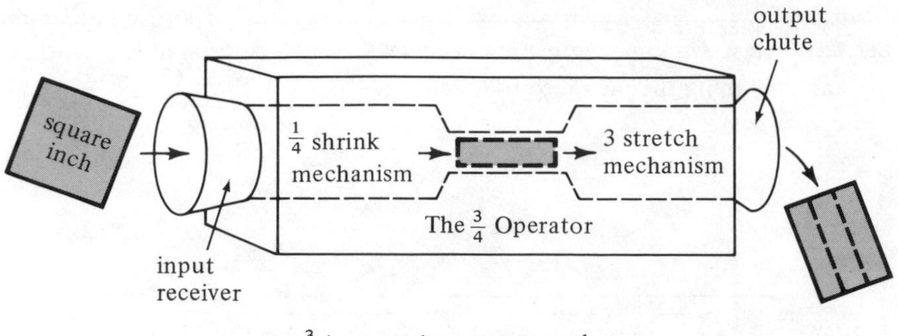

$\frac{3}{4}$ in an active operator role

When two finite sets are being compared for "size" (numerosity), the same basic ideas are present even if we are not used to thinking of this as a measurement situation.

EXAMPLE The statement, "There are two-fifths as many goats as sheep" brings to mind a picture of the sort

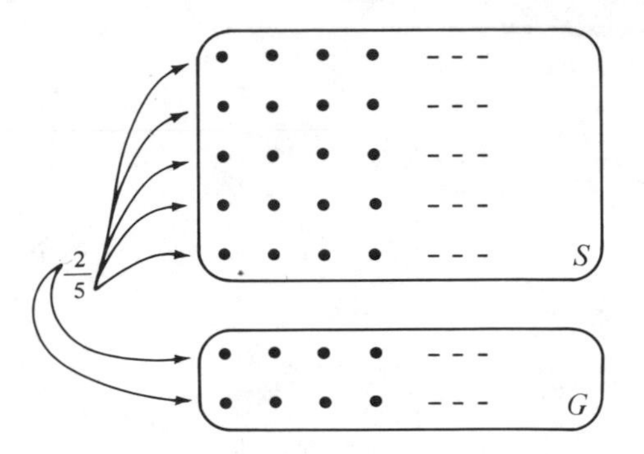

Now the partitioning of the two sets is not into congruent parts, but rather into **equivalent**, or equinumerous, subsets. The shrinking-stretching view is still tenable if shrinking is thought of as a thinning-out process ("keep 1 out of every 5") and stretching as a replicating process ("now duplicate each element").

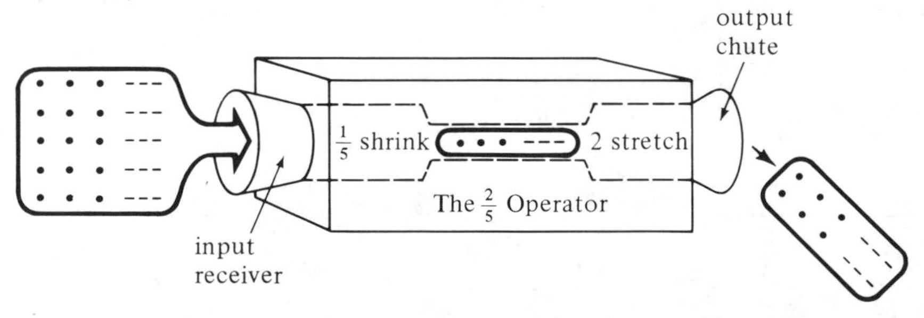

EXAMPLE The statement "two-thirds of the class is girls" is similarly interpretable even though one set is a subset of the other.

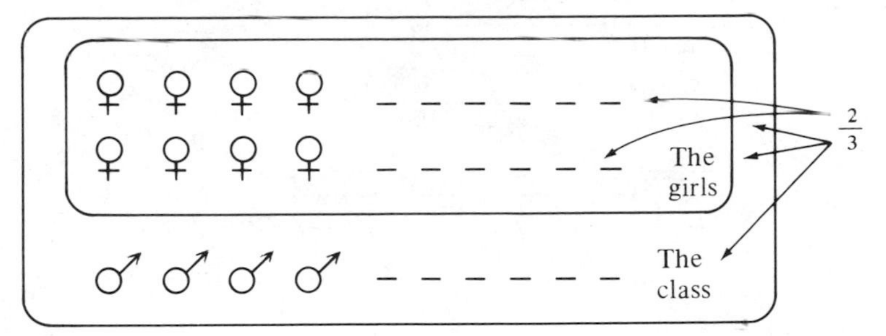

This "measuring" of a subset by comparing it to a superset (a set that includes it) is commonplace. We do it whenever we say such things as

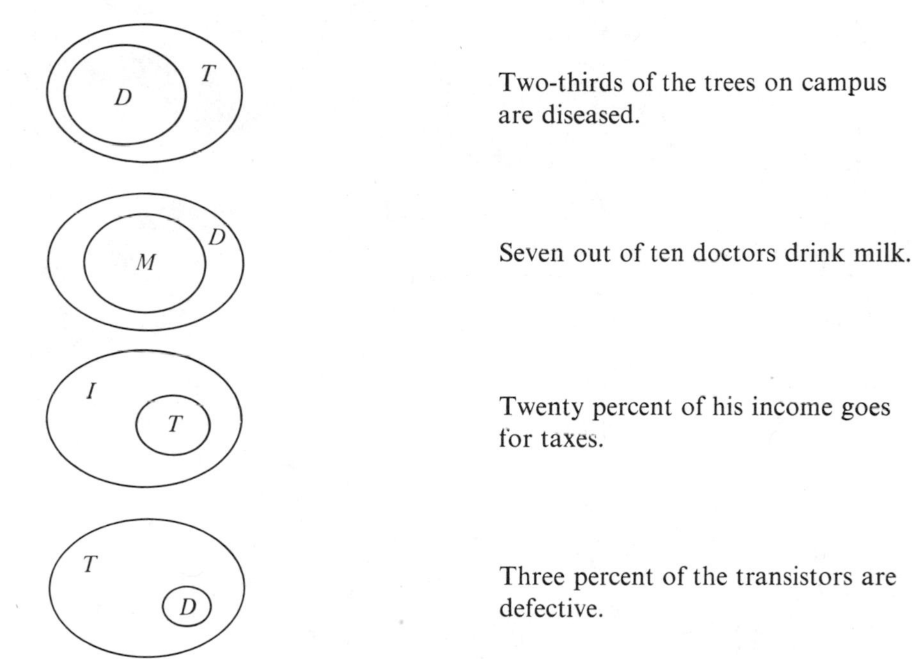

It is this kind of measuring on which elementary probability is based.

EXERCISES

1. Sketch outputs in the following operator diagrams.

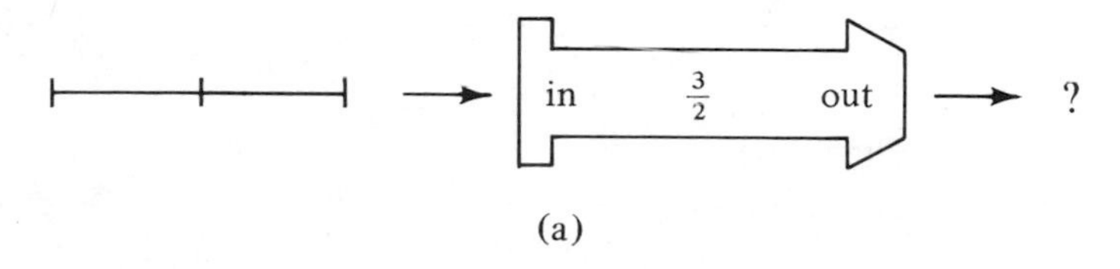

(a)

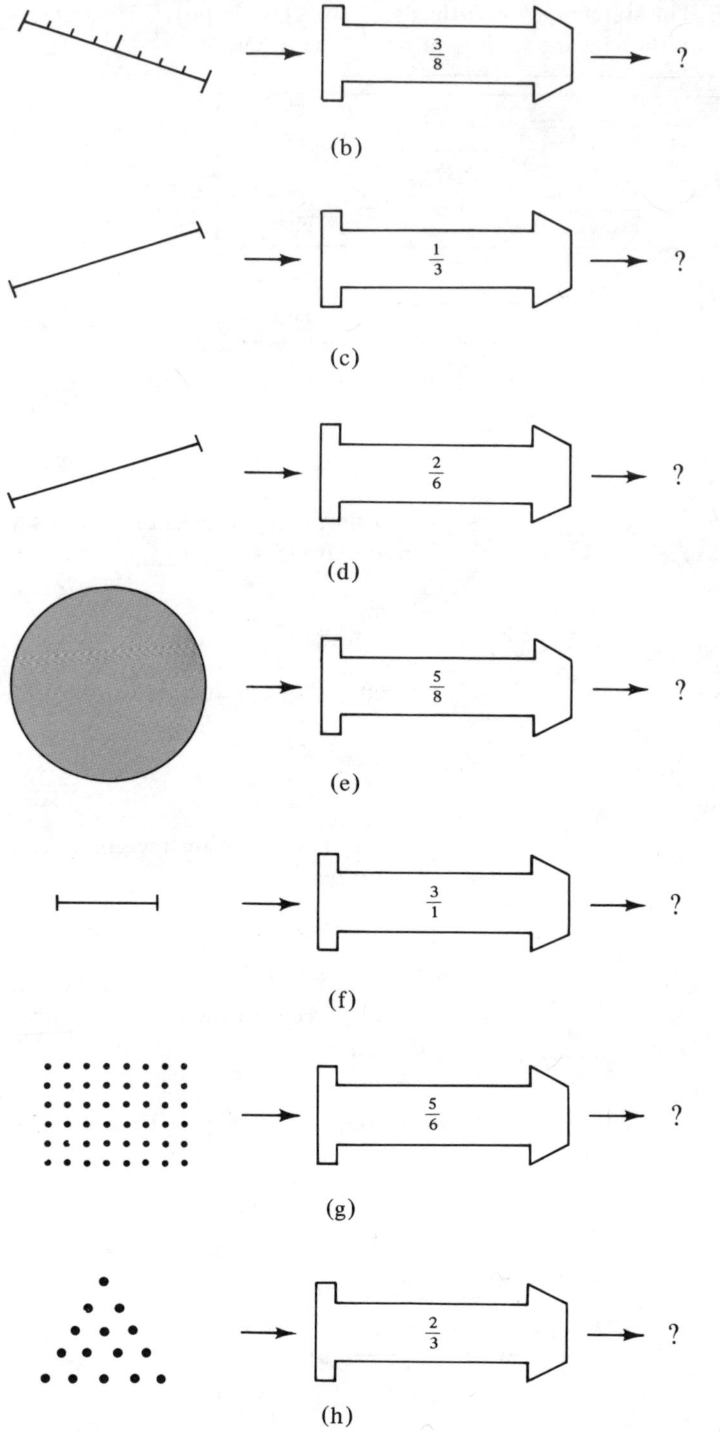
$\frac{3}{8}$
?
(b)
$\frac{1}{3}$
?
(c)
$\frac{2}{6}$
?
(d)
$\frac{5}{8}$
?
(e)
$\frac{3}{1}$
?
(f)
$\frac{5}{6}$
?
(g)
$\frac{2}{3}$
?
(h)

2. (a) What is *in*appropriate about this operator diagram?

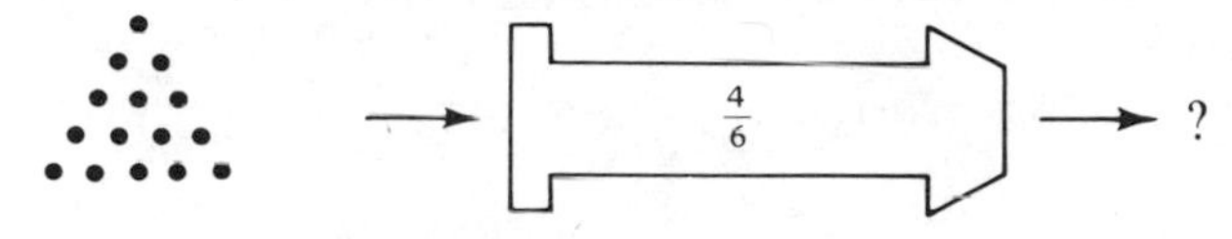

(b) What do you think of this statement:

"Eighty percent of the Great Lakes are contiguous with Michigan."

3. In this exercise operators are stylized as arrows. Fill in whatever is missing: input, output, or operator.

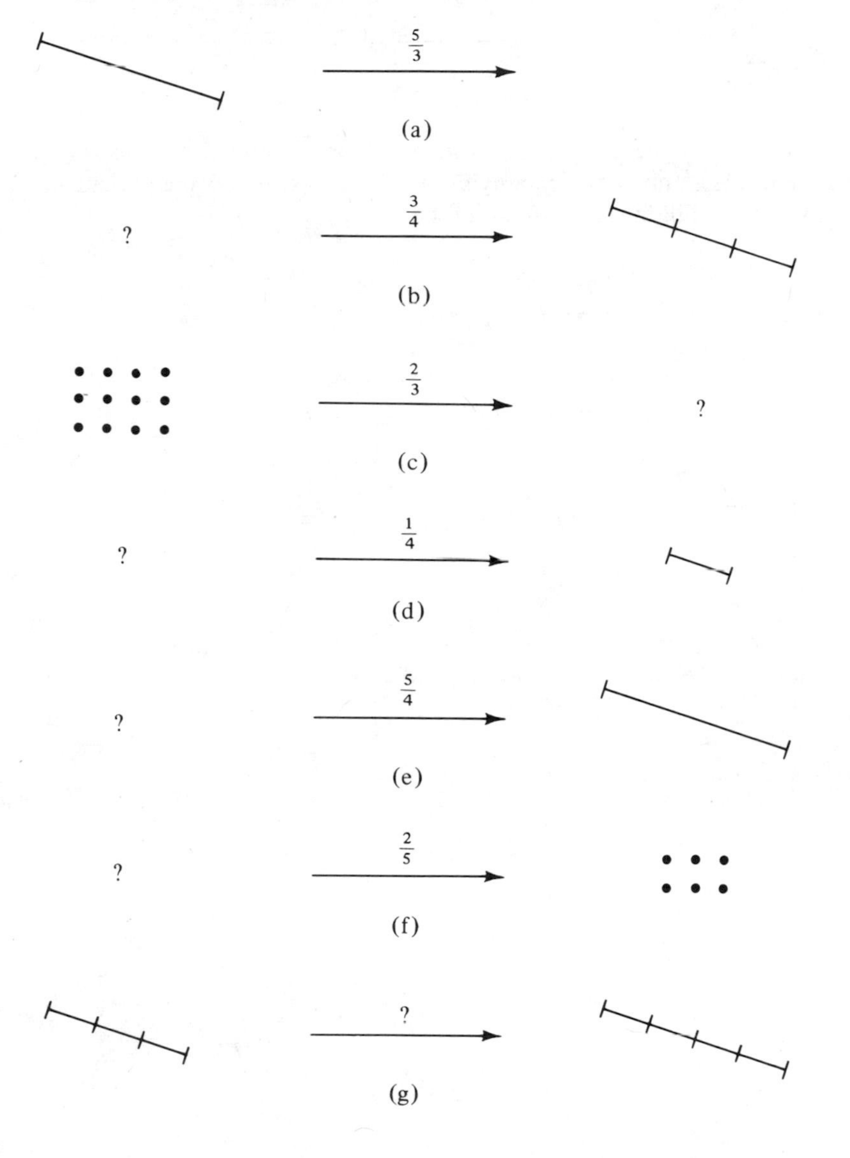

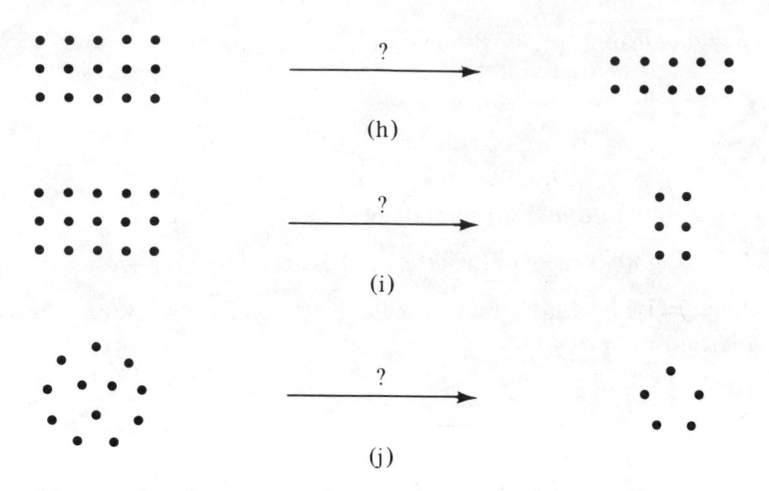

4. For this exercise the arrows represent operators, A and B represent sets, and the whole numbers accompanying A or B represent how many elements are in A or B. Fill in the missing numbers.

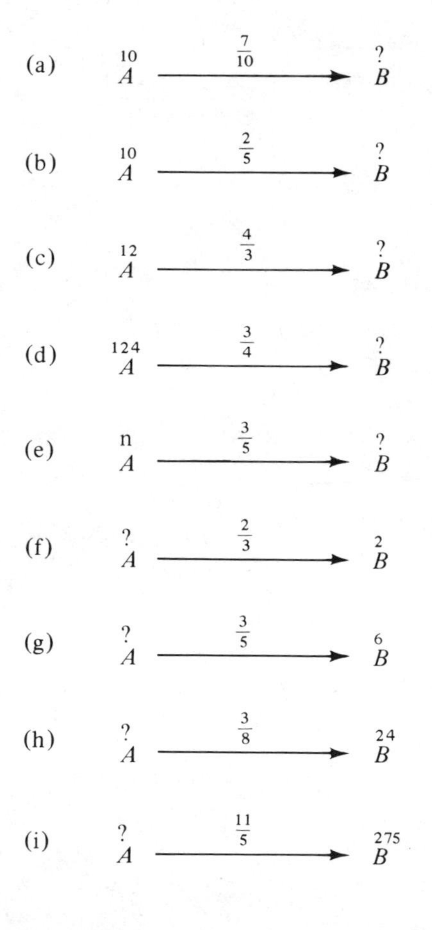

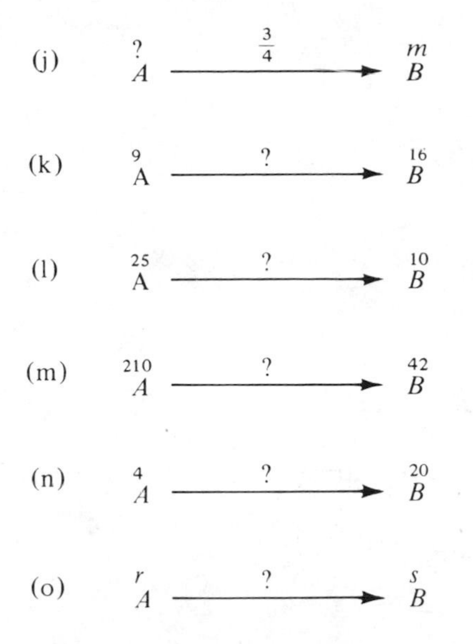

5. Many word problems can be represented by diagrams such as those in Exercise 4 and then solved.

EXAMPLE Three-eighths of the marbles are blue. There are 72 marbles. How many are blue?

Diagram

$$\overset{72}{M} \xrightarrow{\frac{3}{8}} \overset{?}{B}$$

Solution $\frac{3}{8} \cdot 72 = 27$ blue marbles.

Draw diagrams and solve.

(a) Two-thirds of the councilmen are Democrats. There are nine councilmen. How many are Democrats?

(b) In the election, one-third of the eligible voters turned out and 12,600 votes were cast. How many eligible voters are there?

(c) Two out of every five students in the class are juniors. There are 18 juniors. How many students are in the class?

(d) The ratio of males to females in a herd of bison is 2 to 9. There are six males. How many bison are in the herd?

6. For this exercise, U is the 18-element set shown below. Each subset of U can be assigned a rational number that compares it with U. For example, the subset A is assigned the rational number $\frac{7}{18}$. We refer to this number as "the measure of A with respect to U," and write

$$m_U(A) = \frac{7}{18}$$

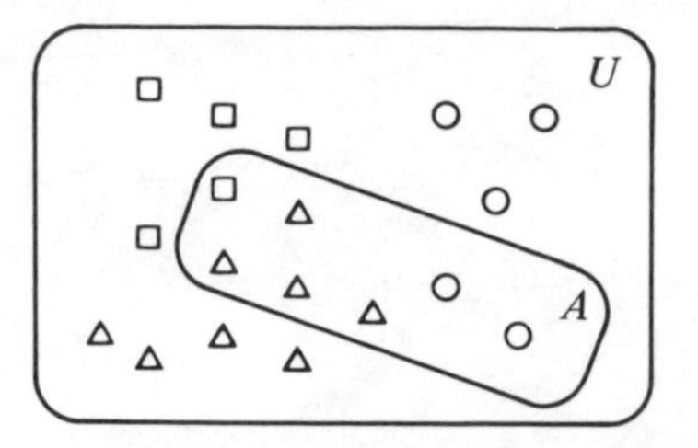

(a) If T is the subset of triangles, $m_U(T) =$ ________.
(b) If C is the subset of circles, $m_U(C) =$ ________.
(c) If S is the subset of squares, $m_U(S) =$ ________.
(d) If P is the subset of polygons, $m_U(P) =$ ________.
We are thinking of m_U as a measure function. Its domain (input set) is the collection of all subsets of U. Here is a partial arrow diagram:

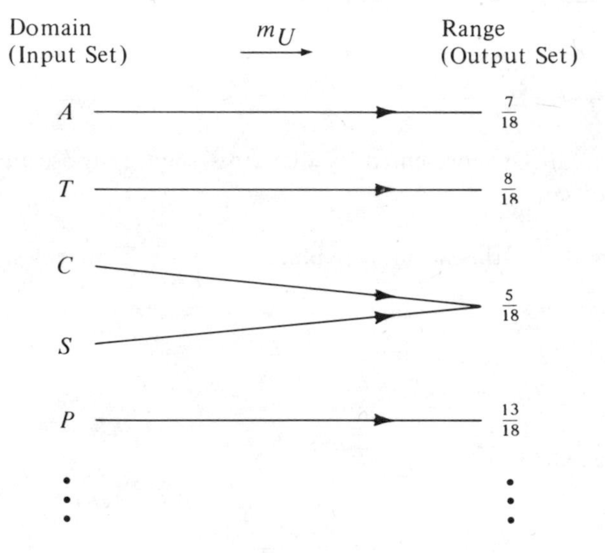

(e) What is the largest possible output? What input leads to it?
(f) What is the smallest possible output? What input leads to it?
(g) How is the range of this measure function m like [unlike] the ranges of the geometric measure functions encountered in Chapter 4?
(h) Does m_U satisfy the "additivity property" described in Chapter 4? Illustrate by completing this chain:

$$m_U(T \cup S) = m_U(P) = \tfrac{13}{18} = \tfrac{8}{18} + \tfrac{5}{18} = \underline{\qquad\qquad}$$

(i) The measure functions of Chapter 4 all have the property that if two of their inputs are *congruent* then they lead to a common output. In symbols,

$$\text{if } A \cong B, \text{ then } m(A) = m(B)$$

In the present context, congruence is not a useful relation between elements (subsets of U); something else is. Complete this statement:

If subset A __________ subset B,
then $m_U(A) = m_U(B)$

7. Think of the paragraph you are presently reading as a set of letters. The first sentence has fifty-six letters in it. Of these, twenty-one are vowels. "Y" is not counted as a vowel. We say that the **relative frequency** of vowels in the first sentence is twenty-one fifty-sixths. This number is the measure of the subset of vowels, where the set of all of the letters has been used as the unit. Complete the following table.

	RELATIVE FREQUENCY OF VOWELS (ROUNDED TO THE NEAREST HUNDREDTH)
in the first sentence	$\frac{21}{56} \doteq .38$
in the first 2 sentences	$\frac{33}{94} \doteq .35$
in the first 3 sentences	
in the first 4 sentences	
in the first 5 sentences	
in the first 6 sentences	
in all 7 sentences	

6.2 EXPERIMENTS: OUTCOMES AND THEIR PROBABILITIES

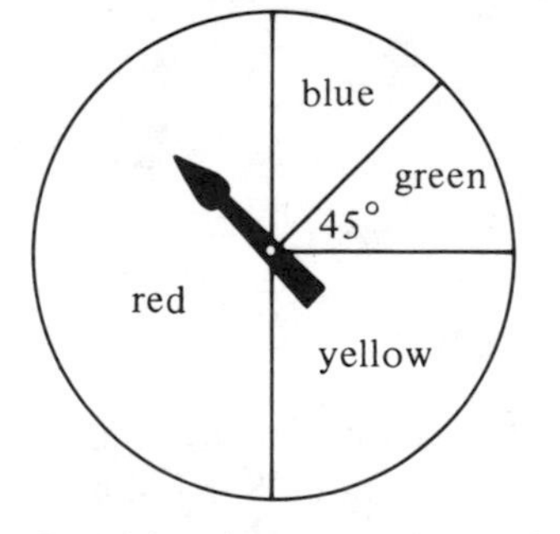

Consider the **random experiment** of spinning this spinner. Each spin is called a **trial**. On each trial there are four possible **outcomes**. The point can stop on red, blue, green, or yellow. The set of all possible outcomes

$\{R, B, G, Y\}$

is called the **outcome set**, or "sample space," for the experiment.

By simply looking at the spinner you probably agree that red is "more likely" than yellow, blue is "less likely" than yellow, while blue and green are "equally likely" outcomes. This is about the same situation we were in back in Chapter 4 when we began our work on measurement. Without the use of numbers we could often make simple decisions of the sort "larger–smaller–same size." or "more numerous–less numerous–equally numerous." Now, as then, we are dissatisfied with such meager information. We want to assign a *number* to each outcome that, in some sense, will measure its likelihood.

The clue for how to do this is in the word "measure." To assign a numerical likelihood to an outcome, we need to compare it with some sort of unit outcome. One way to do it, in this experiment, would be to choose blue as the unit and then decide, by staring at the spinner, that

$$m_B(G) = 1, \quad m_B(Y) = 2, \quad m_B(R) = 4$$

and, of course,

$$m_B(B) = 1$$

Another way would be to choose red as unit:

$$m_R(B) = \tfrac{1}{4}, \quad m_R(G) = \tfrac{1}{4}, \quad m_R(Y) = \tfrac{1}{2}, \quad m_R(R) = 1$$

Yellow could also be chosen:

$$m_Y(B) = \tfrac{1}{2}, \quad m_Y(G) = \tfrac{1}{2}, \quad m_Y(Y) = 1, \quad m_Y(R) = 2$$

The problem is that no single outcome stands out clearly as the most natural one to choose as unit, and until a standard unit is agreed upon we have a Babel of likelihoods for each outcome.

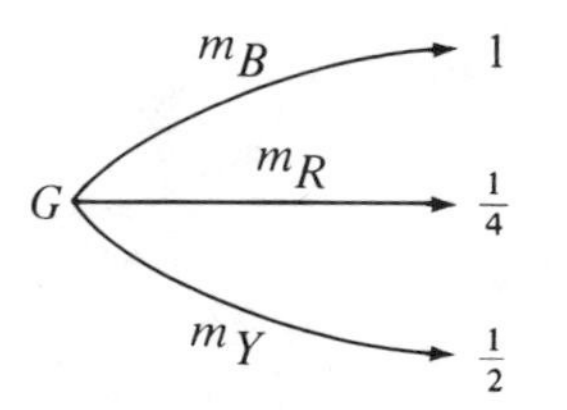

An agreement has been reached on how to choose the unit for this or any other experiment. It is based on the simple observation: Something is bound to happen, so why not measure each outcome against this certainty. Using this certainty as unit, each outcome is assigned a number called its **probability**. For this experiment,

$p(R) = \tfrac{1}{2}$ ("the probability of R is $\tfrac{1}{2}$")

$p(Y) = \tfrac{1}{4}, \quad p(G) = \tfrac{1}{8}, \quad p(B) = \tfrac{1}{8}$

Another way to think about the probability of an outcome is as the relative frequency of that outcome in a large number of trials. It seems reasonable that in a hundred spins of our spinner, the relative frequency of reds would be about $\frac{1}{2}$, that in a thousand spins the relative frequency of reds would be even closer to $\frac{1}{2}$, and that if we imagined more and more trials the relative frequencies of reds would "converge" on $\frac{1}{2}$. Similarly, the relative frequencies of blue, green, and yellow in a large number of trials would be about $\frac{1}{8}$, $\frac{1}{8}$, and $\frac{1}{4}$.

Below is a record of 100 trials of this spinner experiment.

Y B R R Y R B R Y Y R R G R Y R R B R R
G Y R R B G G R Y R B R B Y Y R R R R G
R R R Y R Y R R R Y Y R R Y B Y R G Y R
B B Y R B Y Y Y R R G R R G Y R R B R Y
R Y R R B R B Y G R R G Y Y R R R B G Y

The relative frequencies of red, yellow, blue, and green in the 100 trials are $\frac{48}{100}$, $\frac{27}{100}$, $\frac{14}{100}$, $\frac{11}{100}$, respectively. If we view the record as a set of 100 elements, then $\frac{48}{100}$ is the measure of the subset of R's, where the unit of measure is the entire set. Similarly, $\frac{27}{100}$ is the measure, with respect to the whole set, of the subset of Y's. (*Note*: $\frac{48}{100} \doteq \frac{1}{2}$, $\frac{27}{100} \doteq \frac{1}{4}$, $\frac{14}{100} \doteq \frac{1}{8}$, $\frac{11}{100} \doteq \frac{1}{8}$.)

The two essential aspects of this spinner experiment, the outcome set and the assignment of probabilities to the outcomes, can be described in a single **probability diagram**. A probability diagram is nothing but an arrow diagram for the **probability function** p.

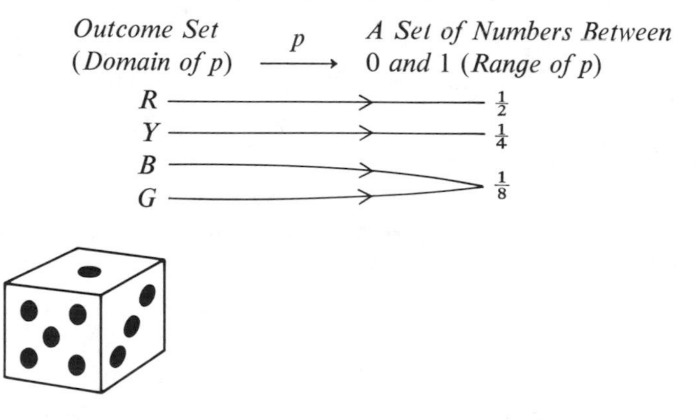

EXAMPLE The experiment consists of rolling a die and recording the number of dots that turn up. The outcome set is $\{1, 2, 3, 4, 5, 6\}$. Each outcome has probability $\frac{1}{6}$. The probability diagram for the experiment can be drawn this way,

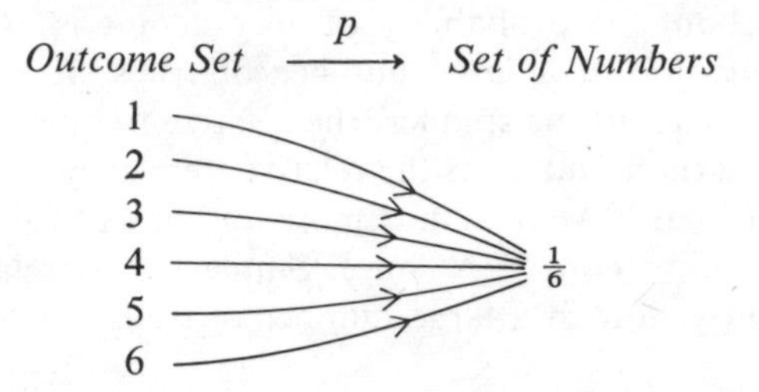

or this way,

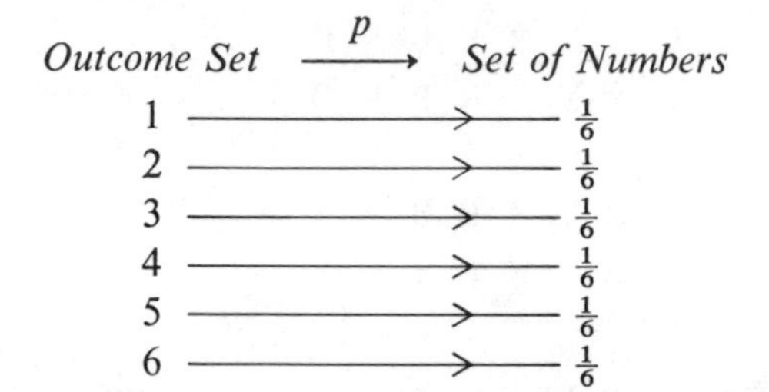

The second drawing violates the canon of mathematical style that, whenever possible, in listing a set one should avoid listing the same element more than once. But it has the virtue of showing very clearly that:

The sum of the probabilities of all of the outcomes of an experiment is 1.

You should get used to both kinds of probability diagrams.

EXERCISES

1. The experiment consists of spinning this spinner.

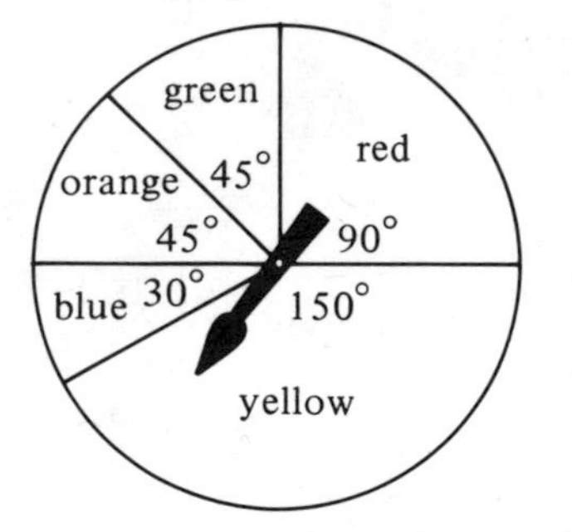

(a) List the outcome set.

(b) If the spinner were spun a million times, about *how many* times would you expect red to turn up? Would you be surprised if 251,423 reds turned up?

(c) In a large number of spins, about what *fraction* of the time would you expect red to turn up? (That is, what do you think the relative frequency of red would be in a large number of trials?) Would you be surprised if, in a million trials, red came up three-eighths of the time?

(d) What probability would you assign to the outcome, Red?

(e) Complete this probability diagram:

$$\textit{Outcome Set} \xrightarrow{\;p\;} \textit{Set of Numbers}$$

(f) Check that the sum of the probabilities of all the outcomes is 1.

2. The experiment consists of spinning this spinner.

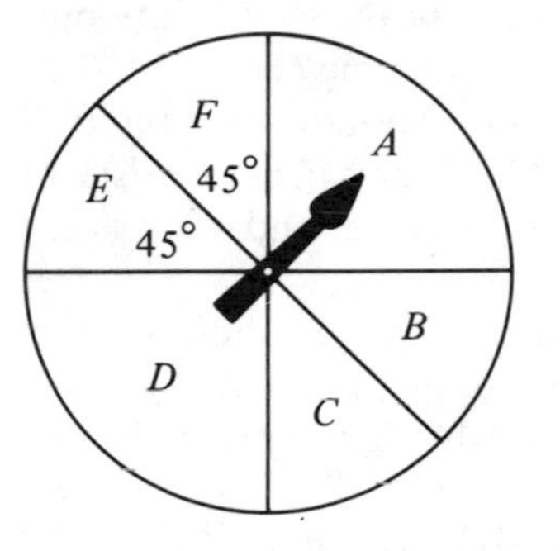

Complete the probability diagram. Check your work by summing the probabilities

$$\textit{Outcome Set} \xrightarrow{\;p\;} \textit{Set of Numbers}$$

3. The experiment consists of flipping a coin.

(a) There are two outcomes. What are they?

(b) Draw a probability diagram for this experiment.

4. In an urn are 7 white and 3 black balls. The experiment is to reach in blindfolded, draw out a ball, unmask and record its color. Draw a probability diagram for this experiment.

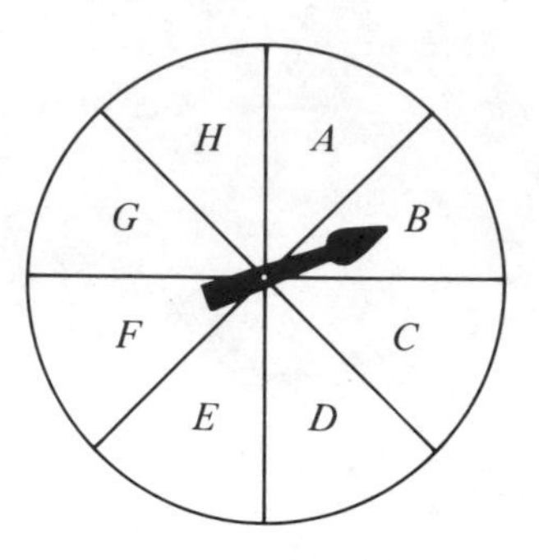

5. The experiment is the same as in Exercise 4, but now there are 3 black, 7 white, and 6 red balls in the urn. Draw a probability diagram.

6. The experiment consists of spinning the spinner above. How many outcomes are there? What is the probability of each outcome?

7. Each letter of the (English) alphabet is written on an individual scrap of paper and the scraps are put into a hat. The experiment consists of drawing out one scrap. How many outcomes are there, and what is the probability of each?

8. The experiment consists of drawing one card from a well-shuffled standard deck. The sample space is {2C, 2D, 2H, 2S, 3C, ..., KS, AC, AD, AH, AS}. How many outcomes are there, and what is the probability of each?

9. For the experiments described so far there has always been one "natural" outcome set, and we referred to it as *the* outcome set. For many experiments several possible outcome sets suggest themselves. For example, if the balls in Exercise 4 were numbered, then two different outcome sets would suggest themselves:

$$\{B, W\} \quad \text{and} \quad \{1, 2, 3, 4, 5, 6, 7, 8, 9, 10\}$$

Draw a probability diagram for this second outcome set.

10. Any set of "happenings" that satisfies the two criteria below can be used as the outcome set for an experiment.

> **(i)** On each trial at least one of the happenings takes place.
> **(ii)** On no trial do two or more of the happenings take place.

Which of these sets of happenings qualify as outcome sets for the experiment of drawing a ball from the urn in Exercise 9?

(a) {odd, even}
(b) {less than 5, more than 5}
(c) {less than 6, more than 4}
(d) {even, prime}

11. The experiment consists of picking a letter, at random, out of this text.

(a) Describe a procedure for doing this.
(b) Using {vowel, consonant} as sample space, draw a probability diagram for the experiment. (Use the results from Exercise 7 p. 230.)
(c) Describe a different outcome set for this same experiment, and describe how you would assign probabilities to its outcomes.

12. A machine produces parts, and an inspector checks if they are good (G) or defective (D). Here is a record of his first 20 observations.

$$G\ G\ D\ G\ D\ G\ G\ G\ G\ D\ D\ G\ D\ G\ D\ G\ G\ D\ D\ G$$

What probability would you assign to the next part's being defective?

13. Of the first 1000 customers at a candy store, 253 were boys under 13, 198 were girls under 13, 307 were boys 13–18, 146 were girls 13–18, and the rest were adults. A customer walks in. What probability (approximately) would you assign to that customer's being

(a) a boy under 13
(b) a girl under 13
(c) a boy 13–18
(d) a girl 13–18
(e) an adult

14. Referring to the record, p. 227, of 100 spins of a spinner, find the following to two decimal places:

	FIRST 20 TRIALS	FIRST 40	FIRST 60	FIRST 80	ALL 100
relative frequency of reds					

15. Flip a coin 100 times and record an H or a T after each toss, for example, $H\ T\ H\ H\ T\ H\ T\ T\ T\ H$ .... Using your data, fill in the table:

	NUMBER OF HEADS	RELATIVE FREQUENCY OF HEADS TO THE NEAREST HUNDREDTH	DIFFERENCE BETWEEN NUMBER OF HEADS AND NUMBER OF TAILS
first 10 flips			
first 20			
first 30			
first 40			
first 50			
first 60			
first 70			
first 80			
first 90			
all 100			

Comment on this statement: "Swamping rather than compensation drives the relative frequency of heads toward $\frac{1}{2}$."

16. The experiment consists of spinning this spinner.

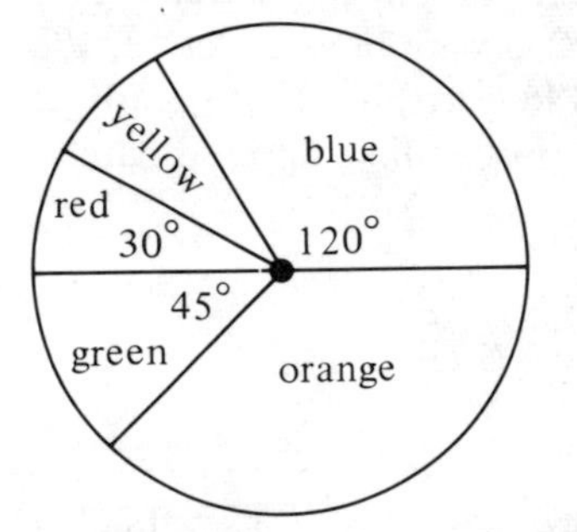

(a) Draw a probability diagram.

(b) Check that the sum of the probabilities of all outcomes is 1.

(c) In appropriate boxes on this "tree," fill in the approximate number of reds, yellows, blues, oranges, and greens you would expect.

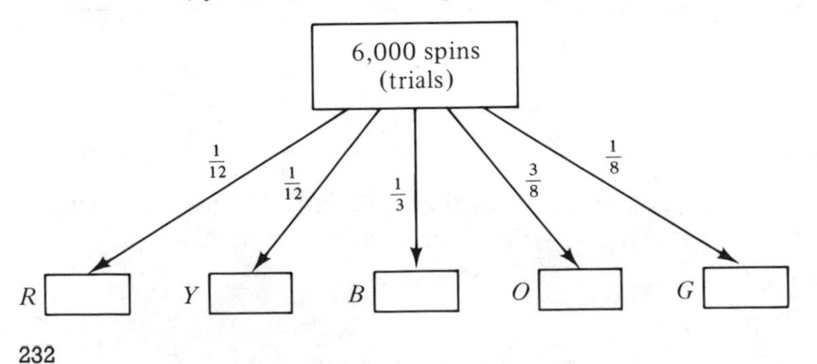

(d) You can think of the probabilities on the branches of the tree as operators. Describe this portion of the tree in terms of stretching and shrinking and inputs and outputs.

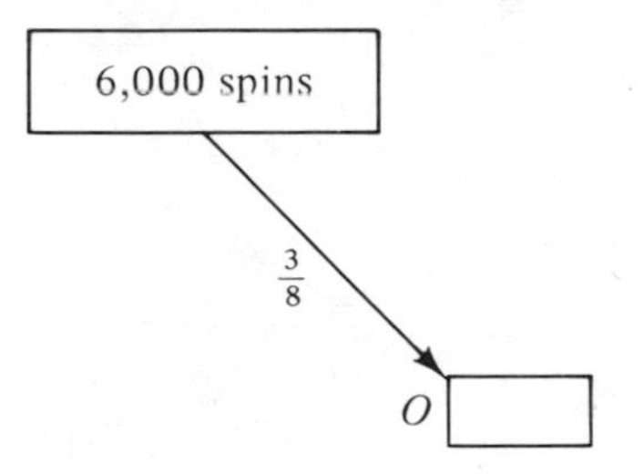

17. We have spent a lot of time on spinner experiments because any random experiment can be **simulated** by a spinner experiment. For example, the toss-a-die experiment can be simulated by this spinner,

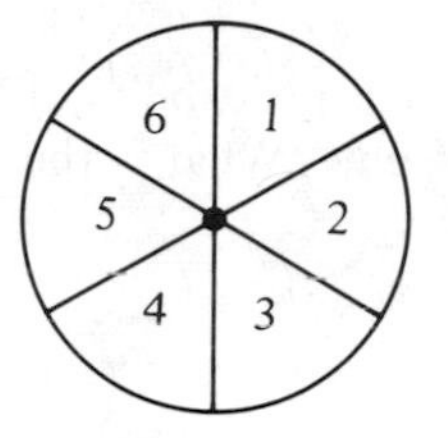

the flip-a-coin experiment by this spinner,

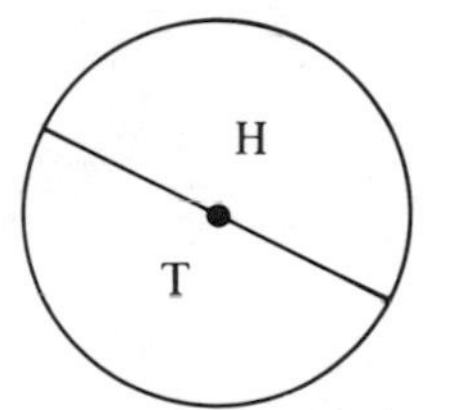

and the draw-a-ball experiment of Exercise 4 by this spinner,

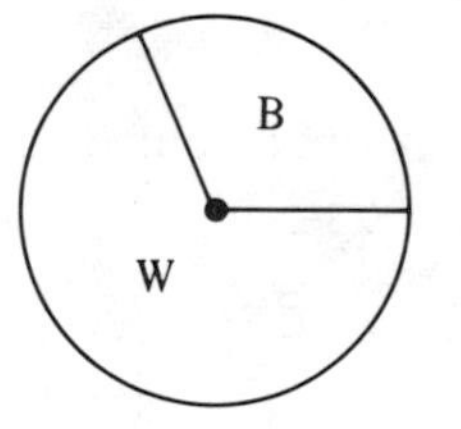

Draw or describe a spinner to simulate the experiments of
(a) Exercise 5
(b) Exercise 7
(c) Exercise 11(b)

18. Simulate the spinner experiment of Exercise 16 by an experiment using colored balls and an urn.

19. A partial probability diagram for a spinner experiment is given here.

Decide on the probability of green and sketch the spinner.

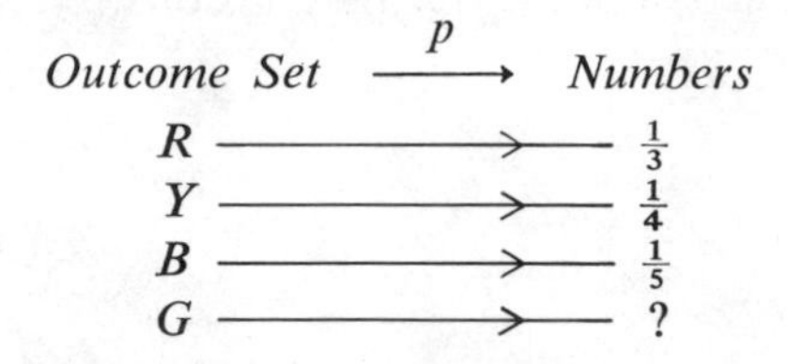

20. Find x and sketch the spinner.

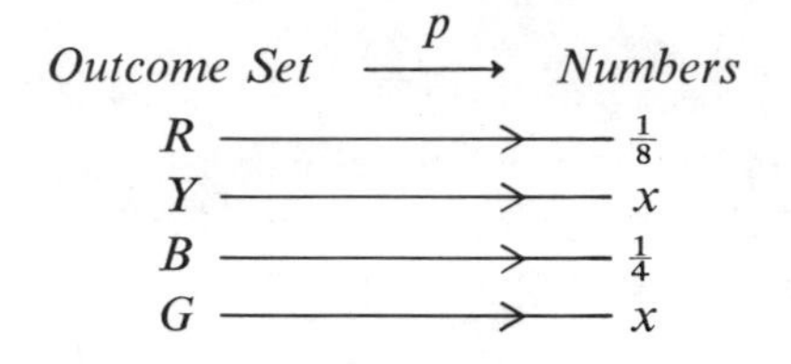

21. A spinner has three colors on it—red, yellow, and blue. The probability of red is twice that of yellow but only two-thirds that of blue. What is the probability of each color?

6.3 EVENTS AND THEIR PROBABILITIES

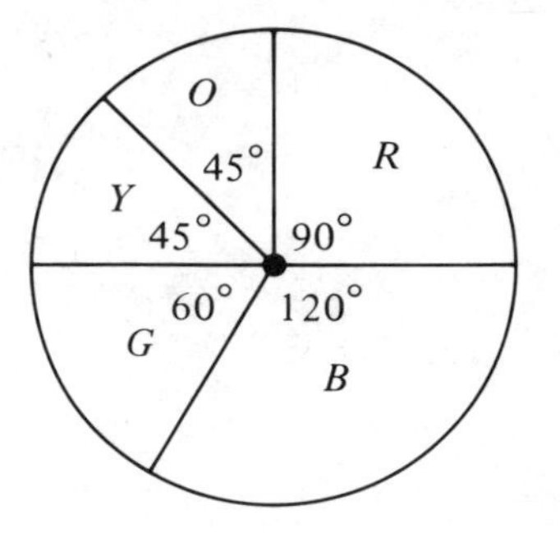

For the experiment of spinning this spinner, there is a "natural" outcome set as indicated in the probability diagram:

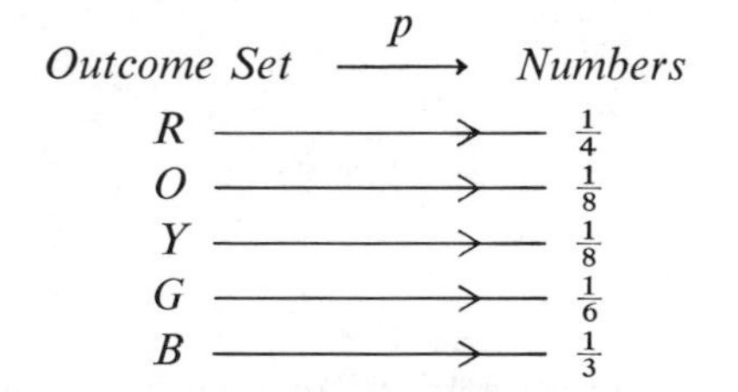

Suppose now that someone asked, "What is the probability of spinning a primary color?" We could scrap the natural outcome set, bring in the new outcome set,

{primary color, not primary color}

and determine the new probability function corresponding to this new outcome set. In deciding what probability to assign to "primary color," we might reason this way. "The primary colors are red, yellow, blue. Since red comes up $\frac{1}{4}$ of the time, blue $\frac{1}{3}$ of the time, and yellow $\frac{1}{8}$ of the time, it follows that a primary color comes up $\frac{1}{4} + \frac{1}{3} + \frac{1}{8} = \frac{17}{24}$ of the time. The probability of 'primary color' is $\frac{17}{24}$."

This reasoning suggests that there is no point in scrapping the old outcome set. The old probability assignments—$p(R) = \frac{1}{4}$, $p(B) = \frac{1}{3}$, $p(Y) = \frac{1}{8}$—were very useful. What we ought to do is make use of the probability function p to assign probabilities not only to the outcomes R, Y, B, O, G themselves, but also to such *subsets* of the outcome set as

"primary color" $= \{R, Y, B\}$
"warm color" $= \{R, O, Y\}$
"cool color" $= \{B, G\}$
"Swedish flag color" $= \{B, Y\}$
...

Any subset of the outcome set is called an **event**.

The procedure for assigning a probability to each event is the obvious one.

$$P(\{R, Y, B\}) = p(R) + p(Y) + p(B) = \tfrac{1}{4} + \tfrac{1}{3} + \tfrac{1}{8} = \tfrac{17}{24}$$

More generally,

The probability of an event is the sum of the probabilities of the outcomes making up that event.

Thus

$$P(\text{warm color}) = P(\{R, O, Y\}) = p(R) + p(O) + p(Y)$$
$$= \tfrac{1}{4} + \tfrac{1}{8} + \tfrac{1}{8} = \tfrac{1}{2}$$
$$P(\text{Swedish flag color}) = P(\{B, Y\}) = p(B) + p(Y) = \tfrac{1}{3} + \tfrac{1}{8} = \tfrac{11}{24}$$

Perhaps you are bothered by the use of both P and p to signify "probability." The reason is that we actually have two probability functions. The original one, p (sometimes called a point probability function), has as its domain the outcome set, $\{R, O, Y, G, B\}$. The new function P has as its domain the set of all 32 subsets of the outcome set:

$$\varnothing, \{R\}, \{O\}, \{Y\}, \{G\}, \{B\}, \{R, O\}, \{R, Y\}, \ldots, \{R, O, Y, G, B\}$$

It is tempting (and not terribly misleading) to think that we have simply enlarged the domain of the probability function p in much the same way as we enlarged the domain of the length (in inches) function ℓ. Having decided on how to assign lengths to segments, we extended the domain of ℓ to the set of all polygonal curves by addition:

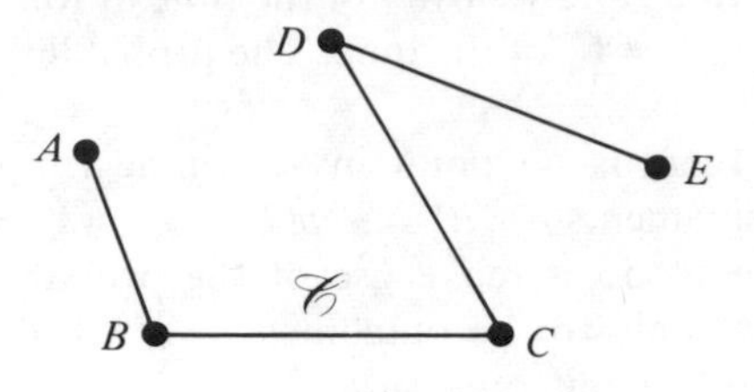

$$\ell(\mathscr{C}) = \ell(\overline{AB}) + \ell(\overline{BC}) + \ell(\overline{CD}) + \ell(\overline{DE})$$

Now, having assigned probabilities to outcomes, we go on to assign probabilities to entire sets of outcomes, again using addition:

$$P(\{R, Y, B\}) = p(R) + p(Y) + p(B)$$

In the case of lengths this was a true enlargement of the domain of ℓ, because each segment is also a polygonal curve. In the present context there is a subtle difference. We have not actually enlarged the domain of p. Technically, we have replaced it by an entirely different domain. Technically, no outcome is also an event:

$$R \neq \{R\}, \quad Y \neq \{Y\}, \quad \ldots$$

It is for this reason that we use a slightly different symbol, P, for the probability function whose domain is the set of all events (subsets of the sample space). Both p and P are called probability functions, however, because even though $\{R\} \neq R$, still $P(\{R\}) = p(R)$. (Why?)

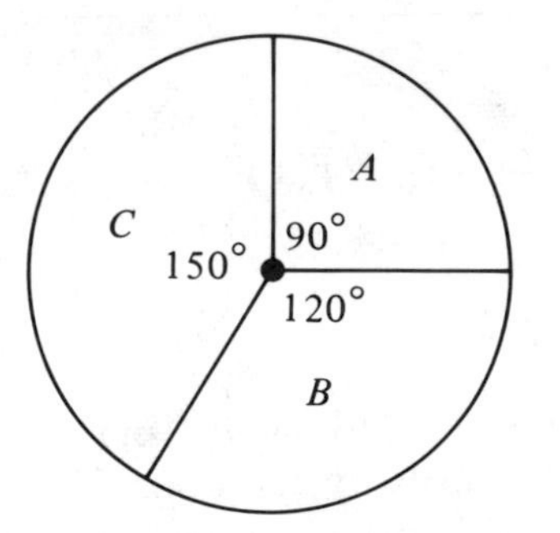

EXAMPLE The experiment consists of spinning this spinner. There are three outcomes: A, B, C. There are eight events:

$$\varnothing, \{A\}, \{B\}, \{C\}, \{A, B\}, \{A, C\}, \{B, C\}, \{A, B, C\}$$

The first is called the **impossible event**; the last is called the **certain event**. The assignment of probabilities is as follows.

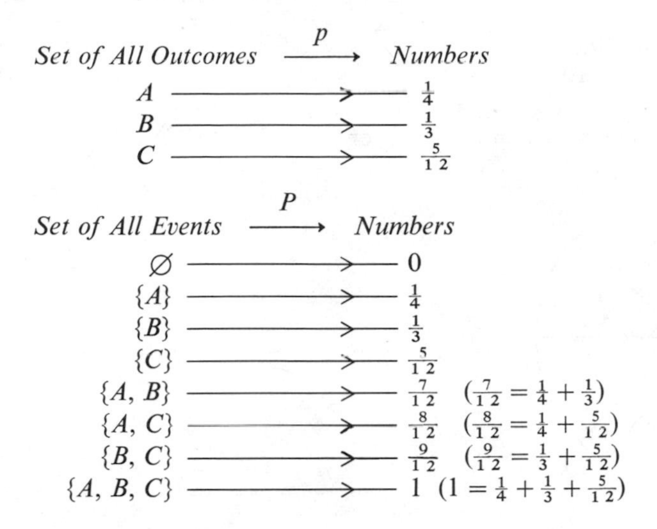

EXERCISES

1. The experiment consists of spinning this spinner.

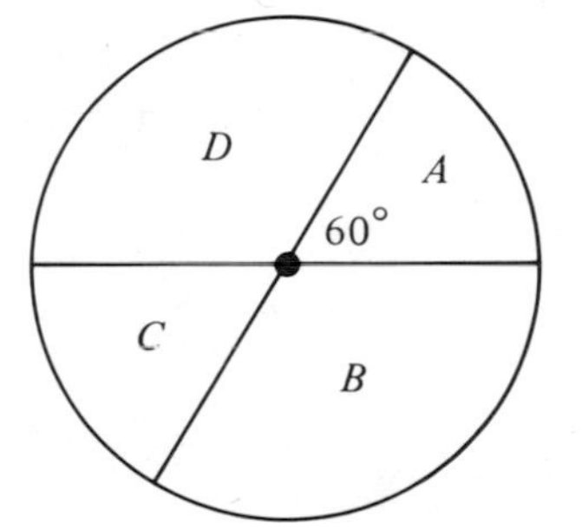

(a) List all outcomes and draw an arrow diagram for the function p.
(b) List all events and draw an arrow diagram for the function P.
(c) Does P have the additivity property of measure functions? Does p?

2. The experiment consists of drawing one ball from an urn containing ten balls numbered 1 through 10. List in roster notation each event that is described in words below. Then find its probability.
(a) an odd number is drawn
(b) a perfect square is drawn
(c) a prime is drawn
(d) a number less than 8 is drawn
(e) a negative integer is drawn
(f) a whole number is drawn
(g) an integral power of 2 is drawn
(h) a divisor of 6 is drawn

3. For the same experiment as in Exercise 2, give a verbal description of each of these events. Also assign each event its probability.
(a) $\{3, 6, 9\}$
(b) $\{2, 4, 6, 8, 10\}$
(c) $\{5, 6, 7, 8, 9, 10\}$
(d) $\{1\}$

4. The experiment is to spin this spinner.

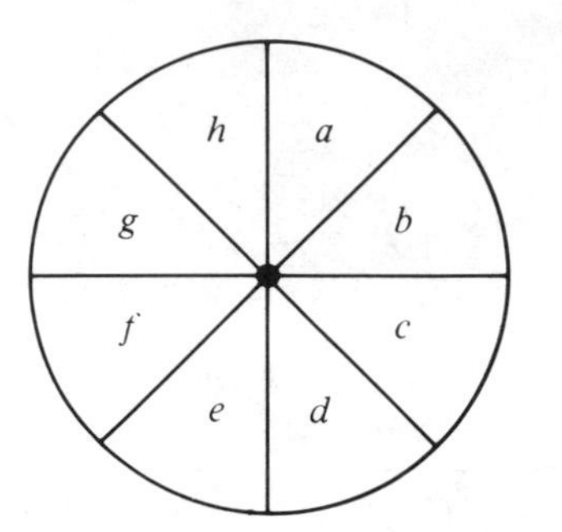

Two events, E and F, have been indicated on the Venn diagram of the

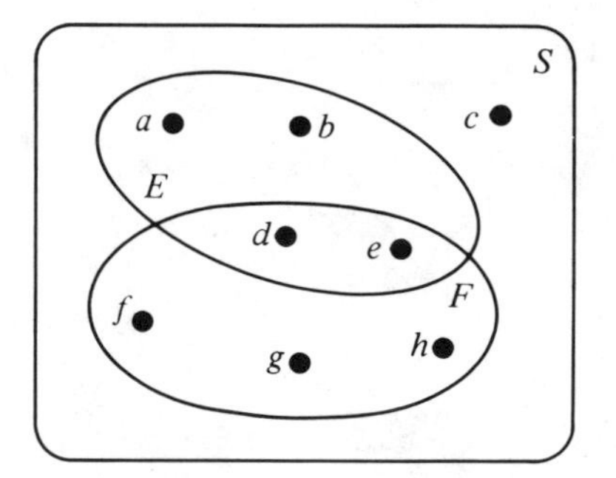

outcome set S. Calculate the following probabilities.

(a) $p(a)$	**(b)** $p(h)$
(c) $P(E)$	**(d)** $P(F)$
(e) $P(E \cup F)$	**(f)** $P(E \cap F)$
(g) $P(F')$	**(h)** $P(E \cap F')$
(i) $P(\{d\})$	**(j)** $p(d)$
(k) $P(E' \cap F')$	**(l)** $P(E \cup F \cup \{c\})$
(m) $P(F - E)$	**(n)** $P(\varnothing)$

5. For the experiment of Exercise 2, let E be the event of drawing an even number and let T be the event of drawing an integral power of 2. Complete the Venn diagram for the outcome set S and the subsets E and T.

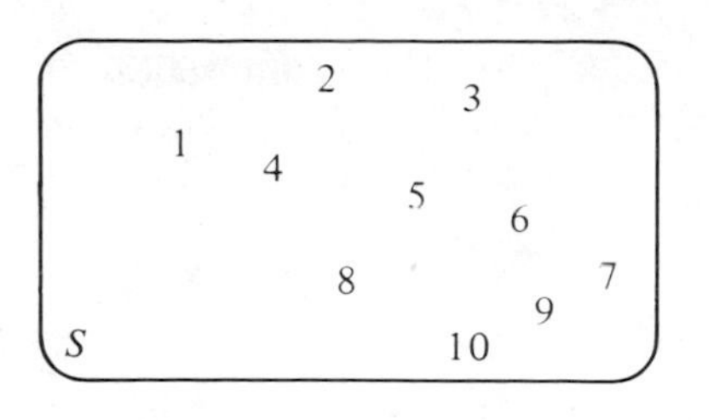

Now calculate these probabilities:
(a) $p(9)$
(b) the probability of drawing an 8
(c) $P(E)$
(d) the probability of drawing an even number or an integral power of 2
(e) $P(T - E)$
(f) the probability of drawing an even number that is not an integral power of 2
(g) the probability of drawing a number that is not an integral power of 2
(h) $P(E \cup T')$
(i) $P((E \cup T)')$
(j) the probability of drawing an odd number that is an integral power of 2.

6. The experiment consists of drawing two balls at once from an urn containing five balls numbered 1 through 5.
(a) Complete the listing of the outcome set begun below:

$$S = \{\{1, 2\}, \{1, 3\}, \{1, 4\}, \{1, 5\}, \ldots\}$$

What is the probability of
(b) each outcome
(c) the event, "the sum of the two numbers drawn is 6"
(d) the event, "the sum of the two numbers drawn is odd"
(e) the event, "the product of the two numbers drawn is odd"
(f) the event, "the difference between the numbers drawn is 3"

7. When an experiment has a large number of outcomes, it becomes important to be able to think about various events without actually listing them. The experiment consists of drawing one card at random from a standard 52-card deck. Find the probability of drawing:
(a) the ace of hearts
(b) the seven of diamonds
(c) a four
(d) a heart
(e) a club or a heart
(f) a seven or a heart
(g) an ace or jack or ten
(h) not a heart
(i) higher than a four and lower than a nine
(j) lower than a four or higher than a nine
(k) a four or a nine

8. The experiment consists of rolling a pair of dice, one red and one green.

The outcomes are "ordered pairs" of numbers. For example, (3, 5) represents the outcome of a 3 on the red die and a 5 on the green, while (5, 3) represents a 5 on the red die and a 3 on the green. The outcome set is represented schematically below.

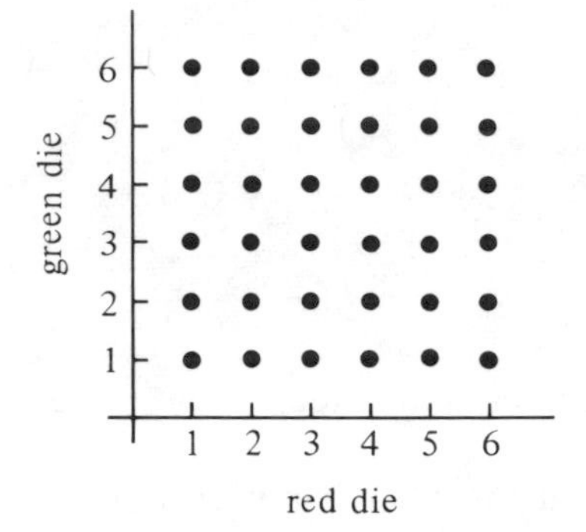

(a) Which dot represents (3, 5)? Which (5, 3)?
(b) How many outcomes are there?
(c) What is the probability of each outcome?
(d) Loop the event, "a sum of 8 is thrown." What is its probability?
(e) Loop the event, "a number less than 4 appears on the red die." What is its probability?
(f) Shade the event "a sum of 8 is thrown and a number less than 4 appears on the red die." What is its probability?
(g) What is the probability of throwing a sum of 8 or a number less than 4 on the red die?
(h) What is the probability of not throwing a sum of 8?
(i) What sum has the greatest probability?

9. Read and explain:

$$P(E') = 1 - P(E) \text{ for any event } E \text{ in any random experiment}$$

10. **Random variables** can be used to abbreviate the description of various events. For the experiment of Exercise 8, let

R represent the number on the red die
G represent the number on the green die
S represent the sum of the numbers on the two dice
M represent the product of the numbers on the two dice

The symbols "$P(S = 9)$" are read "the probability that the sum is nine." The calculation of $P(S = 9)$ can be thought of as follows: "$S = 9$ is shorthand for the open sentence, 'the sum of the components of the ordered pair ________ is 9.' Its domain is the outcome set. Its truth set is $\{(3, 6), (4, 5), (5, 4), (6, 3)\}$. This truth set is the event whose probability we are to calculate. By counting, this probability is $\frac{4}{36}$. Answer: $P(S = 9) = \frac{4}{36}$. ('The probability of rolling a sum of nine is $\frac{4}{36}$.')" Find the following:

(a) $P(S < 4)$ **(b)** $P(R > 4)$
(c) $P(M = 12)$ **(d)** $P(2 < G < 6)$
(e) $P(R = G)$ **(f)** $P(M = R)$
(g) $P(M = S)$ **(h)** $P(R = S)$
(i) $P(M \text{ is odd})$ **(i)** $P(R > G)$

11. Fill in this chart, which compares the terminology of geometry and probability.

	GEOMETRY	PROBABILITY
big set	space	
elements		outcomes
subsets	geometric figures	

12. The number of events in an experiment with n outcomes is the number of subsets of an n-element set. Fill in the chart and decide what that number is.

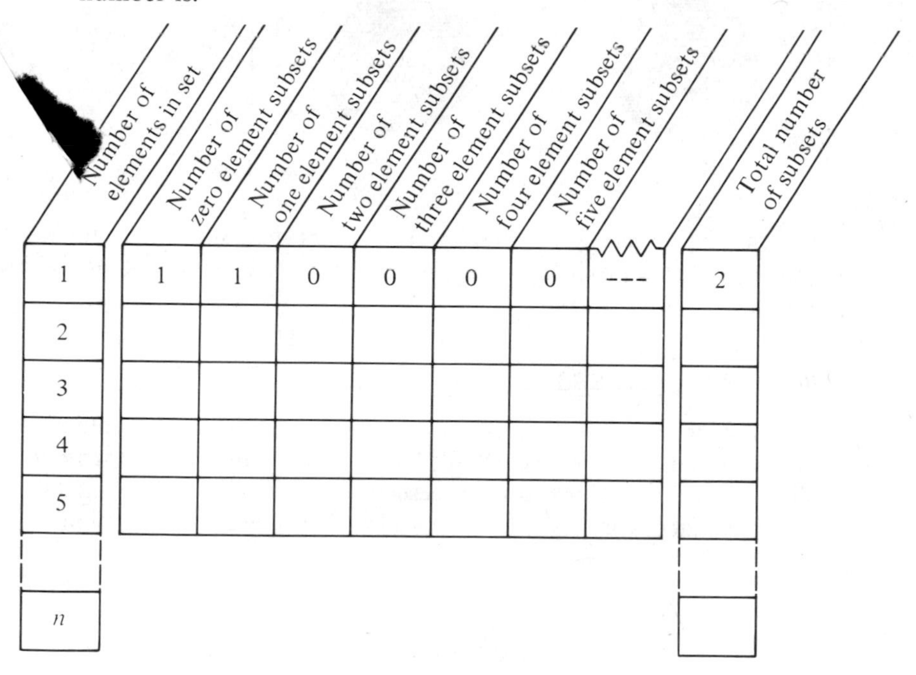

Number of elements in set	Number of zero element subsets	Number of one element subsets	Number of two element subsets	Number of three element subsets	Number of four element subsets	Number of five element subsets		Total number of subsets
1	1	1	0	0	0	0	---	2
2								
3								
4								
5								
n								

***13.** (Research) A basic property of the probability function p is that the sum of the p-probabilities of all outcomes is 1. Try to decide what can be said about the sum of the P-probabilities of all events. (The coin-toss experiment and the example on pp. 236–237 are good places to begin your research.)

*14. A common gambling game at carnivals consists of tossing a dime onto a board, a portion of which is shown. If the dime stays entirely within the black circle, you win. If it touches white, you lose. In the figure, d represents the diameter of a dime.

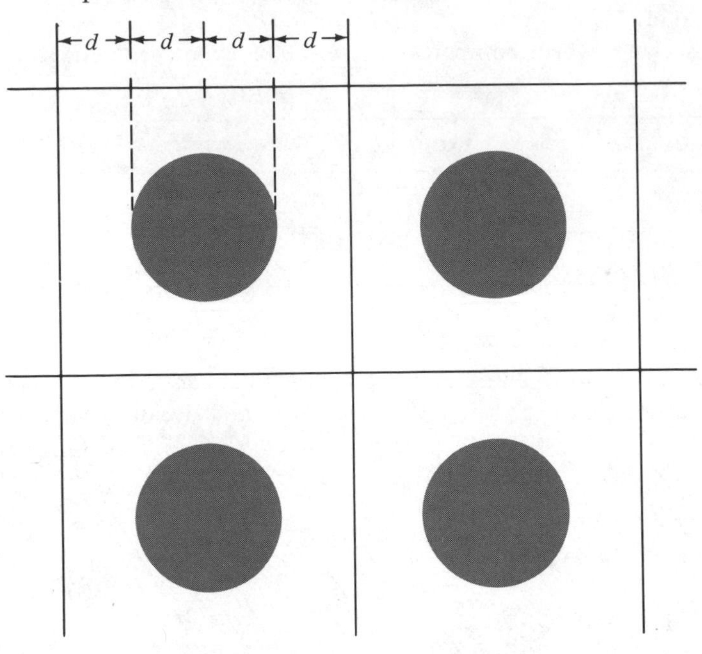

Ignoring any effects of aiming, what is the probability of winning? If there is to be no profit from the game, how much should the prize be worth?

6.4 PROBABILITY TREES

So far our experiments have been single-stage experiments—spinning a spinner, tossing a die, flipping a coin. We look now at a multistage experiment in which two acts are performed in succession.

The experiment consists of spinning this spinner twice in succession.

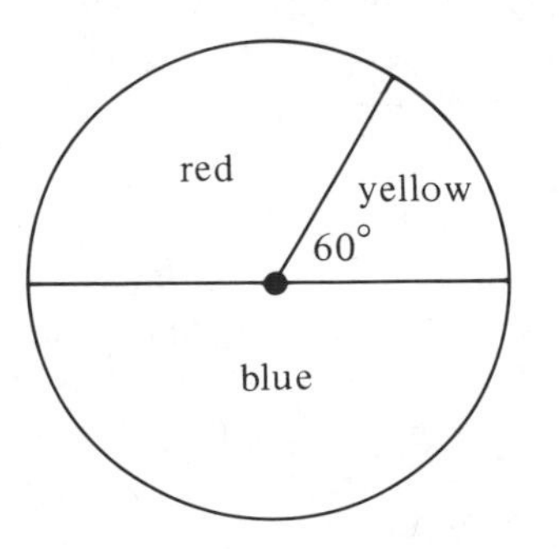

One outcome is "first blue, then red." We denote this outcome by the ordered pair (B, R).

A systematic way of finding *all* outcomes is to draw a **probability tree**. (Probability trees, like factor trees and family trees, usually grow upside down.)

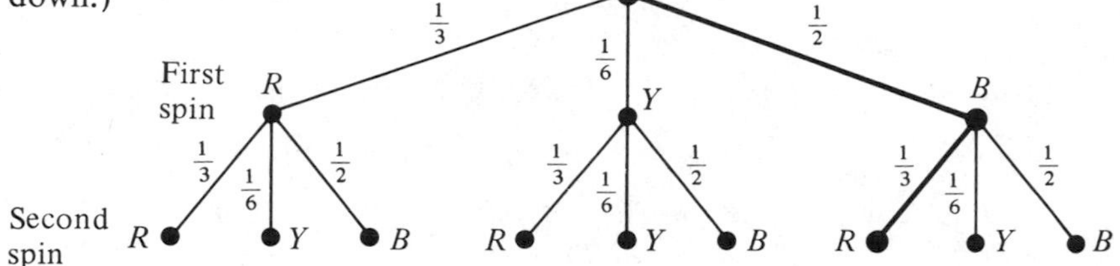

Each path on the tree represents an outcome. For example, the dark path represents the outcome (B, R), "first blue, then red." (Can you find the path corresponding to the outcome (R, B)?) A listing of the outcome set in roster notation can be made up by tracing all paths down the tree:

$$\{(R, R), (R, Y), (R, B), (Y, R), (Y, Y), (Y, B), (B, R), (B, Y), (B, B)\}$$

Having listed the sample space, the next step is to assign a probability to each outcome. Again, it is helpful to look at the tree. The fractions on the branches are suggested by the sizes of the sectors: On any spin, be it the first or the second, the probability of spinning a red is $\frac{1}{3}$, the probability of spinning a yellow is $\frac{1}{6}$, and the probability of spinning a blue is $\frac{1}{2}$. If the experiment were repeated, say 360 times, you would expect to get a blue on the first spin about $\frac{1}{2} \cdot 360$ or 180 times. Of the approximately 180 times a blue came up on the first spin you would expect to get a red on the second spin about $\frac{1}{3} \cdot 180$ or 60 times. Thus, the probability of the outcome (B, R) should be $\frac{60}{360} = \frac{1}{6}$. But that same *probability could have been found by multiplying the fractions along the path* corresponding to (B, R):

$$p((B, R)) = \tfrac{1}{6} = \tfrac{1}{2} \cdot \tfrac{1}{3} = p(B) \cdot p(R)$$

("One-third of one-half of the time you will spin a blue and then a red.")

The probabilities of the other outcomes are found similarly, and a probability diagram can be drawn,

Outcome Set $\xrightarrow{p}$	*Numbers*	
(R, R) ⟶	$\frac{1}{9}$	$(\frac{1}{9} = \frac{1}{3} \cdot \frac{1}{3})$
(R, Y) ⟶	$\frac{1}{18}$	$(\frac{1}{18} = \frac{1}{3} \cdot \frac{1}{6})$
(R, B) ⟶	$\frac{1}{6}$	$(\frac{1}{6} = \frac{1}{3} \cdot \frac{1}{2})$
(Y, R) ⟶	$\frac{1}{18}$	$(\frac{1}{18} = \frac{1}{6} \cdot \frac{1}{3})$
(Y, Y) ⟶	$\frac{1}{36}$	$(\frac{1}{36} = \frac{1}{6} \cdot \frac{1}{6})$
(Y, B) ⟶	$\frac{1}{12}$	$(\frac{1}{12} = \frac{1}{6} \cdot \frac{1}{2})$
(B, R) ⟶	$\frac{1}{6}$	$(\frac{1}{6} = \frac{1}{2} \cdot \frac{1}{3})$
(B, Y) ⟶	$\frac{1}{12}$	$(\frac{1}{12} = \frac{1}{2} \cdot \frac{1}{6})$
(B, B) ⟶	$\frac{1}{4}$	$(\frac{1}{4} = \frac{1}{2} \cdot \frac{1}{2})$

Using these outcome probabilities as a basis, it is now possible to compute the probability of any event.

$$P(\text{at least one red}) = P(\{(R, R), (R, Y), (R, B), (Y, R), (B, R)\})$$
$$= \tfrac{1}{9} + \tfrac{1}{18} + \tfrac{1}{6} + \tfrac{1}{18} + \tfrac{1}{6} = \tfrac{10}{18}$$
$$P(\text{the same color twice}) = P(\{(R, R), (Y, Y), (B, B)\})$$
$$= \tfrac{1}{9} + \tfrac{1}{36} + \tfrac{1}{4} = \tfrac{14}{36}$$
$$P(\text{exactly one blue}) = P(\{(R, B), (Y, B), (B, R), (B, Y)\})$$
$$= \tfrac{1}{6} + \tfrac{1}{12} + \tfrac{1}{6} + \tfrac{1}{12} = \tfrac{6}{12}$$

EXERCISES

1. There are 3 red balls and 2 black balls in an urn. A three-stage experiment consists of drawing a ball at random, recording its color, returning it to the urn, drawing a second ball at random, recording its color, returning it to the urn, drawing a third ball at random, recording its color, and returning it to the urn. A typical outcome is "first black, then red, then red," which we denote by the ordered triple (B, R, R). Use the probability tree below to solve the following problems.

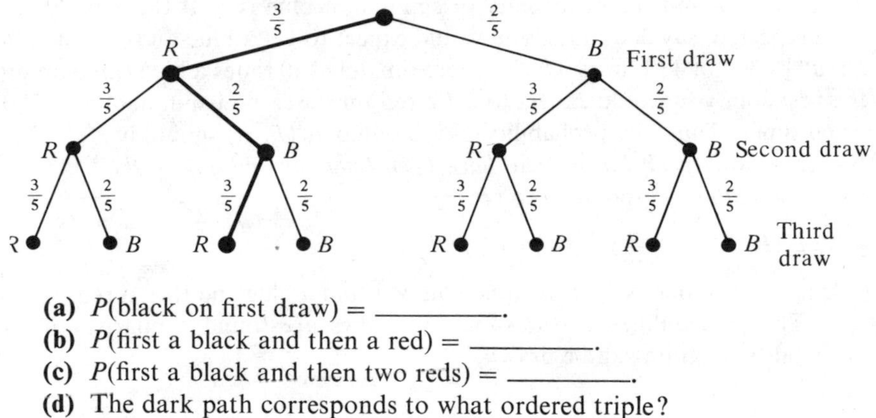

(a) $P(\text{black on first draw}) =$ __________.
(b) $P(\text{first a black and then a red}) =$ __________.
(c) $P(\text{first a black and then two reds}) =$ __________.
(d) The dark path corresponds to what ordered triple?
(e) What path corresponds to the ordered triple (B, B, R)?
(f) Complete the listing of the sample space begun here:

$$S = \{(R, R, R), (R, R, B), \ldots\}$$

(g) Assign a probability to each outcome in S.
(h) Find the sum of the probabilities in part (g). [This is a check on your arithmetic in (g).]
(i) $P(\text{exactly two reds}) =$ __________.
(j) $P(\text{at least two blacks}) =$ __________.

(k) P(never drawing the same color twice in succession) = ________.
(l) P(a black on the third draw) = ________.
(m) P(black on the first and third draws) = ________.

2. If we snip off the bottom eight branches on our tree, then we have a representation of the two-stage experiment of drawing two balls in succession (replacing the first one before drawing the second).
   (a) Complete the listing begun here of the sample space T for this two-stage experiment

$$T = \{(R, R), \ldots\}$$

   (b) Assign a probability to each outcome in T.
   (c) Find the sum of the probabilities in part (b).
   (d) P(drawing the same color twice) = ________.
   (e) P(drawing at least one black) = ________.
   (f) P(drawing at most one red) = ________.

3. The experiment is that of Exercise 2: to draw two balls in succession (replacing the first before drawing the second) from an urn containing three red and two black balls. One explanation of *why* probabilities multiply—for example, why $p((R, B)) = \frac{3}{5} \cdot \frac{2}{5} = \frac{6}{25}$—can be based on a view of the fractions on the tree as operators.

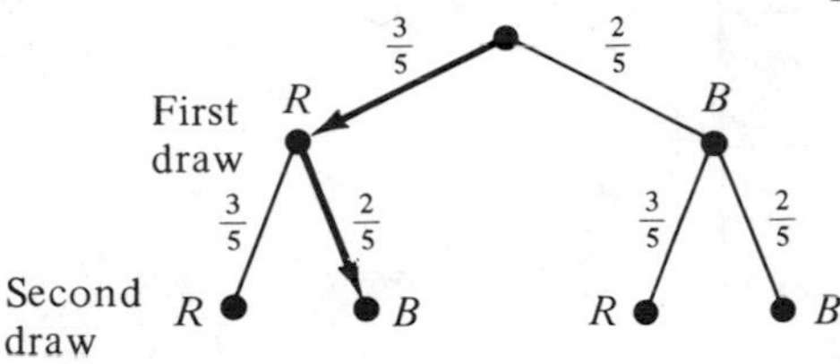

(Note the two arrow heads that have been hung on the tree.) Fill in the boxes.

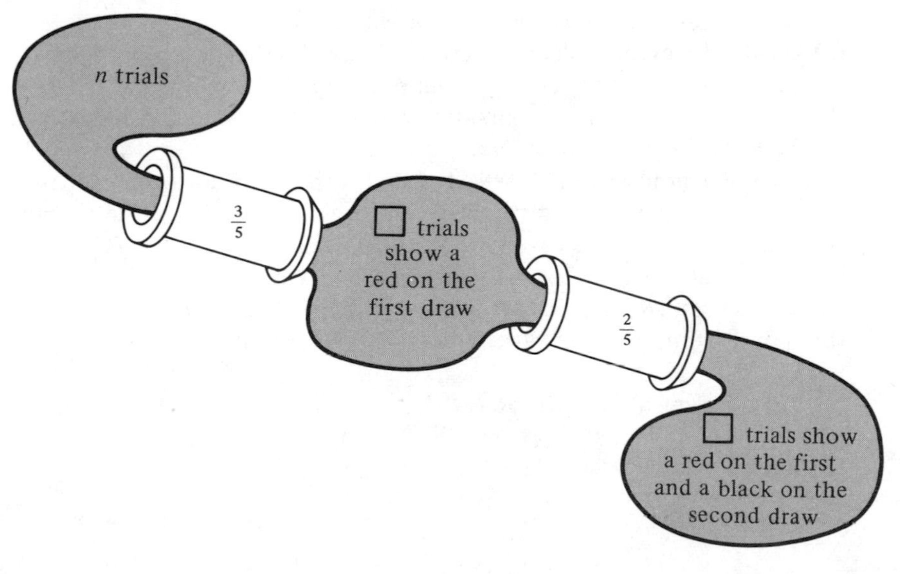

Now simplify the fraction,

$$\frac{\text{number of trials that show a red then a black}}{\text{number of trials}}$$

4. Here is another illustration of *why* $p((R, B)) = \frac{3}{5} \cdot \frac{2}{5} = \frac{6}{25}$. It is based on replacing the experiment of Exercise 3 with a related one having equally likely outcomes. The new experiment is to draw two balls in succession (replacing the first before drawing the second) from an urn containing three *numbered* red balls—R_1, R_2, R_3—and two *numbered* black balls—B_1, B_2. The 25 equally likely outcomes can be listed in a square array. The probability of an event is then the fraction of the square occupied by the outcomes making up that event.

Second draw

First draw	R_1	R_2	R_3	B_1	B_2
R_1	(R_1, R_1)	(R_1, R_2)	(R_1, R_3)	(R_1, B_1)	(R_1, B_2)
R_2	(R_2, R_1)	(R_2, R_2)	(R_2, R_3)	(R_2, B_1)	(R_2, B_2)
R_3	(R_3, R_1)	(R_3, R_2)	(R_3, R_3)	(R_3, B_1)	(R_3, B_2)
B_1	(B_1, R_1)	(B_1, R_2)	(B_1, R_3)	(B_1, B_1)	(B_1, B_2)
B_2	(B_2, R_1)	(B_2, R_2)	(B_2, R_3)	(B_2, B_1)	(B_2, B_2)

(a) Locate the event, "red on the first draw." What is its probability (area with respect to the large square as unit)?
(b) Locate the event "black on second draw." What is its probability (area with respect to the large square as unit)?
(c) Locate the event "red on first draw and black on second draw."
(d) Explain why $p((R, B)) = \frac{3}{5} \cdot \frac{2}{5}$.

5. The **cartesian product** of two sets A and B is the set of all ordered pairs that can be formed having the **first component** in set A and the **second component** in set B. This cartesian product is denoted by "$A \times B$."
(a) Show that the sample space pictured in Exercise 4 is the cartesian product set, $\{R_1, R_2, R_3, B_1, B_2\} \times \{R_1, R_2, R_3, B_1, B_2\}$.
(b) Show that the event "first red then black" is the subset $\{R_1, R_2, R_3\} \times \{B_1, B_2\}$.
(c) List all elements in $\{R_1, R_2, R_3\} \times \{R_1, R_2, R_3\}$ and then describe this event in words.
(d) Write the event "two black" as a cartesian product.

6. The experiment consists of flipping a coin three times in succession.

(a) Draw a probability tree for the experiment. Be sure to label the segments with fractions.

(b) Darken the path representing the outcome (H, T, H).

(c) List the outcome set.

(d) Assign a probability to each outcome.

(e) P(exactly two heads) = ________.

(f) P(more heads than tails) = ________.

7. In an urn are two red, five blue, and three yellow balls. The experiment consists of drawing two balls in succession, the first being replaced before the second is drawn.

(a) Draw a probability tree.

(b) List a sample space.

(c) Assign a probability to each outcome.

(d) P(two balls of same color) = ________.

(e) P(no blue balls) = ________.

(f) This experiment is equivalent to one involving two successive spins of an appropriate spinner. Sketch the spinner.

8. Describe a two-stage experiment equivalent to the experiment of rolling a pair of dice, one red and one green. (See Exercise 8, pp. 239–240.)

9. In an urn are five red and three white balls. The experiment is to draw one ball from the urn and then, *without replacing the first ball,* draw a second ball.

(a) Fill in probabilities on the branches of this tree.

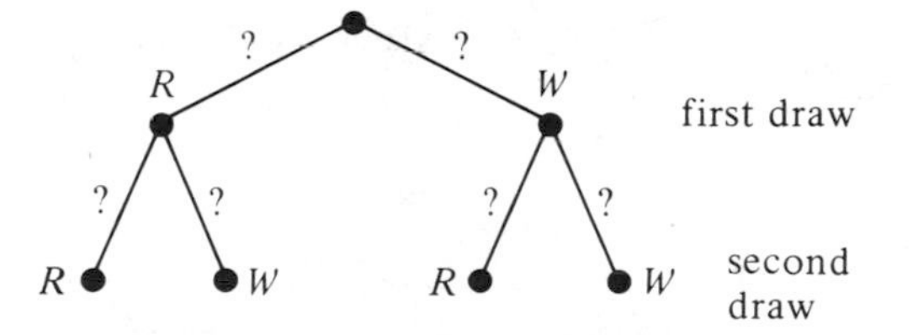

(b) List a sample space.

(c) Assign a probability to each outcome.

(d) Add up the probabilities in part (c).

(e) P(two red balls) = ________.

(f) P(two white balls) = ________.

(g) P(two balls of the same color) = ________.

(h) P(two balls of different colors) = ________.

(i) Is this experiment equivalent to one involving two successive spins of an appropriate spinner?

10. The experiment is to draw three balls in succession, without replacement, from an urn containing five red and three white balls.

(a) Draw a tree for the experiment.

(b) $p((W, R, W))$ = ________.

(c) P(exactly two red balls) = ________.

(d) P(more red than white) = ________.
(e) P(three white) = ________.

11. In an urn are two red, five blue, and three yellow balls. The experiment is to draw three balls in succession, without replacement.
(a) Draw a probability tree.
(b) $p((B, Y, R))$ = ________.
(c) P(two or more blues) = ________.
(d) P(exactly two different colors) = ________.
(e) P(three reds) = ________.

The rule for multiplying probabilities that has been discussed in this section can be stated in this general form:

MULTIPLICATION RULE FOR PROBABILITIES

Suppose that E and F are events and that the probability of E is e. Suppose also that the probability that F will occur after E has occurred is f. Then the probability of "first E then F" is the product $e \cdot f$.

Use this rule to solve the following problems.

12. The experiment is to draw a card from a standard deck and then roll a die. What is the probability of
(a) drawing a spade and rolling a four
(b) drawing a red card and rolling a six
(c) drawing a five and rolling a five

13. The experiment is as follows. First flip a coin. If it comes up heads, roll a die. If it comes up tails, spin this spinner.

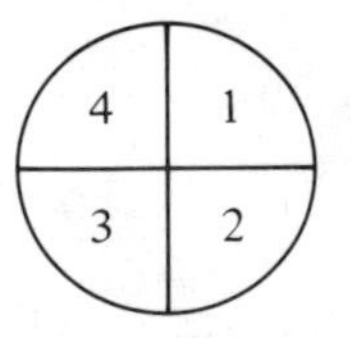

(a) P(getting a 6) = ________.
(b) P(getting a 3) = ________.

6.5 TREE DIAGRAMS IN PROBLEM SOLVING

A very helpful way of thinking about multiplication of rational numbers is as **composition** (hooking together) of operators.

EXAMPLE PROBLEM In a certain city of 1,000,000 people, eligible voters (E) make up $\frac{3}{5}$ of the population (P), and $\frac{4}{7}$ of the eligible voters actually vote (V). What fraction of the total population actually votes?

Solution First consider the diagram,

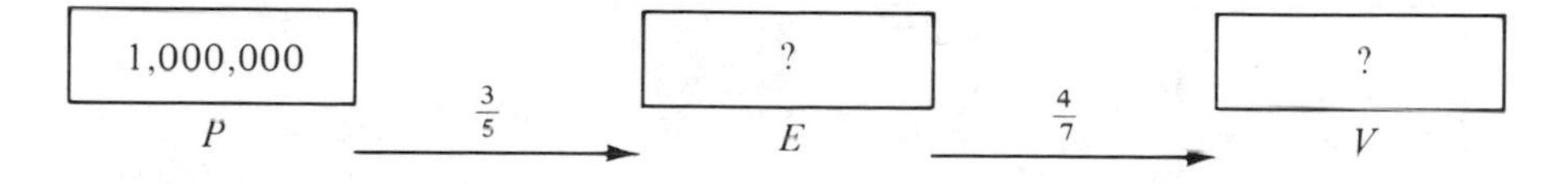

in which the arrows represent operators (machines). The set E contains $\frac{3}{5} \cdot 1{,}000{,}000$ people; the set V contains $\frac{4}{7} \cdot \frac{3}{5} \cdot 1{,}000{,}000$ people. Now consider the diagram

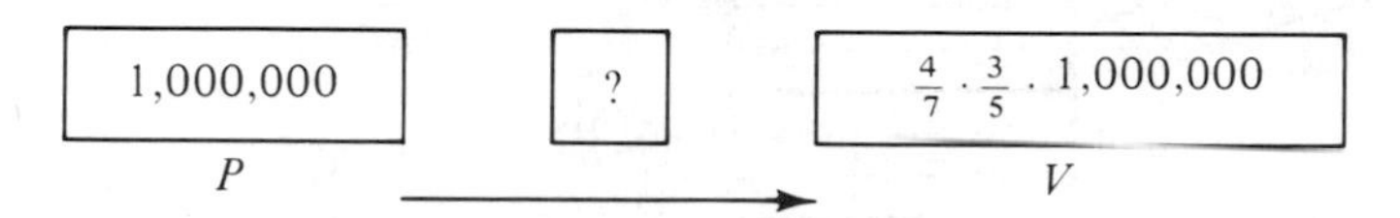

The operator is clearly $\frac{3}{5} \cdot \frac{4}{7}$.

Answer $\frac{3}{5} \cdot \frac{4}{7} = \frac{12}{35}$ of the total population votes.

EXAMPLE PROBLEM Two-fifths of the fish in the lake are out of season, and three-fourths of the fish that are in season are under the legal size limit. What fraction of the fish in the lake are legal to catch?

Solution Draw the diagram (tree) and hook up the operators $\frac{3}{5}$ and $\frac{1}{4}$.

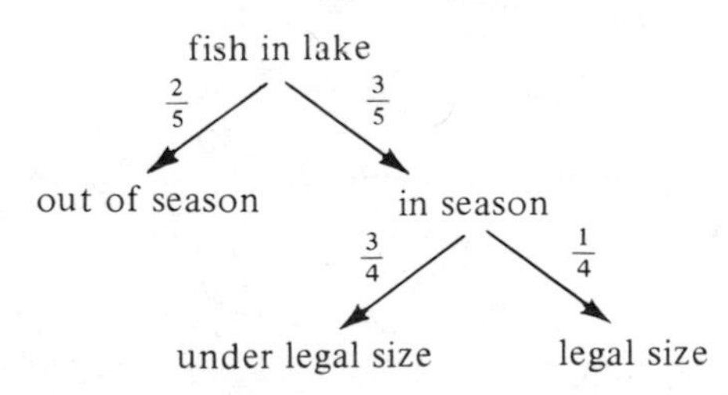

Answer $\frac{3}{5} \cdot \frac{1}{4} = \frac{3}{20}$ of the fish in the lake are legal to catch.

As we noted in the previous section, this view of multiplication can explain the technique of multiplying on a probability tree. But its applicability extends far beyond probability. Even the multiplication technique for finding the area of a rectangle can be thought of in terms of hooking up operators.

PROBLEM Find the area of a $\frac{5}{8}$ in. by $\frac{3}{2}$ in. rectangle.

Solution

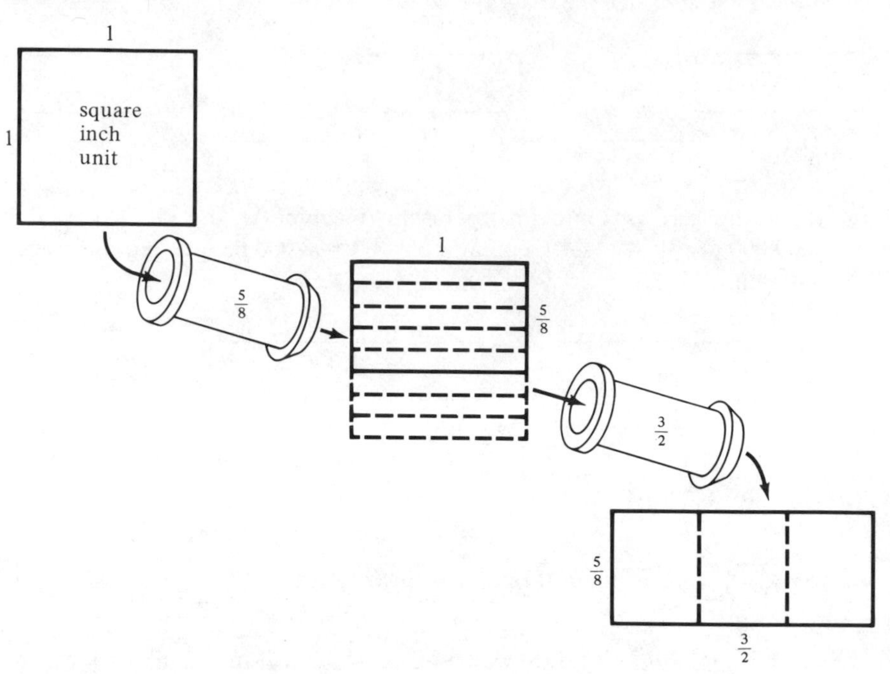

Area with respect to the square inch unit $= \frac{5}{8} \cdot \frac{3}{2}$.

Use tree diagrams and the technique of multiplying down a branch to solve the following problems, not all of which are stated in the language of probability.

EXERCISES

1. Using traffic counters the highway department has determined that $\frac{2}{5}$ of all traffic on Main St. takes Oak Drive

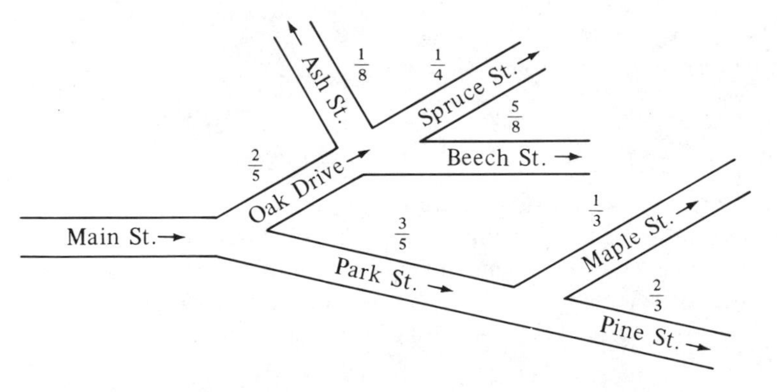

and the rest takes Park St., $\frac{2}{3}$ of the traffic on Park St. takes Pine Drive, and the rest takes Maple St., and so on.

(a) Of every 120 cars traveling up Main St., how many end up on Spruce St.?
(b) What fraction of the Main St. traffic ends up on Spruce St.?
(c) Of every 90 cars traveling up Main St., how many end up on Pine St.?
(d) What fraction of the Main St. traffic ends up on Pine St.?
What fraction of Main St. traffic ends upon
(e) Maple St.?
(f) Beech St.?
(g) Ash St.?

2. A golf ball machine produces 3 defective balls for every 12 good ones. An inspector spots 4 out of 5 defective balls. He also throws out 1 of every 20 good balls by mistake.
(a) Interpret the tree diagram and fill in fractions on its branches.

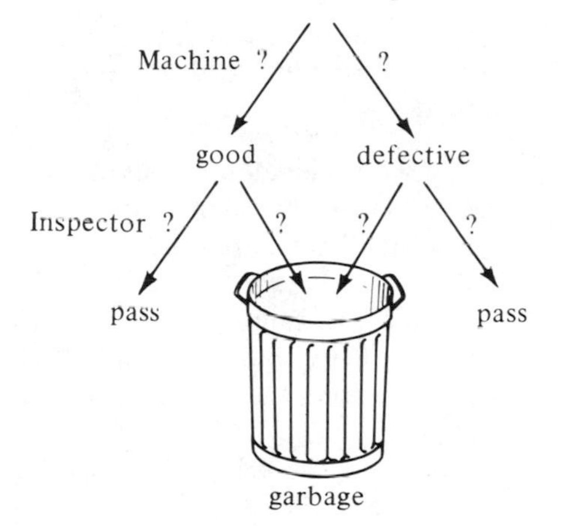

(b) What fraction of the golf balls *sold* by this company is defective?

3. Interpret, then use the tree diagram to answer the questions that follow. At a certain college the failure rate is $\frac{1}{4}$ for freshmen, $\frac{1}{6}$ for sophomores, $\frac{1}{10}$ for juniors, $\frac{1}{25}$ for seniors.

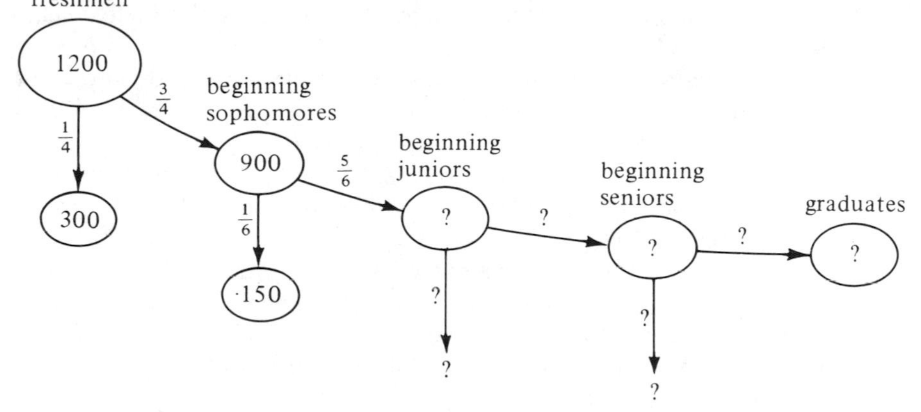

(a) How many students in an entering class of 1200 will flunk out during the first three years?
(b) How many of an entering class of 1200 freshmen can expect to graduate?
(c) What is the probability that an entering freshman will graduate?
(d) What is the probability that a beginning sophomore will not graduate?
(e) What is the probability that a beginning junior will graduate?

4. Seventy percent of the farmers in Cobb County raise only corn. The rest raise only soybeans. Sixty percent of the corn farmers and 80 percent of the soybean farmers own their own farms. What percent of the farmers in Cobb County own their own farms?

5. In order for a pass play to succeed,

(1) the quarterback has to have time to throw the ball,
(2) the quarterback has to get the ball to the intended receiver, and
(3) the receiver has to catch the ball.

Suppose that the quarterback gets tackled with the ball on 2 out of every 15 pass plays, that he throws wildly 3 out of 8 throws, and that the receiver drops 2 out of every 5 balls thrown to him. Draw a tree and find what fraction of the pass plays are successful.

6. If Joe fails the final exam, the probability of his failing the course will be $\frac{1}{3}$. If he passes that exam, the probability of his failing the course will be $\frac{1}{10}$. Joe figures that the probability of his passing the exam is $\frac{3}{4}$. What is the probability Joe will fail the course?

7. If Joe misses the school bus after school, the probability of his having peanut butter sandwiches for supper is $\frac{1}{2}$. If he catches the bus, the probability of his having peanut butter sandwiches for supper is $\frac{1}{20}$. Joe figures that the probability of catching the bus is $\frac{4}{5}$. What is the probability that he will have peanut butter sandwiches for supper?

8. At the school dance Joe sits out $\frac{1}{2}$ of the slow dances and $\frac{9}{10}$ of the fast ones. If the band plays $\frac{2}{3}$ fast dances and $\frac{1}{3}$ slow, what fraction of the dances does Joe sit out?

9. A sewage treatment plant treats sewage in three stages. At the first stage 90 percent of all phosphates are removed. At the second stage 75 percent of the remaining phosphates are removed. At the third stage 50 percent of the remaining phosphates are removed. What percent of phosphates are removed from the sewage by this plant?

10. An inventor devises a two-stage sewage treatment process that removes $\frac{8}{9}$ of the phosphates at each stage. Is his process more or less effective in removing phosphates than the three-stage process of Exercise 9?

11. You drive through six intersections controlled by traffic lights on your way into town. You estimate that for each traffic light the probability that you will be stopped by a red light is $\frac{1}{2}$.
(a) What is the probability that you will be stopped by no traffic light?

(b) What is the probability that you will be stopped by exactly one of the six lights?

12. Lights are timed so that if you miss one the probability that you will make the next is $\frac{3}{4}$. If you make a light, however, the probability that you will make the next one remains $\frac{1}{2}$. You meet four such lights. The probability of making the first one is $\frac{1}{2}$.

(a) What is the probability that you will make all the lights?

(b) What is the probability that you will miss just one?

(c) What is the probability that you will miss them all?

13. Forty percent of the people afflicted with a certain disease show visible symptoms. Eight percent of the population show visible symptoms of the disease. No one without the disease shows symptoms. What percent, to the nearest whole percent, of the apparently unafflicted population actually has the disease? (*Hint*: Use this tree, find x and y, and then return to the original question.)

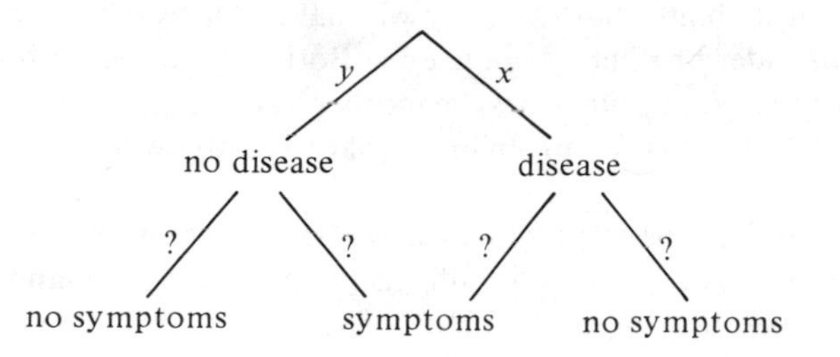

14. A certain automobile plant produces 200 automobiles every working day. Each car is inspected as it comes off the line and is passed, sent back for mechanical repairs, or sent back for paint touch-up. If $\frac{1}{5}$ of all cars produced are sent back for mechanical repairs and $\frac{1}{8}$ of all cars produced are sent back for paint touch-up, about how many cars per day are sent back for some reason?

15. The weather forecast gives the probability of rain before noon as $\frac{1}{7}$ and the probability of rain after noon as $\frac{1}{3}$. What is the probability that it will rain some time during the day?

16. Eye color in humans is a genetic trait. There are two types of eye color genes, brown and blue. Every person carries two such genes. If he carries two blue genes, his eyes will be blue. If he carries either one or two brown genes, his eyes will be brown The brown gene is said to be "dominant," the blue gene "recessive." Each parent contributes one (randomly chosen) gene to the offspring. Thus, all the children of a father with two blue genes and a mother with two brown genes will have one blue and one brown gene, and hence brown eyes.

Suppose, however, that the father has a blue and a brown gene and so does the mother. The eye color of their children is now governed by the laws of probability.

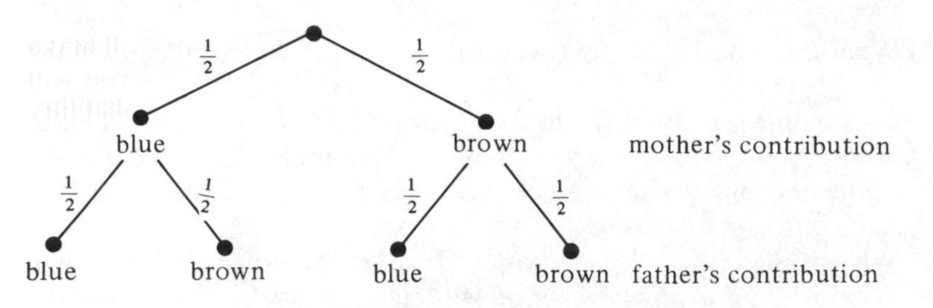

The possible gene combinations for the child (paths) are

(blue, blue), (blue, brown), (brown, blue), (brown, brown)

Each has probability $\frac{1}{2} \cdot \frac{1}{2} = \frac{1}{4}$. The probability that the child has brown eyes (at least one brown gene) is thus $\frac{3}{4}$, and the probability that he has blue eyes (two blue genes) is $\frac{1}{4}$.

(a) A baby is born. The mother has blue eyes. The father has brown eyes, but the paternal grandfather has blue eyes. What genes does the father have? What is the probability that the baby will have blue eyes?

(b) A baby is born. His older brother has blue eyes. Both his mother and father have brown eyes. What genes does the mother have? What genes does the father have? What is the probability that the baby will have blue eyes?

(c) A brown-eyed man and a blue-eyed woman have had six children, all with brown eyes. What is the probability of this happening if the man carries one blue and one brown gene? (Think of spinning this spinner six times.)

What genes do you think the father carries? Can you be certain of your answer?

6.6 COUNTING PERMUTATIONS

In an urn are ten balls numbered 1 through 10. The experiment consists of drawing three balls in succession without replacement.

PROBLEM What is the probability of drawing, in some order, the balls numbered 1, 2, and 3?

The straightforward way to attack this problem would be to

1. List an outcome set.
2. Assign each outcome a probability.

3. List the outcomes comprising the event in question and add up their probabilities.

We being with step 1.

A natural outcome set for the experiment consists of ordered triples such as (1, 7, 4), (2, 9, 3), ... The usual way to list the outcome set of a multistage experiment such as this one is to draw a tree. A portion of a tree for this experiment is sketched below.

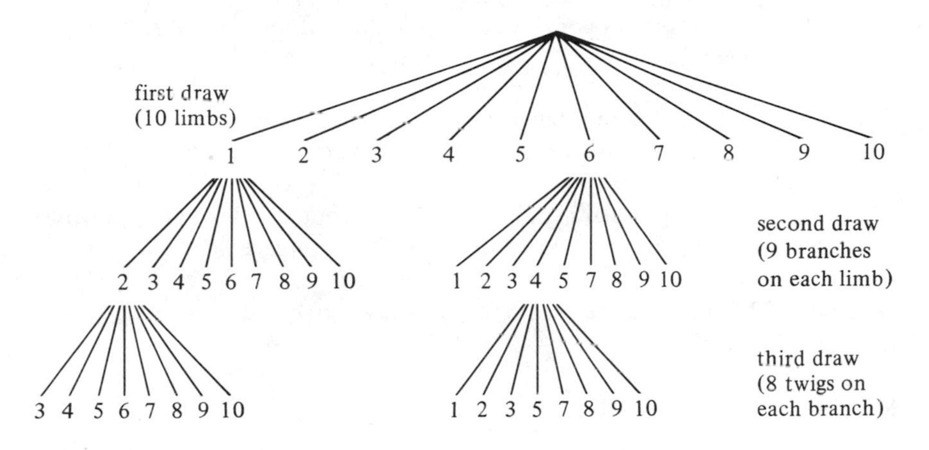

Time and space limitations prevent drawing the entire tree or listing the entire outcome set, but fortunately that is not necessary. We can see that there are $10 \cdot 9 \cdot 8$ paths down the tree; that is, there are $10 \cdot 9 \cdot 8 = 720$ ordered triples in the outcome set

$$\{(1, 2, 3), (1, 2, 4), \ldots, (6, 4, 3), (6, 4, 5), (6, 4, 7), \ldots, (10, 9, 8)\}$$

Now we move on to step (2). Clearly all outcomes in the set above are equally likely. Since there are 720 of them and since their probabilities must add up to 1, each must have probability $\frac{1}{720}$. (Can you see how $\frac{1}{720}$ might also be thought of as $\frac{1}{10} \cdot \frac{1}{9} \cdot \frac{1}{8}$?)

Now we are ready for step (3). The event in question consists of all those ordered triples (outcomes) that are made up from a 1, a 2, and a 3. There are six such outcomes:

$$(1, 2, 3), (1, 3, 2), (2, 1, 3), (2, 3, 1), (3, 1, 2), (3, 2, 1)$$

which for this problem we think of as **favorable outcomes**. (Can you visualize these favorable paths on the tree?) Thus, the probability of the event in question is

$$6 \cdot \frac{1}{720} = \frac{6}{720} = \frac{\text{number of favorable outcomes}}{\text{total number of outcomes}}$$

We have illustrated two points by this example. The first is this:

In an experiment with equally likely outcomes, the probability of an event E is the quotient,

$$\frac{\text{number of favorable outcomes}}{\text{total number of outcomes}}$$

where an outcome is considered "favorable" if and only if it belongs to the event E.

The second is this: For an experiment (or an event) with a great many outcomes, it is essential to be able to "count" the outcomes without actually listing them.

Consider now a similar experiment. An urn contains 100 balls numbered 1 through 100. Four balls are to be drawn without replacement.

PROBLEM What is the probability of drawing the numbers 1, 2, 3, 4 in some order?

Solution The probability is

$$\frac{\text{the number of ordered 4-tuples you can make using the four numbers } 1, 2, 3, 4}{\text{the number of ordered 4-tuples you can make using four of the numbers } 1, 2, \ldots, 100}$$

so immediately we are faced with two counting problems. We consider the denominator first.

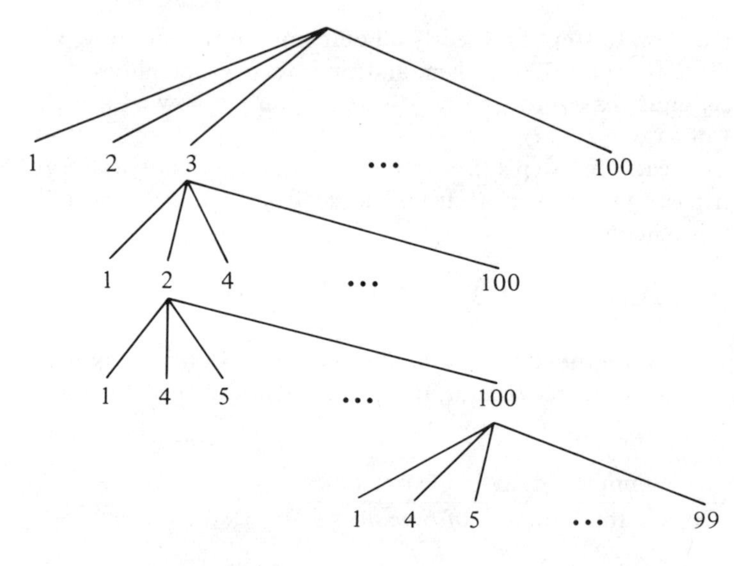

On a tree for this experiment there are 100 limbs. On each limb are 99 branches, on each branch 98 twigs, and on each twig 97 smaller twigs. Thus, there are $100 \cdot 99 \cdot 98 \cdot 97$ paths down the tree. The denominator is $100 \cdot 99 \cdot 98 \cdot 97$.

The numerator is the number of ways of arranging the numbers 1, 2, 3, 4. These arrangements, or **permutations**, can also be visualized as paths on a tree.

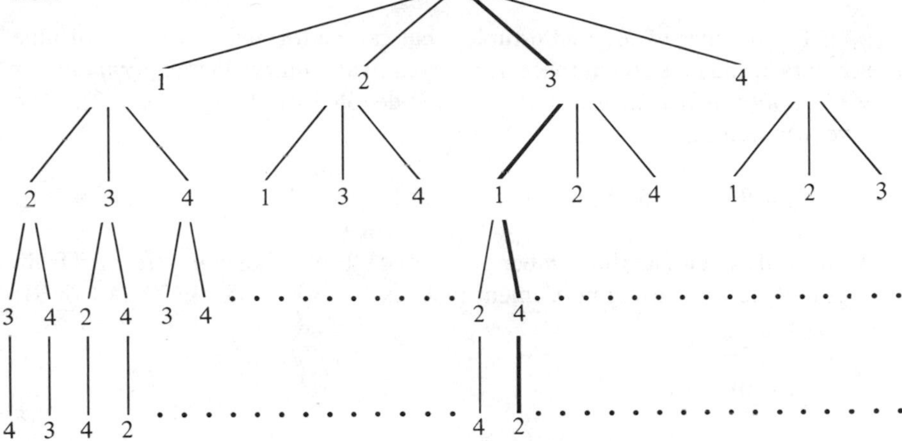

There are $4 \cdot 3 \cdot 2 \cdot 1$ permutations of the numbers 1, 2, 3, 4. (The dark path corresponds to what permutation? Is this tree a part of the larger tree for the entire experiment?)

Answer The probability of drawing 1, 2, 3, 4 in some order is

$$\frac{4 \cdot 3 \cdot 2 \cdot 1}{100 \cdot 99 \cdot 98 \cdot 97} \doteq \frac{25}{100^4} = \frac{1}{4{,}000{,}000}$$

The repeated product $4 \cdot 3 \cdot 2 \cdot 1$ is abbreviated 4! and read "four factorial." More generally, for any natural number n,

$$n! \text{ ("}n\text{ factorial")} = n \cdot (n-1) \cdot (n-2) \cdot \cdots \cdot 3 \cdot 2 \cdot 1$$

EXAMPLE How many permutations are there of the set

$$\{A, B, C, D, E, F, G, H, I\}$$

Answer (Picture a tree!) There are

$$9 \cdot 8 \cdot 7 \cdot 6 \cdot 5 \cdot 4 \cdot 3 \cdot 2 \cdot 1 = 9!$$

permutations of the given set of nine elements.

EXAMPLE How many ordered 6-tuples can be formed from the set

$$\{A, B, C, D, E, F, G, H, I\}$$

using no element twice?

Answer (Picture a tree.) $9 \cdot 8 \cdot 7 \cdot 6 \cdot 5 \cdot 4$

The number of ordered 6-tuples that can be formed from a set of nine elements, using no element twice, is referred to as "*the number of permutations of nine things taken six at a time*" and is denoted by the symbol "P_6^9." We have just seen that

$$P_6^9 = 9 \cdot 8 \cdot 7 \cdot 6 \cdot 5 \cdot 4$$

We have also seen that the number of ordered 9-tuples that can be formed from a set of 9 elements using no element twice is $9 \cdot 8 \cdot 7 \cdot 6 \cdot 5 \cdot 4 \cdot 3 \cdot 2 \cdot 1$ or $9!$; that is,

$$P_9^9 = 9!$$

EXERCISES

Write the answers to Exercises 1–8 as products of natural numbers.

1. How many ordered pairs can be made up from the 26 letters of the alphabet if no letter is used twice?

2. How many ordered triples can be made up from the digits 0, 1, 2, 3, 4, 5, 6, 7, 8, 9 if no digit is used twice?

3. A club has 30 members. How many slates of officers—president, vice president, secretary, treasurer—are possible if no person is allowed to hold two offices at once?

4. How many permutations are there of the set $\{A, B, C, D, E\}$?

5. In how many ways can 8 people stand in line at a supermarket check-out aisle?

6. How many different batting orders are possible for a baseball team of 9 players?

7. A Little League team has 15 players. (Of course only 9 can play at a time.) How many different batting orders could the coach turn in to the umpire at the start of a game?

8. A baseball league has 6 teams. How many "final standings" are possible for this league? (Ties are dissolved by a flip of a coin.)

9. Write out as a product:

(a) P_2^7 **(b)** P_3^{12}

(c) P_4^6 **(d)** P_6^6

(e) P_4^{100} **(f)** P_8^{75}

(g) P_2^n **(h)** P_3^n

(i) P_8^n (j) P_n^n
(k) 4! (l) 9!

10. Express the answer to each of Exercises 1–8 using a symbol of the form P_k^n.

11. Consider 100 points in a plane, no three of which are collinear.
   (a) How many rays do they determine?
   (b) How many lines do they determine?
   (c) How many angles do they determine?
   (d) How many triangles do they determine?

Some people prefer to remember the following verbalized rule rather than always to visualize a tree.

THE PRODUCT RULE FOR COUNTING

If you can perform a first act in n_1 ways and after that is done you can perform a second act in n_2 ways and after that is done you can perform a third act in n_3 ways, and so on, then you can perform all the acts in succession in $n_1 \cdot n_2 \cdot n_3 \cdot \cdots$ ways.

---

12. If you can choose among 5 people for president and 6 other people for vice president, how many choices have you for the pair of officers?

13. Joe has 6 shirts and 4 pairs of slacks. How many different suits of clothes can he wear to school?

14. A golfer has 4 woods, 8 irons, and 2 putters. He brags that he finished the first hole by hitting a wood, an iron, and a putt. In how many different ways could he have done this?

15. **(a)** How many ordered triples can be made up from the 26 letters of the alphabet if no letter is used twice?
   (b) How many ordered triples can be made up from the letters of the alphabet if letters can be used more than once?

16. How many possible social security numbers are there?

17. License "numbers" in a certain state consist of 2 letters (not necessarily different) followed by 4 digits (not necessarily different). How many license numbers are possible?

18. In a hat are the numbers 1, 2, 3, 4, 5, 6, 7, 8, 9. An experiment consists of drawing out 3 numbers (without replacement) and letting the first be the hundreds digit, the second the tens digit, and the third the units digit of a 3-digit number.
   (a) How many outcomes are there in the sample space {123, 124, 125, ...}?
   (b) What is the probability of each outcome?
   (c) How many outcomes have 5 as the hundreds digit?
   (d) What is the probability that in one trial a number between 500 and 600 will be formed?
   (e) What is the probability that a number over 800 will be formed?
   (f) What is the probability that a number under 500 will be formed?

19. The experiment consists of first drawing a spade from the 13 spades and then a heart from the 13 hearts.
  (a) List a few outcomes and then decide how many outcomes there are.
  (b) What is the probability of drawing the king of spades and the three of hearts?
  (c) What is the probability that both cards will be face cards? (Consider jacks, queens, kings, and aces as face cards.)
  (d) What is the probability that both cards will be lower than an eight? (Consider aces as high cards.)
  (e) What is the probability of not drawing a seven?
  (f) What is the probability of drawing at least one seven?
  (g) Add the probabilities in (e) and (f).

20. You are dealt 4 cards from a standard deck. What is the probability that you will get 4 aces?

21. You are dealt 4 cards in succession from a standard deck, but now you put each card back in the deck and the deck is reshuffled before you are dealt the next card. What is the probability of 4 aces?

22. You are dealt 5 cards from a standard deck (without replacement). What is the probability of 4 aces?

6.7 COUNTING COMBINATIONS

If a club with 20 members wants to choose a president, vice president, secretary, and treasurer, and if no one is to hold two offices, then there are

$$P_4^{20} = 20 \cdot 19 \cdot 18 \cdot 17$$

possible slates of officers: 20 choices for president, 19 for vice president, 18 for secretary, and 17 for treasurer. (Can you visualize a tree?) The slates of officers are *ordered 4-tuples.*

$$\begin{matrix} P & V & S & T \\ (\text{Ann}, & \text{Bob}, & \text{Carol}, & \text{Dan}) \end{matrix} \neq \begin{matrix} P & V & S & T \\ (\text{Dan}, & \text{Ann}, & \text{Carol}, & \text{Bob}) \end{matrix}$$

Suppose now that the club wants to choose a 4-member committee to clean up the clubhouse after the meeting. How many such committees are possible? Clearly, the order in which people are chosen for this committee is immaterial.

$$\{\text{Ann, Bob, Carol, Dan}\} = \{\text{Dan, Ann, Carol, Bob}\}$$

The problem here is simply to count how many different (unordered) 4-*element subsets* there are in a 20-element set. Using what we know about permutations, that is not difficult.

Each 4-element subset can be arranged into 4! different ordered 4-tuples. Thus, each such subset would be represented 4! times on the list of all possible ordered 4-tuples. Since there are $20 \cdot 19 \cdot 18 \cdot 17$ ordered 4-tuples (officer slates), there must be

$$\frac{20 \cdot 19 \cdot 18 \cdot 17}{4!}$$

unordered 4-element subsets (committees).

EXAMPLE How many three-element subsets are there in the set $\{A, B, C, D, E\}$?

Solution There are $5 \cdot 4 \cdot 3$ ordered triples, but each (unordered) 3-element subset is represented by $3! = 3 \cdot 2 \cdot 1$ such ordered triples. Thus, there are,

$$\frac{5 \cdot 4 \cdot 3}{3 \cdot 2 \cdot 1}$$

3-element subsets.

A subset (unordered) is sometimes referred to as a **combination** to distinguish it from a permutation, which always refers to some kind of ordered -tuple. The number of 3-element subsets of a 5-element set is referred to as "*the number of combinations of* 5 *things taken* 3 *at a time*," and is denoted by "C_3^5." We have just seen that

$$C_3^5 = \frac{P_3^5}{3!}$$

and

$$C_4^{20} = \frac{P_4^{20}}{4!}$$

EXAMPLE How many segments do 10 points determine?

Solution We can choose $10 \cdot 9 = P_2^{10}$ ordered pairs of points as end points. But (A, B) and (B, A) determine the same segment. We must divide by 2. There are

$$\frac{10 \cdot 9}{2} = \frac{P_2^{10}}{2!} = C_2^{10}$$

segments determined by the ten points.

EXAMPLE Interpret and evaluate C_5^{14}.

Solution C_5^{14} is the number of combinations of 14 things taken 5 at a time; that is, the number of 5-element subsets of a 14-element set.

$$C_5^{14} = \frac{P_5^{14}}{5!} = \frac{14 \cdot 13 \cdot \not{12} \cdot 11 \cdot \not{10}}{\not{5} \cdot \not{4} \cdot \not{3} \cdot \not{2} \cdot 1} = 2002$$

EXERCISES

Write the answers to Exercises 1–5 as whole numbers.

1. How many 2-element sets can be formed using the 26 letters of our alphabet?

2. How many 3-element subsets are there of the set {0, 1, 2, 3, 4, 5, 6, 7, 8, 9}?

3. How many different 5-card poker hands can be made up from a standard 52-card deck?

4. How many triangles are determined by 10 points, no 3 of which are collinear?

5. Twelve baseball teams are to be split up into two 6-team leagues. In how many ways can this be done?

6. Evaluate each symbol as in the example.

EXAMPLE

$$C_3^9 = \frac{P_3^9}{3!} = \frac{\overset{3}{\not{9}} \cdot \overset{4}{\not{8}} \cdot 7}{\not{3} \cdot \not{2} \cdot 1} = 84$$

(a) C_4^6 **(b)** C_2^5
(c) C_5^{10} **(d)** C_6^7
(e) C_1^7 **(f)** C_4^{11}
(g) C_7^{11} **(h)** C_3^{100}
(i) C_4^4 **(j)** C_4^3

7. When C_{17}^{29} is expressed as a quotient, the denominator is ________. The numerator is a product of ________ factors, the largest of which is ________ and the smallest of which is ________.

8. Complete the following.

(a) P_k^n represents the number of ordered ________-tuples that can be made up from a ________ element set using no element twice. $P_k^n =$ ________.

(b) C_k^n represents the number of .... $C_k^n =$ ________.

9. (a) C_0^5 represents the number of ________ element subsets of a ________ element set, and thus $C_0^5 =$ ________.

(b) Explain why P_0^5 should be assigned the value 1.

(c) What value should be assigned to 0! so that the general formula

$$C_k^n = \frac{P_k^n}{k!}$$

will continue to hold even in the case $k = 0$?

Express the answers to Exercises 10–14 in terms of symbols of type C_k^n or P_k^n.

10. In how many different ways can 4 people sit in a 4-seat automobile?

11. In a group of 7 people, 3 are to ride in a first car and 4 in a second. In how many ways can the group be split up?

12. How many different seating arrangements are possible for the people in Exercise 11?

13. Ten lines are drawn, of which no two are parallel and no three are concurrent. How many crossing points are formed?

14. How many choices have you if you must choose five days out of the week on which you will work?

15. An experiment consists of drawing 3 numbers (at once) from a hat containing the numbers 1, 2, 3, 4, 5, 6, 7, 8, 9, 10.

(a) List two more outcomes in the outcome set

$$\{\{1, 2, 3\}, \{1, 2, 4\}, \{1, 2, 5\}, \ldots\}$$

(b) How many outcomes in this outcome set?
(c) What is the probability of each outcome?
(d) How many outcomes consist of three even numbers?
(e) What is the probability of drawing three even numbers?
(f) What is the probability of drawing three numbers over 6?
(g) What is the probability of drawing three numbers under 7?
(h) What is the probability of drawing three numbers with sum 10?

16. An experiment consists of drawing two numbers (at once) from a hat containing the numbers 1, 2, 3, 4, 5, 6, 7.

(a) Decide on an outcome set and list several outcomes.
(b) What is the probability of drawing two odd numbers?
(c) What is the probability of drawing two even numbers?
(d) What is the probability of drawing a sum of 8?
(e) What is the probability of drawing a sum greater than 7?

17. An experiment consists of randomly choosing four different letters of the alphabet. Express each probability below in terms of symbols of the type C_k^n. Do not "simplify."

(a) P(all four are vowels)
(b) P(all four are consonants)
(c) P(none follow j in the alphabet)
(d) P(all precede e in the alphabet)
(e) P(at least one is a vowel)

18. The experiment consists of drawing two balls (without replacement) from an urn containing 5 red and 3 green balls.

(a) View this as a two-stage experiment and draw a probability tree.
(b) What is the probability of drawing two red balls?
(c) In how many ways can two balls be chosen from among eight?
(d) In how many ways can two balls be chosen from among five?
(e) Use your answers to (c) and (d) to check your answer to (b).

19. Write an expression for the probability that the first two entrants are both among the 100 lucky winners in a sweepstakes with 1 million entrants.

20. An experiment consists of flipping a coin ten times in succession. Each outcome is thus an ordered 10-tuple such as *HHTHTTTHTT* or *HTHTT HHTTH*. One way to represent this experiment would be by a tree diagram. Such a tree would have ten levels and $2^{10} = 1024$ tiny twigs at the bottom. A neater way of picturing the experiment uses the idea of a "random walk." Imagine a particle that begins at the origin in the figure and walks one unit right whenever a head is tossed and one unit up whenever a tail is tossed. The dotted walk represents the outcome *HHTH TTTHTT*, the dark walk represents the outcome *HTHTTHHTTH*.

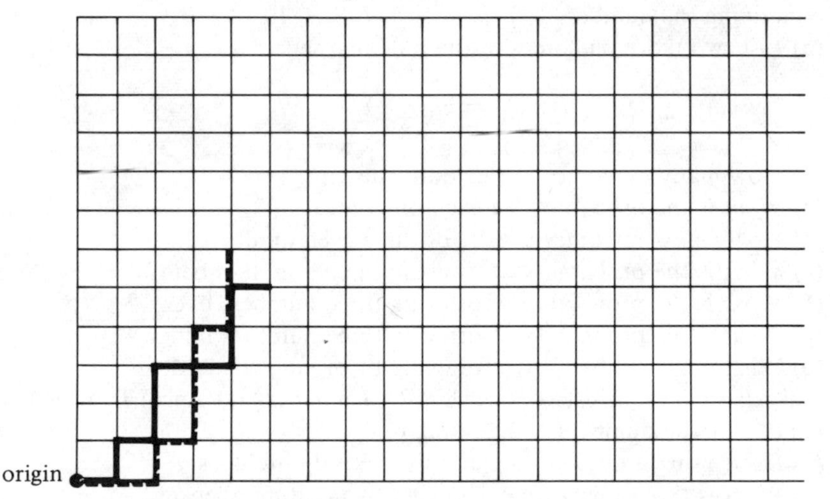

Draw walks for these outcomes.

(a) *HHTHTHHHTH* **(b)** *THHHTHTHHH*
(c) *HHHHTHTTHH* **(d)** *TTTHHHHHHH*

All of these walks should have terminated at the point (7, 3), which is seven units right (7 heads) and 3 units up (3 tails).

(e) Of the total number of 1024 walks, how many terminate at point (7, 3)? (*Hint*: You have ten blanks, __ __ __ __ __ __ __ __ __ __. How many ways are there of choosing the 3 blanks to fill with *T*'s?)

On your drawing, draw walks representing these outcomes.

(f) *THHTTTHTTH*
(g) *HTTHHTHTTT*
(h) *THTHTTHTHT*
(i) How many walks terminate at point (4, 6)?

(j) On the figure below, label the other dark points, and determine how many walks go to each one.

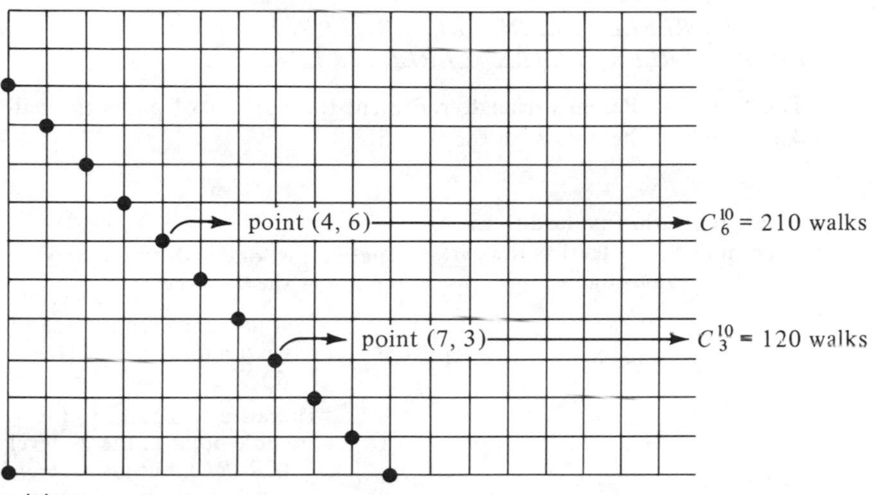

(k) Compute to three decimal places the probabilities of getting 10 heads, 9 heads, 8 heads, . . . , 0 heads.

(l) What is the probability of getting 8 or more heads?

(m) What is the probability of getting more than 7 heads or more than 7 tails on 10 tosses of an honest coin?

6.8 PASCAL'S TRIANGLE

Angular obstructions are attached to an inclined board as shown: one at the 0-level, two at the 1-level, three at the 2-level, and so on. A small ball is released through the chute at the top of the board. Each time it hits the point of one of the obstructions, as it rolls down the boards, it can go either left or right.

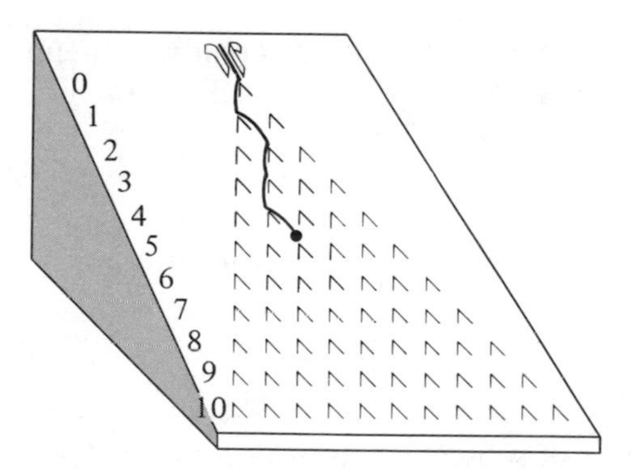

Figure 1.

The dark path can be represented by the symbol *LRLLR*. It is one of *ten* different paths that would take the ball to the same position at the 5-level.

RRLLL RLRLL RLLRL RLLLR LRRLL
LRLRL LRLLR LLRRL LLRLR LLLRR

The entries in **Pascal's triangle** represent the number of paths the ball can take to each of the points on the board.

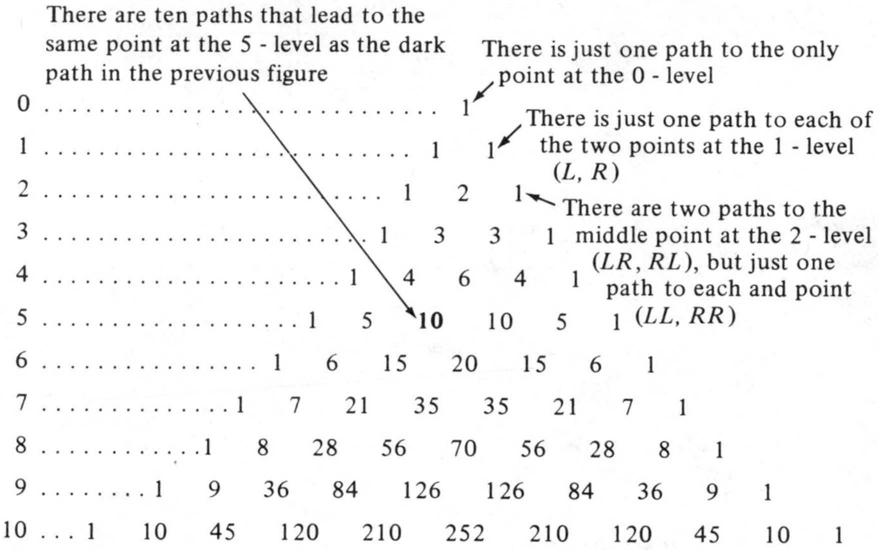

There are ten paths that lead to the same point at the 5 - level as the dark path in the previous figure

There is just one path to the only point at the 0 - level

There is just one path to each of the two points at the 1 - level (*L*, *R*)

There are two paths to the middle point at the 2 - level (*LR*, *RL*), but just one path to each and point (*LL*, *RR*)

0 . . . 1
1 . . . 1 1
2 . . . 1 2 1
3 . . . 1 3 3 1
4 . . . 1 4 6 4 1
5 . . . 1 5 **10** 10 5 1
6 . . . 1 6 15 20 15 6 1
7 . . . 1 7 21 35 35 21 7 1
8 . . . 1 8 28 56 70 56 28 8 1
9 . . . 1 9 36 84 126 126 84 36 9 1
10 . . . 1 10 45 120 210 252 210 120 45 10 1

Figure 2. Pascal's triangle.

Each entry in Pascal's triangle can be computed in a very simple way from entries above it. Can you discover the rule? Can you explain, in terms of counting paths, why the numbers in Pascal's triangle should be related in this way?

Pascal's triangle has another interpretation, which makes it very useful in probability. All of its numbers are combination numbers. For example, the numbers at the 3-level,

1 3 3 1

are just

$$C_0^3 \quad C_1^3 \quad C_2^3 \quad C_3^3$$

which we know represent the number of subsets of various sizes of a three-element set.

$\{a, b, c\} \rightarrow \{\ \}$	$C_0^3 = 1$	0-element subset
$\{a\}, \{b\}, \{c\}$	$C_1^3 = 3$	1-element subsets
$\{a, b\}, \{a, c\}, \{b, c\}$	$C_2^3 = 3$	2-element subsets
$\{a, b, c\}$	$C_3^3 = 1$	3-element subset

The numbers at the 4-level,

$$1 \quad 4 \quad 6 \quad 4 \quad 1$$

are just the combination numbers

$$C_0^4 \quad C_1^4 \quad C_2^4 \quad C_3^4 \quad C_4^4$$

For example,

$$C_2^4 = \frac{P_2^4}{2!} = \frac{4 \cdot 3}{2 \cdot 1} = 6$$

Another way of writing Pascal's triangle, then, is

0 .... C_0^0

1 .... $C_0^1 \quad C_1^1$

2 .... $C_0^2 \quad C_1^2 \quad C_2^2$

3 .... $C_0^3 \quad C_1^3 \quad C_2^3 \quad C_3^3$

4 .... $C_0^4 \quad C_1^4 \quad C_2^4 \quad C_3^4 \quad C_4^4$

5 .... $C_0^5 \quad C_1^5 \quad \mathbf{C_2^5} \quad C_3^5 \quad C_4^5 \quad C_5^5$

Figure 3. Pascal's triangle.

Earlier, however, we found the same dark entry, $C_2^5 = 10$, by counting paths. The reason why there are C_2^5 paths that lead to the same point as the dark path in Figure 1 is because each such path can be described by filling 2 of the 5 blanks __ __ __ __ __ with R's (and the rest with L's). There are as many paths like this as there are ways of choosing 2 things from among 5, namely C_2^5.

EXERCISES

Use Pascal's triangle (Figure 2) to answer Exercises 1–7 below.

1. How many 2-element subsets has an 8-element set?

2. How many 6-element subsets has a 9-element set?

3. How many 4-man committees can be chosen from 10 persons?

4. How many 4-man Ping Pong teams can be chosen from 7 persons?

5. In how many ways can you choose 5 problems to work on a 7-problem test?

6. In how many ways can you choose 2 problems not to work on a 7-problem test?

7. Ten players enter a chess tournament. How many different pairs could meet in the championship match?

8. Evaluate each symbol below using the formula

$$C_r^n = \frac{P_r^n}{r!}$$

and then compare with the appropriate entry in Figure 2.

EXAMPLE

$$C_2^5 = \frac{P_2^5}{2!} = \frac{5 \cdot 4}{2 \cdot 1} = 10$$

which is the 2-entry at the 5-level in Figure 2. (We begin our counting of levels and entries with 0, not 1.)

(a) C_3^6 **(b)** C_3^4 **(c)** C_8^9 **(d)** C_1^7 **(e)** C_0^7

9. How does Pascal's triangle show that $C_0^n = C_n^n$, $C_1^n = C_{n-1}^n$, $C_2^n = C_{n-2}^n$, and so on?

10. **(a)** Explain *how* each entry in Pascal's triangle (Figure 2) can be calculated from entries above it.

(b) Explain *why*, in terms of counting paths, the numbers in Pascal's triangle should be related in this way.

(c) Complete the following:

$$C_r^n + C_{r+1}^n = C_?^?$$

11. **(a)** Add up all entries in Pascal's triangle that are at the 0-level, *ibid.* the 1-level, *ibid.* the 2-level, and so on.

(b) What do you think is the sum of all entries at the n-level?

(c) Use your answer to (b) to decide how many subsets an n-element set has. [Look back at Exercise 12, page 241.]

12. The "algebraic expression" $a + b$ is called a "*bi*nomial" because it is the sum of *two* "terms." Continue the "expansion" of powers of this binomial begun below and look for a pattern:

$$(a + b)^0 = 1$$

$$(a + b)^1 = 1a + 1b$$

$$(a + b)^2 = 1a^2 + 2ab + 1b^2$$

$(a+b)^3 =$

$(a+b)^4 =$

$(a+b)^5 =$

Can you explain *why* this pattern shows up?

13. Corroborate you answer to Exercise 11(b) by setting $a = 1$ and $b = 1$ in the binomial expansions in Exercise 12.

14. Suppose the apparatus of Figure 1 is constructed so that at each point the probability of the ball going right is the same as the probability of its going left. Compute the probabilities of these events.
  (a) The ball hits the center point at the 2-level.
  (b) The ball hits the center point at the 4-level.
  (c) The ball hits an end point at the 3-level.
  (d) The ball hits an end point at the 3-level or below.
  (e) The ball hits one of the three middle points at the 10-level.

15. The experiment of flipping a coin eight times can be simulated by the experiment of rolling a ball down the apparatus in Figure 1.
  (a) What is the probability of 4 heads and 4 tails?
  (b) What is the probability of more heads than tails?
  (c) What is the probability of more than twice as many of one kind as of the other?

16. Each of four people is asked to think of a number from among 1, 2, . . . , 9.
  (a) Describe a spinner experiment that simulates this experiment.
  (b) What is the probability that each person thinks of a different number?
  (c) Would you be willing to bet that at least two of the people would think of the same number?

17. Joe wants to raise goldfish in his pond, but cannot tell the difference between males and females. How many goldfish should he buy if he wants to be at least "90 percent certain" of getting fish of both sexes?

6.9 SOLUTION OF POLYA'S PROBLEM (OPTIONAL)

Back in Chapter 1 we posed three related questions, collectively known as Polya's problem.

1. Into how many pieces is a line separated by n points?
2. Into how many regions can a plane be separated by n lines?
3. Into how many chunks can space be separated by n planes?

In this section we give complete answers to these questions. Our solution to the problem is based on Pascal's triangle and particularly on this property of it:

Each entry that is not at the end of a row is the sum of the two entries above it.

Two versions of Pascal's triangle are reproduced below. You should find them useful in answering the questions that follow.

$$
\begin{array}{c}
1 \\
1 \quad 1 \\
1 \quad 2 \quad 1 \\
1 \quad 3 \quad 3 \quad 1 \\
1 \quad 4 \quad 6 \quad 4 \quad 1 \\
1 \quad 5 \quad 10 \quad 10 \quad 5 \quad 1 \\
\cdots\cdots\cdots\cdots\cdots\cdots \\
1 \quad (n-1) \quad \dfrac{(n-1)(n-2)}{2} \quad \cdots \quad (n-1) \quad 1 \\
1 \quad n \quad \dfrac{n(n-1)}{2} \quad \dfrac{n(n-1)(n-2)}{3 \cdot 2 \cdot 1} \quad \cdots \quad n \quad 1 \\
\cdots\cdots\cdots\cdots\cdots\cdots\cdots\cdots\cdots
\end{array}
$$

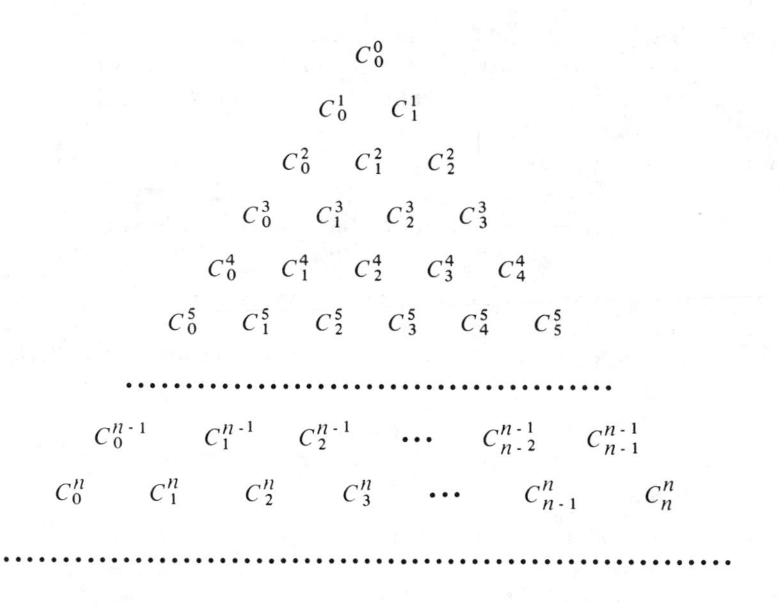

QUESTION 1 The figures that follow illustrate how each entry in the second diagonal of Pascal's triangle can be viewed as a sum of entries from the first diagonal.

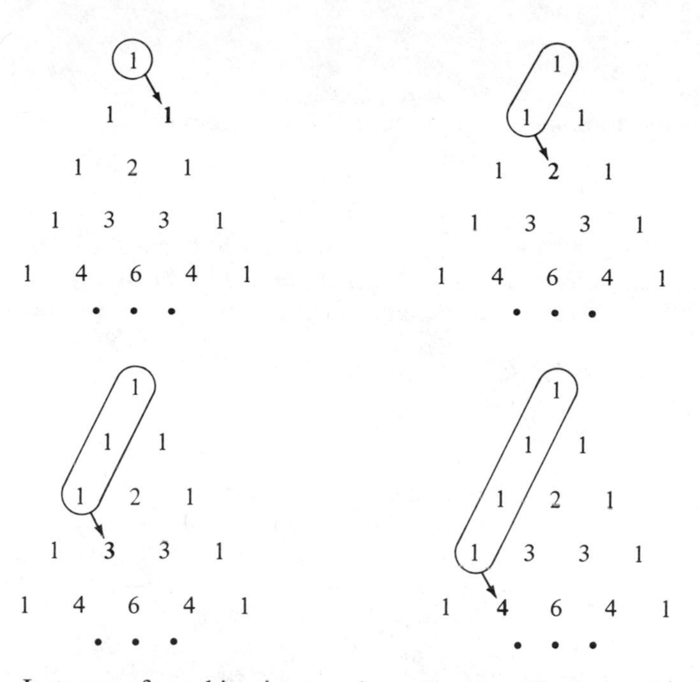

In terms of combination numbers, the general relationship is

$$C_0^0 + C_0^1 + C_0^2 + C_0^3 + \cdots + C_0^{n-1} = C_?^?$$

QUESTION 2 Discover how each entry in the third diagonal can be viewed as a sum of entries from the second diagonal. Explain how the diagrams below suggest *why* $1 + 2 + 3 + 4 = 10$.

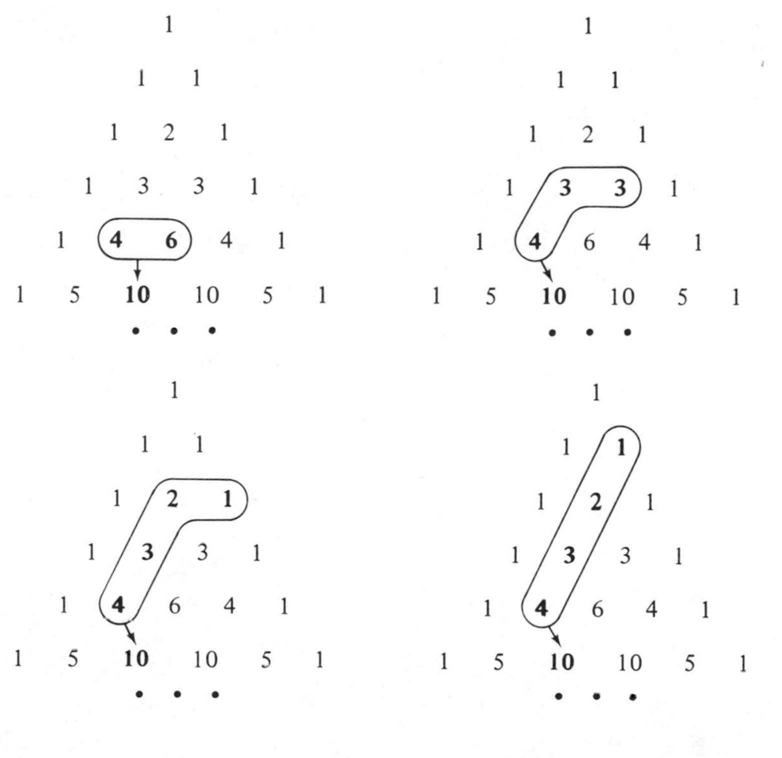

In terms of the combination numbers, the general relationship is

$$C_1^1 + C_1^2 + C_1^3 + \cdots + C_1^{n-1} = C_?^?$$

QUESTION 3 Discover how each entry in the fourth diagonal can be viewed as a sum of entries from the third diagonal. Can you explain *why* the relationship holds? In terms of the combination numbers, the general relationship is

$$C_2^2 + C_2^3 + C_2^4 + \cdots + C_2^{n-1} = C_?^?$$

QUESTION 4 (Polya's Problem—Part 1) Let $\ell(n)$ be the number of pieces into which n points separate a line. Clearly $\ell(0) = 1$. Now

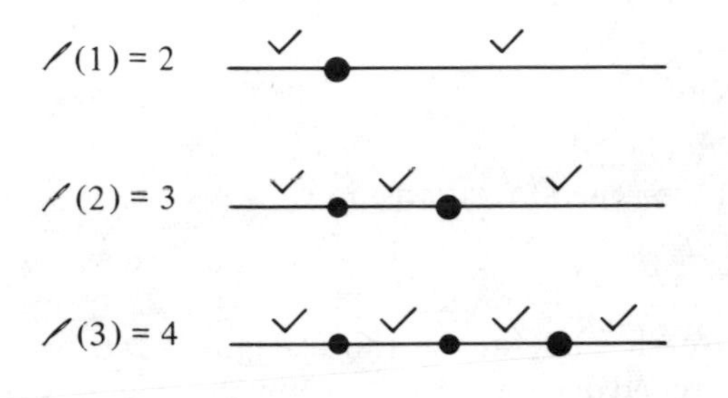

and in general,

$$\ell(n) = 1 + n$$

which we choose to write as

$$\ell(n) = C_0^n + C_1^n$$

What happens if we set $n = 0$ in the above formula? Simplify $\ell(0)$, $\ell(1)$, $\ell(2)$, $\ell(3)$, $\ell(4)$.

QUESTION 5 (Polya's Problem—Part 2). Let $p(n)$ be the number of pieces into which n lines can separate a plane. Clearly, $p(0) = 1$. Now

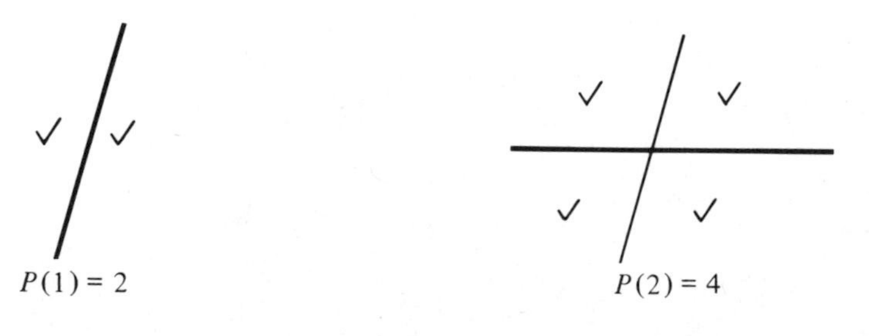

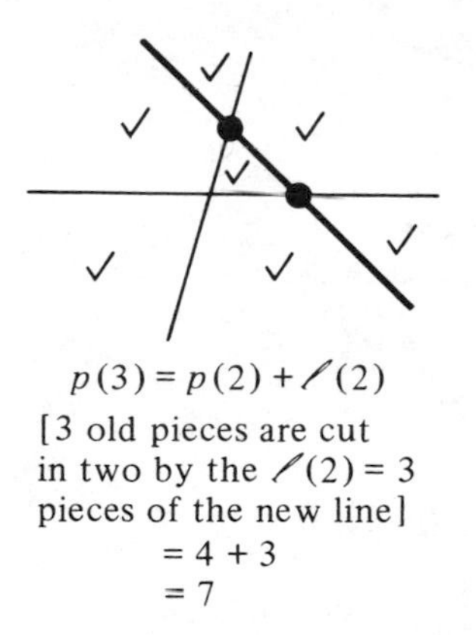

$p(3) = p(2) + \ell(2)$
[3 old pieces are cut in two by the $\ell(2) = 3$ pieces of the new line]
$= 4 + 3$
$= 7$

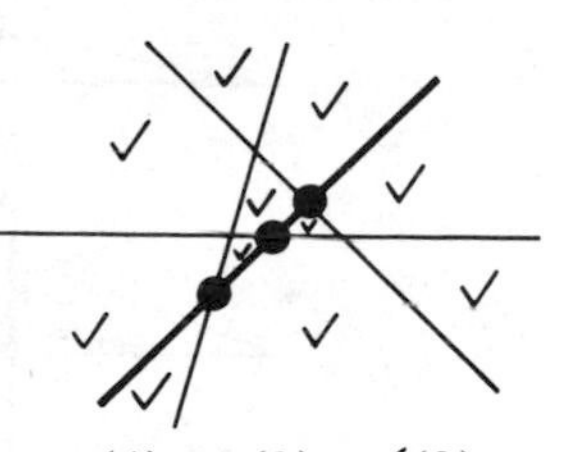

$p(4) = p(3) + \ell(3)$
[4 old pieces are cut in two by the $\ell(3) = 4$ pieces of the new line]
$= 7 + 4$
$= 11$

In general, for any k,

$$p(k) = p(k-1) + \ell(k-1)$$

because the kth line is drawn so that it hits all $k - 1$ of the previous lines in distinct points. These $k - 1$ points on the kth line separate it into $\ell(k-1)$ pieces, and each of these pieces cuts in two one of the regions formed by the first $k - 1$ lines. Using this relation over and over, we obtain

$$\begin{aligned} p(n) &= p(n-1) + \ell(n-1) \\ &= p(n-2) + \ell(n-2) + \ell(n-1) \qquad \text{Why?} \\ &= p(n-3) + \ell(n-3) + \ell(n-2) + \ell(n-1) \\ &\cdots \\ &= p(0) + \ell(0) + \ell(1) + \ell(2) + \cdots + \ell(n-2) + \ell(n-1) \end{aligned}$$

Now, using the formula for $\ell(n)$ and the fact that $p(0) = 1$,

$$\begin{aligned} p(n) &= 1 + [C_0^0 + C_1^0] + [C_0^1 + C_1^1] + [C_0^2 + C_1^2] \\ &\quad + \cdots + [C_0^{n-2} + C_1^{n-2}] + [C_0^{n-1} + C_1^{n-1}] \\ &= 1 + [C_0^0 + C_0^1 + C_0^2 + \cdots + C_0^{n-2} + C_0^{n-1}] \\ &\quad + [C_1^0 + C_1^1 + C_1^2 + \cdots + C_1^{n-2} + C_1^{n-1}] \\ &= 1 + C_1^n + C_2^n \qquad \text{Why?} \end{aligned}$$

which we choose to write in the form

$$p(n) = C_0^n + C_1^n + C_2^n$$

Simplify $p(0)$, $p(1)$, $p(2)$, $p(3)$, $p(4)$, $p(5)$.

QUESTION 6 (Polya's Problem—Part 3) Let $s(n)$ be the number of chunks into which n planes can separate space. Clearly, $s(0) = 1$. Now

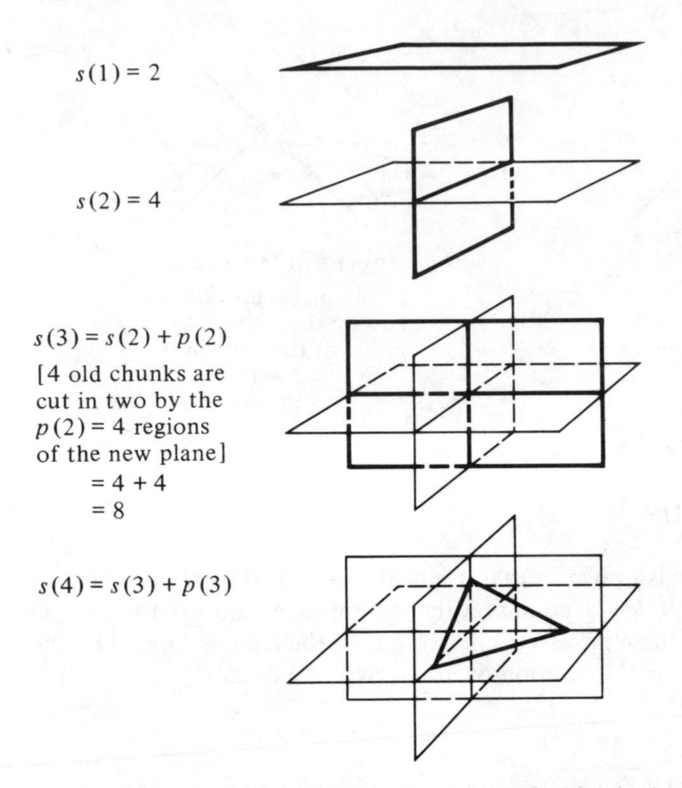

because the new fourth plane (only a piece of which is shown) is drawn so that it meets the first three planes in three nonconcurrent lines, no two of which are parallel. These three lines on the fourth plane separate it into $p(3) = 7$ pieces and each of these pieces cuts an old chunk in two. Thus,

$$s(4) = s(3) + p(3) = 8 + 7 = 15$$

Explain why, in general,

$$s(k) = s(k - 1) + p(k - 1)$$

for any k.

Using the above relation over and over, we obtain

$$\begin{aligned} s(n) &= s(n-1) + p(n-1) \\ &= s(n-2) + p(n-2) + p(n-1) \qquad \text{Why?} \\ &= s(n-3) + p(n-3) + p(n-2) + p(n-1) \\ &\cdots \\ &= s(0) + p(0) + p(1) + p(2) + \cdots + p(n-1) \end{aligned}$$

Now, using the fact that $s(0) = 1$ and the formula for $p(n)$,

$$\begin{aligned}
s(n) &= 1 + [C_0^0 + C_1^0 + C_2^0] + [C_0^1 + C_1^1 + C_2^1] \\
&\quad + [C_0^2 + C_1^2 + C_2^2] + \cdots + [C_0^{n-1} + C_1^{n-1} + C_2^{n-1}] \\
&= 1 + [C_0^0 + C_0^1 + C_0^2 + \cdots + C_0^{n-1}] \\
&\quad + [C_1^0 + C_1^1 + C_1^2 + \cdots + C_1^{n-1}] \\
&\quad + [C_2^0 + C_2^1 + C_2^2 + \cdots + C_2^{n-1}] \\
&= 1 + C_1^n + C_2^n + C_3^n \qquad \text{Why?}
\end{aligned}$$

which we choose to write in the form

$$s(n) = C_0^n + C_1^n + C_2^n + C_3^n$$

Simplify $s(0)$, $s(1)$, $s(2)$, $s(3)$, $s(4)$, $s(5)$.

QUESTION 7 You slice through a block of cheese ten times in a random fashion. How many pieces of cheese result?

VOCABULARY

6.1 Comparison, Measurement, and Positive Rational Numbers

operator (→ as machine)
equivalent
relative frequency

6.2 Experiments: Outcomes and Their Probabilities

random experiment
outcome
probability of an outcome
probability function, p
trial
outcome set (sample space)
probability diagram
simulate

6.3 Events and Their Probabilities

event
probability of an event, P
impossible event
certain event
random variables

6.4 Probability Trees

probability tree
cartesian product

first component
second component
multiplication rule for probabilities

6.5 Tree Diagrams in Problem Solving

composition (of operators)

6.5 Counting Permutations

favorable outcomes
permutation
factorial ($n!$)
number of permutations of n things taken k at a time, P_k^n
product rule for counting .

6.7 Counting Combinations

combination
number of combinations of n things taken k at a time, C_k^n
random walk

6.8 Pascal's Triangle

Pascal's triangle

GEOMETRIC CONSTRUCTIONS AND DRAWINGS

OVERVIEW

Educational psychologists today generally agree that manipulative experience is a desirable base on which to build mathematical concepts Consequently, mathematics instruction in the modern elementary school places heavy stress on a wide range of manipulative materials: attribute blocks, cuisenaire rods, Dienes blocks, tangrams, geoboards, mirror cards, balance scales, probability spinners, and so on. It is not surprising that this movement should also embrace such hoary devices as the straightedge and compass. Once the tools of high school geometry, the straightedge and compass are now used extensively in the K-6 curriculum to make meaningful the abstract concepts of geometry. For example, the abstract concept of perpendicularity becomes much more concrete once the child is able to construct a perpendicular to a line with a straightedge and compass.

This chapter has several goals. One is simply to review the basic techniques of construction and drawing that are presently in use in the elementary schools. Another is to make more concrete the concepts of slide, turn, and flip, which may have been unfamiliar when first introduced. A third is to make plausible, via constructions, the congruence conditions for triangles, which will serve as foundation for much of the work in similarity and trigonometry to follow.

In Section 7.1 we make the distinction between constructing and drawing, and we indicate very briefly the two different schools of geometric thought to which these different techniques belong.

Sections 7.2–7.5 cover the basic geometric constructions. Each construction is accompanied by a technique for drawing the same kind of figure. (Constructions are given top billing only because we have done so much work already with the other drawing instruments, ruler and protractor.) Among other skills, by the end of Section 7.5 one has the ability to construct (in principle) the rational number line.

In Section 7.6 the basic SAS, ASA, and SSS congruence conditions are made plausible via construction techniques. In Section 7.7 the situation is reversed: The congruence conditions are assumed and the construction techniques are then justified in terms of those conditions.

7.1 CONSTRUCTIONS AND DRAWINGS

The ancient Greeks were the first people to organize geometric information into a unified subject of study. Literal translations of Euclid's geometry books have been used as texts for 2000 years. Many geometry texts of only a generation ago followed Euclid's path very closely. Only in recent years have significant changes been made in the teaching of geometry. These changes were largely inspired by the great modern mathematicians, David Hilbert (1862–1943) and George David Birkhoff (1884–1944).

We encountered some of Hilbert's important ideas when we studied "separation" and "betweenness" in Chapter 1. Hilbert's efforts were directed toward tightening up the logical structure of Euclid's approach by including further axioms which explicate certain properties that had been passed over as intuitively obvious by Euclid. A sample of one of Hilbert's extra axioms is this:

> For any two distinct points A and B there is always a point C between A and B, and a point D such that B is between A and D.

Birkhoff's approach was more revolutionary. He proposed replacing the Euclid-Hilbert axioms by a radically different set of axioms suggested by the *ruler* and the *protractor*. A sample of one of Birkhoff's axioms is this so-called "ruler axiom":

> For each line there exists a one-to-one correspondence between the points of the line and the set of all real numbers such that the distance between each two points of the line is the absolute value of the difference of the real numbers that correspond to the two points.

Euclid and the other Greek geometers made much greater use of the **straightedge** (a ruler tipped upside down so none of the markings show) and the **compass**.

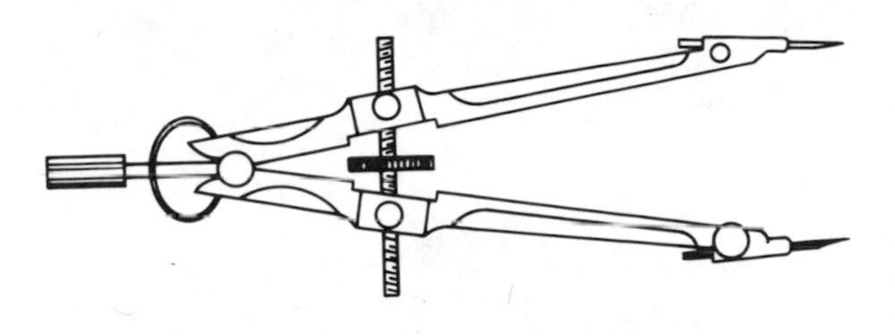

compass

(A "Euclidean compass," like the straightedge, has no markings on it. Your compass may have a scale on it indicating the distance from tip to tip You should ignore that scale in this chapter.) They were interested in seeing just what kinds of figures they could depict using only those two instruments. We refer to such pictures as **geometric constructions**.

A construction is a picture of a geometric figure made with only straightedge and compass.

You may remember a geometric construction of a regular hexagon.

1. Draw a circle with a *compass*.
2. Without changing the opening of the compass, put the metal tip on the circle and "swing off an arc" to cut the circle.

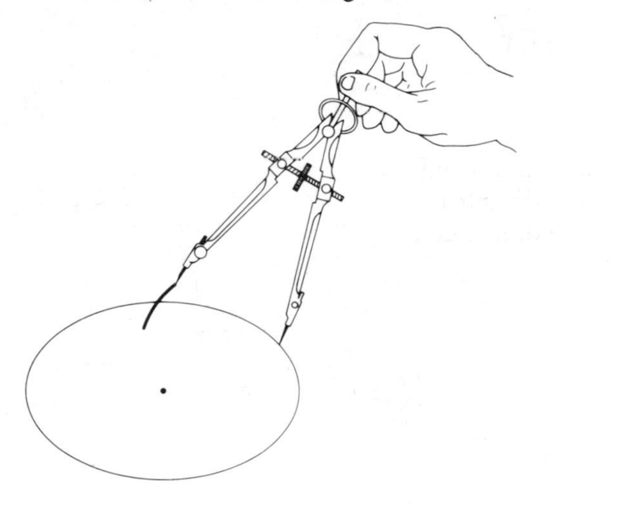

3. Move the compass so that the metal tip is at the point of intersection of the circle and the arc. Swing off another arc.
4. Move the compass to the new intersection point and swing another arc, and so on.
5. When you have gone completely around the circle, join the six points with line segments using the *straightedge*.

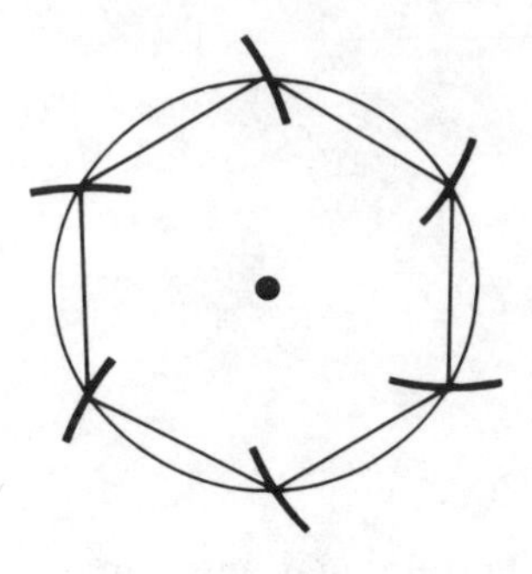

This is how the ancient Greeks constructed a regular hexagon.

You may also remember how to draw a regular hexagon using a *protractor* (and compass and straightedge).

1. Draw a circle with a compass.
2. Notice that if a hexagon were "inscribed" in the circle and if each of its six vertices were joined to the center of the circle, then six congruent angles would be arranged around the center of the circle. Each, then, would have to measure $360°/6 = 60°$.

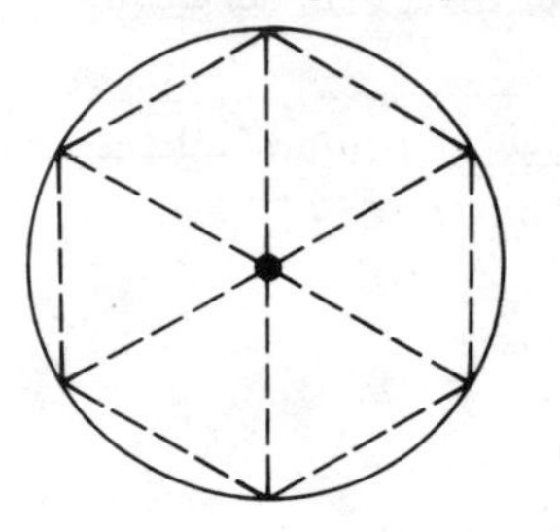

3. Draw a radius, using the straightedge.
4. Use protractor and straightedge to draw a $60°$ angle having the original radius as one of its sides.

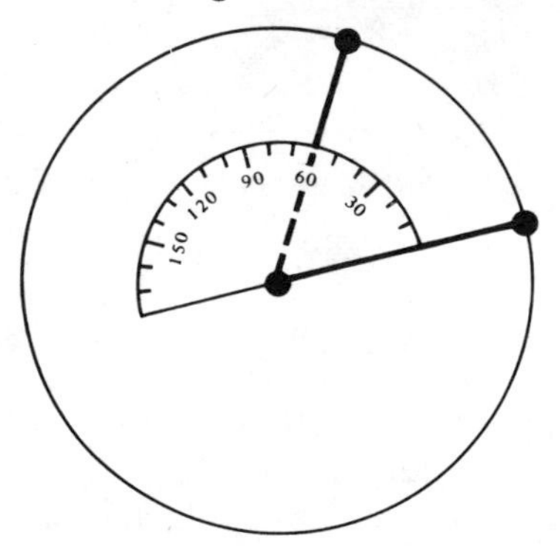

5. Move the protractor to the second radius drawn, draw another $60°$ angle, and so on.
6. Finally, join the six points on the circle with segments.

This procedure is more in the modern spirit of Birkhoff.

You probably prefer the straightedge-compass construction. It is quicker, easier, and more exact, but only because we drew hexagons. What if we had wanted to draw a regular nonagon? Using a protractor we would proceed exactly as before, except now we would want nine, instead of six, congruent angles arranged around the center of the circle. Now each would have to have measure $360°/9 = 40°$.

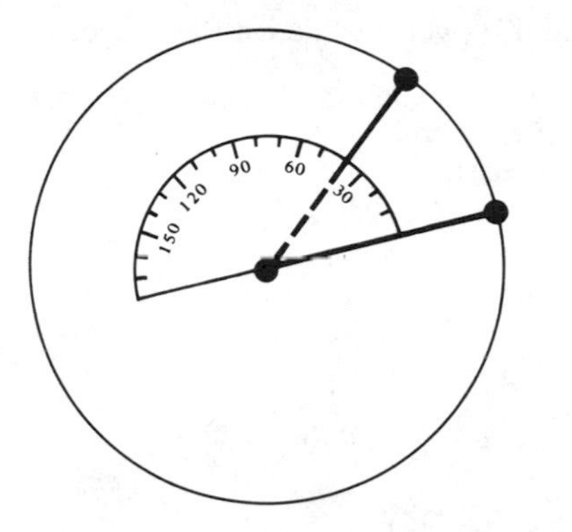

Using only a straightedge and compass we would *never* be able to construct a regular nonagon! This construction has been *proved* to be *impossible*.

In this chapter we shall study the most basic straightedge-compass constructions. You might like to imagine that you are learning an ancient handcraft. Drawing geometric figures with only a compass and straightedge is a little like making pottery with only a potter's wheel, a knife, and your bare hands. Sometimes you turn out a beautiful piece (like a regular hexagon) that makes a machine-made (protractor-made) copy look clumsy. Other times (as in the case of the nonagon), your simple equipment just cannot do what a machine can.

CAUTION In this chapter we are very finicky about language. When you are asked to *construct* a figure you are restricted to using just two geometric tools, straightedge and compass. When you are asked to *draw* a figure you are free to use ruler and protractor as well as straightedge and compass.

EXERCISES

1. Sketch a convex octagon and join one vertex to all other nonadjacent vertices with segments.

(a) Into how many triangles has the octagon been chopped?

(b) What is the sum of the measures of the interior angles of the octagon?

2. (a) The sum of the measures of the interior angles of a convex n-gon is ________

(b) Each interior angle of a regular n-gon has measure ________

3. **(a)** Draw a horizontal segment 2 in. long in the middle of a clean sheet of paper.
   (b) Using only a ruler and protractor, draw a regular pentagon having this segment as one of its sides.
   (c) Repeat part (a). Then, using only a straightedge, compass, and protractor, draw a regular pentagon having this segment as one of its sides.

A polygon is said to be **inscribed** in a circle if its vertices lie on the circle.

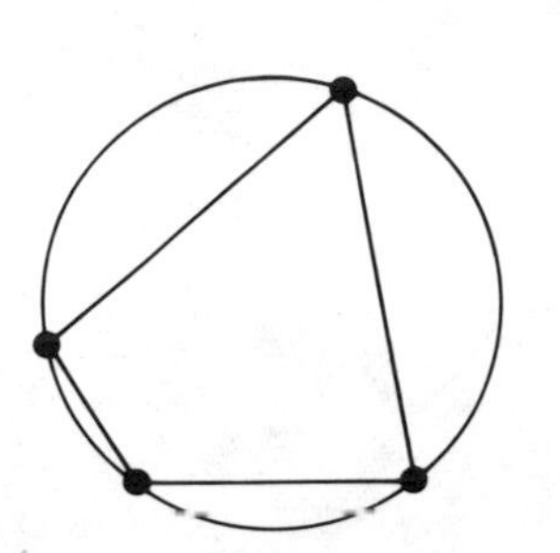

a quadrilateral inscribed in a circle

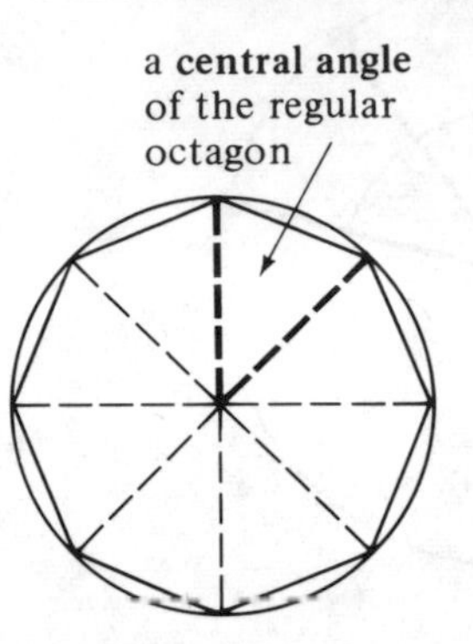

a regular octagon inscribed in a circle

4. A regular octagon is inscribed in a circle and each vertex is joined to the center of the circle as shown. What is the measure of each of the eight central angles?
5. Use straightedge, compass, and protractor to inscribe a regular octagon in a circle.
6. Interpret the schematic diagram below.

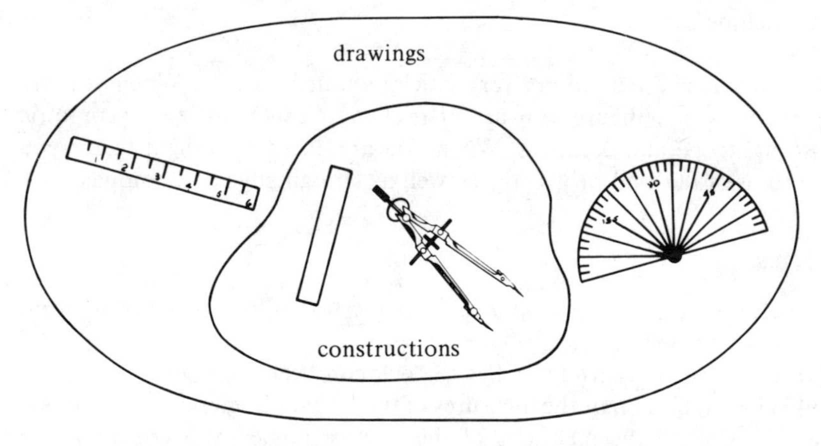

7. **(a)** If you were presented with a regular octagon, how would you go about drawing the circle in which it is inscribed?
   (b) Is this drawing in fact a construction?

(c) Repeat (a) and (b) for a regular pentagon.
(d) What do you think is meant by the **center** of a regular polygon?

8. (a) Write an algebraic expression for the measure of a *central angle* of a regular n-gon.
(b) Write an algebraic expression for the measure of an *interior angle* of a regular n-gon.
(c) Find the value of n for which a regular n-gon has its central angles and interior angles of the same size.

***9.** For this exercise let your unit of measure be a quarter of an inch. In the middle of a large sheet of paper mark two points, A and B, exactly 6 units apart. Now draw the family of 12 circles with center at A and radii 1, 2, ..., 12 units. Draw the same sort of family of circles with center at B.
(a) Darken $\{P \mid PA = 3 \text{ and } PB = 5\}$. That is, darken the set of all points P such that the distance from P to A is 3 and the distance from P to B is 5.
(b) Darken $\{P \mid PA = 4 \text{ and } PB = 4\}$.
(c) Darken $\{P \mid PA = 7 \text{ and } PB = 1\}$.
(d) Graph $\{P \mid PA + PB = 8\}$.
Using a different color for each one, graph these other sets:
(e) $\{P \mid PA + PB = 7\}$
(f) $\{P \mid PA + PB = 10\}$
(g) $\{P \mid PA + PB = 12\}$
(h) $\{P \mid PA + PB = 6\}$
(i) $\{P \mid PA + PB = 5\}$
The figures you graphed in parts (d)–(g) are called **ellipses**, and the points A and B are called their **foci**. Kepler's first law of planetary motion states that the orbit of each planet is an ellipse with the Sun at one focus. If points A and B above represent the Sun and the Earth's "other" focus, then the Earth's orbit would be very nearly the graph of $\{P \mid PA + PB = 360\}$.
(j) What is the approximate shape of the Earth's orbit?

***10.** Draw points A and B and the two families of circles as in Exercise 9. Graph the following figures, each in a different color:
(a) $\{P \mid PA - PB = 1\}$
(b) $\{P \mid PA - PB = 2\}$
(c) $\{P \mid PA - PB = 3\}$
(d) $\{P \mid PA - PB = 4\}$
(e) $\{P \mid PA - PB = 5\}$
(f) $\{P \mid PA - PB = 6\}$
(g) $\{P \mid PA - PB = 7\}$
(h) $\{P \mid PA - PB = 0\}$
Each figure you graphed in parts (a)–(e) is called a **branch** of a **hyperbola**. A complete hyperbola consists of two branches. Graph the following complete hyperbola.
(i) $\{P \mid |PA - PB| = 4\}$

7.2 COPYING FIGURES

The simplest way to copy a geometric figure, and the way that is most closely related to the superposition meaning of congruence, is to place a sheet of translucent paper over the figure and trace. A (congruent) copy of any polygon can also be drawn very easily using a ruler and protractor.

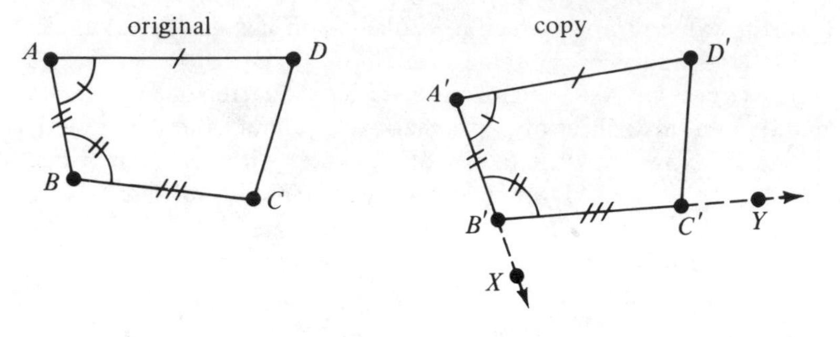

1. Measure $\overline{AD}$ (with the ruler) and draw a segment $\overline{A'D'}$ of the same length (with the ruler). Why will $\overline{AD}$ and $\overline{A'D'}$ be congruent?
2. Measure $\angle DAB$ (with the protractor) and draw an angle $\angle D'A'X$ of the same measure (with protractor and ruler). Why will $\angle DAB$ and $\angle D'A'X$ be congruent?
3. Measure $\overline{AB}$ and mark a point B' on ray $\overrightarrow{A'X}$ so that $\overline{AB}$ and $\overline{A'B'}$ have the same length.
4. Measure $\angle ABC$ and draw an angle, $\angle A'B'Y$, of the same measure.
5. Measure $\overline{BC}$ and mark a point C' on ray $\overrightarrow{B'Y}$ so that $\overline{BC}$ and $\overline{B'C'}$ have the same length.
6. Draw $C'D'$.

We would like to be able to *construct* a (congruent) copy of any polygon using only straightedge and compass. If you study the steps we followed in copying quadrilateral $ABCD$, with ruler and protractor, you will probably see that the hard problem of copying a polygon breaks down into many easy problems of just two kinds:

1. copying segments
2. copying angles

Both of these things can be done with only a compass and straightedge.

The procedure for constructing a (congruent) copy of a given line segment is left as an exercise.

To construct a (congruent) copy of a given angle, $\angle O$

1. Draw a ray $\overrightarrow{O'X}$

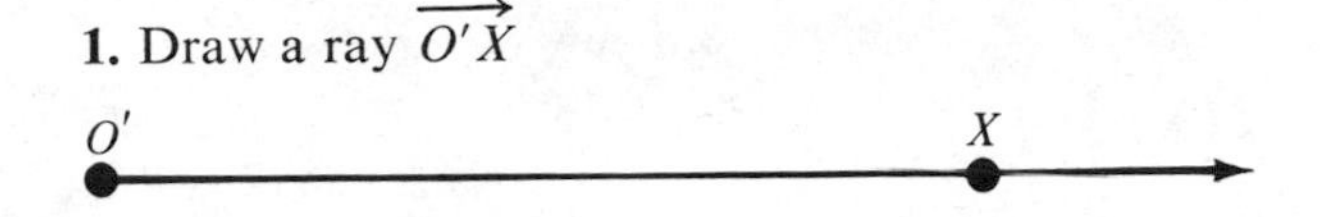

2. Place the metal tip of the compass at the vertex O of the original angle and swing an arc, of convenient radius, to cut both sides of the angle. Let A and B be the points of intersection.

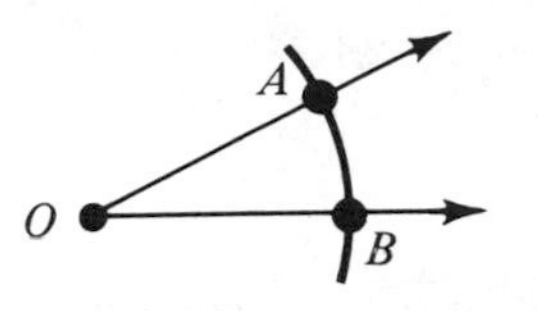

3. Without changing the opening of the compass, place the metal tip at O' and swing an arc. Let B' be the point of intersection of arc with ray.

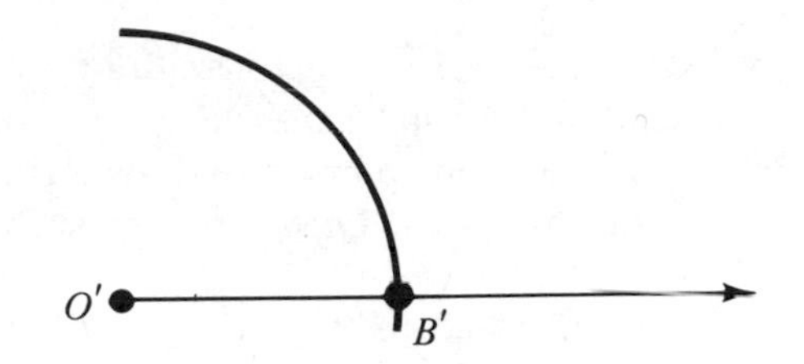

4. Reset the compass so that one tip is at A and the other at B.

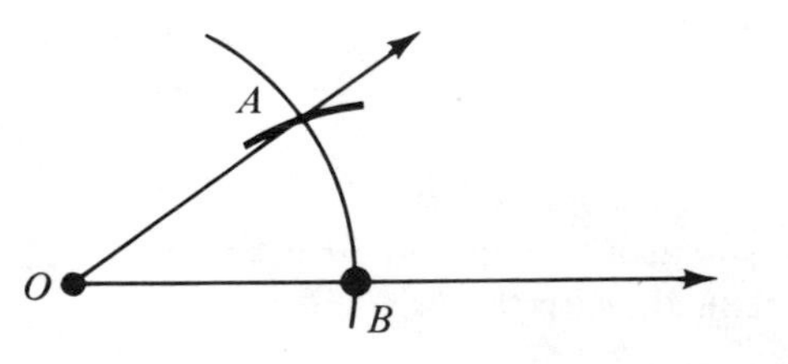

5. Without changing the opening of the compass, place the metal tip at B' and swing an arc to cut the first arc. Let A' be the point of intersection of these arcs.

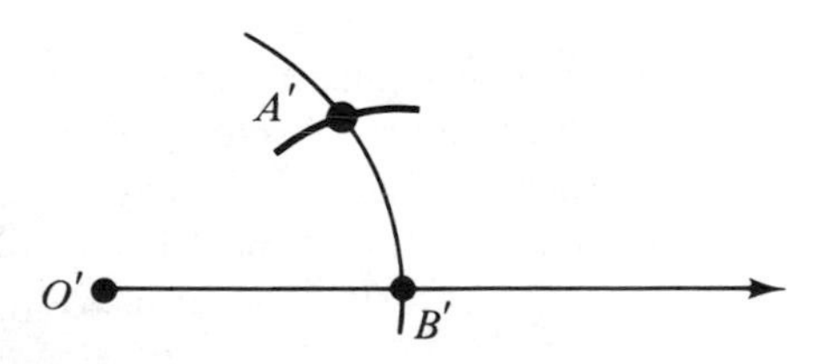

6. Draw ray $\overrightarrow{O'A'}$.

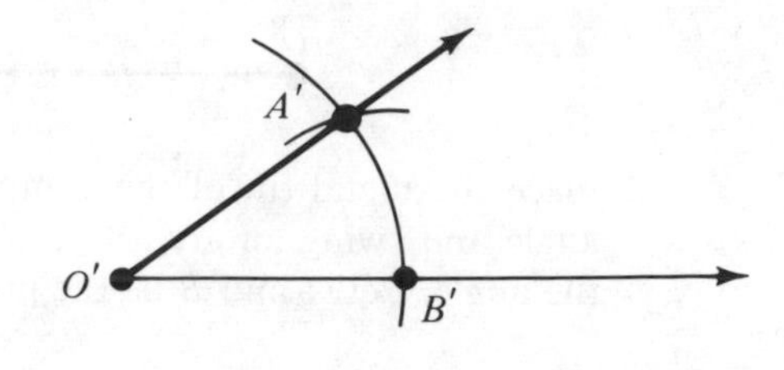

$\angle A'O'B' \cong \angle AOB.$

EXERCISES

1. Write careful, step-by-step instructions for how to construct a congruent copy of the given line segment $\overline{AB}$.

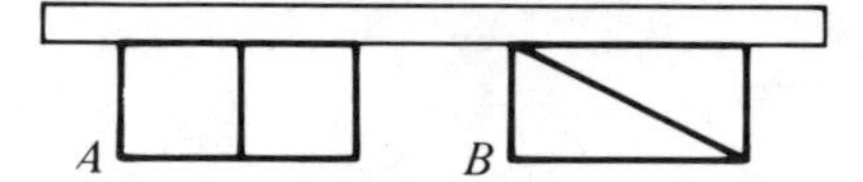

2. Use ruler and protractor to draw a triangle having two adjacent sides of lengths 2 in. and 3 in. which form a 54° angle.

3. Use ruler and compass to draw a triangle whose sides have lengths 2 in., 3 in., and 4 in. Draw another such triangle. Do the two seem to be congruent?

4. Use ruler and compass to draw a quadrilateral whose sides have lengths 2 in., 3 in., 4 in., 5 in. Draw another such quadrilateral. Do the two seem to be congruent?

5. Which is the better support for the platform, A or B? Why?

A B

6. Draw a copy of the figure shown using ruler and protractor. Check that your copy can be made to coincide with the original.

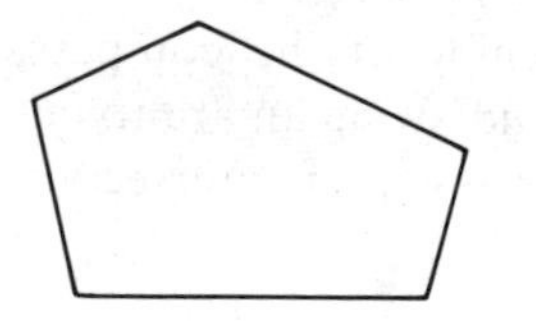

7. We made five measurements in copying the quadrilateral on p. 000 with ruler and protractor. We measured three sides and two angles in the order SASAS.

(a) Could we have copied the quadrilateral by beginning with an angle instead of a side and measuring ASASA?

(b) How many measurements would have to be made to copy this pentagon?

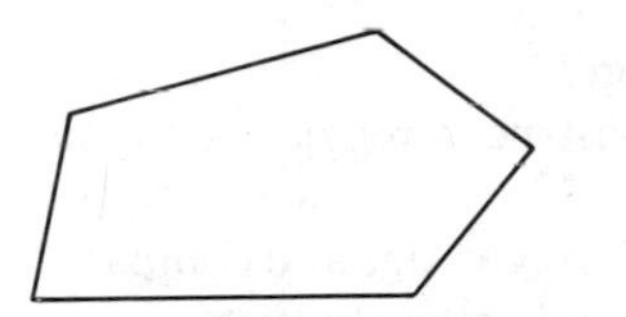

(c) How many to copy a hexagon?
(d) Fill in this chart:

POLYGON	NUMBER OF RULER AND PROTRACTOR MEASUREMENTS MADE IN COPYING IT
triangle	______
quadrilateral	5
pentagon	______
hexagon	______
heptagon	______
octagon	______
...	...
n-gon	______
...	...
______	197

8. For each geometric figure described, do four things: (i) *Draw* it using ruler and protractor; (ii) *construct* a copy of your drawing using straightedge and compass; (iii) *check* with ruler and protractor that the copy you constructed has the same measurements as the original; (iv) *check* that your figures can be made to coincide.
(a) 3-in. segment
(b) 65° angle
(c) a triangle having a pair of sides of lengths 2 in. and 3 in. meeting in a right angle

9. Draw an angle $\angle AOB$ and try this construction for copying it.

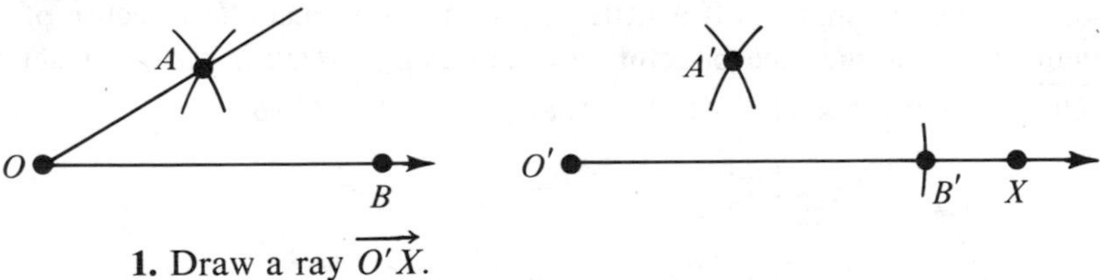

1. Draw a ray $\overrightarrow{O'X}$.
2. Mark B' on $\overrightarrow{O'X}$ so that $OB = O'B'$.
3. Swing an arc, from B', with radius BA.
4. Swing an arc, from O', with radius OA.
5. Let A' be the point of intersection of the arcs and draw $\overrightarrow{OA'}$.

10. Suppose that you were handed a 20° angle.
  (a) Could you then construct a 40° angle?
  (b) Could you then construct a regular nonagon?
  (c) We stated (p. 281) that it is impossible to construct a regular nonagon. It is possible to construct a 20° angle?
  (d) Is it possible to trisect a 60° angle using only straightedge and compass?

11. Construct the image of the figure $\mathscr{A}$ under the indicated rotation.

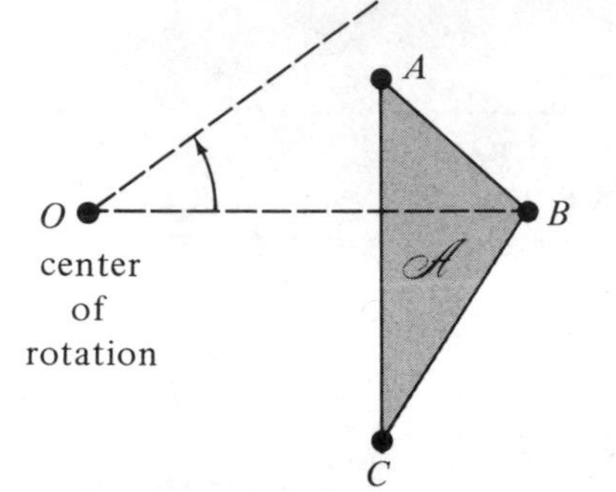

12. Draw the image of the figure $\mathscr{B}$ under a counterclockwise turn, about P, of 80°.

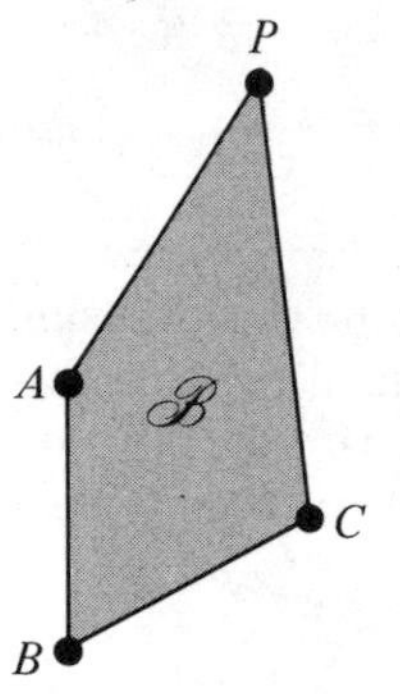

7.3 BISECTING ANGLES

It was easy to construct a copy of a segment, but a little harder to construct a copy of an angle. Just the opposite is true of bisecting. It is easy to construct the bisector of an angle, and a little harder to construct the bisector of a segment. To construct the **bisector** of an angle, $\angle AOB$, is to construct a ray, $\overrightarrow{OC}$, in the interior of $\angle AOB$, so that $\angle AOC \cong \angle COB$.

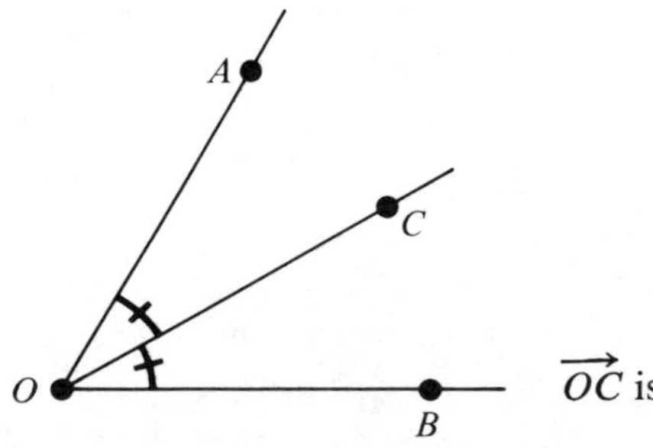

$\overrightarrow{OC}$ is the bisector of $\angle AOB$.

The construction of $\overrightarrow{OC}$ takes just four steps.

EXAMPLE *Construct* the bisector of $\angle AOB$.

1. With the metal tip of the compass at O, swing an arc to intersect both $\overrightarrow{OA}$ and $\overrightarrow{OB}$.

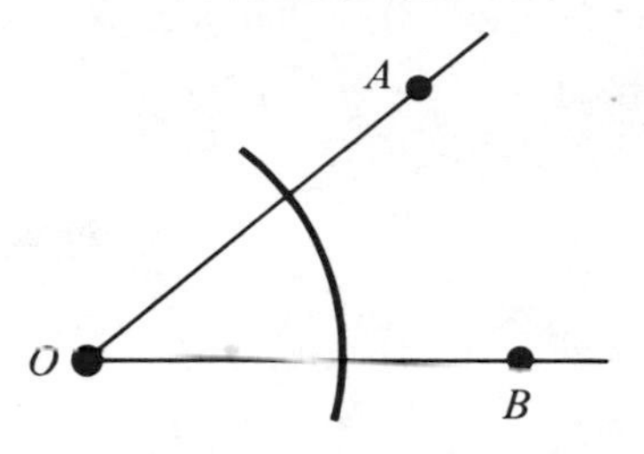

(Now you can change the opening of the compass if you like. We have done so here.)

2. Place the metal tip of the compass at one of the intersection points and swing an arc in the interior of $\angle AOB$.

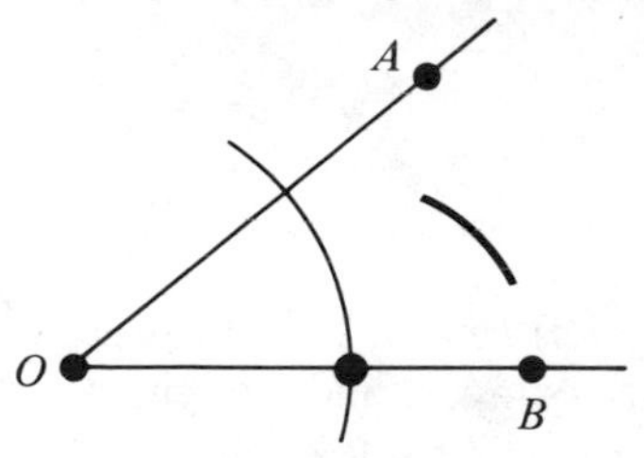

3. Without changing the opening of the compass, place its metal tip at the other intersection point and swing an arc to cut the arc drawn in step 2. Let C be the point of intersection

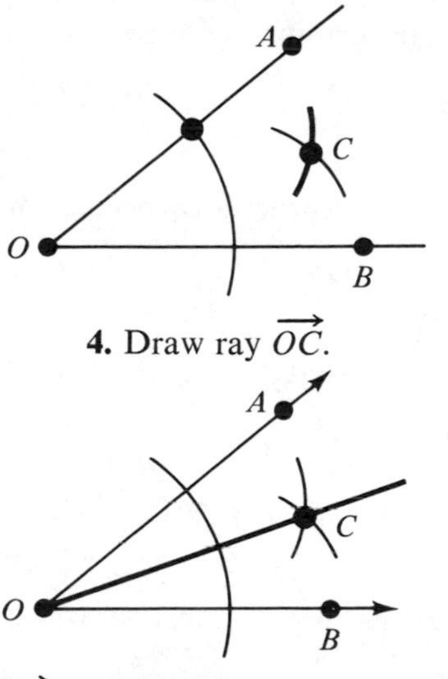

4. Draw ray $\overrightarrow{OC}$.

$\overrightarrow{OC}$ bisects $\angle AOB$.

If we are willing to use a protractor, bisecting an angle is even easier.

EXAMPLE *Draw* the bisector of $\angle AOB$.

1. Measure $\angle AOB$ with the protractor. $m\,\angle AOB = 64°$.
2. Divide its measure by two. $\frac{64}{2} = 32$.
3. Use the protractor to find a point C (inside $\angle AOB$) so that $m\,\angle COB = 32°$.

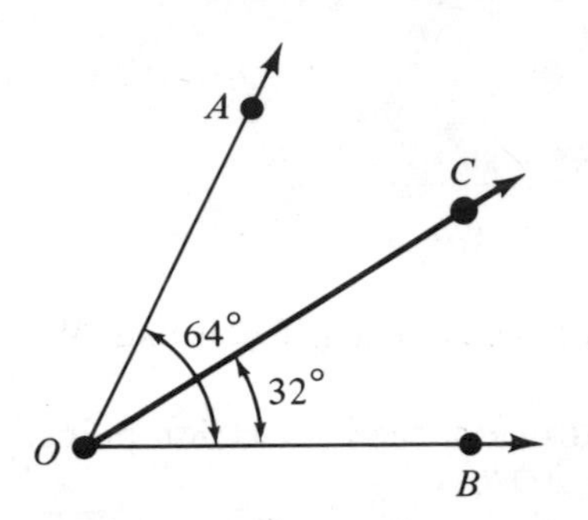

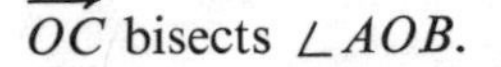
$\overrightarrow{OC}$ bisects $\angle AOB$.

EXERCISES

1. For each angle described below, do three things: (i) *Draw* it using protractor and straightedge; (ii) *construct* its bisector with compass and straightedge; (iii) *check* with the protractor that the angles on either side of the bisector have the same measure.
   (a) 70° **(b)** 170° **(c)** 180°
2. Construct a line perpendicular to ℓ at A. Recall the definition: m is perpendicular to ℓ if m and ℓ intersect so that adjacent angles are congruent.

3. Write out a step-by-step description of how to construct a perpendicular to a line ℓ through a point A on ℓ.
4. **(a)** Construct a 45° angle.
   (b) Describe how to construct a regular octagon.
5. Without using a protractor find the measures, in degrees, of each of these angles:
   (a) $\angle EOB$
   (b) $\angle AOB$
   (c) $\angle COE$
   (d) $\angle DOB$

   (See the figure at the top of p. 291).
6. Without doing any measuring or constructing, decide through which marks on the protractor in these rays will pass:

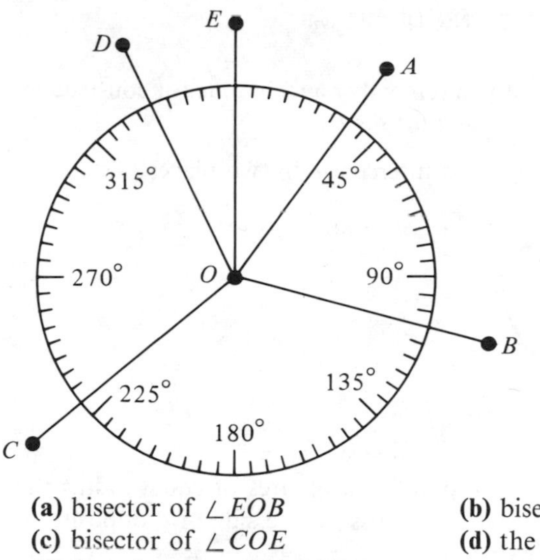

(a) bisector of $\angle EOB$

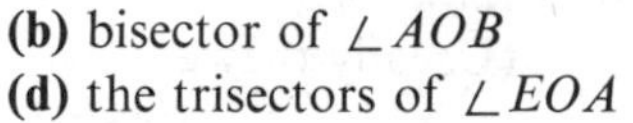

(b) bisector of $\angle AOB$

(c) bisector of $\angle COE$

(d) the trisectors of $\angle EOA$

(e) the trisectors of $\angle COD$

7. Sketch an angle of about 80°. Now construct an angle that is $\frac{3}{8}$ as large.

8. (A "practical application" of angle bisection) If you are ever lost, but have a wristwatch and the sun is shining, here is a way to find approximate directions. Hold the watch so the face is up and parallel to the ground. Point the hour hand in the direction of the sun. The bisector of the angle between the hour hand and "12 o'clock" on your watch will point south. For this method to work, in what direction should the sun be at 12 noon? 9 a.m.? 6 a.m.? 3 p.m.? 6 p.m.? Describe how to tell time with a compass on a sunny day.

9. Draw an angle on a sheet of scratch paper. Find its bisector by folding the paper.

7.4 BISECTING SEGMENTS; PERPENDICULARS

In the previous section we developed a technique for constructing a *perpendicular to a line through a point* ***on*** *the line.*

1. Think of ℓ as a 180° angle with vertex at A.

2. Bisect that angle in the usual way: Swing arcs from A to intersect ℓ on either side of A; open the compass a little further and from these intersection points swing arcs that cross; draw the line through A and this crossing point. This line is perpendicular to ℓ.

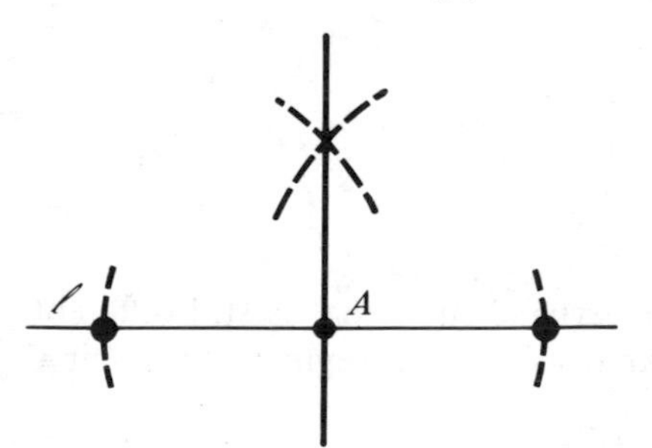

Before reading further, see if you remember a technique for constructing a *perpendicular to a line through a point **off** the line.*

1. From point A, swing an arc to intercept ℓ in two places.

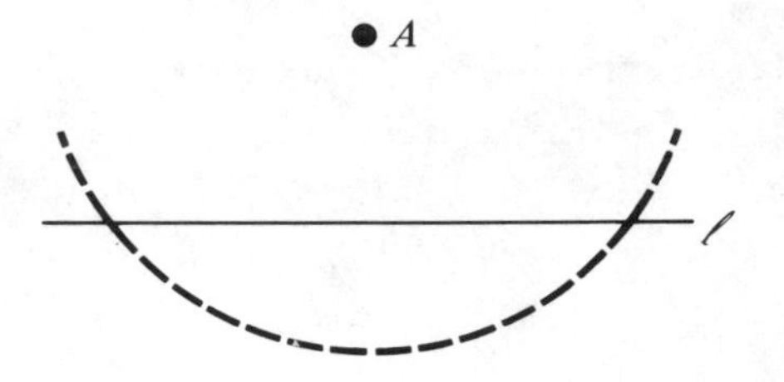

2. From these two intersection points, swing arcs of equal radius that cross. (Usually they are made to cross on the side of ℓ opposite A.)

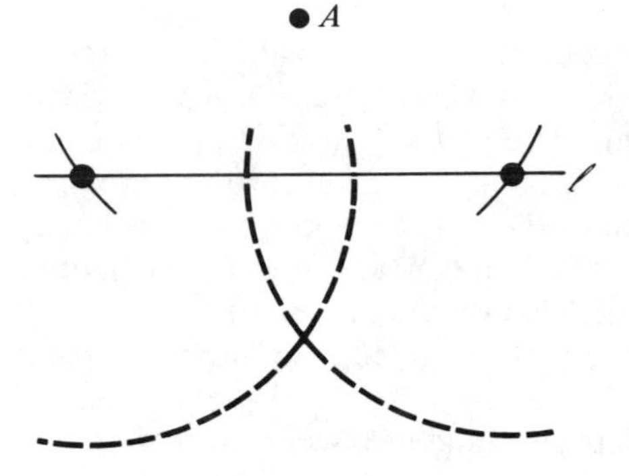

3. Draw the line through A and this crossing point. This line is perpendicular to ℓ.

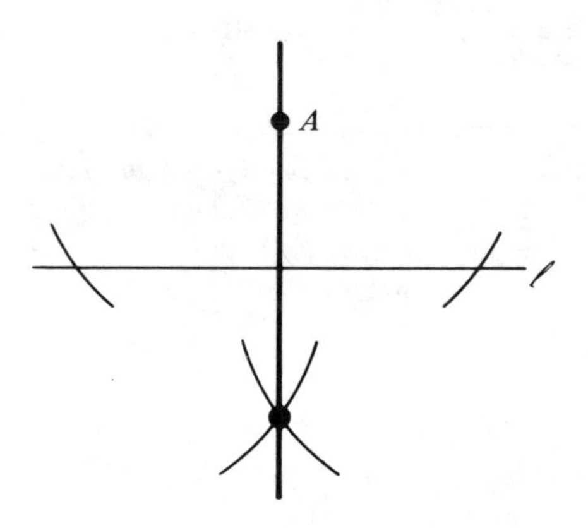

There is a third construction of a perpendicular that is still different. It is the construction of a perpendicular bisector of a segment. A point C

between A and B is called the **midpoint** of $\overline{AB}$ if $\overline{AC}$ is congruent to CB.

A C B

C is the midpoint of $\overline{AB}$.

Any line through this midpoint (except the line $\overleftrightarrow{AB}$) is called a **bisector** of $\overline{AB}$. Among these bisectors is just one line perpendicular to $\overline{AB}$. It is called the **perpendicular bisector** of $\overline{AB}$.

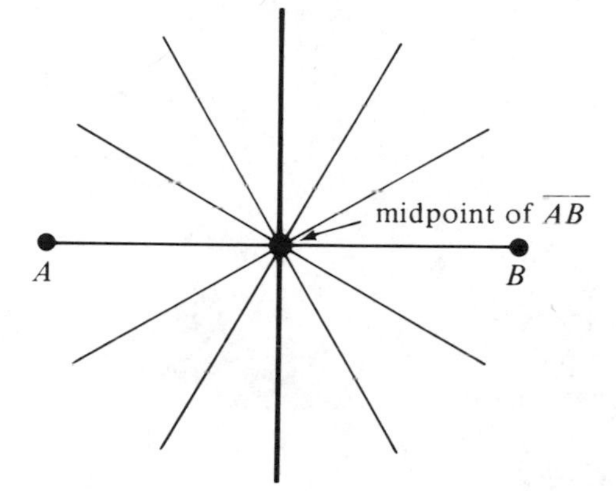

Many bisectors of $\overline{AB}$—just one perpendicular bisector.

If the midpoint of segment $\overline{AB}$ has already been located, then constructing the perpendicular bisector is easy. All you need to do is follow the first procedure, the one for constructing a perpendicular to a line from a point on the line. The importance of the third construction, which follows, is that *it locates the midpoint* of $\overline{AB}$ for you.

1. From A and B draw arcs (of the same radius) to intersect on both sides of $\overline{AB}$. (The radius will have to be more than half the length of $\overline{AB}$.)

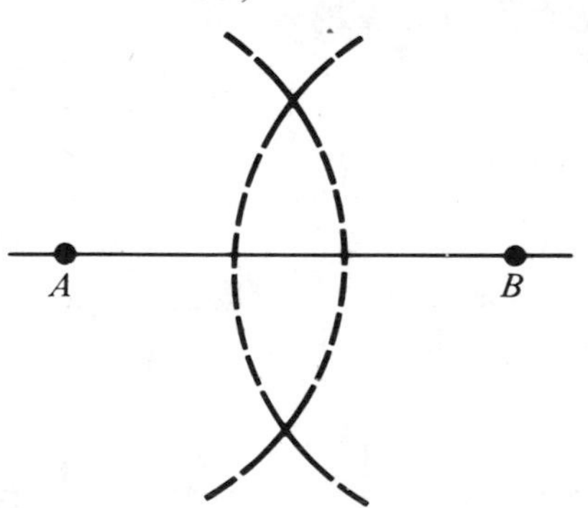

2. Draw the line through the two intersection points. This line is the perpendicular bisector of $\overline{AB}$.

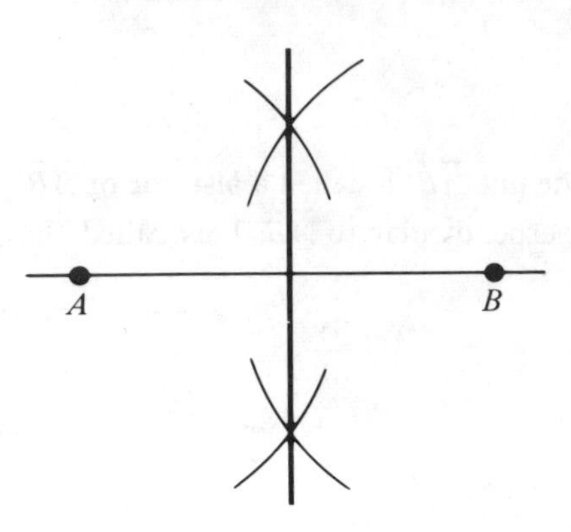

EXERCISES

1. Using a ruler and protractor, *draw*
   (a) a perpendicular to a line at a point on the line
   (b) a perpendicular to a line from a point off the line
   (c) the perpendicular bisector of a segment
2. *Construct*:

   C

   A B

   (a) the perpendicular to $\overleftrightarrow{AB}$ at A
   (b) the perpendicular to $\overleftrightarrow{AB}$ from C
   (c) the perpendicular bisector of $\overline{AB}$
   How are the lines you constructed related? How can you *check* that each is perpendicular to $\overleftrightarrow{AB}$?
3. (a) If two lines in a plane fail to intersect, must they be parallel?
   (b) If two lines in space fail to intersect, must they be parallel?
   (c) If ℓ, m, n are distinct lines in a plane and $m \perp \ell$ and $n \perp \ell$, must $m \parallel n$?
   (d) If ℓ, m, n are distinct lines in space and $m \perp \ell$ and $n \perp \ell$, must $m \parallel n$?
4. Do you think that between any two points on a line, no matter how close together, there is another point? Explain.
5. Without using a ruler, find the measure in inches of each of these segments. (The inch has been used as the unit on the number line.)

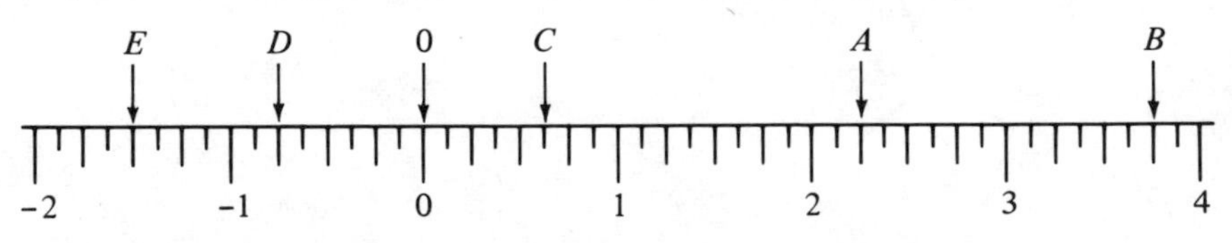

(a) $\overline{OA}$ **(b)** $\overline{CB}$ **(c)** $\overline{EO}$ **(d)** $\overline{AD}$

6. Without doing any measuring or constructing, decide at which marks on the number line these points will fall.

(a) midpoint of $\overline{OA}$
(b) midpoint of $\overline{CB}$
(c) midpoint of $\overline{AD}$
(d) trisectors of $\overline{OA}$
(e) trisectors of $\overline{CB}$
(f) trisectors of $\overline{AD}$

7. Mark two points, A and B, about 3 in. apart near the middle of a clean sheet of paper.

(a) Open your compass to 2 in. and draw the set of all points that are 2 in. from A.
(b) Draw the set of all points 2 in. from B.
(c) Darken those points that are 2 in. from *both* A and B.
(d) Repeat (a), (b), and (c) with a compass opening of 3 in. instead of 2 in.
(e) Repeat (a), (b), and (c) with a compass opening of 4 in. instead of 2 in.
(f) Describe the set of all points that are **equidistant** from A and B.

8. Two pebbles are dropped simultaneously, but several feet apart, into a still pond.

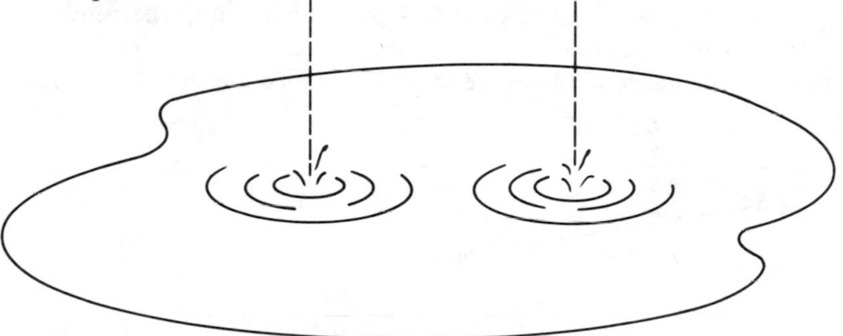

(a) Describe the point at which their first rings will first meet.
(b) Describe the set of *all* points in which the first rings will meet as they travel across the pond.

9. Describe the shortest path from a point to a line.

10. A man lifts a weight with the apparatus shown below, consisting of two posts, a rope, a pulley, and a sliding hook. Describe the path followed by the weight as it is raised.

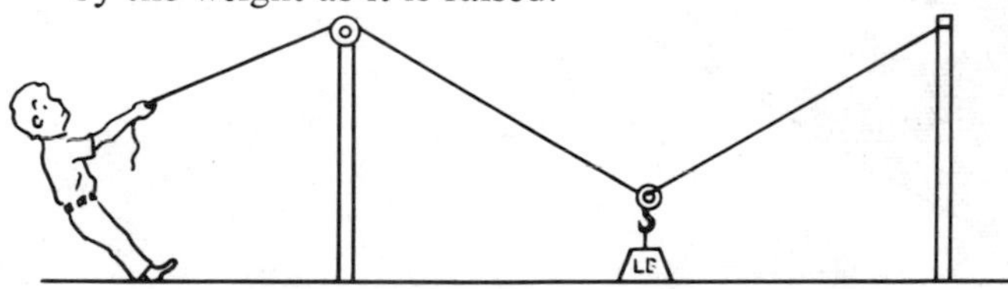

11. Draw a 4-in. segment with a ruler. Now construct a segment of length $1\frac{3}{4}$ in.

12. Construct a figure similar to this one but having a perimeter half as great.

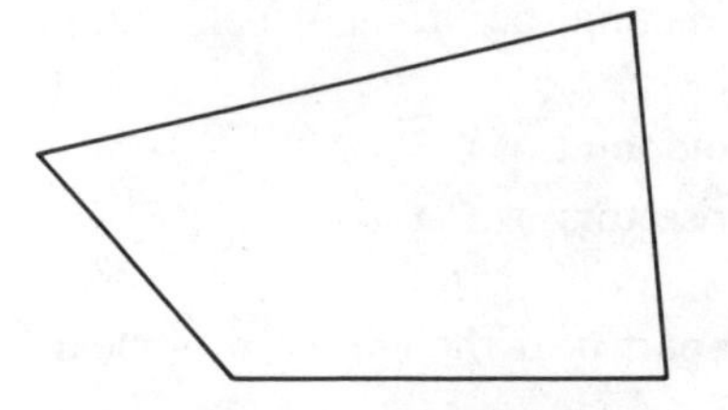

13. Using the constructions for perpendiculars described in this section, construct a line through A parallel to ℓ.

A

ℓ

14. **(a)** On a sheet of scratch paper, draw a line ℓ and a point A on ℓ. Locate the perpendicular to ℓ through A by paper folding.

(b) Repeat part (a) with the point A *off* ℓ.

(c) Draw a segment on a sheet of scratch paper and locate its perpendicular bisector by paper folding.

15. Construct the image of the figure $\mathscr{A}$ under a reflection (flip) in the line ℓ.

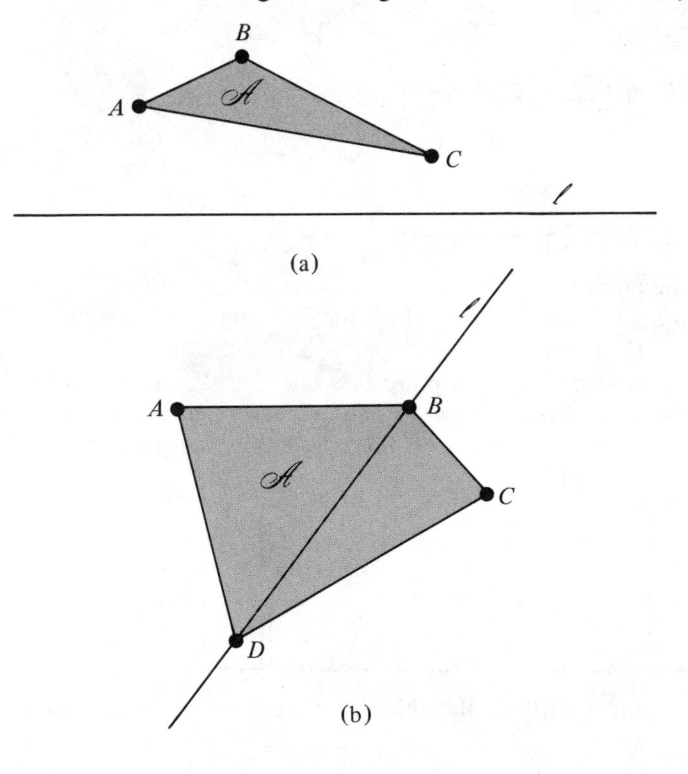

(a)

(b)

16. (An Op Art Project) (i) On a clean sheet of paper, draw a large triangle no two sides of which are congruent. Draw or construct the midpoints of each side of the triangle. Join each vertex of the triangle to the midpoint of the opposite side. Do the three segments you just drew seem to be concurrent? (ii) Using a different colored pencil and the same triangle as in step (i), draw or construct the bisectors of the three angles. Do these bisectors seem to be concurrent? Put the colored pencil in your compass, place the metal tip at their point of concurrence, and draw the largest circle that stays inside the triangle. (iii) Using still a different colored pencil and the same triangle, draw or construct the perpendicular bisectors of the three sides. Do these perpendicular bisectors seem to be concurrent? Put the colored pencil in your compass, place the metal tip at their point of concurrence, and draw the smallest circle that stays outside the triangle.

7.5 PARALLELS; PARTITIONING A SEGMENT

You learned one way to construct a line parallel to a given line through a point not on it.

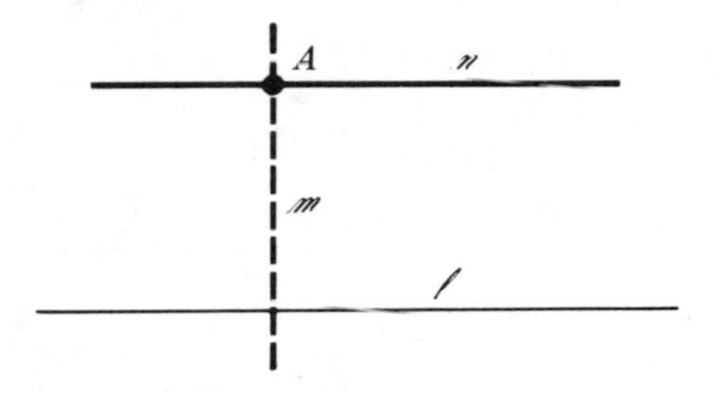

1. Construct a line m through A perpendicular to ℓ.
2. Construct a line n perpendicular to m at A.

Lines ℓ and n could not possibly intersect, because if they did a triangle with two right angles would be formed. (What is wrong with that?) Therefore lines ℓ and n must be parallel.

If you remember this one basic fact about parallels, a simpler construction is possible.

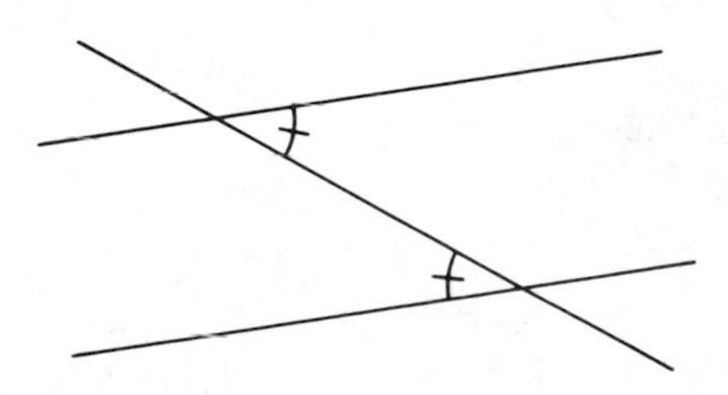

If when two lines are cut by a transversal the alternate interior angles are congruent, then the two lines are parallel.

Given a line ℓ and a point A not on ℓ. To construct a line through A parallel to ℓ:

1. Draw any line m through A that crosses ℓ. Say m crosses ℓ at B and one of the angles formed is called "$\angle 1$."

2. Copy $\angle 1$ on the opposite side of m using A as vertex and $\overrightarrow{AB}$ as one of the sides. The other side will be parallel to ℓ, since alternate interior angles are congruent.

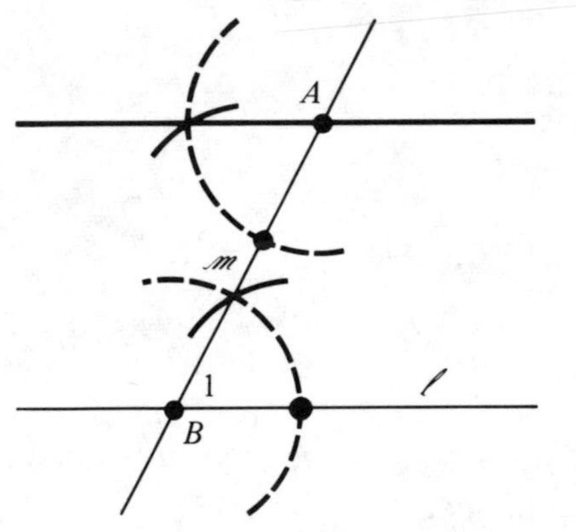

EXERCISES

1. In the figure below, name all pairs of

(a) vertical angles
(b) adjacent angles

(c) alternate interior angles
(d) corresponding angles
(e) alternate exterior angles

2. In the figure above, name all angles congruent to $\angle 1$; to $\angle 2$. (The horizontal lines are parallel.)
3. Construct a copy of $\angle ABC$ on the other side of $\overleftrightarrow{AB}$ using A as a vertex and $\overrightarrow{AB}$ as one side.

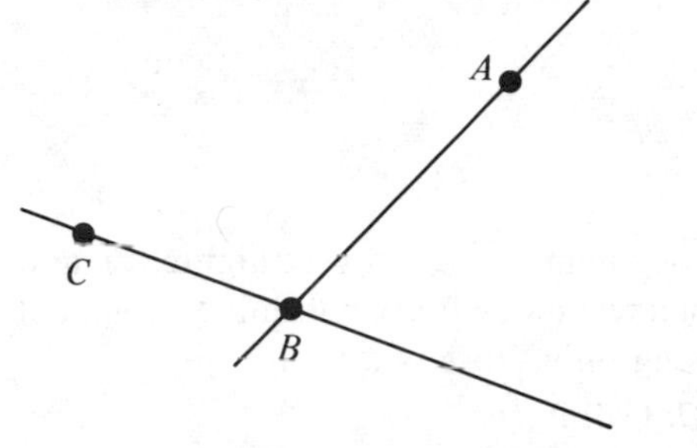

4. Construct a parallel to ℓ through A.

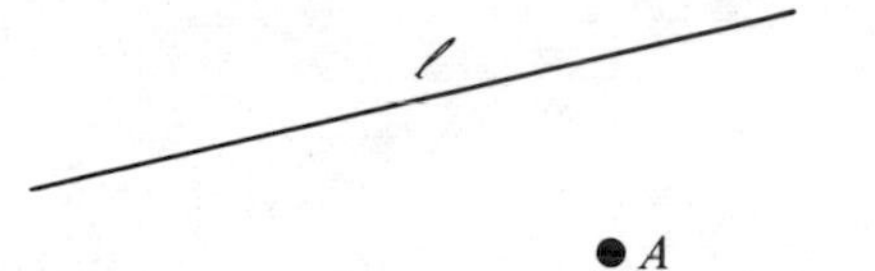

5. Explain how you would draw a parallel to ℓ through A using straightedge and protractor.
6. Invent a procedure for constructing a parallel to ℓ through A based on the fact:

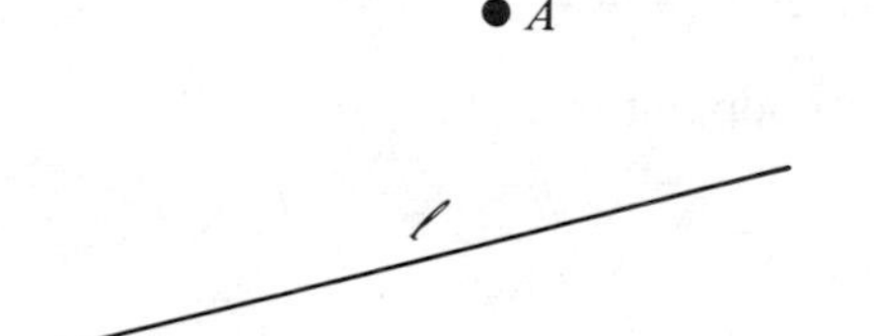

> If corresponding angles are congruent when two lines are cut by a transversal, then the two lines are parallel.

7. Locate the parallel to a given line through a given point by paper folding.
8. Recall that a slide has both length and direction, and can be described by a "vector." Construct the image of the figure $\mathscr{A}$ under the slide given by vector **v**.

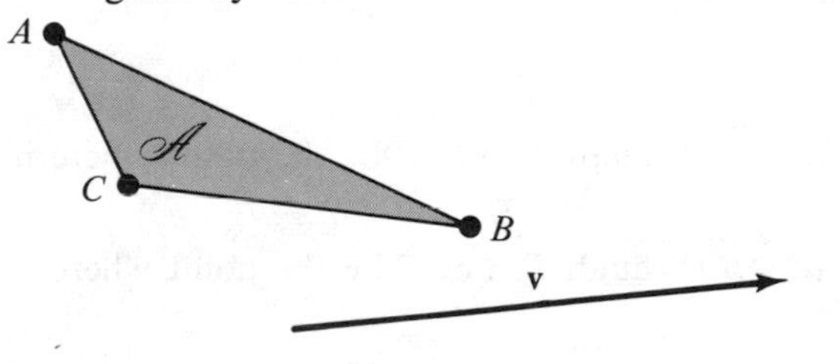

(a)

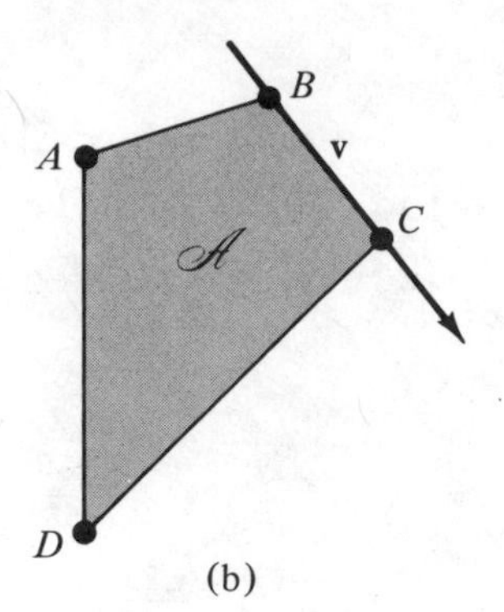

(b)

You already know how to *bisect* a segment using only straightedge and compass. The next few exercises suggest a technique for partitioning a segment into *any number* of congruent parts using only those two tools.

9. (a) Construct the midpoint of $\overline{AC}$. Call it D.

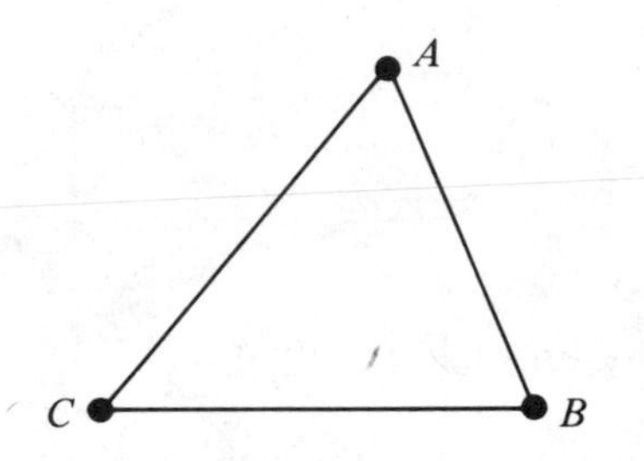

(b) Construct a line through D parallel to $\overline{CB}$. Let E be the point where it crosses $\overline{AB}$.

(c) Check with a ruler that E is the midpoint of $\overline{AB}$.

10. In the figure, $\overline{AD} \cong \overline{DE} \cong \overline{EC}$.

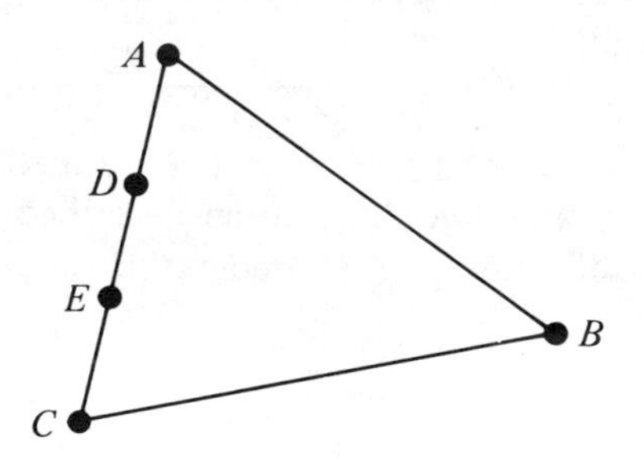

(a) Construct a parallel to $\overleftrightarrow{AB}$ through D. Let F be the point where it intersects $\overline{CB}$.

(b) Construct a parallel to $\overleftrightarrow{AB}$ through E. Let G be the point where it intersects $\overline{CB}$.

(c) Check with ruler or compass that $\overline{CG} \cong \overline{GF} \cong \overline{FB}$.

11. Complete: "If $\overline{AK} \parallel \overline{BJ} \parallel \overline{CI} \parallel \overline{DH} \parallel \overline{EG}$, and $\overline{FG} \cong \overline{GH} \cong \overline{HI} \cong \overline{IJ} \cong \overline{JK}$, then. . . .

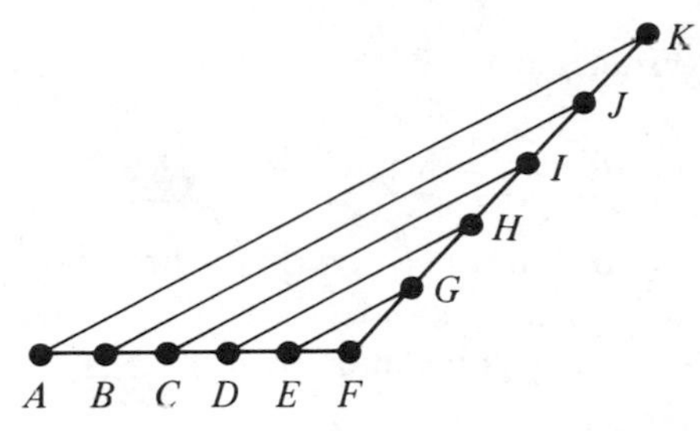

12. Complete this description of how to partition a given segment, $\overline{AF}$, into five congruent parts using only straightedge and compass. (The figure can be used as a guide.) (i) Use straightedge to draw a ray from F. Any ray not along $\overleftrightarrow{AF}$ is all right. (ii) Use compass to swing off, on this ray, five congruent segments (of any length you please): $\overline{FG}$, $\overline{GH}$, $\overline{HI}$, $\overline{IJ}$, $\overline{JK}$. (iii) Use straightedge to draw $\overline{AK}$. . . .

13. Draw a segment and partition it into three congruent pieces using only straightedge and compass.

14. Draw a segment and partition it into seven congruent pieces using straightedge, compass, and protractor—but no ruler.

15. Explain how, in principle, one could construct points on the nascent number line below corresponding to the numbers:

0 1

(a) $\frac{1}{5}$ **(b)** $\frac{1}{10}$ **(c)** $\frac{7}{10}$ **(d)** $-\frac{13}{10}$
(e) m/n, where m and n are any natural numbers

16. Name with numbers those points on the number line that would partition each segment into the requested number of congruent parts.

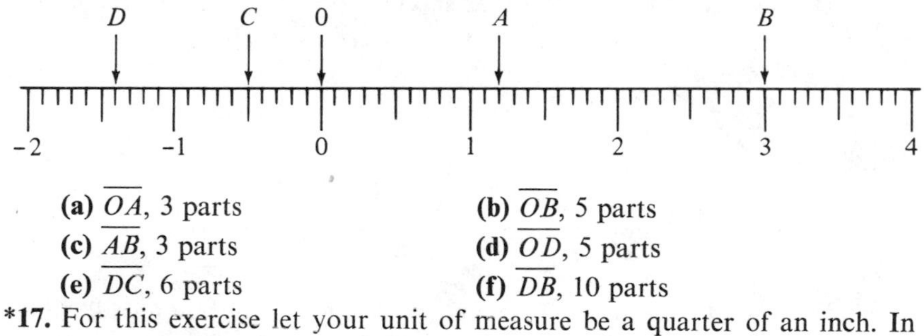

(a) $\overline{OA}$, 3 parts
(b) $\overline{OB}$, 5 parts
(c) $\overline{AB}$, 3 parts
(d) $\overline{OD}$, 5 parts
(e) $\overline{DC}$, 6 parts
(f) $\overline{DB}$, 10 parts

***17.** For this exercise let your unit of measure be a quarter of an inch. In the middle of a large sheet of paper, draw a line ℓ and a point A exactly 4 units away from it (whatever that means). Now draw the family of 12 circles with center at A and radii 1, 2, . . . , 12 units. Also draw the family of 12 lines parallel to ℓ, on A's side of ℓ, and at distances 1, 2, . . . ,

12 units from ℓ (whatever that means). Now graph the following figure, which is known as a **parabola**:

$$\{P \mid \text{distance from } P \text{ to } A = \text{distance from } P \text{ to } \ell\}$$

7.6 CONGRUENT TRIANGLES

The last straightedge-compass constructions we shall work on are those for copying triangles. If you are handed a triangle and asked to construct a (congruent) copy of it, there are several ways you can do it.

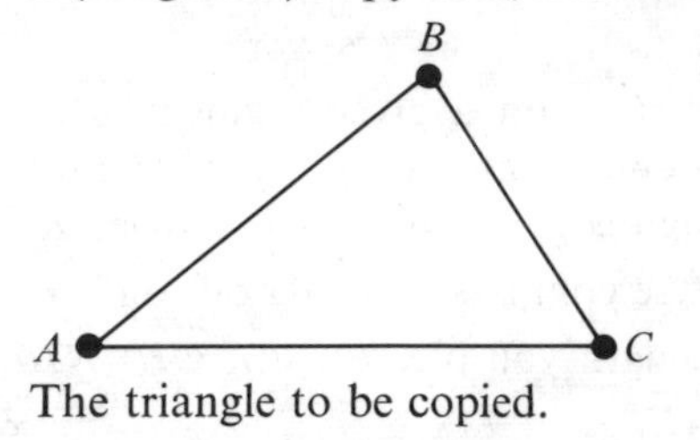

The triangle to be copied.

I. (i) Construct a copy of $\overline{AC}$.

A' ——— C'

(ii) Construct a copy of $\angle A$ using A' as vertex and $A'C'$ as one side.

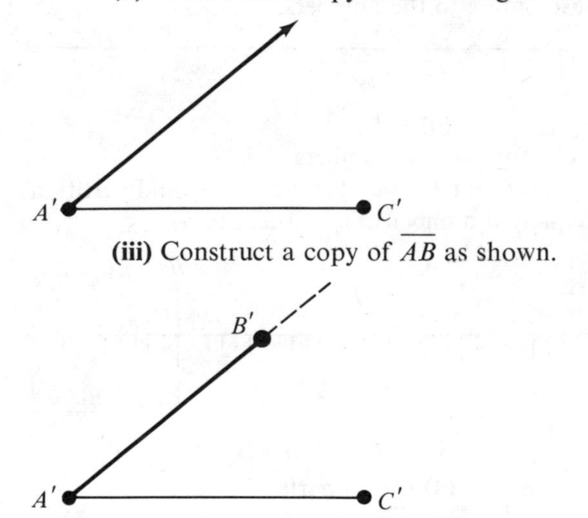

(iii) Construct a copy of $\overline{AB}$ as shown.

There is no need to copy $\angle C$ or $\angle B$ or $\overline{BC}$. By simply joining B' to C' a congruent copy of $\triangle ABC$ will result. It will happen automatically that the other "corresponding parts" are congruent:

$$\overline{B'C'} \cong \overline{BC}, \quad \angle B' \cong \angle B, \quad \angle C' \cong \angle C$$

To copy a triangle, then, it is enough to copy (any) two sides and the angle between them. This property of triangles is known as the side-angle-side (SAS) condition for congruence.

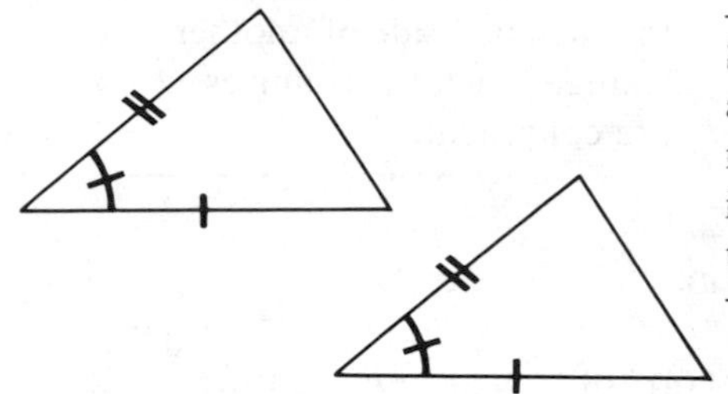

SAS: If two sides and the included angle of one triangle are congruent respectively to two sides and the included angle of another triangle, then the triangles are congruent.

Triangles congruent by SAS condition.

II. Another way to construct a copy of triangle ABC is as follows.

(i) Construct a copy of $\overline{AC}$.

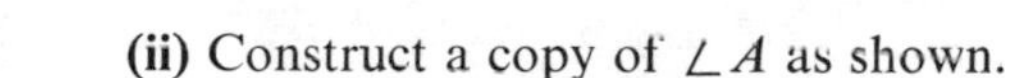

(ii) Construct a copy of $\angle A$ as shown.

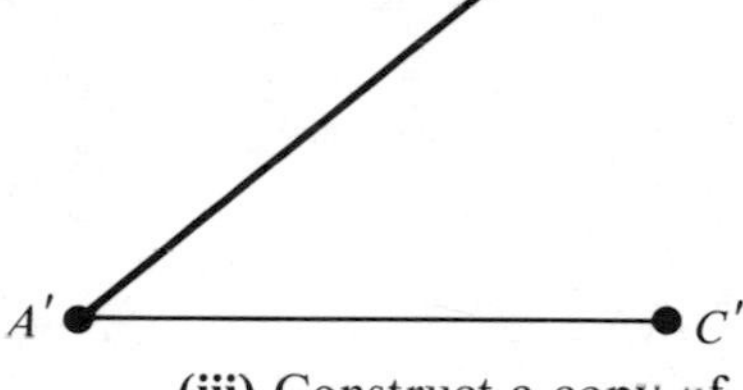

(iii) Construct a copy of $\angle C$ as shown.

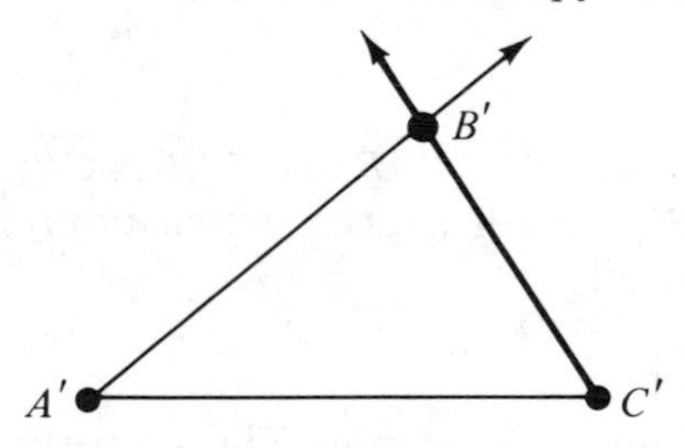

The rays from A' and C' will intersect at a point B', and $\triangle A'B'C'$ will be congruent to $\triangle ABC$. It will happen automatically that the other "corresponding parts" are congruent:

$$\overline{A'B'} \cong \overline{AB}, \quad \overline{B'C'} \cong \overline{BC}, \quad \angle B' \cong \angle B$$

To copy a triangle, then, it is enough to copy (any) two angles and the side between them. This property of triangles is known as the angle-side-angle (ASA) condition for congruence.

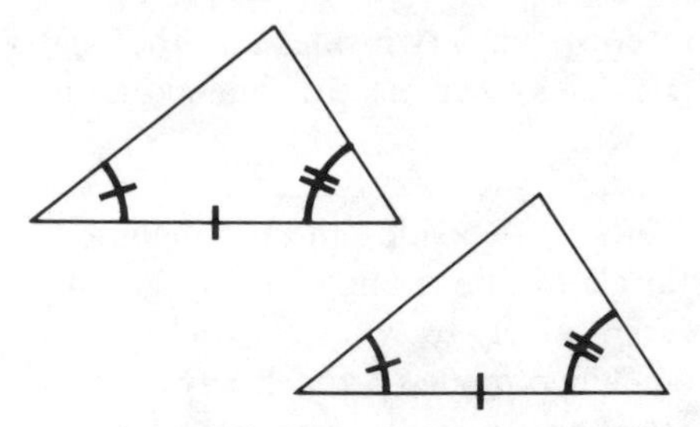

ASA: If two angles and the included side of one triangle are congruent respectively to two angles and the included side of another triangle, then the triangles are congruent.

Triangles congruent by ASA condition.

III. A third way to construct a copy of triangle ABC is this:

(i) Construct a copy of $\overline{AC}$.

A' C'

(ii) Swing an arc of radius AB from A'.

A' C'

(iii) Swing an arc of radius CB from C'.

B'

A' C'

The arcs will intersect at a point B' for which $\overline{A'B'} \cong \overline{AB}$ and $\overline{C'B'} \cong \overline{CB}$, and $\triangle A'B'C'$ will be congruent to $\triangle ABC$. It will happen automatically that the "corresponding angles" are congruent;

$$\angle A' \cong \angle A, \quad \angle B' \cong \angle B, \quad \angle C' \cong \angle C.$$

To copy a triangle, then, it is enough to copy all three sides. This property of triangles is known as the side-side-side (SSS) condition for congruence.

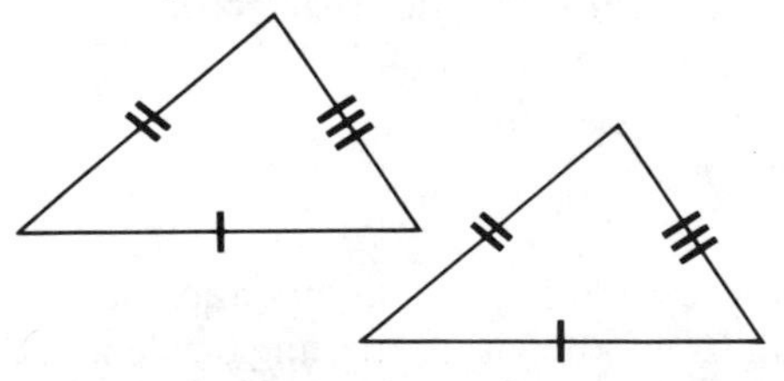

SSS: If three sides of one triangle are congruent respectively to three sides of another triangle, then the two triangles are congruent.

Triangles congruent by SSS condition.

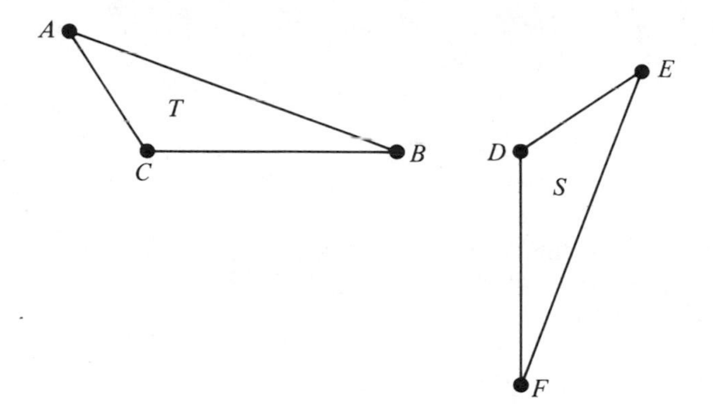

To sharpen the concept of "corresponding parts," let us look at congruence of triangles in terms of rigid motions. To say that triangle T is congruent to triangle S is to say that there is a rigid motion—some sequence of slides, turns, flips—that will make triangle T coincide with triangle S. In the figure, a 90° clockwise turn about C followed by a slide 3 in. right will do the trick. Under this rigid motion the vertices A, B, C of triangle T and the vertices D, E, F of triangle S will be matched up as follows:

$$A \leftrightarrow E, \quad B \leftrightarrow F, \quad C \leftrightarrow D$$

Any one of these matched pairs of vertices is called a pair of **corresponding vertices** with respect to this particular rigid motion.

The match-up of vertices induces a match-up of angles and sides as well.

$$\angle A \leftrightarrow \angle E, \quad \angle B \leftrightarrow \angle F, \quad \angle C \leftrightarrow \angle D$$

$$\overline{AB} \leftrightarrow \overline{EF}, \quad \overline{BC} \leftrightarrow \overline{FD}, \quad \overline{CA} \leftrightarrow \overline{DE}$$

Any one of these six pairs is said to be a pair of **corresponding parts** with respect to the rigid motion. In summary,

---

Corresponding vertices (sides, angles) of congruent triangles are those pairs of vertices (sides, angles) that will be matched up by a rigid motion that makes the triangles coincide.

If two triangles are congruent, then, of course, there are six pairs of congruent corresponding parts.

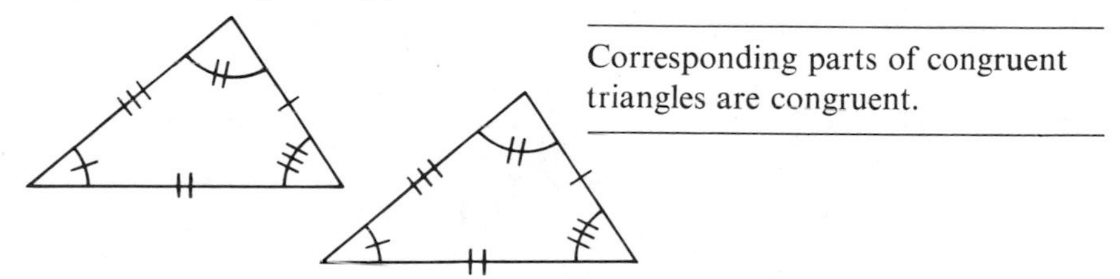

Corresponding parts of congruent triangles are congruent.

The SAS, ASA, and SSS congruence conditions tell us that if in two triangles we can find three strategically located pairs of congruent parts, then the triangles must be congruent and so all the other parts must also match up in congruent pairs.

Notational Convention

There are acceptable and unacceptable ways to report that two triangles are congruent. It is easier to illustrate the notational convention with examples than to try to explain it in words.

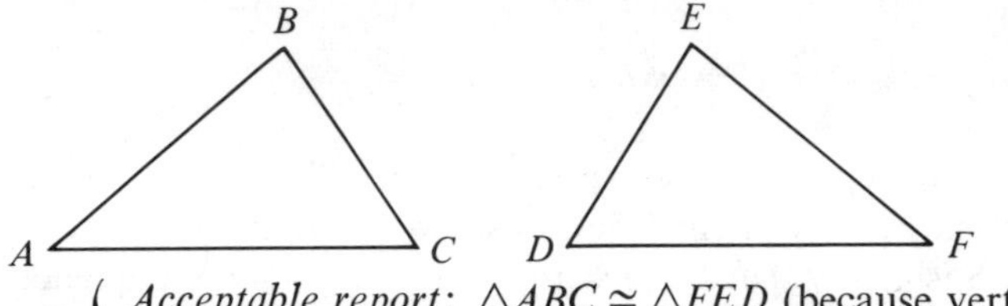

Acceptable report: $\triangle ABC \cong \triangle FED$ (because vertices A, B, C correspond to vertices F, E, D in that order).
Unacceptable report: $\triangle ABC \cong \triangle DEF$ (because there is no rigid motion matching A with D, B with E, and C with F).

Acceptable report: $\triangle CAB \cong \triangle DFE$.
Unacceptable report: $\triangle CAB \cong \triangle FED$.

EXERCISES

1. Construct copies of this triangle using

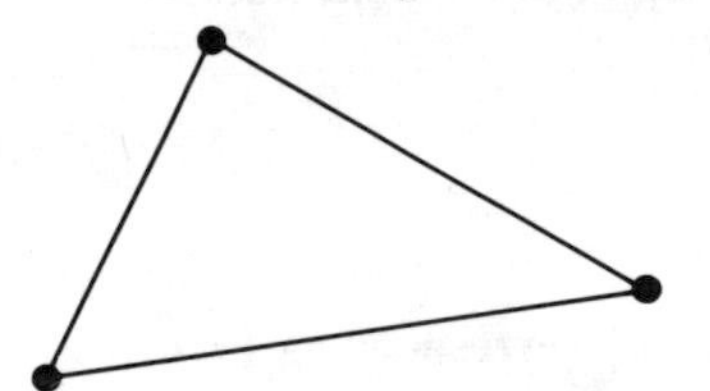

(a) the SAS procedure
(b) the ASA procedure
(c) the SSS procedure
Check that all of your copies can be made to coincide with the original.

2. (a) Why are these triangles congruent?

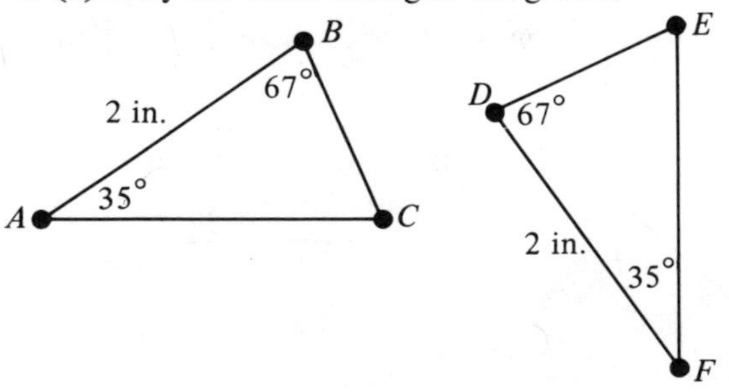

(b) What side corresponds to $\overline{AC}$? What is true of the sides opposite corresponding angles?

(c) Complete: $\triangle ABC \cong \triangle$ ________

3. (a) Why are these two triangles congruent?

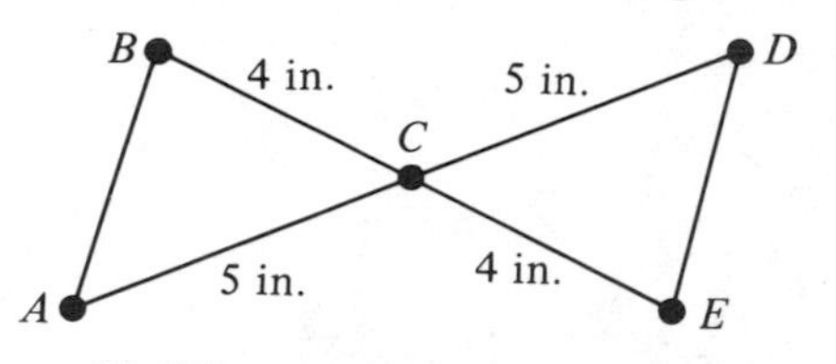

(b) What angle corresponds to $\angle D$? What is true of the angles opposite corresponding sides?

(c) Complete: $\triangle BCA \cong \triangle$ ________

(d) Describe a rigid motion that would make the left triangle coincide with the right one.

4. B and D are the centers of the circles.

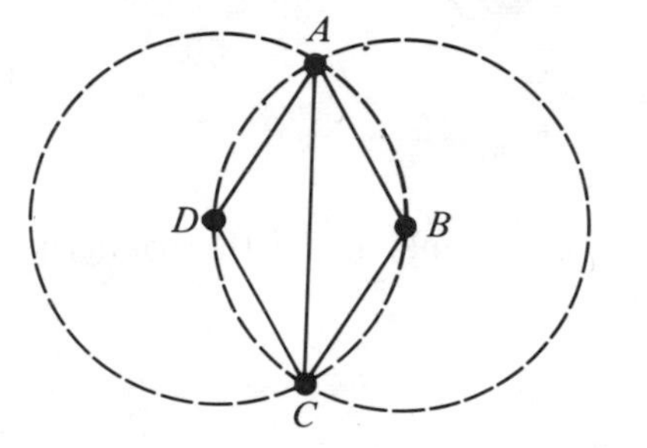

(a) Why are these two triangles congruent?

(b) Describe a flip that will make the right-hand triangle coincide with the left-hand triangle.

(c) Complete: $\triangle ABC \cong \triangle A$ ________

(d) Describe a turn (rotation) that will make the right-hand triangle coincide with the left-hand one.

(e) Complete: $\triangle ABC \cong \triangle C$ ________

5. (a) Write out all the acceptable ways there are for reporting that the triangles of Exercise 2 are congruent.

(b) How many acceptable ways are there for reporting that the triangles in Exercise 4 are congruent?

(c) How many acceptable ways are there for reporting that these two triangles are congruent?

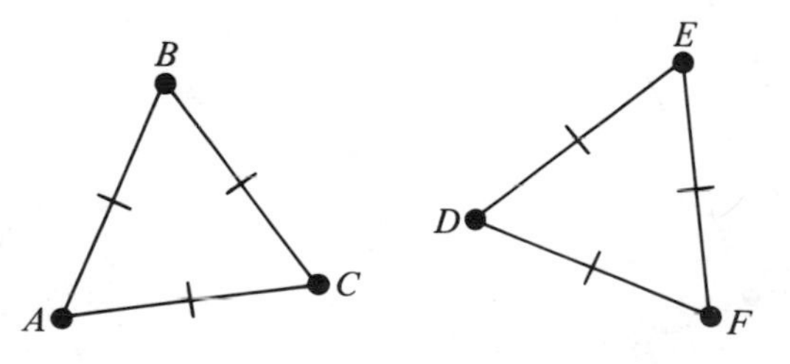

6. State why each pair of triangles is congruent, report the congruence according to the notational convention, and then pair off all corresponding parts. [Be prepared to describe (roughly) a rigid motion that would put one triangle on the other.]

EXAMPLE

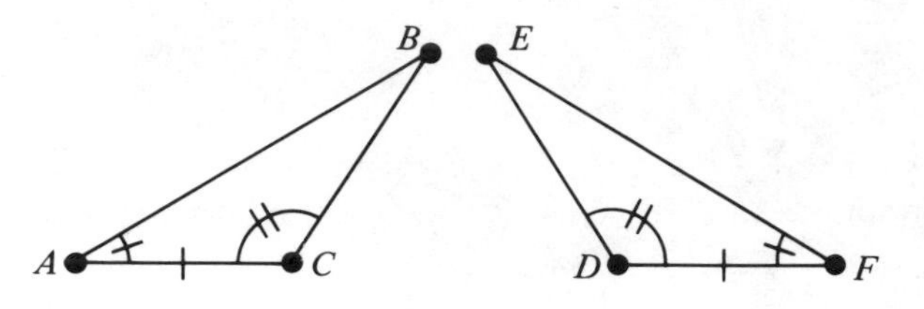

Answer Congruent by ASA. $\triangle ABC \cong \triangle FED$.

$$\angle A \leftrightarrow \angle F \quad \text{so} \quad \overline{BC} \leftrightarrow \overline{DE}$$

$$\angle C \leftrightarrow \angle D \quad \text{so} \quad \overline{AB} \leftrightarrow \overline{EF}$$

$$\overline{AC} \leftrightarrow \overline{DF} \quad \text{so} \quad \angle B \leftrightarrow \angle E$$

Reflection in the perpendicular bisector of $\overline{CD}$ would make the triangles coincide.

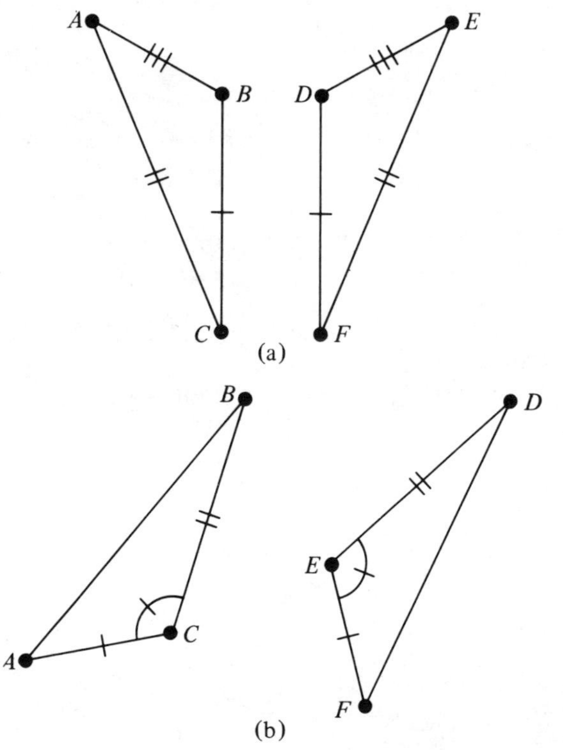

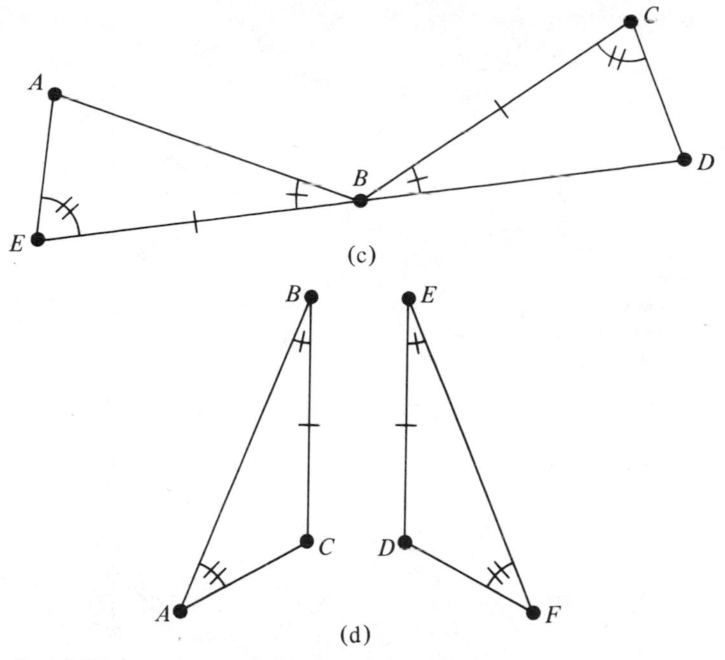

7. (a) Write out carefully an SAA congruence condition for triangles.

(b) Explain how the diagram below shows that there is *no* SSA congruence condition.

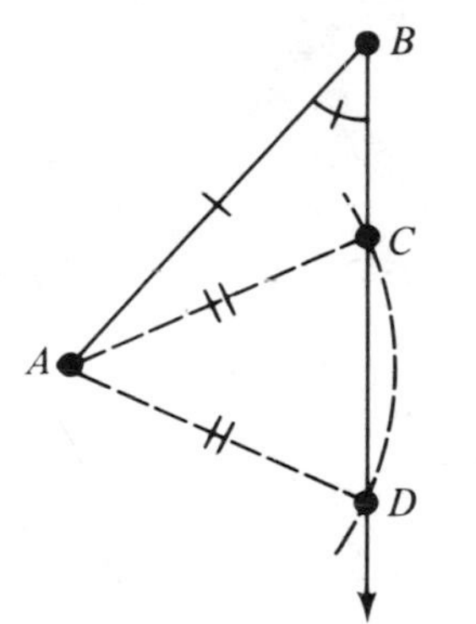

(c) Sketch a diagram that shows there is *no* AAA congruence condition.

8. In each figure, find a pair of congruent triangles, give the congruence condition that assures you they are congruent, and report the congruence using the notational convention.

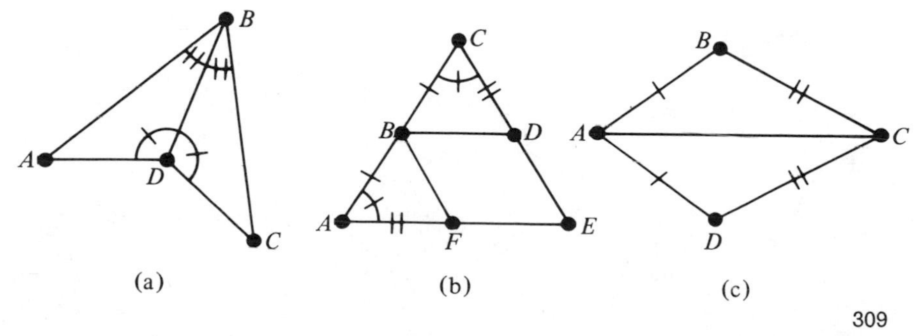

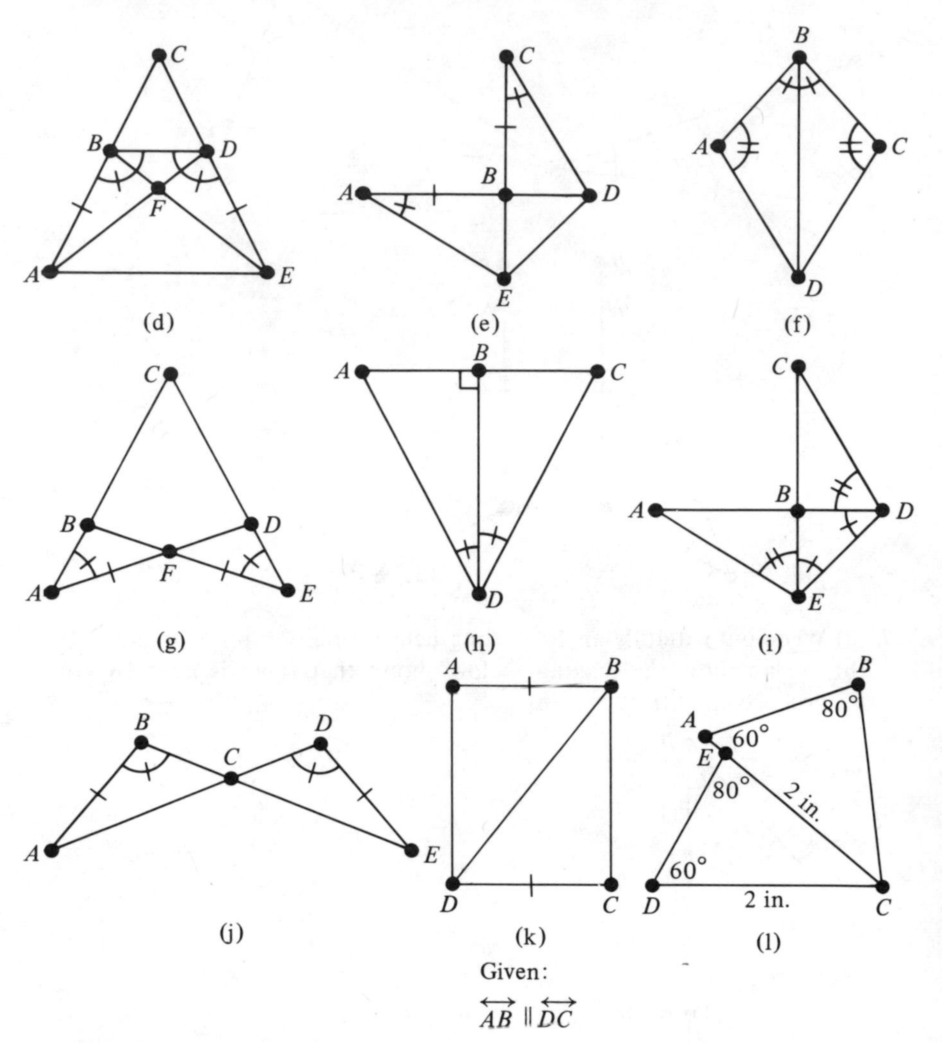

9. Without measuring, fill in the blanks:

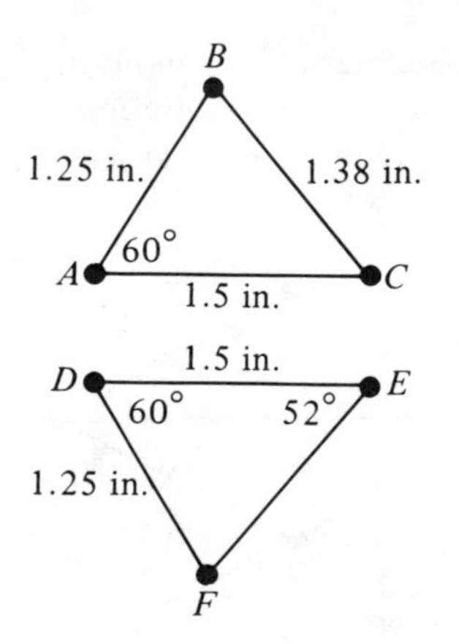

(a) $EF =$ ________
(b) $m \angle F =$ ________
(c) $m \angle C =$ ________
(d) $m \angle B =$ ________

10. Locations on a map are assigned coordinates (ordered pairs) according to the "first right, then up" convention. For example, the approximate coordinates of forest ranger A are (44.4, 66.3).

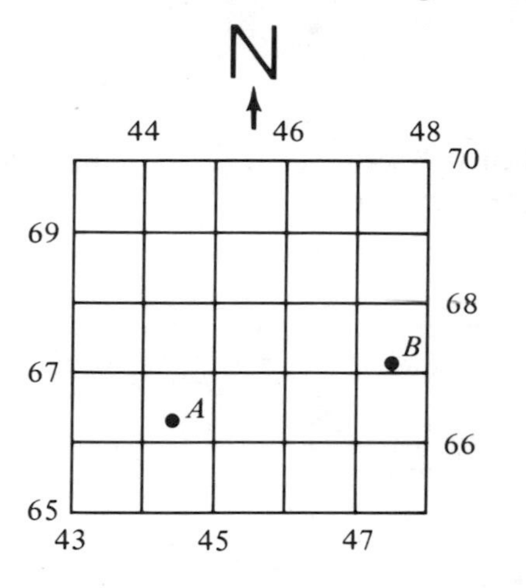

(a) What are the approximate coordinates of forest ranger B?
(b) Forest ranger A sees a flash of lightning 40° east of north. Forest ranger B sees the same flash 15° west of north. Find the approximate coordinates of the lightning strike using your protractor.
(c) Lightning strikes again somewhere to the north, but this time neither ranger is able to read an angle. By counting the elapsed time between flash and thunder, ranger A computes the distance to the strike as 3 miles. Ranger B computes a distance of 2 miles. Find approximate coordinates of the strike. (Each square on the map is 1 mile by 1 mile.)

7.7 CONGRUENT TRIANGLES AND DEDUCTION

The congruence conditions for triangles can be used to explain *why* various straightedge-compass constructions work and *why* a number of other geometric properties should be true. A good deal of time is spent on this sort of deduction in conventional high school geometry courses. The exercise list below contains only a sampling. It is questionable whether any such deduction should be carried out in the elementary school.

EXERCISES

1. The figure suggests how compass and straightedge were used to bisect an angle.

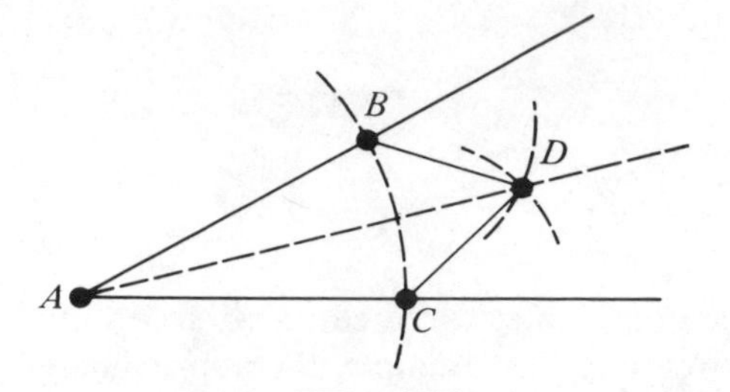

(a) Why is $\overline{AB} \cong \overline{AC}$?
(b) Why is $\overline{BD} \cong \overline{CD}$?
(c) Why is $\triangle ABD \cong \triangle ACD$?
(d) Why is $\angle BAD \cong \angle CAD$?

2. In this construction of the perpendicular bisector of $\overline{AB}$, the compass opening is left fixed so that $\overline{AC} \cong \overline{BC} \cong \overline{BD} \cong \overline{AD}$.

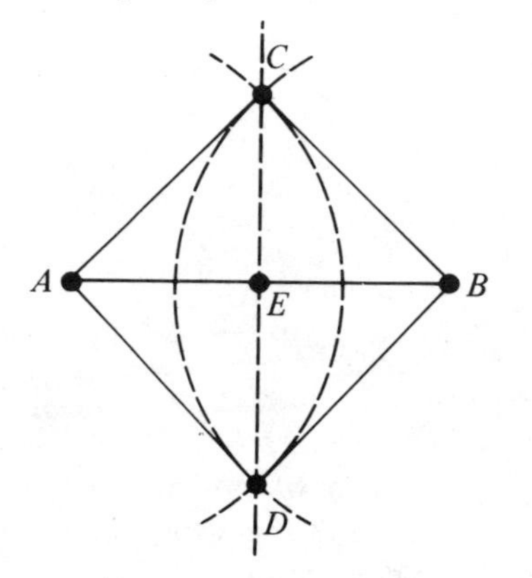

(a) Why is $\triangle ACD \cong \triangle BCD$?
(b) Why is $\angle ACD \cong \angle BCD$?
(c) Why is $\triangle ACE \cong \triangle BCE$?
(d) Why is $\overline{AE} \cong \overline{EB}$?
(e) Why is $\overline{CD} \perp \overline{AB}$? (Recall the definition: Two lines are perpendicular if they meet to form congruent adjacent angles.)

3. A triangle with two congruent sides is called an **isosceles** triangle. The angles opposite the congruent sides in an isosceles triangle are called its **base angles**. To show that the base angles of an isosceles triangle are congruent, let D be the midpoint of $\overline{CB}$ and draw $\overline{AD}$.

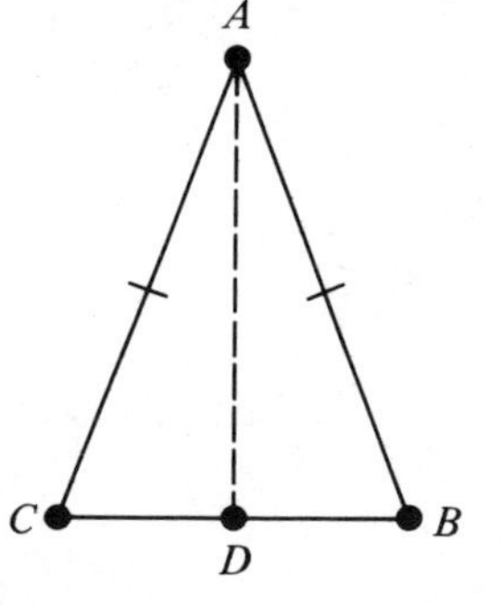

(a) Why is $\triangle ACD \cong \triangle ABD$?
(b) Why is $\angle C \cong \angle B$?

4. Recall that a parallelogram is a quadrilateral whose opposite sides are parallel. For parallelogram $ABCD$, answer the following.

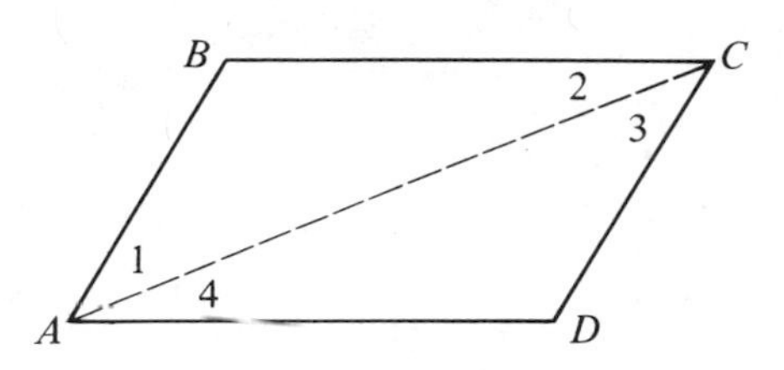

(a) Why is $\angle 1 \cong \angle 3$?
(b) Why is $\angle 2 \cong \angle 4$?
(c) Why is $\triangle ABC \cong \triangle CDA$?
(d) Why is $\overline{AB} \cong \overline{CD}$?
(e) Why is $\overline{BC} \cong \overline{AD}$?
(f) Why is $\angle B \cong \angle D$?
(g) Why is $\angle BAD \cong \angle DCB$?

Parts (d) and (e) above suggest the general statement:

"Opposite sides of a parallelogram are congruent."

(h) What general statement is suggested by parts (f) and (g)?

5. Use the figure below and show that "if opposite sides of a quadrilateral are congruent, then the quadrilateral is a parallelogram."

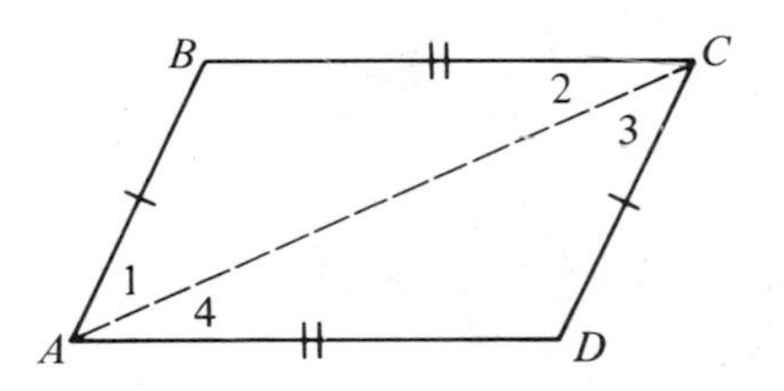

6. Use the figure below and show that "the diagonals of a parallelogram bisect each other."

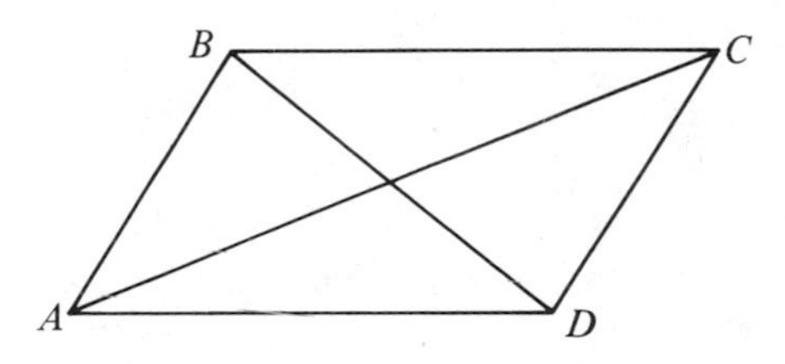

7. The figure below suggests the construction of a perpendicular to ℓ at A. Show that $\overleftrightarrow{DA} \perp \overleftrightarrow{BC}$.

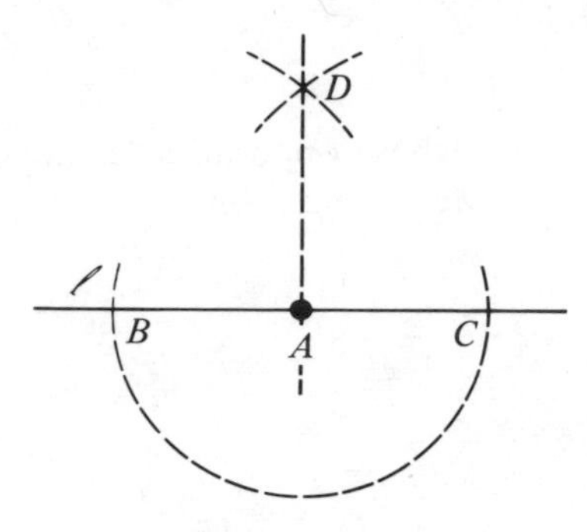

8. The perpendicular bisector of a segment $\overline{AB}$ is a set of points. Let us call it $\mathscr{P}$. And let us denote by $\mathscr{E}$ the set of all points equidistant from A and B. Exercises 7 and 8 (p. 295) suggested that $\mathscr{P} = \mathscr{E}$.

(a) Use the figure below and show that if $X \in \mathscr{P}$, then $X \in \mathscr{E}$.

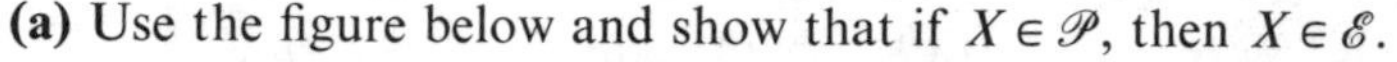

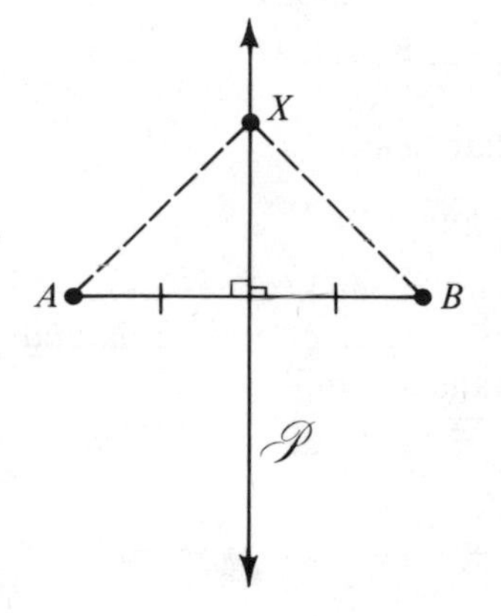

(b) Use the figure below and show that if $X \in \mathscr{E}$, then $X \in \mathscr{P}$.

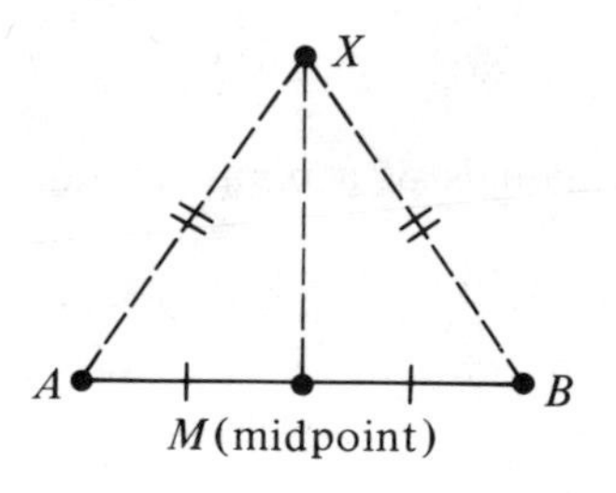

(c) How do (a) and (b) show that $\mathscr{P} = \mathscr{E}$?

9. Using the fact that the perpendicular bisector of a segment $\overline{AB}$ is the set of all points P such that $AP = PB$, prove that the perpendicular bisectors of the three sides of a triangle are concurrent. (*Hint*: Let the perpendicular bisectors of $\overline{BC}$ and $\overline{AC}$ meet at, say, P. Show that P is on the perpendicular bisector of $\overline{AB}$. See the figure the top of p. 315.)

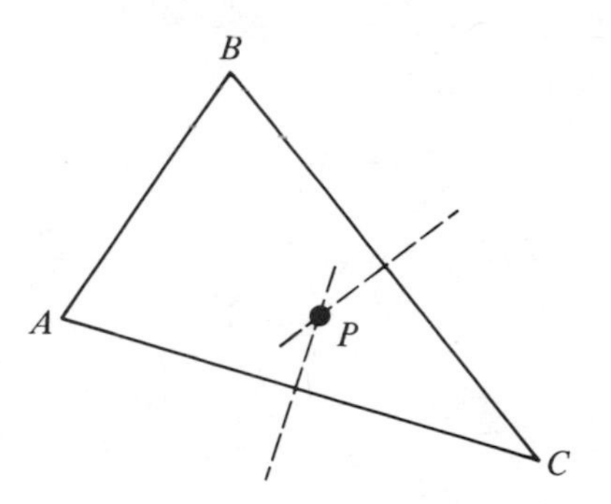

10. The figure below suggests the construction of the trisectors, G and F, of the original segment $\overline{AB}$. In the figure,

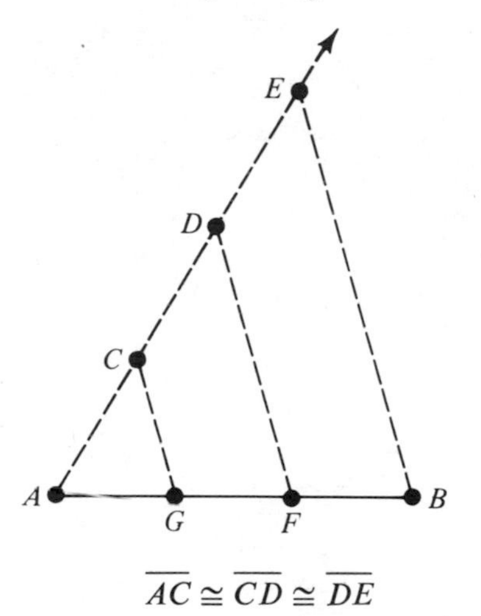

$$\overline{AC} \cong \overline{CD} \cong \overline{DE}$$

and

$$\overline{CG} \parallel \overline{DF} \parallel \overline{EB}$$

The problem is to show that $\overline{AG} \cong \overline{GF} \cong \overline{FB}$. Begin by sketching a line through C parallel to $\overline{AB}$. Let it meet $\overline{DF}$ at H and $\overline{EB}$ at I. Also sketch a line through D parallel to $\overline{AB}$. Let it meet $\overline{EB}$ at J. Use the result of Exercise 4.

11. Draw a figure and show that if two opposite sides of a quadrilateral are both congruent and parallel, then the quadrilateral is a parallelogram.

VOCABULARY

7.1 Constructions and Drawings

straightedge
compass
geometric construction
inscribed
central angle
center of a regular polygon

7.2 Copying Figures

7.3 Bisecting Angles

bisector (of an angle)

7.4 Bisecting Segments; Perpendiculars

midpoint
bisector (of a segment)
perpendicular bisector
equidistant

7.5 Parallels; Partitioning a Segment

7.6 Congruent Triangles

SAS
ASA
SSS
corresponding vertices
corresponding parts
SAA

7.7 Congruent Triangles and Deduction

isosceles triangle
base angles

SIMILARITY, RIGHT TRIANGLES, AND TRIGONOMETRY

OVERVIEW

Most of the mathematical concepts studied in this chapter appear, at present, in the junior high school curriculum. A few of them, however, are already showing up in the elementary curriculum, and this trend can be expected to continue.

The basic purpose of the first three sections is to explain what it means to say that two geometric figures are similar. The general definition is this: Two geometric figures are similar if all distances between corresponding point pairs are proportional. This definition cannot be understood without some prior familiarity with the concepts of ratio and proportion. Consequently Section 8.1 is devoted to a study of those two ideas.

In Section 8.2 the idea of similarity of figures is introduced gradually —by concentrating on triangles and relating the similarity conditions for triangles to the familiar congruence conditions for triangles studied at the end of Chapter 7. In Section 8.3 similarity of more general geometric figures is studied. The approach is very concrete. Projection techniques for making scale drawing are described and related to the similarity conditions for triangles. Experience in actually making such drawings is provided in the exercises. The last portion of the exercise list deals with questions of how the areas (and perimeters, surface areas, and volumes) of similar figures are related.

The answers to these questions are, of course, important in their own right. But as a bonus they provide the basis for a simple proof (Polya's) of the Pythagorean theorem in the next section.

Right triangles provide the central core for the rest of the chapter. From this center we move off in two directions.

First we consider the question of when two right triangles are *congruent*. This leads naturally to the Pythagorean theorem. A geometric plausibility argument is given in addition to the proof (just mentioned) based on areas of similar triangles. In Section 8.5 applications of the Pythagorean theorem are made. This, of course, entails the use of square roots. No algorithm for finding square roots is given, but rather exercise in using a table is provided. A straightedge-compass construction of $\sqrt{2}$ is called for, and arguments for its uniqueness and irrationality are outlined.

Beginning in Section 8.6 we move off in the second direction. By considering the question of when two right triangles are *similar*, we are led to the trigonometric ratios associated with an acute angle of a right triangle. In Section 8.7 the meaning of the trigonometric ratios is brought together with their numerical values, as given in a trig table, to solve problems involving right triangles.

The chapter concludes with a section on the special 45–45–90 and 30–60–90 right triangles. In this section essential use is made of both the Pythagorean theorem and the trigonometric ratios.

8.1 RATIO AND PROPORTION

The word "ratio" is often used when two geometric figures are compared for "size." For example, we say that the ratio of the *area* of region R to the *area* of region S is "4 to 12" (or 1 to 3); we write, "the ratio of the area of R

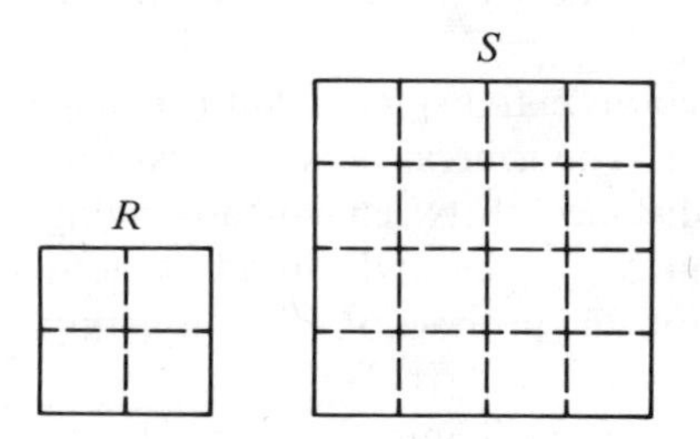

to the area of S is $\frac{4}{12}$." The ratio of the *perimeter* of R to the *perimeter* of S is 4 to 14 ($\frac{4}{14}$) or 2 to 7 ($\frac{2}{7}$). The ratio of *length* of R to the *length* of S is 2 to 4 ($\frac{2}{4}$) or 1 to 2 ($\frac{1}{2}$). The ratio of the *width* of R to the *width* of S is 2 to 3 ($\frac{2}{3}$). For region R the ratio of length to width is 2 to 2 ($\frac{2}{2}$) or 1 to 1 ($\frac{1}{1}$), and for region S the ratio of length to width is 4 to 3 ($\frac{4}{3}$).

In comparing figures such as R and S, it is important to specify exactly which aspects are being compared: area, perimeter, length, or width. For simpler figures it is unnecessary to be so specific. When we say, "the ratio of

$\overline{AB}$ to $\overline{CD}$ is 5 to 3," it is clear that we are referring to their lengths.

A ●—+—+—+—+—● B C ●—+—+—● D

Undoubtedly you have used the language of ratio before in nongeometric situations, and realize that it is just a medium for describing how two sets compare. Any "ratio statement" can be replaced by a "fraction statement." Instead of saying "the ratio of the area of R to the area of S is 4 to 12," we could just as well say "the area of R is four-twelfths the area of S." Instead of saying "the ratio of the length of S to the width of S is 4 to 3," we could say "the length of S is four-thirds the width of S." We have two languages for describing these size relationships.

The language of ratio was begun by the early Greek geometers, but has now been incorporated into everyday usage. A baseball announcer uses ratio language when he says that a batter is "2 for 5." Stockbrokers study the "price to earnings ratio" of common stocks. Doctors watch the ratio of red to white cells in blood. Farmers look for the nitrate to phosphate ratio when they buy fertilizer.

In solving certain types of problems, it is often useful to think in terms of ratios.

EXAMPLE Sam Steady hit 6 home runs and 21 singles in the month of July. If his ratio of home runs to singles remained the same in August, and if he hit 8 home runs in August, how many singles did he hit in August?

Solution 1 We know the ratio of July homers to July singles is the same as the ratio of August homers to August singles. Letting x stand for the number of August singles and writing ratios as fractions, we arrive at the equation

$$\frac{6}{21} = \frac{8}{x}$$

which can be read, "6 is to 21 as 8 is to x." Multiplying both sides by $21x$ yields

$$6 \cdot x = 8 \cdot 21$$

and dividing both sides by 6 gives

$$x = \frac{8 \cdot 21}{6} = 28$$

He hit 28 singles in August.

A statement of equality between two ratios, such as "$\frac{6}{21} = \frac{8}{28}$," is sometimes called a **proportion**.

Solution 2 Sam Steady hits $\frac{21}{6}$ as many singles as homers, so if he hit 8 homers then he must have hit $\frac{21}{6} \cdot 8 = 28$ singles.

EXAMPLE In a certain pet shop the ratio of guppies to puppies is 27 to 4. If there are 12 puppies, how many guppies are there?

Solution If we let x be the number of guppies, then the statement "the number of guppies is to the number of puppies as 27 is to 4" can be rewritten as the equation

$$\frac{x}{12} = \frac{27}{4}$$

Solving for x,

$$\frac{x}{12} = \frac{27}{4} \Leftrightarrow 4x = 12 \cdot 27 \Leftrightarrow x = 81$$

There are 81 guppies.

EXERCISES

A B

C D

1. **(a)** What is the ratio of $\overline{AB}$ to $\overline{CD}$?
   (b) What is the ratio of $\overline{CD}$ to $\overline{AB}$?
   (c) If $AB = 40$ ft, then $CD =$ __________.
   (d) If $AB = 12$ in., then CD = __________.
   (e) If $CD = 15$ cm, then $AB =$ __________.
   (f) If $CD = 10$ m, then $AB =$ __________.
   (g) If $AB + CD = 4$ ft, then $AB =$ __________.
   (h) If $\overline{AB}$ is 6 in. longer than $\overline{CD}$, then $CD =$ __________.
   (i) What is the measure of $\overline{AB}$ with respect to the unit $\overline{CD}$?
   (j) What is the measure of $\overline{CD}$ with respect to the unit $\overline{AB}$?

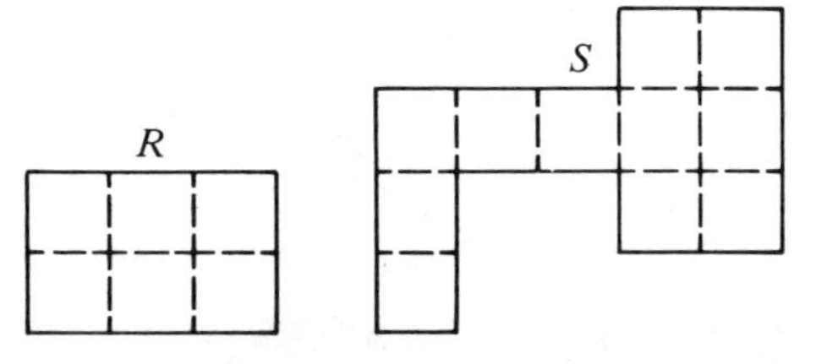

2. **(a)** What is the ratio of the area of R to the area of S?
   (b) What is the ratio of the perimeter of S to the perimeter of R?

(c) If R has perimeter 12 in., what is the perimeter of S?
(d) If R has perimeter 10 cm, what is the perimeter of S?
(e) If S has perimeter 10 ft, what is the perimeter of R?
(f) If R has area 30 acres, what is the area of S?
(g) If S has area 50 km^2, what is the area of R?

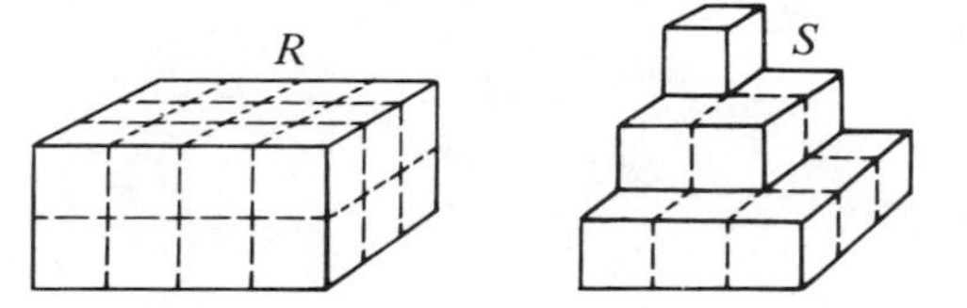

3. (a) What is the ratio of volume of R to volume of S?
(b) What is the ratio of surface area of R to surface area of S?
(c) If R and S are made of the same material and S weighs 10 lb, how much does R weigh?
(d) If 15 cc of paint were required to paint R, how many cubic centimeters are required to paint S?
4. If 5 boys drink 3 qt of lemonade, how much lemonade will 30 boys drink?
5. If a man can drive 260 mi in 5 hr, how long will it take him to drive 400 mi?
6. If corn flakes are sold at 3 boxes for 89¢, how much will 12 boxes cost?
7. If 2 out of every 9 of Mr. Smith's dollars go for taxes and Mr. Smith earns $9450 per year, how much does he pay in taxes per year?
8. In his first 65 official times at bat, a baseball player hits 4 home runs. If he can expect to maintain this hitting pace and come to bat officially 585 times during the season, how many home runs can he expect to hit?
9. A cookie recipe calls for 2 cups of sugar and $1\frac{1}{4}$ cups of shortening. Sue finds that she has only $1\frac{1}{3}$ cups of sugar and decides to use it all and make a smaller batch. How much shortening should she use?
10. Six boys can pick 50 qt of strawberries in a day. A farmer gets an order for 175 qt of strawberries for the next day. How many boys should he hire?
11. For every $4 Joe saves, his brother saves $11. How much will Joe have to save if the two boys want to pool their savings to buy a $60 bicycle?
12. Judy spends 12 min of every hour staring out the window. How much time should she allow herself to do an hour's homework?
13. Translate into an equation: "The areas, A_1 and A_2, of triangles T_1 and T_2 are proportional to the lengths, b_1 and b_2 of their bases.

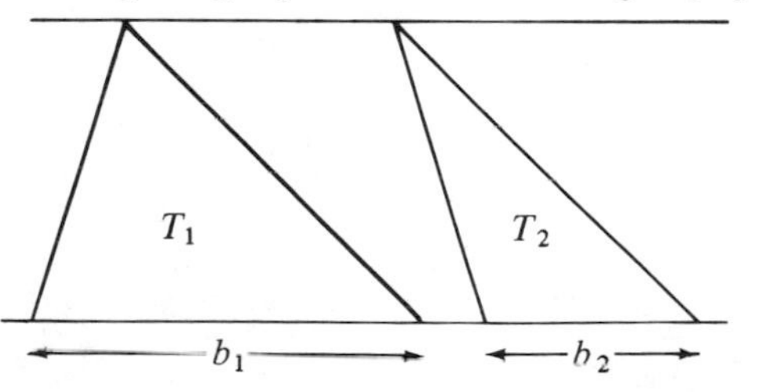

14. Translate into an equation: "The areas, A_1 and A_2, of equilateral triangles T_1 and T_2 are proportional to the squares of the lengths, b_1 and b_2 of their bases."

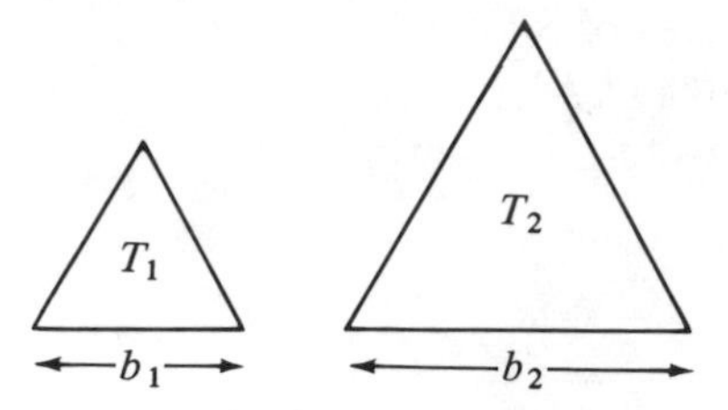

15. For each pair of triangles below, $\triangle ABC$ and $\triangle ADE$, use your protractor to compare $\angle ACB$ and $\angle E$, $\angle ABC$ and $\angle D$; and use your ruler to find the ratio of DE to BC. Do you see some pattern?

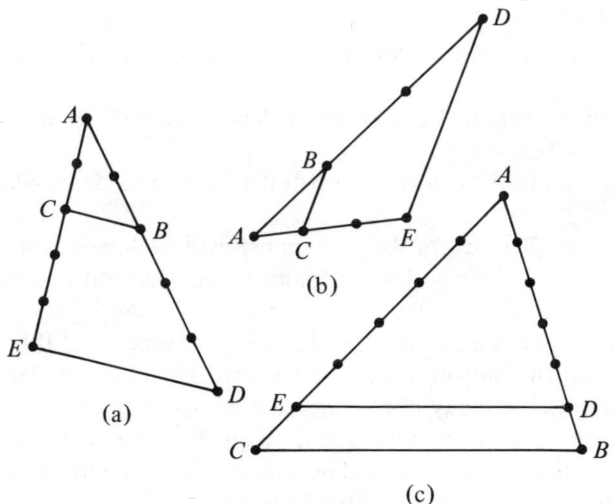

16. In Chapter 5 we noted that the length of the arc of a circle cut off by a central angle is "proportional" to the size of the angle. Write a proportion suggested by this figure and determine the value of a. (Use $\pi \doteq \frac{22}{7}$.)

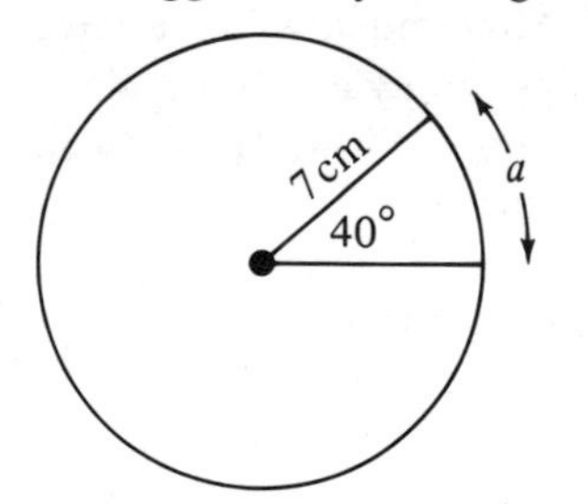

8.2 SIMILAR TRIANGLES

Loosely speaking, two triangles are similar if they "look alike" except perhaps that one is larger than the other. We might also say that two triangles are similar if they have the "same shape," but not necessarily the same size.

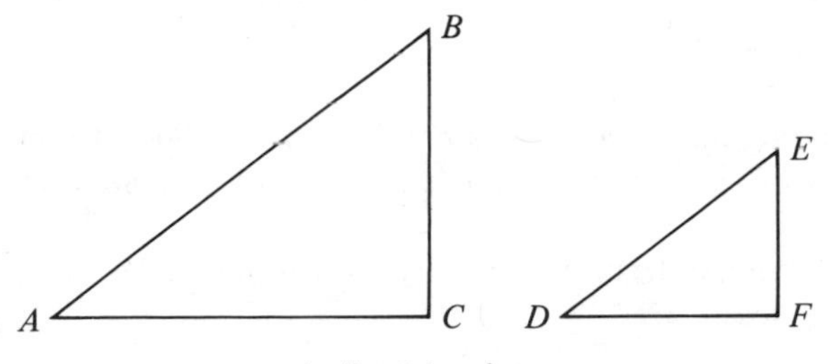

similar triangles

Neither of these tests for similarity is exact enough to be of much use in numerical problems. Fortunately you can discover usable, accurate, foolproof conditions for similarity by measuring "corresponding parts" of the similar triangles, $\triangle ABC$ and $\triangle DEF$. Use protractor and ruler to fill in this chart:

$m\angle A =$ ______	$m\angle B =$ ______	$m\angle C =$ ______
$m\angle D =$ ______	$m\angle E =$ ______	$m\angle F =$ ______
$AC =$ ______	$DF =$ ______	$AC/DF =$ ______
$AB =$ ______	$DE =$ ______	$AB/DE =$ ______
$CB =$ ______	$FE =$ ______	$CB/FE =$ ______

You should have arrived at this conclusion:

> Corresponding angles of *similar* triangles are congruent, and corresponding sides are *proportional* (that is, the ratios of the three pairs of corresponding sides are equal).

This property of *similar* triangles is very much like the property of *congruent* triangles: "Corresponding parts of congruent triangles are congruent." Many of the ideas about congruence and similarity are very much alike.

When we worked with congruent triangles, we could decide which parts corresponded by turning, sliding, and flipping the triangles until they coincided.

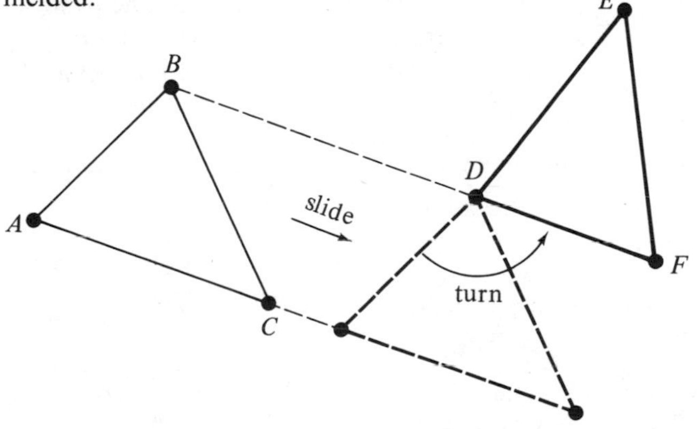

The corresponding vertices are

$$A \leftrightarrow F \qquad B \leftrightarrow D \qquad C \leftrightarrow E$$

from which the corresponding angles and sides can be determined. (If you have matched up all pairs of corresponding angles, how do you match up corresponding sides?)

When we work with similar triangles, usually we cannot make them coincide, but we can come close:

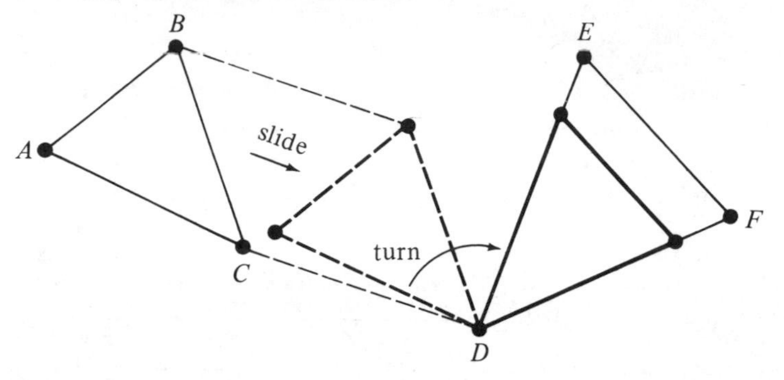

The corresponding vertices are

$$C \leftrightarrow D \qquad A \leftrightarrow E \qquad B \leftrightarrow F$$

from which the corresponding sides and angles can be determined. The common ratio of the sides of $\triangle ABC$ to the sides of $\triangle EFD$, which in the above figure is about $\frac{10}{13}$, is called the **ratio of similarity** of $\triangle ABC$ to $\triangle EFD$. (Notice the order in which the vertices of the second triangle were named.)

To check if two triangles were congruent, we did not have to measure all six pairs of corresponding parts. Three strategically located pairs were enough. (Recall the SAS, ASA, SSS, SAA congruence conditions.) To check if two triangles are similar, it is also unnecessary to measure all six pairs of corresponding parts. There are simple similarity conditions very much like the congruence conditions. For example:

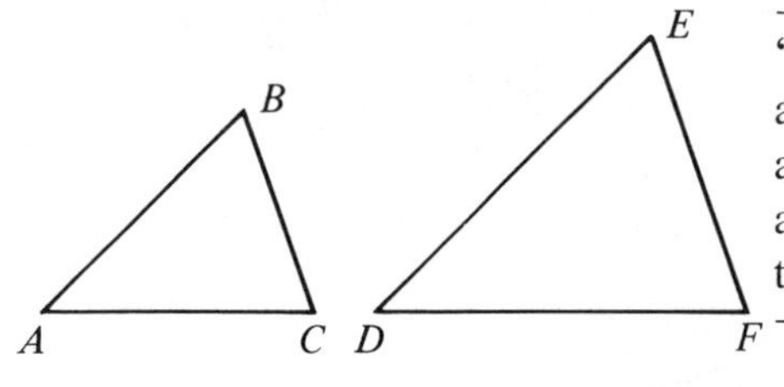

"SAS": If two sides of one triangle are proportional to two sides of another triangle and if the included angles are congruent, then the triangles are similar.

In symbols,

$$\text{If } AB/DE = AC/DF \text{ and } \angle A \cong \angle D, \text{ then } \triangle ABC \sim \triangle DEF.$$

(The symbol " $\sim$ " is read "is similar to." Notice that the congruence symbol " $\cong$ " suggests similarity (same shape) *and* equality (same size). We put the similarity condition, "SAS", in quotes to remind ourselves that it is not quite the same as the congruence condition, SAS.)

EXAMPLE Fill in the missing angle measures.

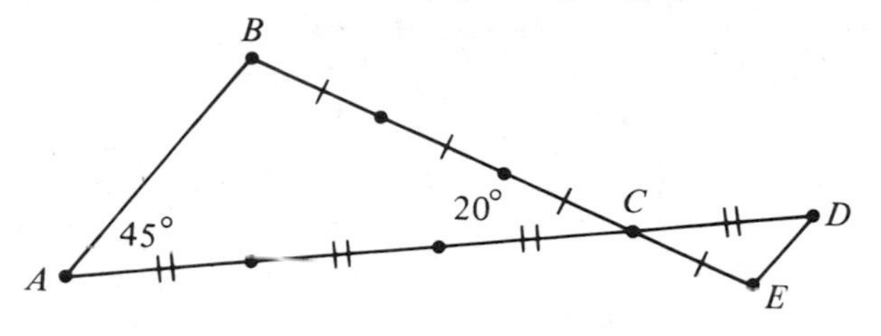

Solution

1. $\triangle ABC \sim \triangle DEC$ because (a) $BC/CE = AC/CD$ since each ratio is $\frac{3}{1}$. (b) $\angle BCA \cong \angle ECD$ since they are vertical angles. Thus the "SAS" condition is satisfied.
2. $m\angle B = 115°$ since $m\angle A + m\angle B + m\angle C$ must $= 180°$.
3. $m\angle D = m\angle A$ $(= 45°)$ because these are corresponding angles of similar triangles. ($\angle D$ corresponds to $\angle A$ since the sides opposite them correspond.)
4. $m\angle E = m\angle B$ $(= 115°)$ for the same reason.

(Can you explain why $\overleftrightarrow{AB} \parallel \overleftrightarrow{DE}$?)

Another similarity condition that is very much like a congruence condition is this:

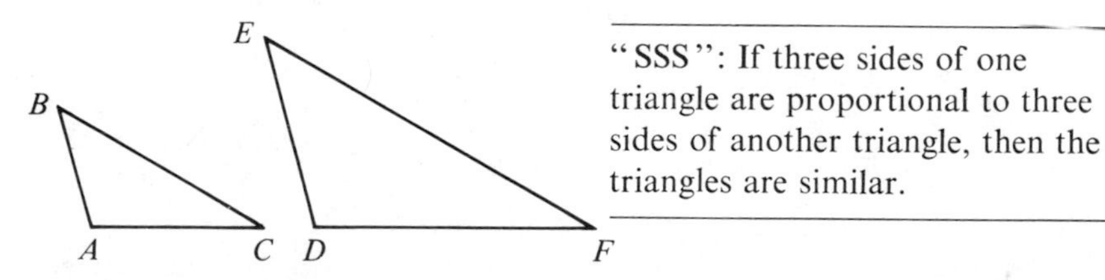

"SSS": If three sides of one triangle are proportional to three sides of another triangle, then the triangles are similar.

In symbols,

If $AB/DE = AC/DF = BC/EF$, then $\triangle ABC \sim \triangle DEF$.

Notice that we have written a "run-on" mathematical sentence, "$AB/DE = AC/DF = BC/EF$." This is to be interpreted in the same way as a run-on inequality such as $2 < x < 5$. Namely, "$AB/DE = AC/DF = BC/EF$ means

"$AB/DE = AC/DF$ *and* $AC/DF = BC/EF$"

and hence, of course, $AB/DE = BC/EF$ too.

EXAMPLE Fill in the missing angle measures:

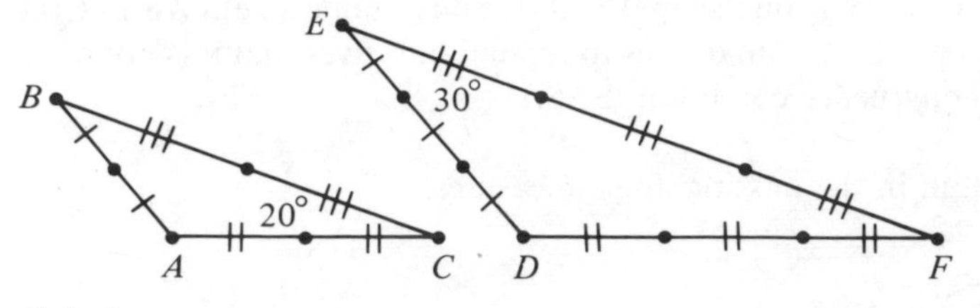

Solution

1. $\triangle ABC \sim \triangle DEF$ by "SSS" since the ratio of each pair of corresponding sides is $\frac{2}{3}$.
2. $m \angle B = m \angle E$ $(= 30°)$ since $\angle B$ and $\angle E$ are corresponding (why?) angles of similar triangles.
3. $m \angle A = 130°$ since $m \angle A + m \angle B + m \angle C$ must $= 180°$.
4. $m \angle D = m \angle A$ $(= 130°)$ corresponding angles.
5. $m \angle F = m \angle C$ $(= 20°)$ corresponding angles.

There is a single, simple similarity condition corresponding to the ASA and SAA congruence conditions.

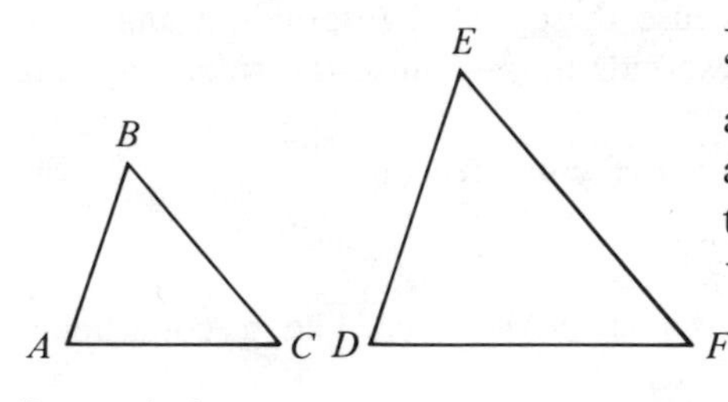

"AA": If two angles of one triangle are congruent respectively to two angles of another triangle, then the triangles are similar.

In symbols,

If $\angle A \cong \angle D$ and $\angle B \cong \angle E$, then $\triangle ABC \sim \triangle DEF$.

EXAMPLE Fill in the missing side lengths:

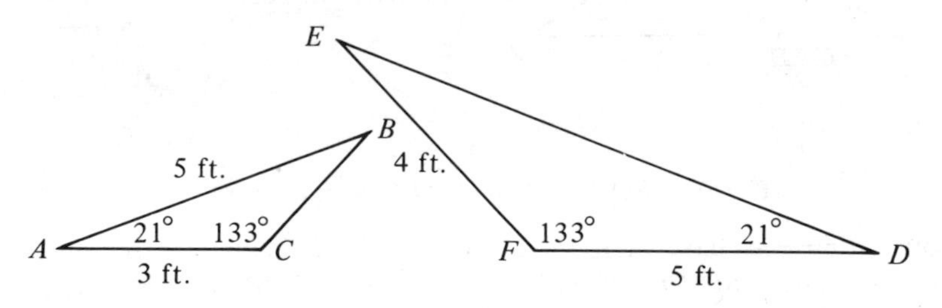

Solution

1. $\triangle ABC \sim \triangle DEF$ by "AA."
2. Since the triangles are similar, corresponding sides (sides opposite corresponding angles) must be proportional:

$$AB/DE = BC/EF = CA/FD$$

Substituting the given lengths,

$$5 \text{ ft}/DE = BC/4 \text{ ft} = 3 \text{ ft}/5 \text{ ft}$$

leads to two equations.

3. $5 \text{ ft}/DE = 3 \text{ ft}/5 \text{ ft} \Rightarrow DE = 5 \cdot \frac{5}{3} \text{ ft} = 8\frac{1}{3} \text{ ft}.$

4. $BC/4 \text{ ft} = 3 \text{ ft}/5 \text{ ft} \Rightarrow BC = 4 \cdot \frac{3}{5} \text{ ft} = 2\frac{2}{5} \text{ ft}.$

EXERCISES

1. Using only a protractor, decide if the following pairs of triangles are similar. If two triangles are similar, report that similarity in such a way that the correspondence between vertices is clear.

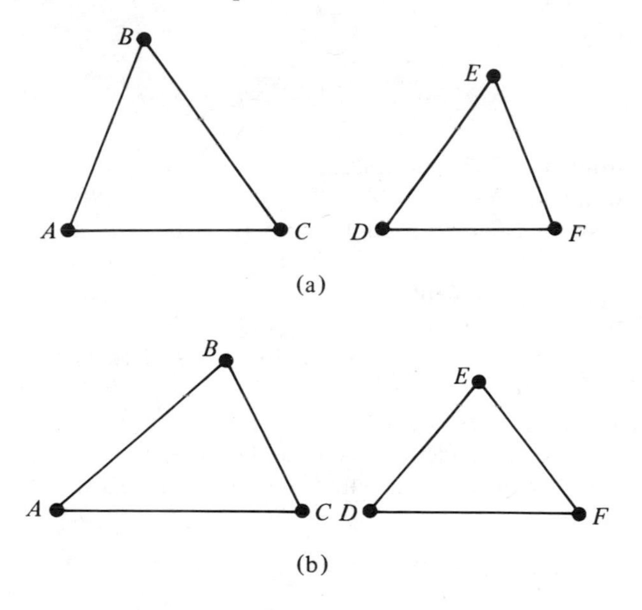

Is it possible to decide if two triangles are congruent using only a protractor?

2. Using only a ruler, decide if the following pairs of triangles are similar.

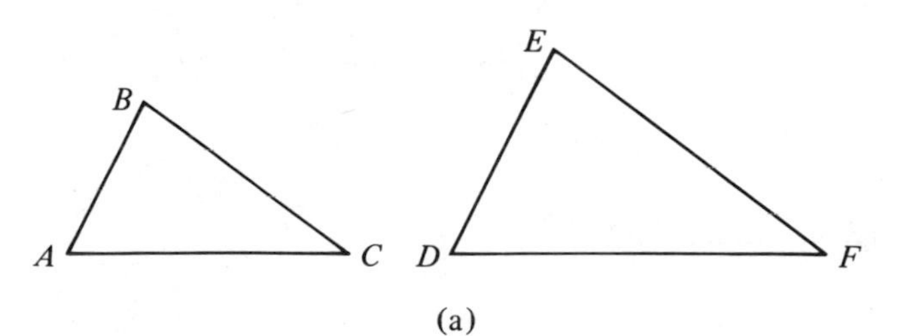

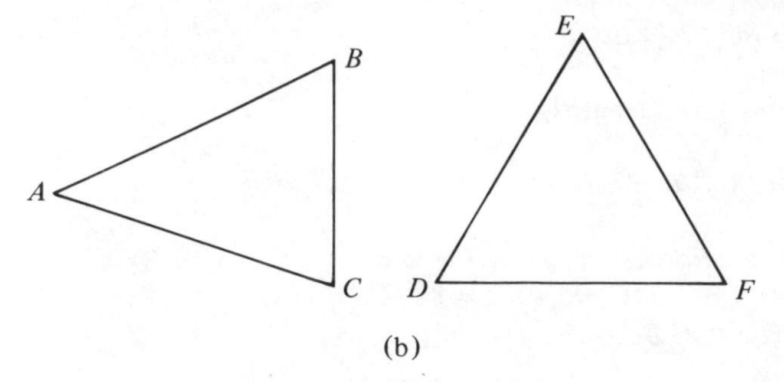

(b)

Is it possible to decide if two triangles are congruent using only a ruler?

3. (a) Construct, by an "SAS" procedure, a triangle similar to $\triangle ABC$ but having twice the perimeter.

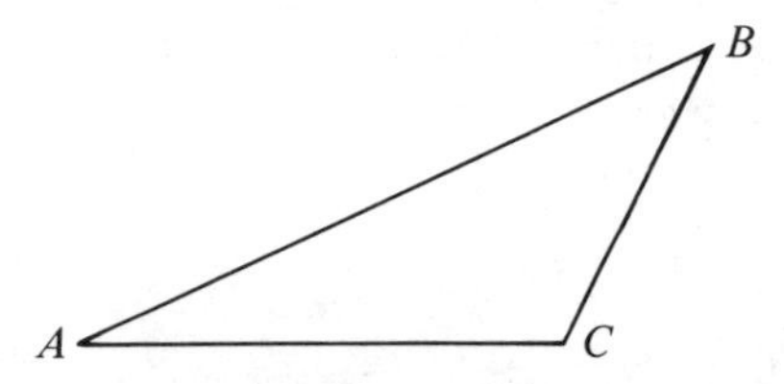

(b) Could you do it using an "SSS" procedure?

(c) Could you do it using an "AA" procedure?

(d) How does the area of the enlarged triangle compare to the area of the original?

4. (a) Must similar triangles be congruent?

(b) Must congruent triangles be similar?

(c) What is the ratio of the lengths of corresponding sides of two congruent triangles?

5. Explain why each pair of triangles is similar. Then report the similarity in such a way that the correspondence between the vertices is clear. Finally, using a ruler only when necessary, find the ratio of similarity of the first triangle to the second.

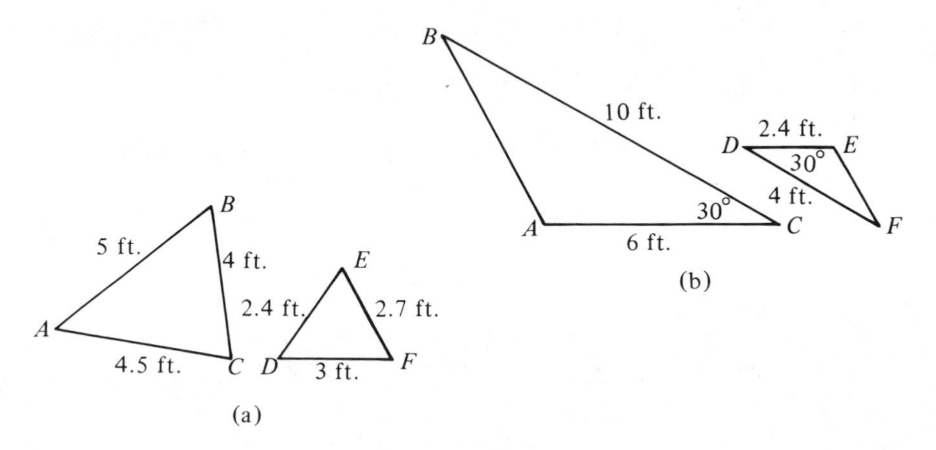

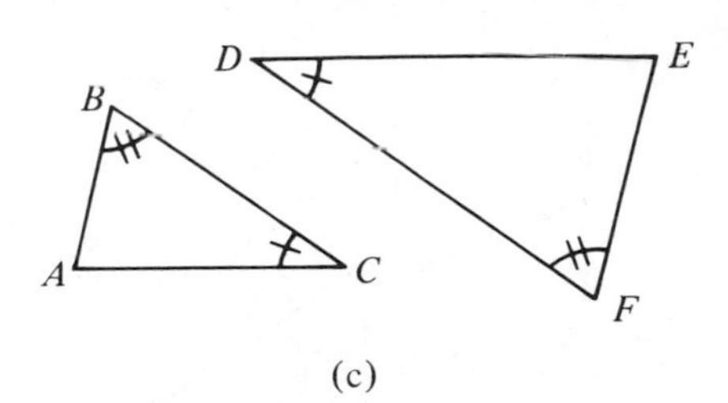

(c)

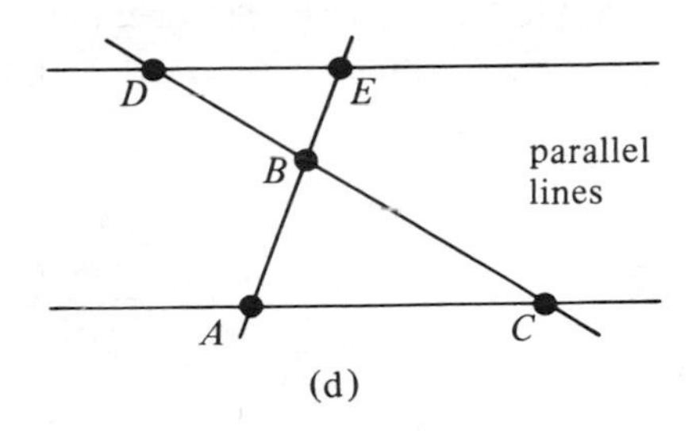

(d)

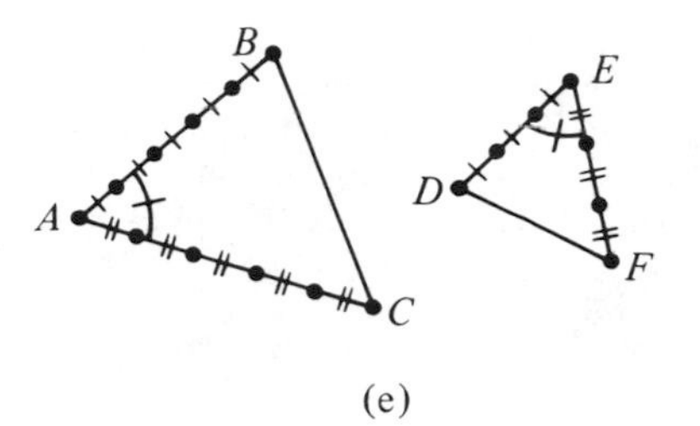

(e)

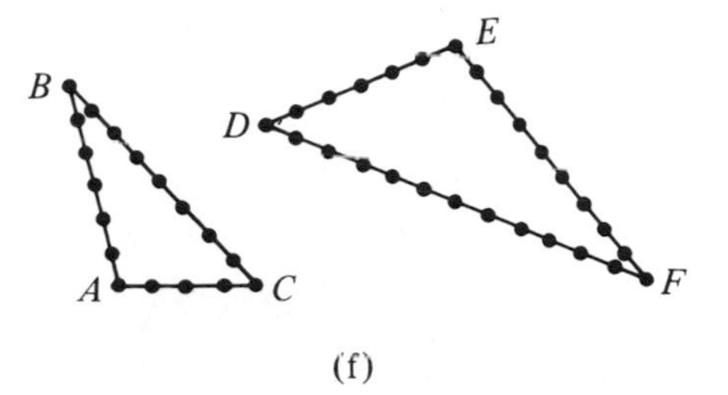

(f)

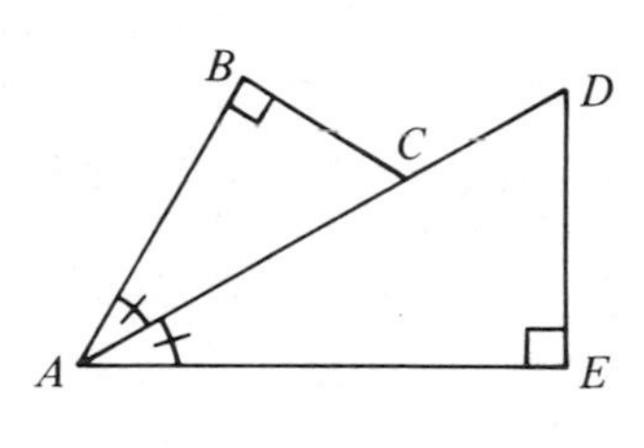

(g)

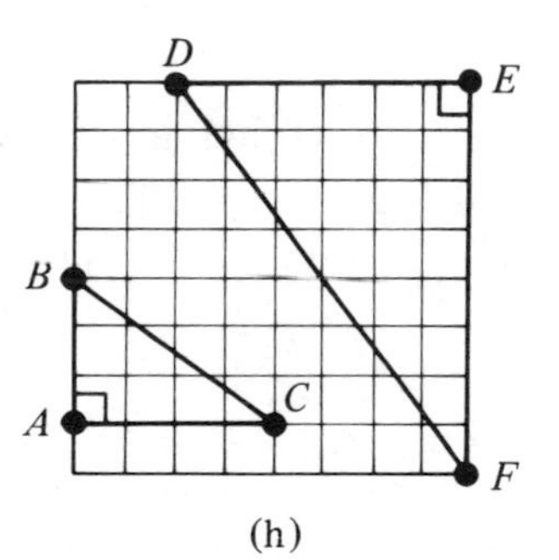

(h)

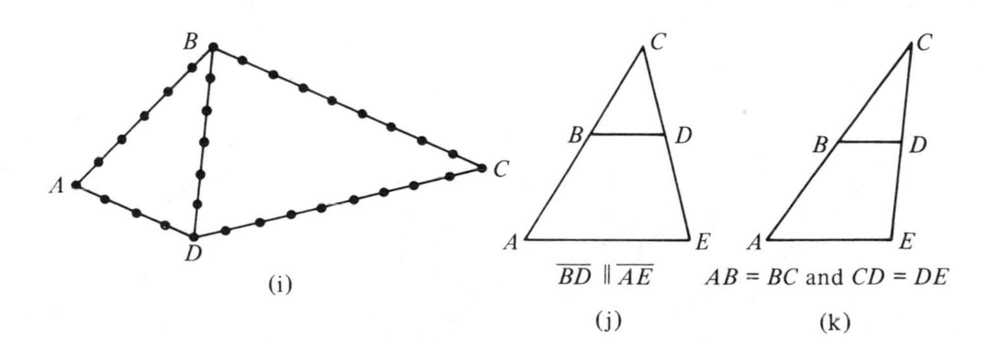

(i)

$\overline{BD} \parallel \overline{AE}$

(j)

$AB = BC$ and $CD = DE$

(k)

6. Find as many pairs of similar triangles as you can and explain why each pair is similar. Express all similarities in the conventional manner.

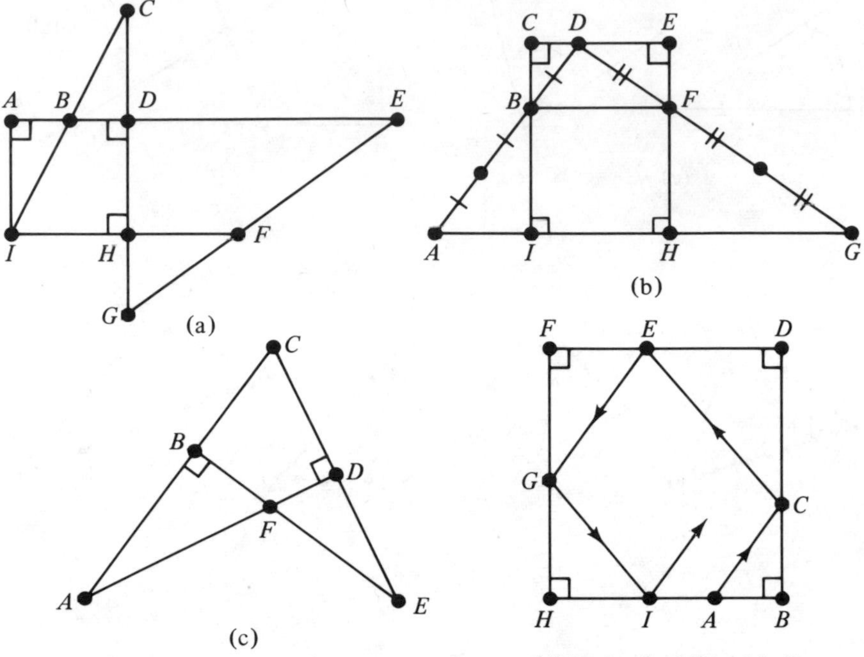

(d) Path of a billiard ball

What assumption are you making about how the ball rebounds off the rail?

In Exercises 7–11, explain why the triangles are similar and then fill in the blanks.

7.

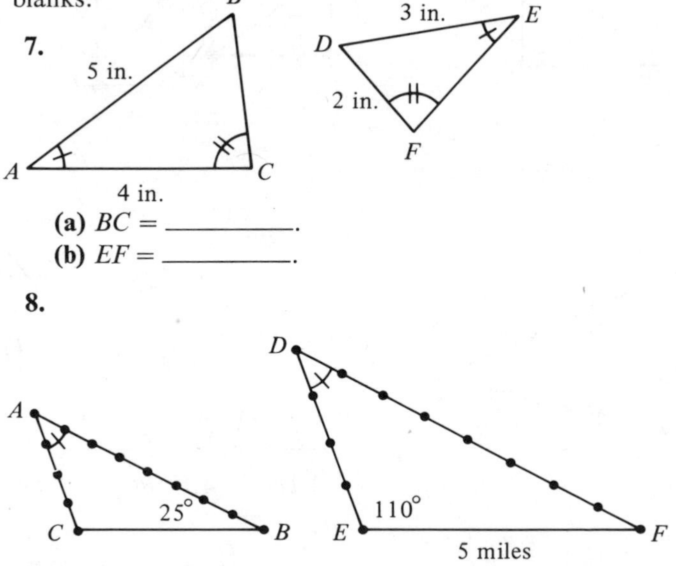

(a) $BC =$ __________.

(b) $EF =$ __________.

8.

(a) $m \angle C =$ ________.
(b) $m \angle A =$ ________.

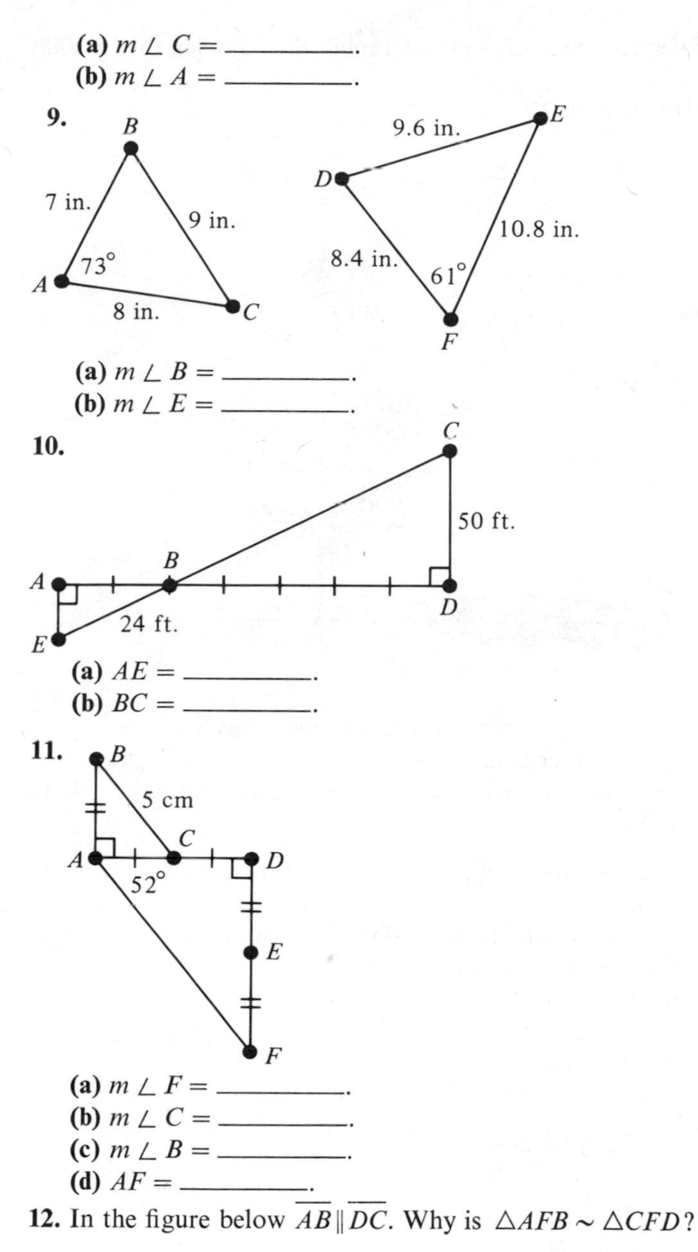

9.

(a) $m \angle B =$ ________.
(b) $m \angle E =$ ________.

10.

(a) $AE =$ ________.
(b) $BC =$ ________.

11.

(a) $m \angle F =$ ________.
(b) $m \angle C =$ ________.
(c) $m \angle B =$ ________.
(d) $AF =$ ________.

12. In the figure below $\overline{AB} \parallel \overline{DC}$. Why is $\triangle AFB \sim \triangle CFD$?

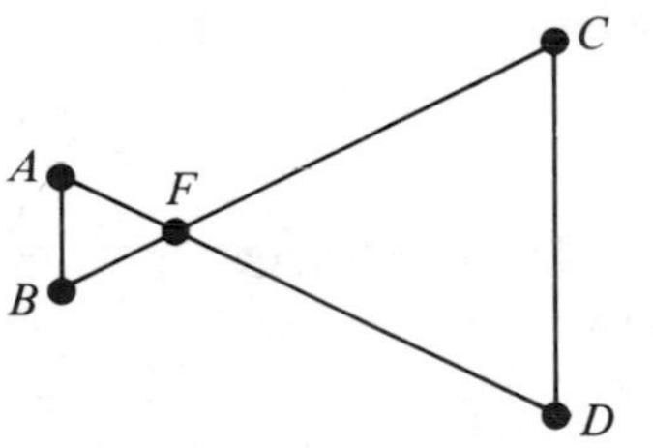

13. Why is the projected image of a photographic slide similar to the slide itself?

14. (a) Why is $\triangle ABC \sim \triangle DEF$?

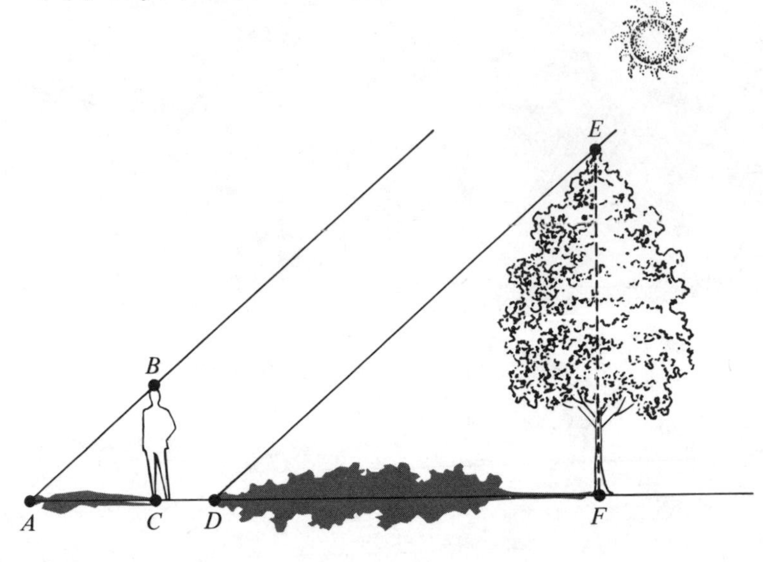

(b) What assumption did you make about the sun's rays?

(c) Did you make any other assumptions?

(d) If the man is 6 ft tall and casts a 7 ft shadow and the tree casts a 42 ft shadow, how tall is the tree?

In Exercises 15–17, use the similarity conditions for triangles as the basis for your explanations.

15. (a) Show that the segment joining the midpoints of two sides of a triangle must be parallel to the third side.

(b) How is BD related to AE?

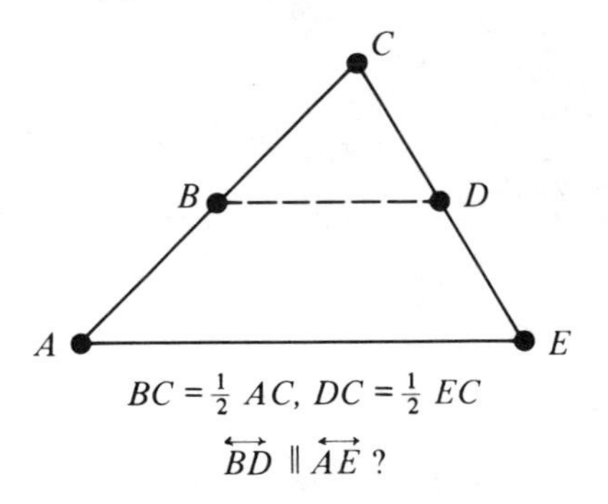

16. (a) Show that a segment that "divides two sides of a triangle proportionally" must be parallel to the third side.

(b) How is BD related to AE?

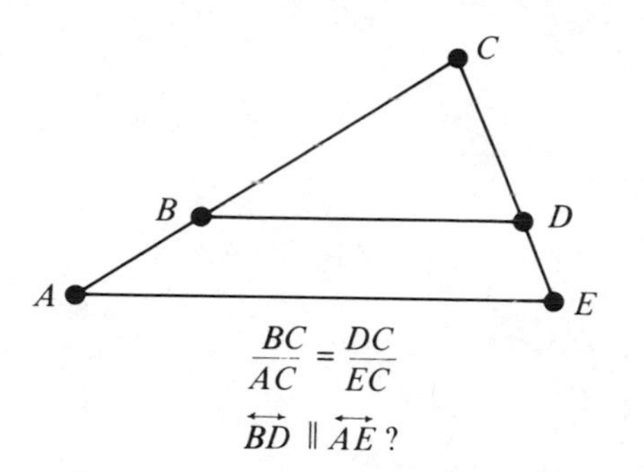

$$\frac{BC}{AC} = \frac{DC}{EC}$$

$$\overleftrightarrow{BD} \parallel \overleftrightarrow{AE}\ ?$$

17. (a) Show that a line drawn through two sides of a triangle parallel to the third side divides those two sides proportionally.

(b) How is BD related to AE?

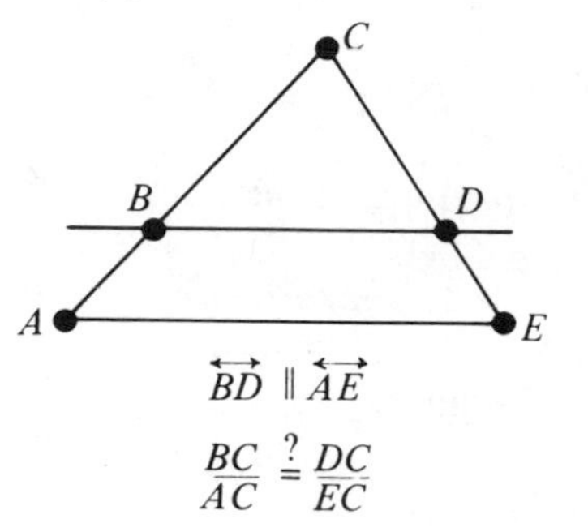

$$\overleftrightarrow{BD} \parallel \overleftrightarrow{AE}$$

$$\frac{BC}{AC} \stackrel{?}{=} \frac{DC}{EC}$$

18. Which of the following pairs of polygons would you consider to be similar?

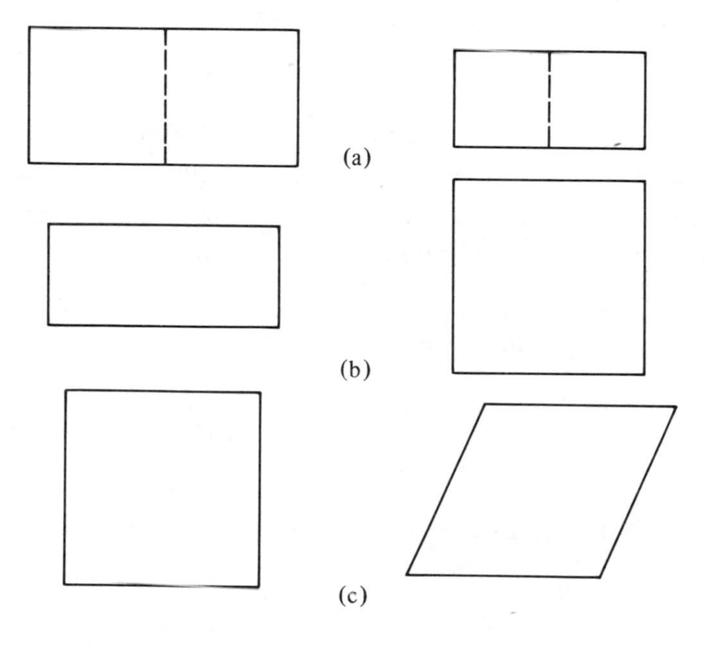

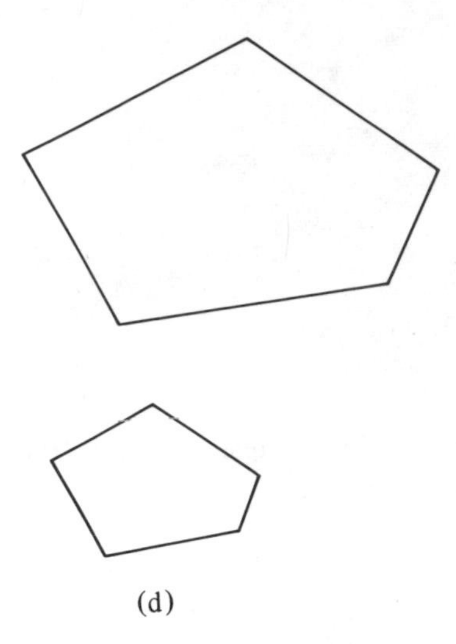

(d)

19. How many pairs of corresponding parts do two similar pentagons have? What is true about these pairs of corresponding parts?

20. Do you think there is an "AAAA" condition for similarity of quadrilaterals? An "SSSS" condition? An "SASAS" condition? Explain.

21. You might have noticed that no definition of similarity for triangles was given. Most contemporary high school geometry courses use a definition that is essentially the property given on p. 323.

> Two triangles are similar if and only if there is a one-to-one correspondence between their vertices such that corresponding angles are congruent and corresponding sides are proportional.

Formulate an analogous definition of congruence for triangles.

8.3 SIMILAR GEOMETRIC FIGURES; SCALE DRAWINGS

We studied similarity of *triangles* in detail because triangles are the simplest polygons. Consequently the theory of similarity of triangles was also not very complicated. There is a much more general notion of similarity.

Two geometric figures are **similar** if all distances between corresponding point pairs are proportional.

If the figures are polygons, then it turns out to be enough to check that the corresponding sides are proportional and the corresponding angles are congruent. If the polygons are triangles, then we know still simpler tests ("SAS," "SSS," "AA").

To understand what we mean when we say that "all distances between corresponding point pairs are proportional," use your ruler on the similar figures below and then simplify the ratios called for.

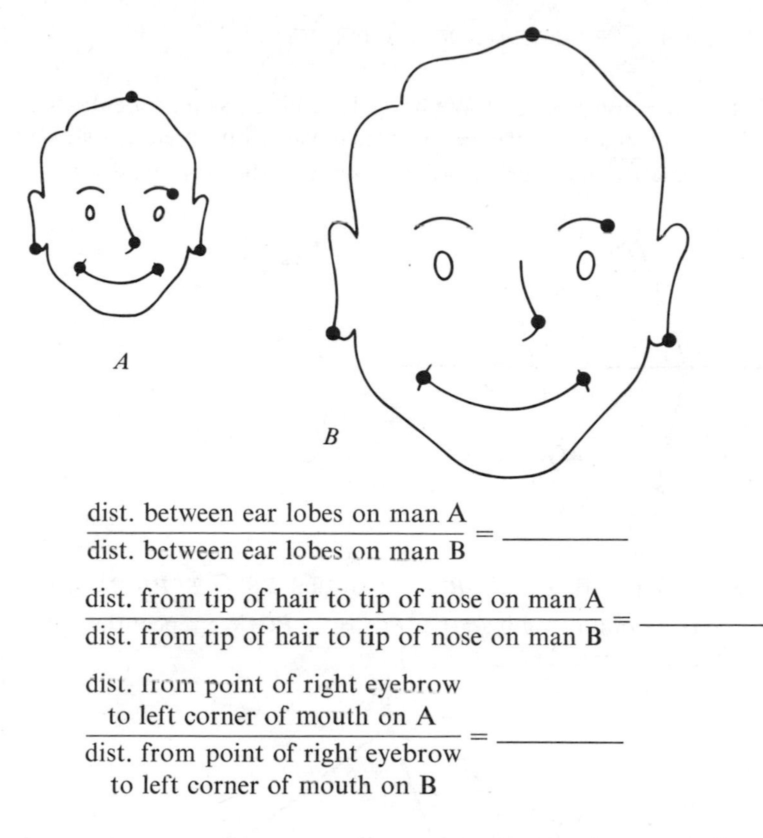

$$\frac{\text{dist. between ear lobes on man A}}{\text{dist. between ear lobes on man B}} = \underline{\qquad\qquad}$$

$$\frac{\text{dist. from tip of hair to tip of nose on man A}}{\text{dist. from tip of hair to tip of nose on man B}} = \underline{\qquad\qquad}$$

$$\frac{\begin{array}{c}\text{dist. from point of right eyebrow}\\ \text{to left corner of mouth on A}\end{array}}{\begin{array}{c}\text{dist. from point of right eyebrow}\\ \text{to left corner of mouth on B}\end{array}} = \underline{\qquad\qquad}$$

Locate two more "corresponding point pairs" and compute the ratio of the distance on A to the distance on B. You should discover that all of the ratios are $\frac{1}{2}$. We say that A is a "scale drawing" of B drawn to the scale $\frac{1}{2}$. We could also say that B is a scale drawing of A drawn to the scale $\frac{2}{1}$.

A **scale drawing** of a figure is another figure, similar to the first. The **scale** of the drawing is the (common) ratio of distances between corresponding point pairs. The scale is just what we referred to as the ratio of similarity in the previous section.

In this section we learn a simple technique for making scale drawings.

Suppose we want to draw a pentagon similar to pentagon *ABCDE* but drawn to the scale $\frac{2}{1}$. [Each "drawing" described below could actually be carried out as a "construction" if one wanted to take the time.]

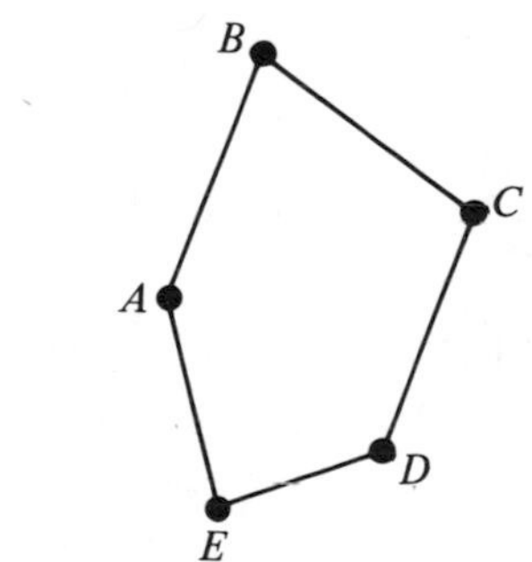

We first pick a "projection point" P. We have chosen it outside the pentagon, but we could have picked it anywhere—inside, outside, on an edge, or at a vertex. Next we draw $\overrightarrow{PA}$ and mark a point A' on it so that $PA' = 2 \cdot PA$.

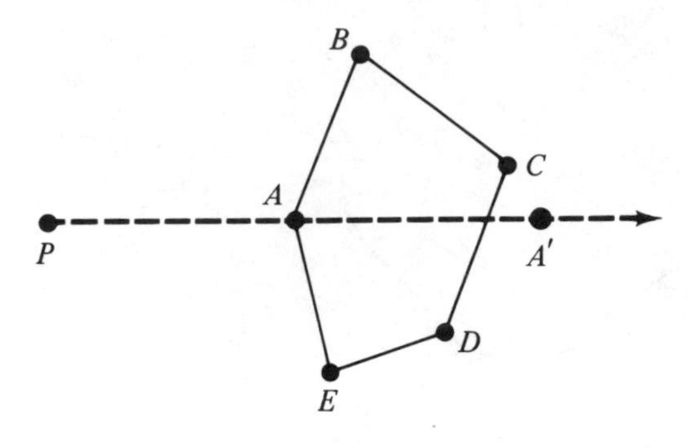

Next we draw $\overrightarrow{PB}$ and mark a point B' on it so that $PB' = 2 \cdot PB$. Next we draw $\overrightarrow{PC}$ and mark a point C' on it so that $PC' = 2 \cdot PC$. We repeat the same procedure at vertices D and E.

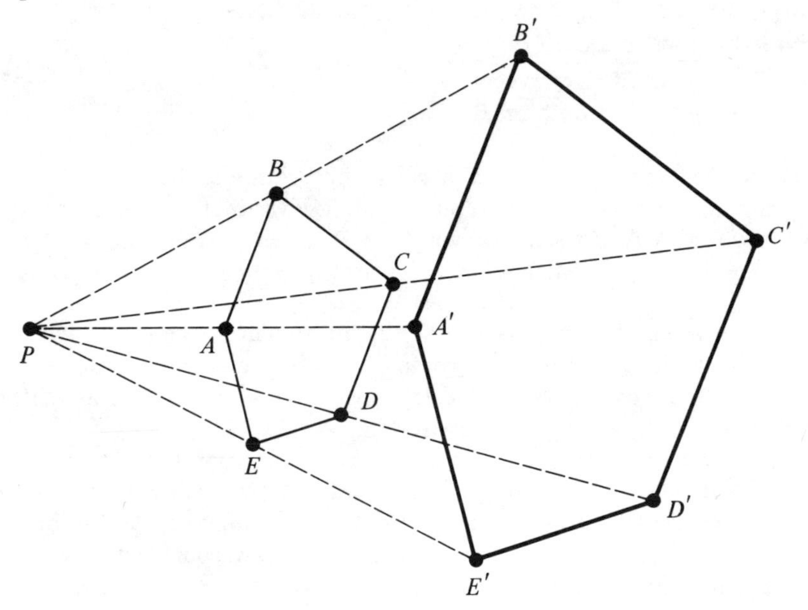

The points A', B', C', D', E' determine a pentagon similar to the original, but having each side twice as long.

We can explain *why* the figures must be similar on the basis of what we know about similar triangles:

1. $\triangle PAB \sim \triangle PA'B'$ by "SAS" ($PA/PA' = PB/PB' = \frac{1}{2}$ and $\angle APB = \angle A'PB'$).

2. Therefore $AB/A'B' = \frac{1}{2}$ (corresponding sides of similar triangles). That is,

$$A'B' = 2 \cdot AB$$

In the same way one could show that

$$B'C' = 2 \cdot BC, C'D' = 2 \cdot CD, D'E' = 2 \cdot DE,$$
$$E'A' = 2 \cdot EA$$

3. $m \angle PBA = m \angle PB'A'$ (corresponding angles of similar triangles).
4. $\triangle PBC \sim \triangle PB'C'$ by "SAS" (as in step 1).
5. Therefore $m \angle PBC = m \angle PB'C'$ (corresponding angles of similar triangles).
6. Now $m \angle ABC = m \angle PBC - m \angle PBA$ and $m \angle A'B'C' = m \angle PB'C' - m \angle PB'A'$. Therefore, by steps 3 and 5, $m \angle ABC = m \angle A'B'C'$. More simply,

$$\angle B \cong \angle B'$$

In the same way one could show that

$$\angle C \cong \angle C', \angle D \cong \angle D', \angle E \cong \angle E', \angle A \cong \angle A'$$

The pentagons are similar because corresponding angles are congruent and corresponding sides are proportional.

If we wish to shrink pentagon $ABCDE$ to a similar pentagon with one-third the perimeter we proceed just as before, but now we locate the points A', B', C', D', E' so that $PA' = \frac{1}{3} \cdot PA$, $PB' = \frac{1}{3} \cdot PB$, $PC' = \frac{1}{3} \cdot PC$, $PD' = \frac{1}{3} \cdot PD$, $PE' = \frac{1}{3} \cdot PE$.

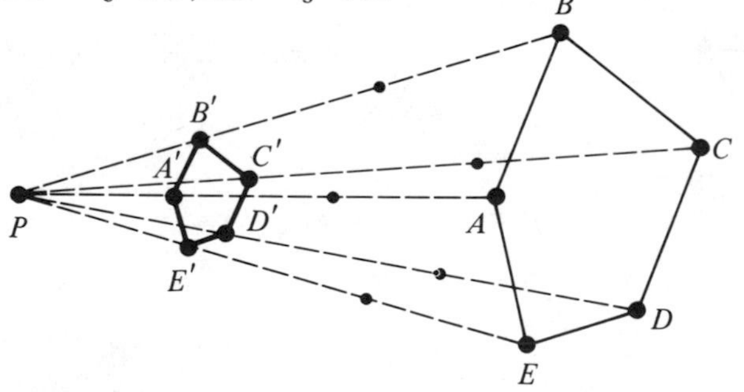

EXERCISES

1. Suppose you are the editor of a dictionary and you have decided that each illustration should be 1 in. high. What will be the approximate scale of your illustrations of the following?

(a) an elephant 10 ft tall
(b) the earth (diameter $\doteq$ 8000 miles)
(c) a ladybug
(d) a U.S. 25-cent piece

2. Make scale drawings with projection point P and scale as indicated.
(a) projection point P outside the quadrilateral; scale $\frac{3}{1}$:

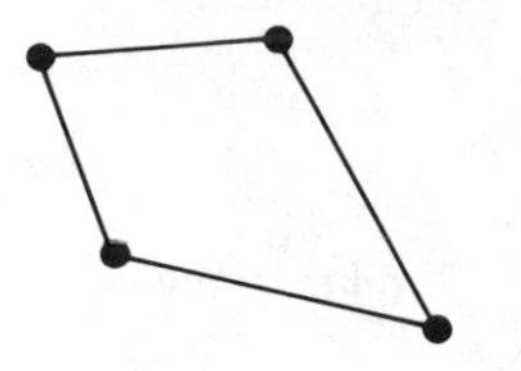

(b) P inside; scale $\frac{1}{2}$:

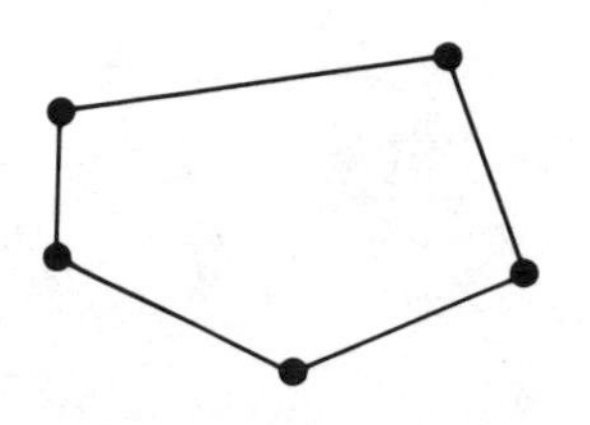

(c) P at a vertex; scale $\frac{2}{3}$:

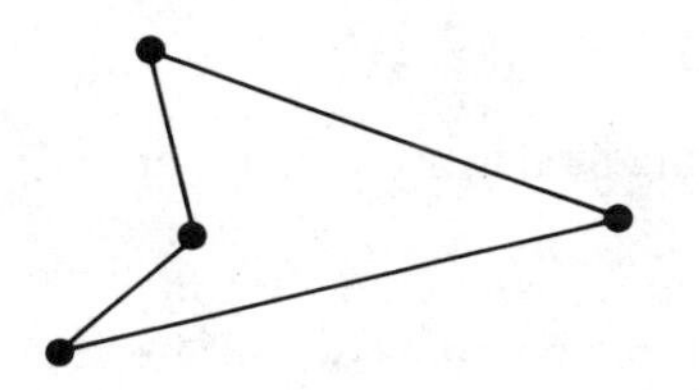

(d) P on an edge but not at a vertex; scale $\frac{3}{2}$:

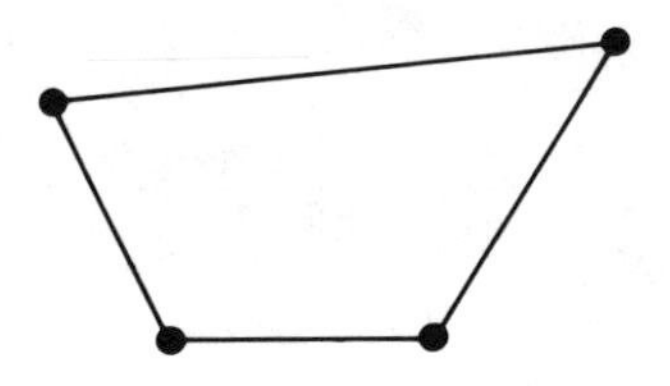

(e) P outside; scale $\frac{1}{2}$:

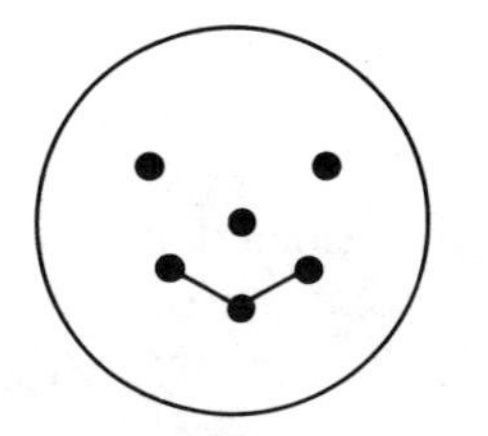

3. Locate the missing projection point and determine the scale (ratio of similarity) for each pair of similar figures below.

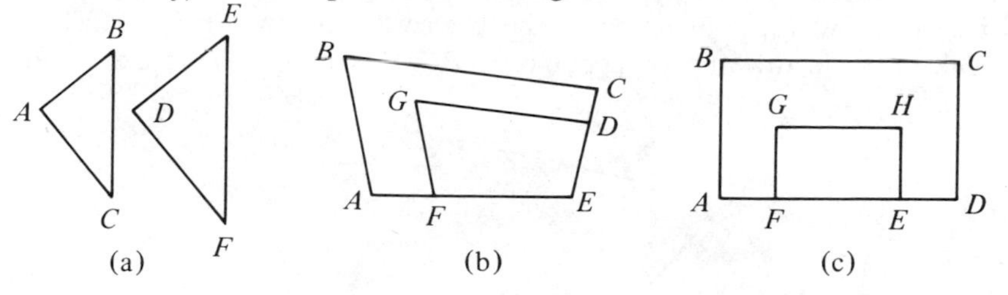

4. (**a**) Can you find a projection point for the pair of similar figures below?

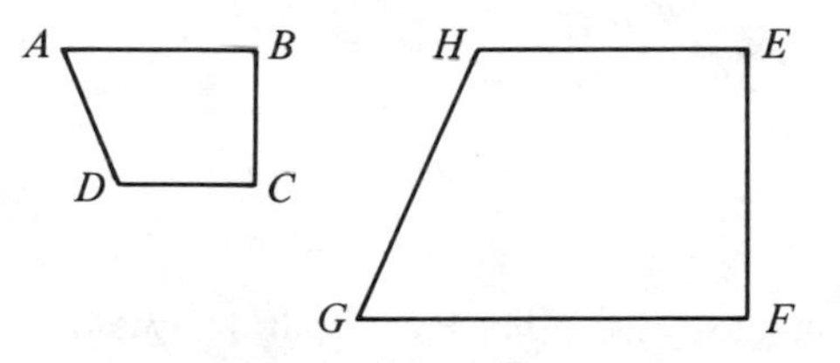

(**b**) Explain how to turn, slide, or flip quadrilateral $EFGH$ so that a projection point can be found.

5. Repeat Exercise 4 for these pairs of figures:

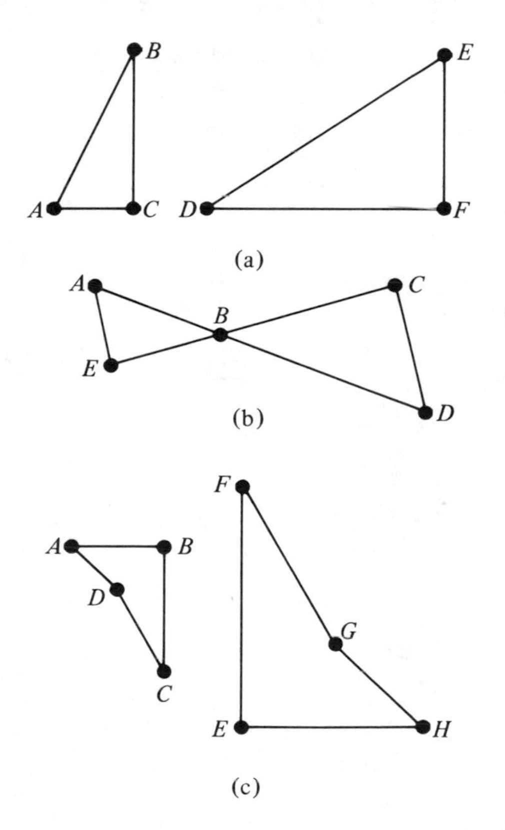

6. Do you think that any pair of similar plane figures can be made to coincide using slides, turns, flips, and then a projection?
7. Explain how you would locate the projection point P from which to make a scale drawing of pentagon $ABCDE$ having FG as one of its sides.

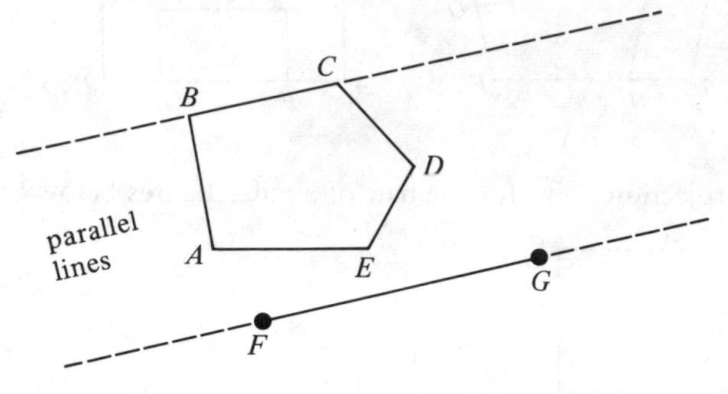

8. The incomplete diagrams below suggest two other techniques for making scale drawings. (The first one is the method used by Michelangelo and other fresco painters to enlarge a small drawing to a full-scale pattern, or *cartoon*, for the fresco.) Study each technique and complete the drawings.

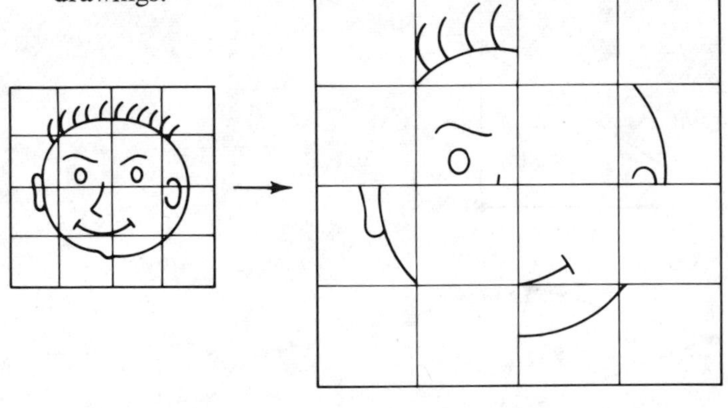

(a)

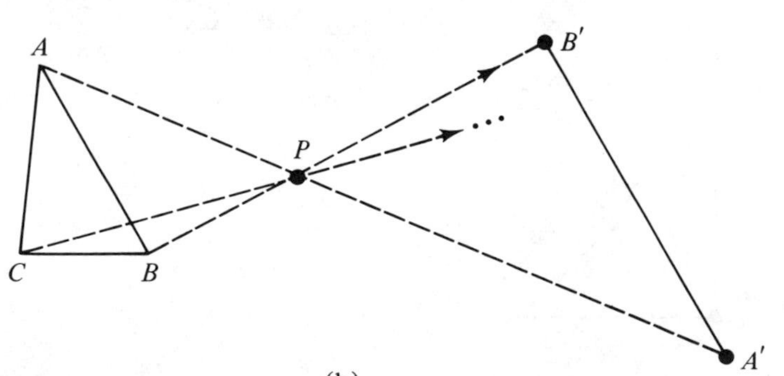

(b)

9. (a) Make a scale drawing F' of figure F to the scale $\frac{3}{2}$ using the projection point P.

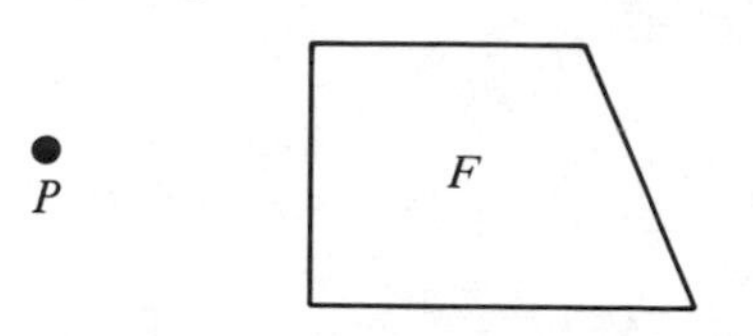

(b) Use your ruler to find the (approximate) perimeters of F and F'. What is the ratio of the perimeter of F' to the perimeter of F?

(c) Calculate the areas of F and F'. What is the ratio of the area of F' to the area of F?

10. Repeat all parts of Exercise 9 for the figure F and point P shown, and the scale $\frac{2}{5}$.

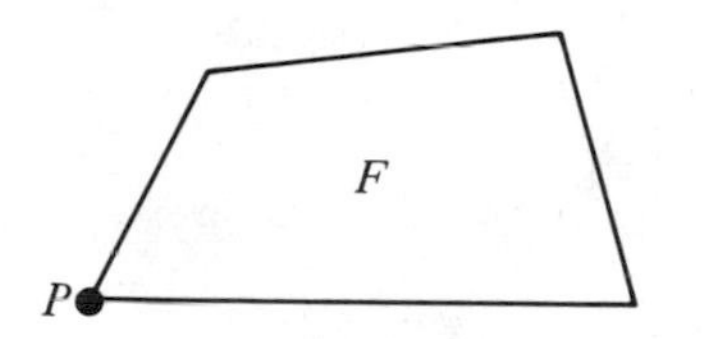

11. Exercises 9 and 10 suggest that:

The ratio of the areas of similar plane regions is equal to the square of the ratio of similarity (scale).

Explain this fact by describing what happens to a 1 in. by 1 in. square inside the figure F of Exercises 9 and 10.

12. How do the perimeters of similar figures compare?

13. The figures in each part below are similar.

(a)

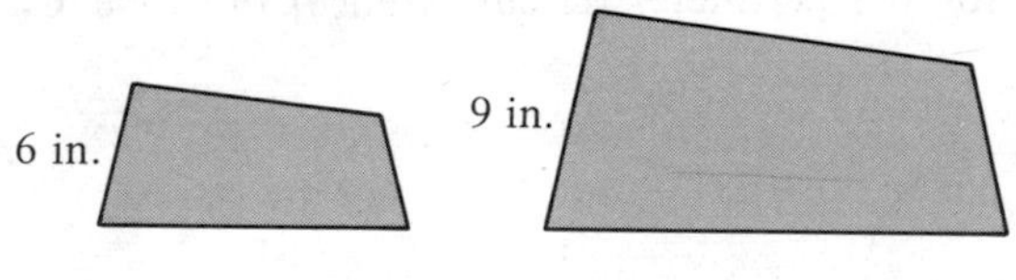

(b)

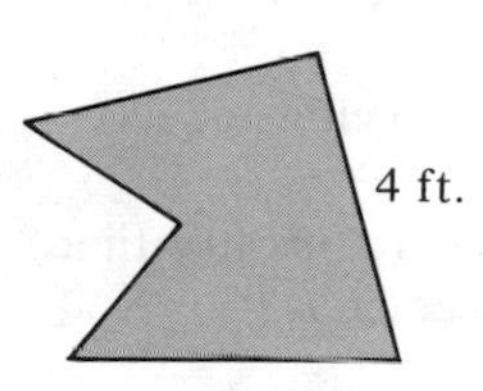

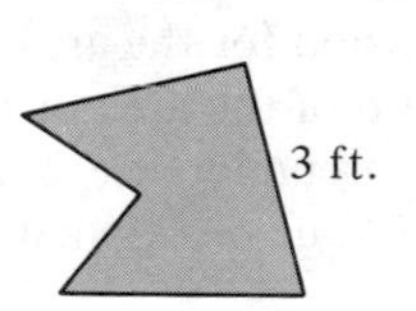

(c)

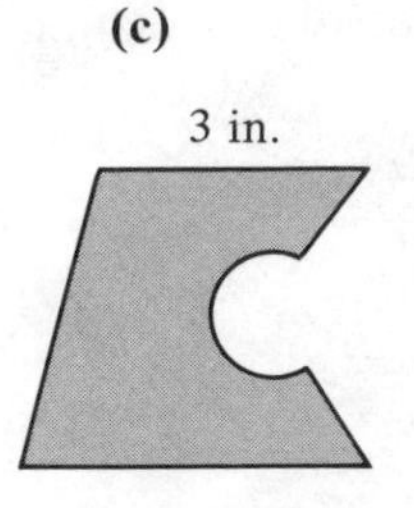

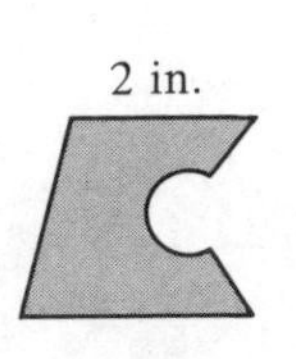

perimeter 36 cm	perimeter ________
area ________	area 20 cm^2

(d)

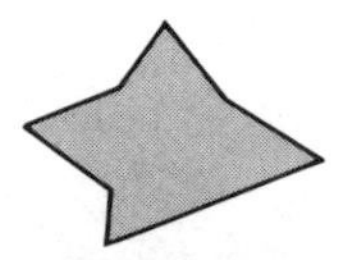

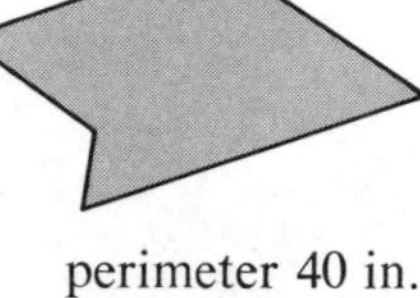

perimeter 30 in.	perimeter 40 in.
area 30 in.^{2}	area ________

(e)

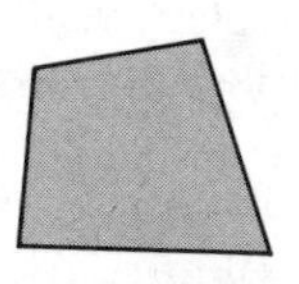

perimeter 21 cm	perimeter ________
area 25 cm^2	area 9 cm^2

14. (a) Write an expression for the perimeter (circumference) of circle C_1. Also circle C_2.

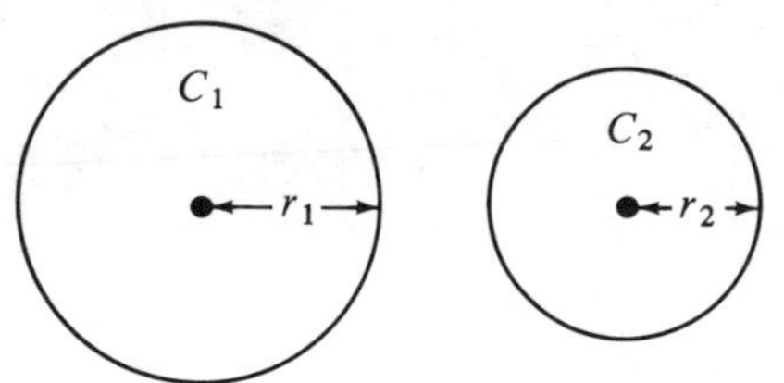

(b) What is the ratio of the perimeter of C_1 to the perimeter of C_2?
(c) What is the scale of C_1 to C_2?
(d) Write an expression for the area inside circle C_1. Also circle C_2.
(e) What is the ratio of the area of C_1 to the area of C_2?

15. (a) If circle C has circumference 2 mi and circle D has radius five times as long as the radius of C, what is the circumference of D?

(b) If circle E has area 3 sq mi and circle F has radius one-fourth the radius of E, what is the area of circle F?

(c) To double the circumference of a circle, the radius must be multiplied by ________.

(d) To double the area of a circle, the radius must be multiplied by ________.

16. (a) Make a (perspective) scale drawing C' of the cube C to the scale $\frac{3}{1}$ using the projection point P.

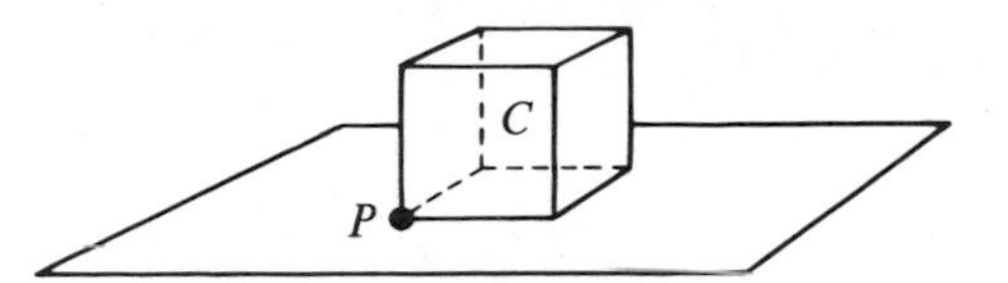

(b) What is the ratio of the length of an edge of C' to an edge of C?

(c) What is the ratio of the total edge length of C' to the total edge length of C?

(d) What is the ratio of the area of a face of C' to the area of a face of C?

(e) What is the ratio of the surface area of C' to the surface area of C?

(f) What is the ratio of the volume of C' to the volume of C?

17. Study Exercise 16 and then decide

(a) how the surface areas of similar solid regions are related.

(b) how the volumes of similar solid regions are related.

18. (a) If sphere S_1 has radius 2 in. and sphere S_2 has radius 10 in., what is the ratio of the surface area of S_2 to the surface area of S_1?

(b) What is the ratio of the volume of S_2 to the volume of S_1?

(c) If the radius of a sphere is tripled, then its surface area is multiplied by ________, and its volume is multiplied by ________.

(d) Which of the factors, r or r^2 or r^3, do you think will appear in the fomula for the surface area of a sphere of radius r?

(e) Same question as in (d) for the volume formula.

19. Joe has a miniature iron cannon that is claimed to be an *exact* replica of one of Napoleon's iron cannons, except that Joe's is 2 in. high while Napoleon's was 6 ft high. Joe's cannon weighs 1 lb. How much did Napoleon's cannon weigh?

20. Two geometric figures are similar if all distances between corresponding point pairs are proportional. Formulate an analogous criterion for congruence of two geometric figures.

8.4 RIGHT TRIANGLES AND THE PYTHAGOREAN THEOREM

Since the sum of the measures of the angles of a triangle is 180° and since a right angle has measure 90°, it follows that no triangle can have more than one right angle

---

A triangle having one right angle is called a **right triangle**.

In fact, if a triangle has one right angle, then both of its other angles are **acute** angles (angles of measure less than 90°).

Right triangles are much used in theory and practice. For that reason a special terminology has developed. The side opposite the right angle is called the **hypotenuse** of the right triangle. The other two sides are called its **legs**.

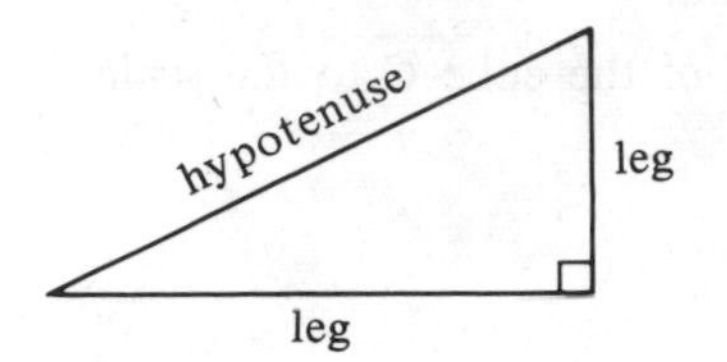

A right triangle.

Since any two right triangles have a pair of angles congruent to begin with (the right angles), deciding when two right triangles are congruent is quite easy. Do you see how the ASA congruence condition applies to each pair of right triangles below?

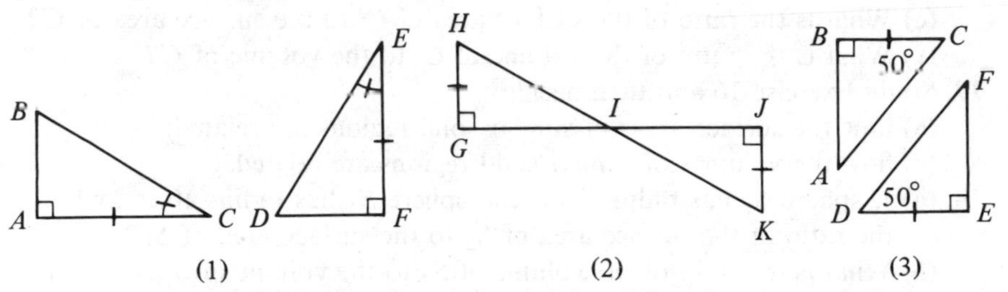

The SAS condition can be rephrased for right triangles as follows:

LL Two right triangles are congruent if the *legs* of one are congruent respectively to the legs of the other.

We give this "leg-leg" condition a name, LL, because we want to refer back to it. Do not clutter up your mind trying to remember it. It is just SAS.

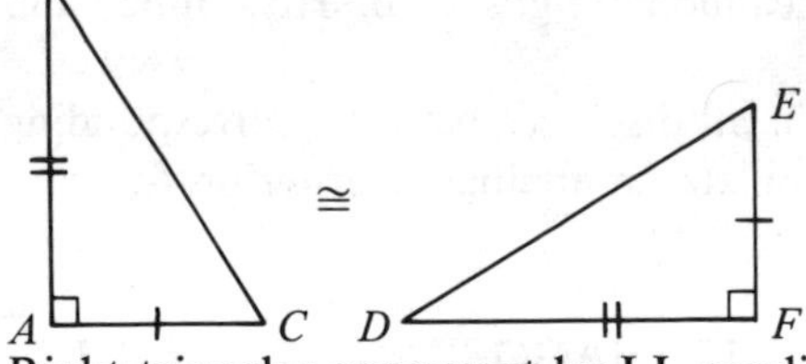

Right triangles congruent by LL condition.

More surprising is the hypotenuse-leg condition.

HL Two right triangles are congruent if the hypotenuse and a leg of one are congruent respectively to the hypotenuse and a leg of the other.

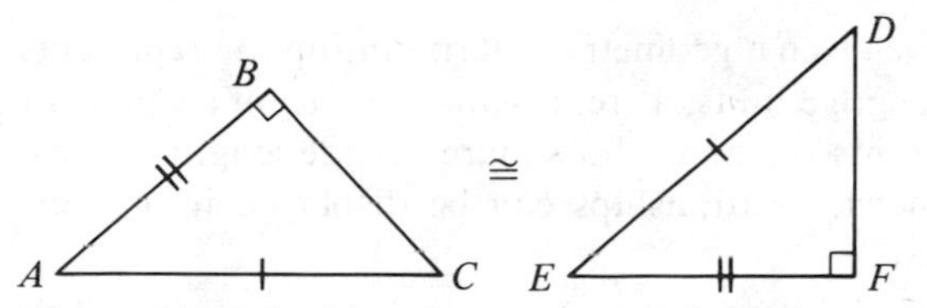

Figure 1. Right triangles congruent by HL condition.

On the surface this looks like an SSA condition, but we know that in general there is no SSA congruence condition! The reason this condition holds goes much deeper. It depends on what is probably the most famous result in geometry, "the Pythagorean theorem."

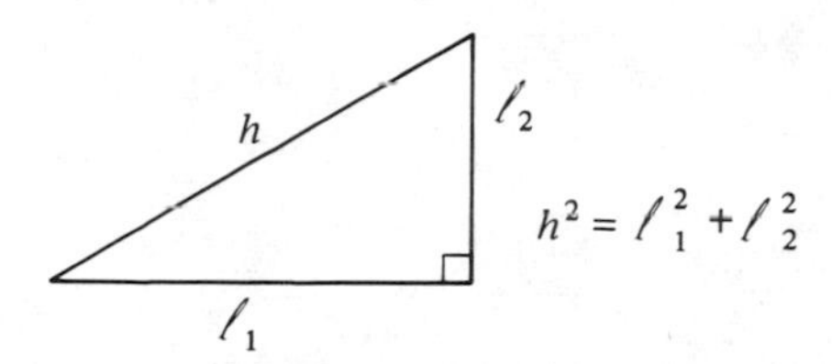

Pythagorean theorem: In a right triangle, the square of the length of the hypotenuse is equal to the sum of the squares of the lengths of the legs.

In Figure 1, then,

$$AB^2 = AC -{}^2 BC^2 = DE^2 - DF^2 = EF^2$$

and so $AB = EF$. In light of the Pythagorean theorem, then, the HL condition is just the SSS or SAS condition in disguise. Thus the HL condition, like the LL condition, is not worth memorizing.

No one knows for sure who first discovered the Pythagorean theorem or how he stumbled on it, but it is thought that the "3–4–5 right triangle" provided the key. The numbers 3, 4, and 5 are related both arithmetically and geometrically. Arithmetically,

$$3^2 + 4^2 = 5^2$$

Geometrically, a triangle with sides of length 3, 4, and 5 units is a right triangle.

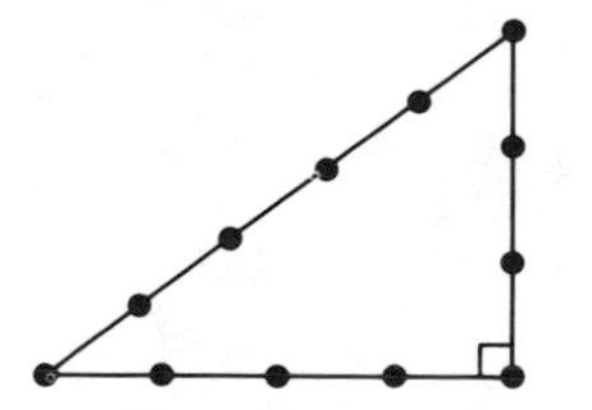

It is thought that the Egyptians knew this fact and used a knotted loop of rope for making square corners. How did they do that?

The arithmetic relation can be given a geometric interpretation: 3^2 represents the area of a square of side length 3 units, 4^2 represents the area of a square of side length 4 units, 5^2 represents the area of a square of side length 5 units. So the arithmetic and geometric relationships can be displayed in a single geometric figure.

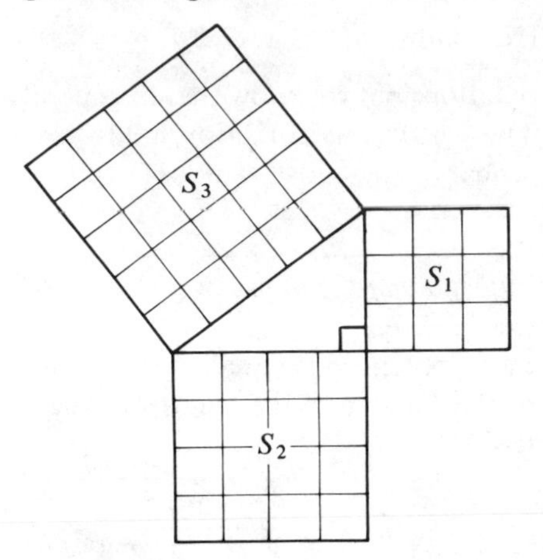

Area of S_1 + area of S_2 = area of S_3.

The person who discovered the Pythagorean theorem was the one who *generalized* this relationship from 3–4–5 right triangles to all right triangles.

Pythagorean theorem If $\angle ACB$ is a right angle, then area of S_1 + area of S_2 = area of S_3.

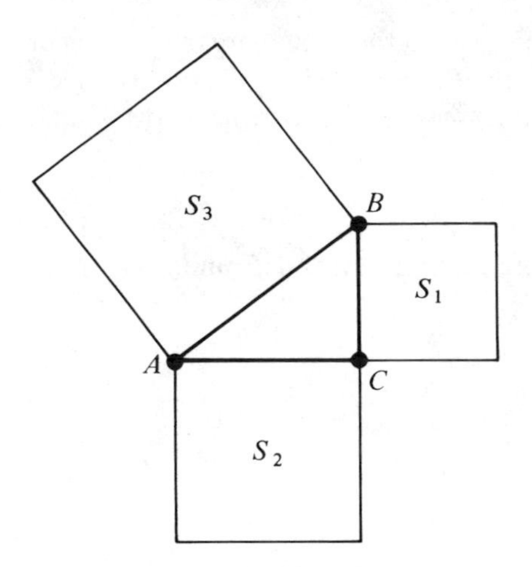

A derivation of this theorem, based on similarity, is called for in the exercises. An intuitive derivation that could be used in the elementary school is to attach four copies of $\triangle ABC$ to S_3, as shown,

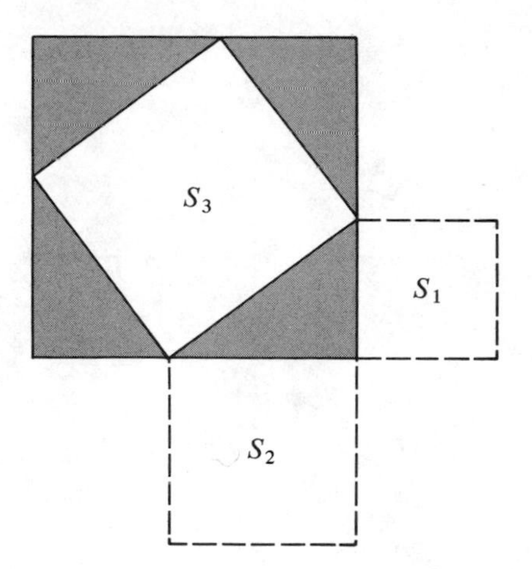

and then to also attach four copies of $\triangle ABC$ to $S_1 \cup S_2$ as shown.

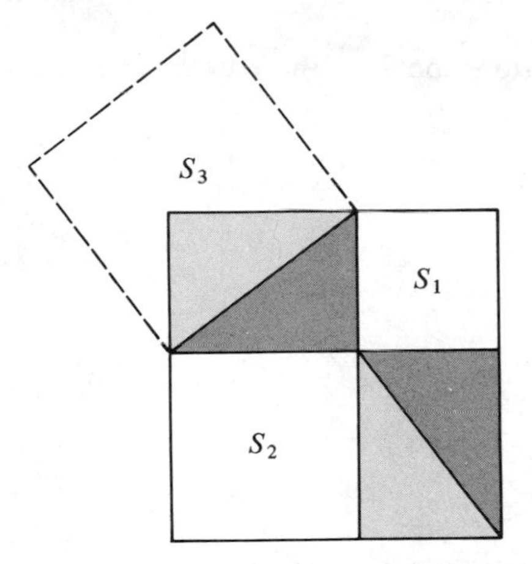

Now, since the two large squares have the same area,

$$\text{area}\,(S_3) + 4 \cdot \text{area}\,(\triangle ABC) = \text{area}\,(S_1) + \text{area}\,(S_2) + 4 \cdot \text{area}\,(\triangle ABC)$$

and so

$$\text{area}\,(S_3) = \text{area}\,(S_1) + \text{area}\,(S_2)$$

EXERCISES

1. Explain why each pair of triangles below is congruent and report the congruence in a way that suggests the correspondence between vertices.

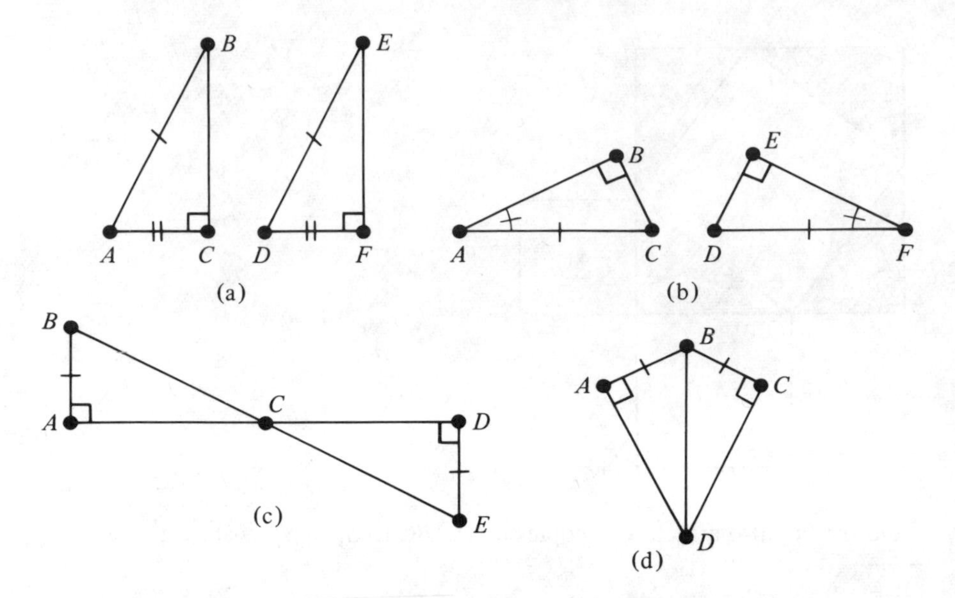

2. Why doesn't the LL or HL condition apply to the pair of triangles below?

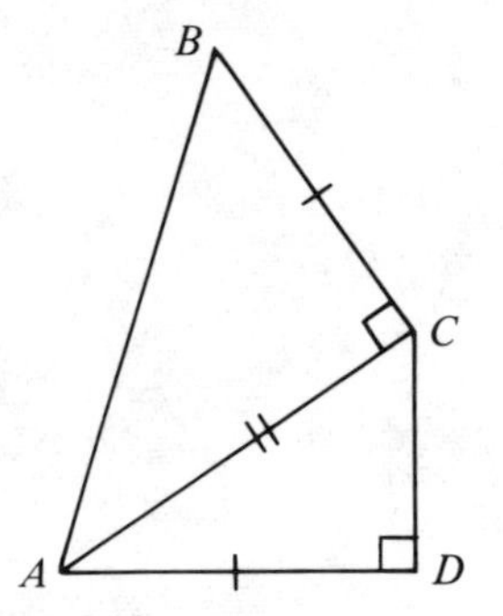

3. Why doesn't the SAA condition apply to the pairs of triangles below?

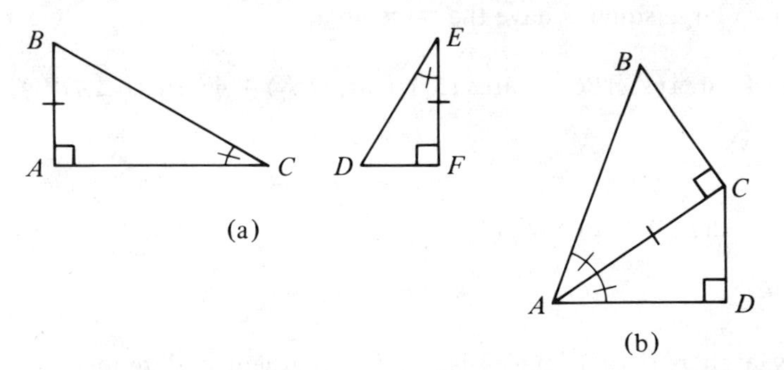

The diagrams in Exercises 4–7 are intentionally distorted.

4. Why is $\triangle ABC$ a right triangle?

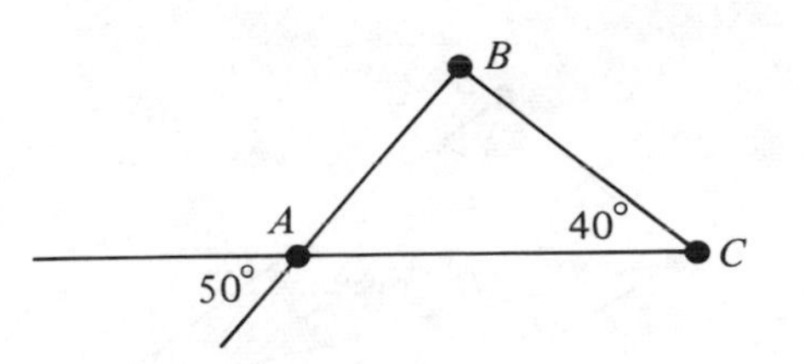

Name its hypotenuse.
Name its legs.

5. Why is $\triangle DEF$ a right triangle?

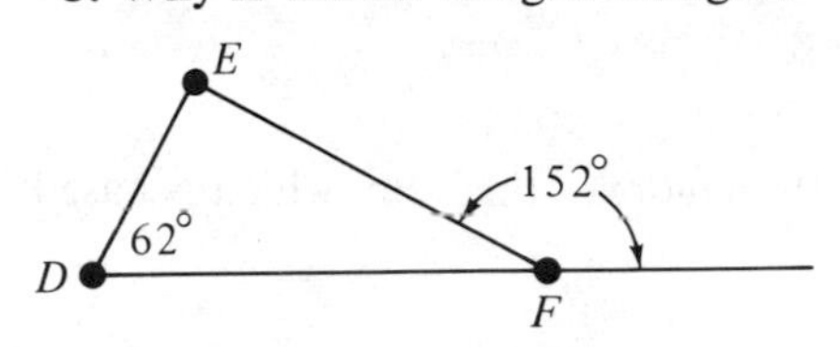

Name its hypotenuse and legs.

6. Why is $\triangle GHI$ a right triangle?

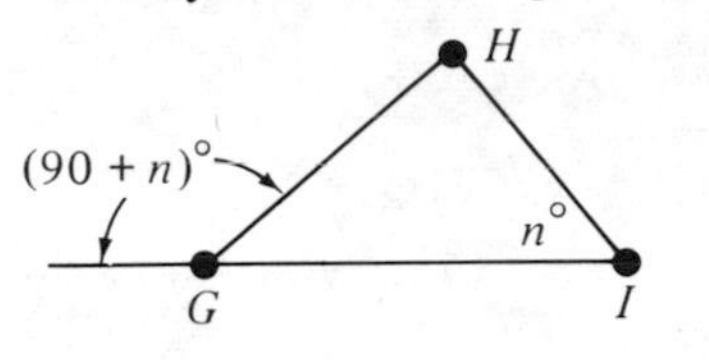

7. Why is $\triangle JKL$ a right triangle?

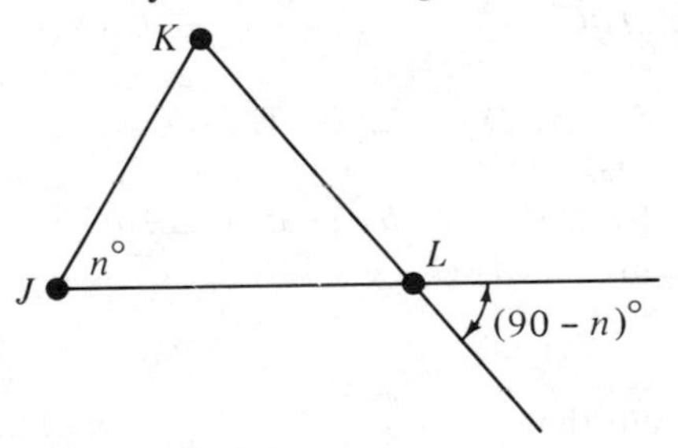

8. Name all right triangles in each diagram below.

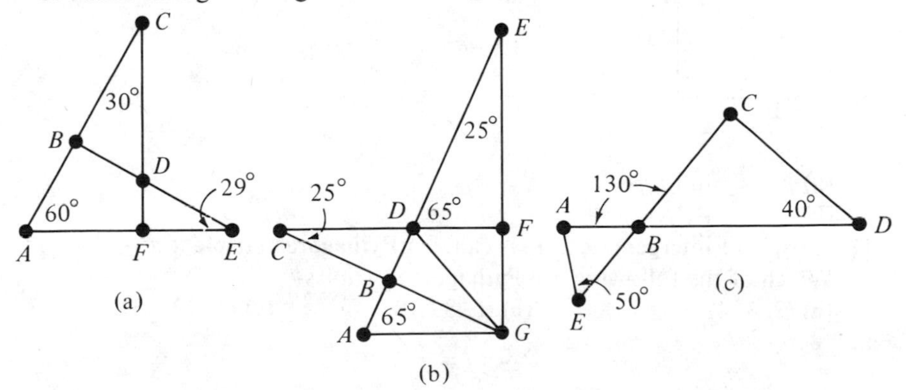

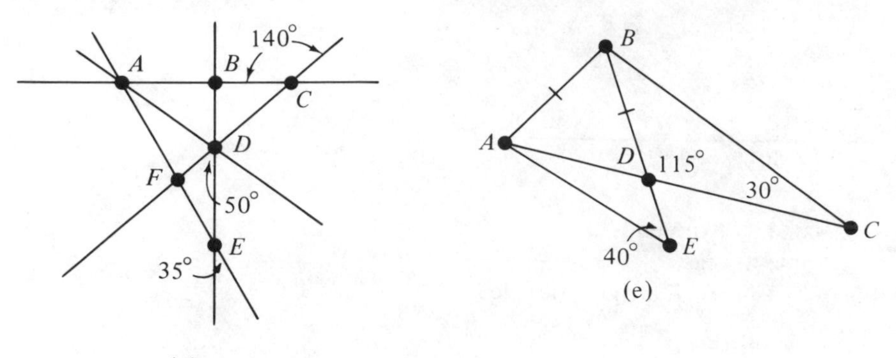

(d)

9. Supply reasons in this proof of the Pythagorean theorem, which is based on similarity and area.

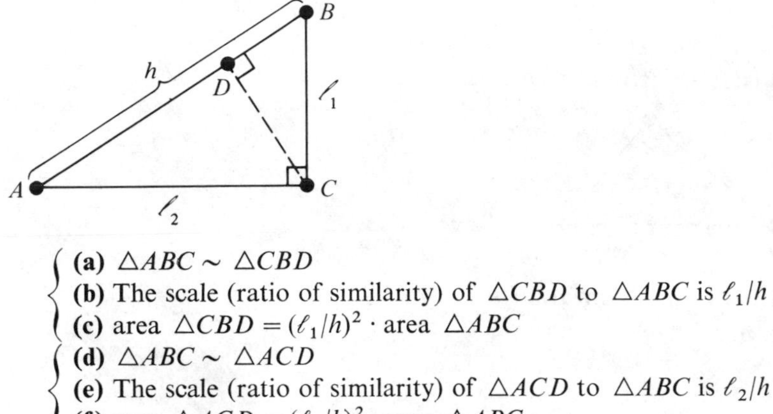

(a) $\triangle ABC \sim \triangle CBD$
(b) The scale (ratio of similarity) of $\triangle CBD$ to $\triangle ABC$ is ℓ_1/h
(c) area $\triangle CBD = (\ell_1/h)^2 \cdot$ area $\triangle ABC$
(d) $\triangle ABC \sim \triangle ACD$
(e) The scale (ratio of similarity) of $\triangle ACD$ to $\triangle ABC$ is ℓ_2/h
(f) area $\triangle ACD = (\ell_2/h)^2 \cdot$ area $\triangle ABC$
(g) area $\triangle CBD +$ area $\triangle ACD = [(\ell_1/h)^2 + (\ell_2/h)^2] \cdot$ area $\triangle ABC$
(h) area $\triangle CBD +$ area $\triangle ACD =$ area $\triangle ABC$
(i) $(\ell_1/h)^2 + (\ell_2/h)^2 = 1$
(j) $\ell_1^2 + \ell_2^2 = h^2$

10. Deduce from the Pythagorean theorem that

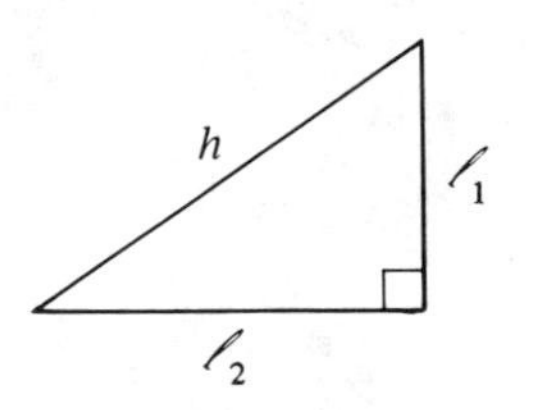

(a) $h > \ell_1$ and $h > \ell_2$
(b) $h < \ell_1 + \ell_2$

11. A triple of intergers (a, b, c) is called a **Pythagorean triple** if $a^2 + b^2 = c^2$. Which of the following are Pythagorean triples?
(a) (2, 3, 4) **(b)** (6, 8, 10) **(c)** (5, 12, 13)

(d) $(5n, 12n, 13n)$ **(e)** $(2n, 3n, 4n)$

12. Two members of a Pythagorean triple are 25 and 7. Find the third.

13. A farmer tells you that it is exactly 9 miles to a gas station as the crow flies, but that you can also get there by going exactly 6 miles East and exactly 7 miles North. Should you believe him?

14. Explain how to show, using only a ruler and some arithmetic, that $\triangle ABC$ below is not a right triangle.

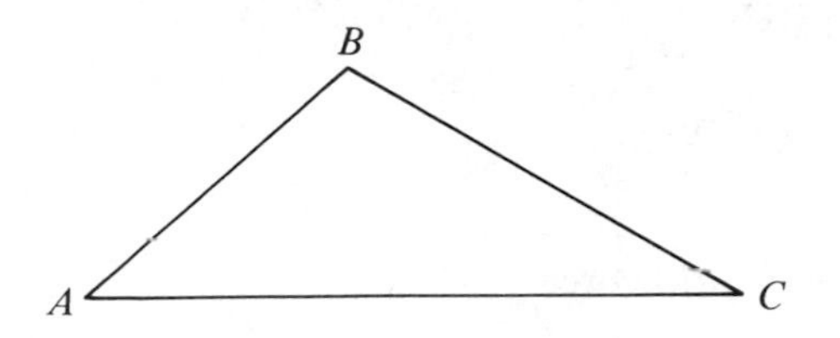

15. **(a)** Do you agree with this statement: "If the square of the length of one side of a triangle equals the sum of the squares of the lengths of the other two sides, then the triangle is a right triangle"?

(b) How does the above statement differ from the Pythagorean theorem?

(c) Explain how to show, using only a protractor, that $a^2 + b^2 \neq c^2$.

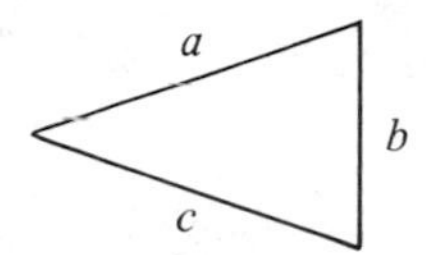

8.5 USING THE PYTHAGOREAN THEOREM; SQUARE ROOTS

To use the Pythagorean theorem to compute a side length of a right triangle, you must be able to find the square root of a number. Recall that if a is any positive number,

> The **square root** of a, written $\sqrt{a}$, is the positive number which when squared equals a.

For example,

$$\sqrt{9} = 3, \quad \sqrt{49} = 7, \quad \sqrt{10{,}000} = 100$$

EXAMPLE If $AB = 12$ in. and $BC = 5$ in., how long is $\overline{AC}$?

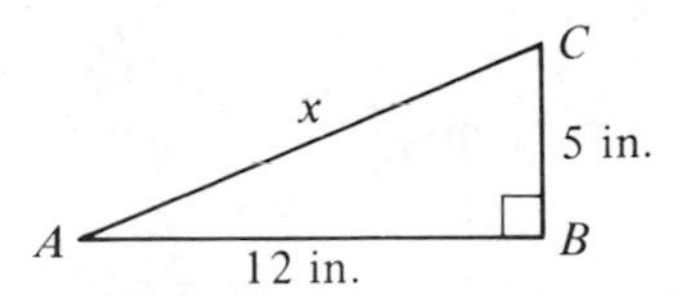

Solution Let $x = AC$. By the Pythagorean theorem,

$$x^2 = 12^2 + 5^2 = 144 + 25 = 169$$

so

$$x = \sqrt{169} = 13 \text{ in.}$$

EXAMPLE If $AB = 24$ cm and $AC = 25$ cm, how long is $\overline{BC}$?

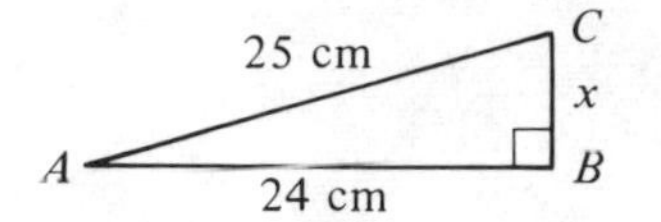

Solution Let $x = BC$. By the Pythagorean theorem,

$$x^2 + 24^2 = 25^2$$

so

$$x^2 = 25^2 - 24^2 = 625 - 576 = 49$$

Hence

$$x = \sqrt{49} = 7 \text{ cm}$$

EXAMPLE If $AB = 5$ in. and $BC = 4$ in., how long is $\overline{AC}$?

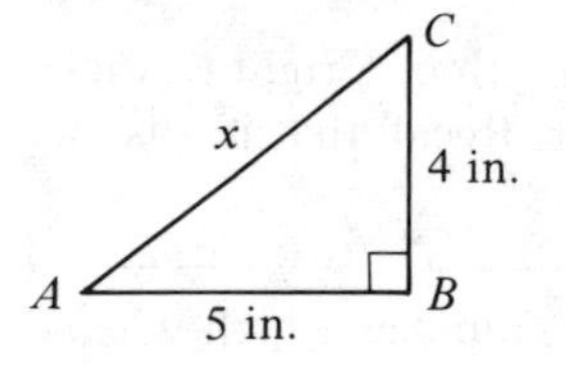

Solution Let $x = AC$. By the Pythagorean theorem,

$$x^2 = 4^2 + 5^2 = 16 + 25 = 41$$

so

$$x = \sqrt{41} \text{ in.}$$

The problem now is to get a decimal approximation to $\sqrt{41}$. There are at least four approaches to the problem.

1. Learn a procedure (algorithm) for computing square roots to as many decimal places as you please. (Perhaps you know one.)

2. Use trial and error to close in on $\sqrt{41}$.

$$6^2 = 36 < 41 \quad \text{and} \quad 7^2 = 49 > 41 \Rightarrow 6 < \sqrt{41} < 7$$

$$6.4^2 = 40.96 < 41 \quad \text{and} \quad 6.5^2 = 42.25 > 41 \Rightarrow$$

$$6.4 < \sqrt{41} < 6.5$$

$$6.40^2 = 40.9600 < 41 \quad \text{and} \quad 6.41^2 = 41.0881 > 41 \Rightarrow$$

$$6.40 < \sqrt{41} < 6.41$$

$$\ldots$$

3. Look up $\sqrt{41}$ in a table of square roots. One such table is given on p. 485.

$$\sqrt{41} \doteq 6.403 \qquad \text{correct to three decimal places}$$

4. Use a slide rule or pocket calculator to find a close approximation to $\sqrt{41}$.

EXERCISES

1. Use the table to find three-place decimal approximations to

(a) $\sqrt{3}$ **(b)** $\sqrt{37}$
(c) $\sqrt{370}$ **(d)** $\sqrt{10}$
(e) $\sqrt{650}$ **(f)** $\sqrt{1000}$

2. Without using the square root table, find

(a) $\sqrt{16}$ **(b)** $\sqrt{.16}$ **(c)** $\sqrt{.0016}$ **(d)** $\sqrt{1600}$
(e) $\sqrt{1600000000}$

3. Use the table to find (approximately)

(a) $\sqrt{160}$ **(b)** $\sqrt{1.6}$ **(c)** $\sqrt{.016}$ **(d)** $\sqrt{16000}$
(e) $\sqrt{3}$ **(f)** $\sqrt{30}$ **(g)** $\sqrt{300}$ **(h)** $\sqrt{3000}$
(i) $\sqrt{.3}$ **(j)** $\sqrt{.03}$ **(k)** $\sqrt{.003}$ **(l)** $\sqrt{.000000003}$
(m) $\sqrt{1.4}$ **(n)** $\sqrt{.27}$ **(o)** $\sqrt{.0036}$ **(p)** $\sqrt{7500}$
(q) $\sqrt{520}$ **(r)** $\sqrt{.013}$ **(s)** $\sqrt{.2}$ **(t)** $\sqrt{750}$

4. Find the unknown distance x to the nearest tenth of a unit. (Use the table of square roots when necessary.)

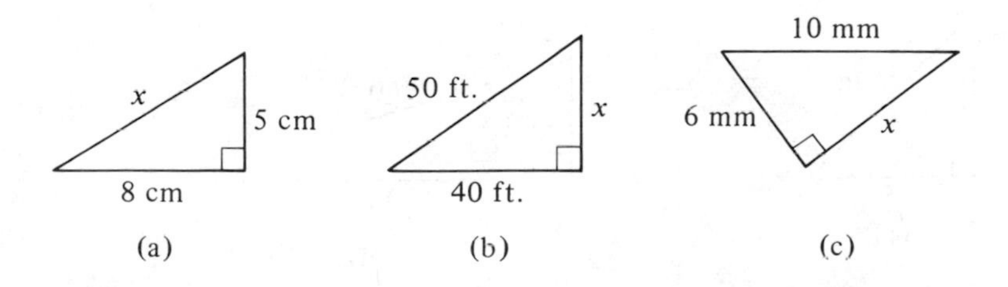

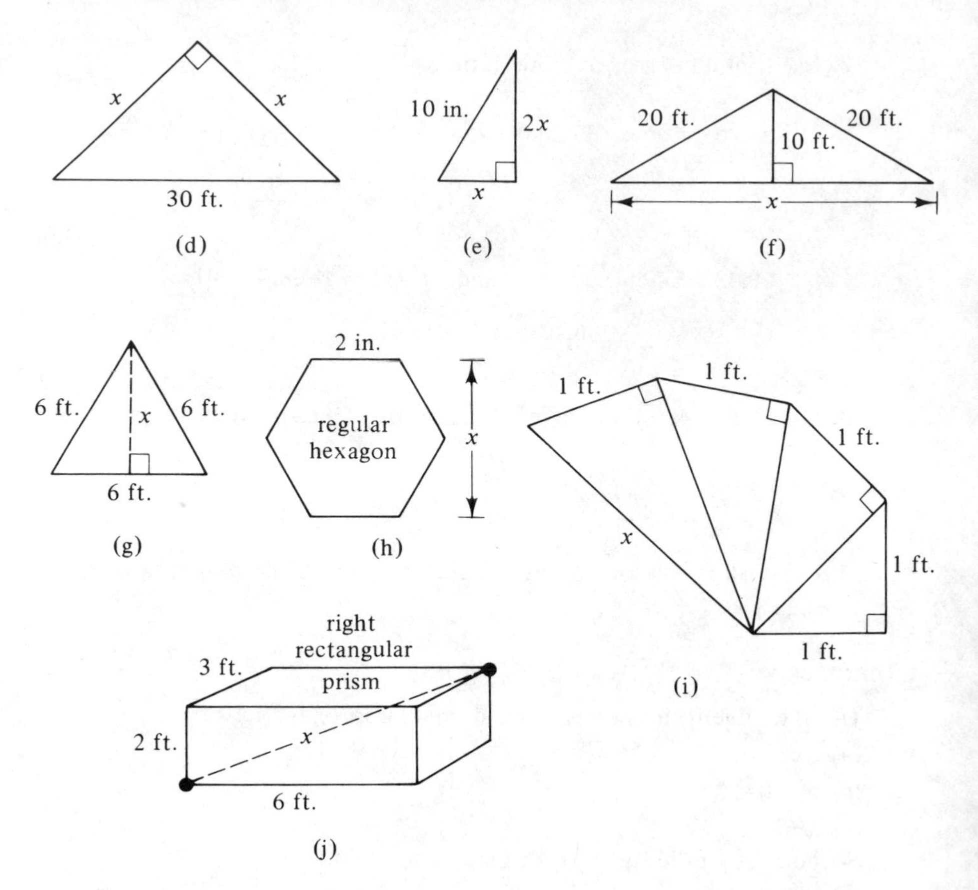

5. Find the unknown distance x and also find the area of the region.

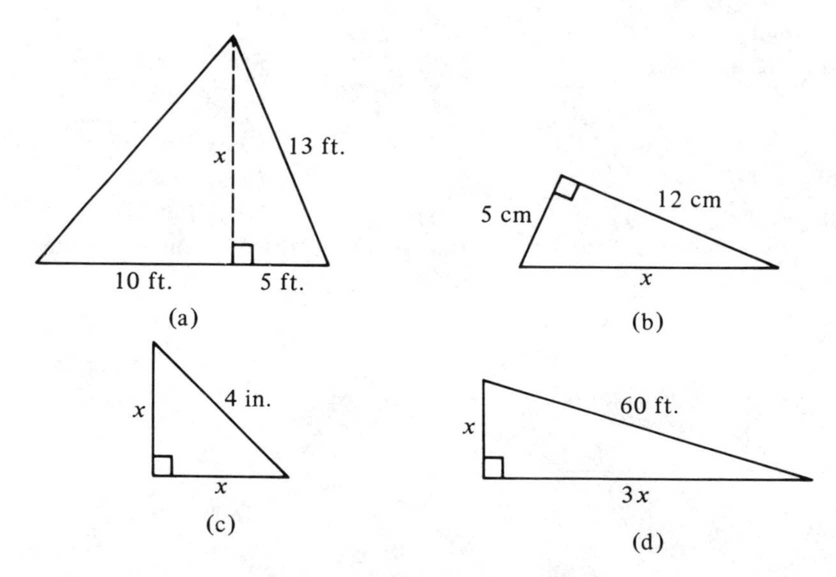

6. Find the areas of the following regions to the nearest whole unit.

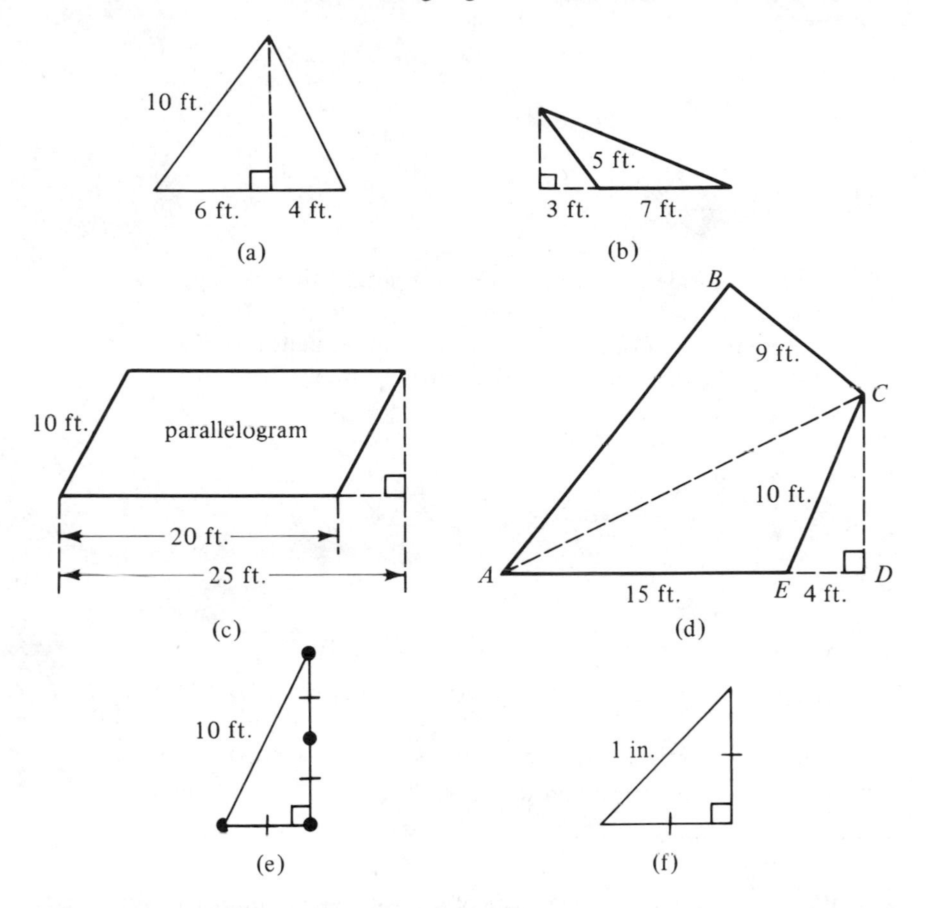

7. If a 16-ft ladder is leaned against a house so that the base of the ladder is 6 ft from the wall, how high on the house will the ladder reach?

8. If each step is 12 in. deep and there is a 9-in. rise between steps, how long is the railing?

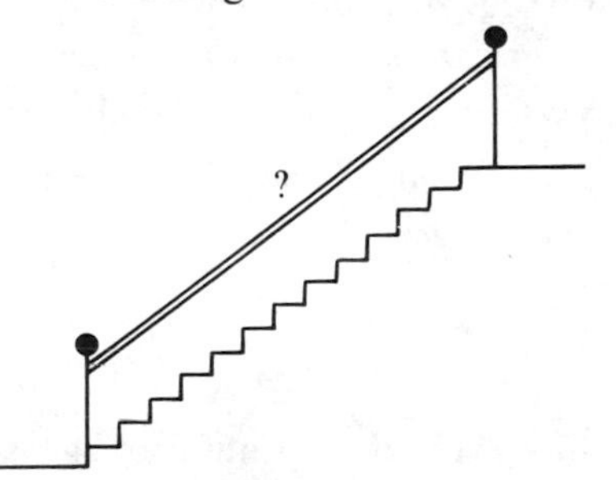

9. What is the radius of the smallest circle that can be "circumscribed" about a square of side length 1 in.? (See the figure on top of p. 356.)

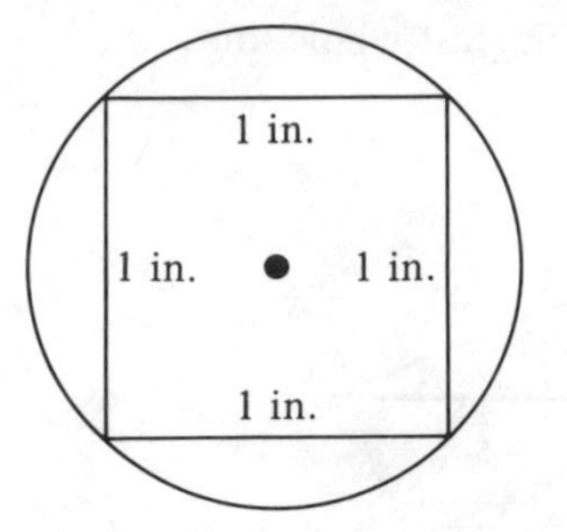

10. What is the side length of the largest square that can be *in*scribed in a circle of radius 1 in.?

11. Suppose that the distance between consecutive nails in a row (or column) on the geoboard is chosen as the unit of distance.

Can you make squares whose areas are 1 square unit? 2? 3? 4? 5? 6? 7? 8? 9? 10?

12. Using only straightedge and compass, construct points on the number line below that correspond to $\sqrt{2}$, $\sqrt{3}$, $\sqrt{5}$.

0 1

***13.** We glibly assumed the existence of a positive real number whose square is 2 and denoted it by $\sqrt{2}$. Exercise 12 lends further credence to the existence of such a real number. It also suggests its uniqueness. Suppose that a and b are both positive real numbers and $a^2 = 2$ and $b^2 = 2$. Show algebraically that a and b must be equal. (*Hint*: "Factor" $a^2 - b^2$.)

***14.** If we assume that every positive integer can be expressed uniquely as a product of primes, then we can show that $\sqrt{2}$ is an irrational number. Suppose, to the contrary, that $\sqrt{2}$ is rational. That is, suppose that $\sqrt{2} = m/n$, where m and n are positive integers. Then

$$2 = \frac{m^2}{n^2} \quad \text{so} \quad 2n^2 = m^2$$

Now the prime factorization of m^2 involves an even number of factors (why?) and the prime factorization of $2n^2$ involves an odd number of factors (why?), but this is impossible (why?). Thus $\sqrt{2}$ is irrational.

***15.** Assuming that there is at least one positive real number whose square is 3, and following the plan of Exercise 13, prove that there is just one such number.

***16.** Following the plan of Exercise 14, prove that $\sqrt{3}$ is irrational.

***17.** Assuming that there is a positive real number whose cube is 2, prove that it is unique.

***18.** Prove that $\sqrt[3]{2}$ is irrational.

***19.** In one corner of the ceiling of a 10 ft by 14 ft room (with an 8 ft ceiling) sits a bug. In the remote corner of the floor is a bread crumb. What is the length of the shortest path from the bug to the crumb? (The bug cannot fly.)

8.6 SIMILAR RIGHT TRIANGLES

One of the similarity conditions for triangles was the "AA" condition: If two angles of one triangle are congruent respectively to two angles of another triangle, then the triangles are similar. For right triangles this condition simplifies still further:

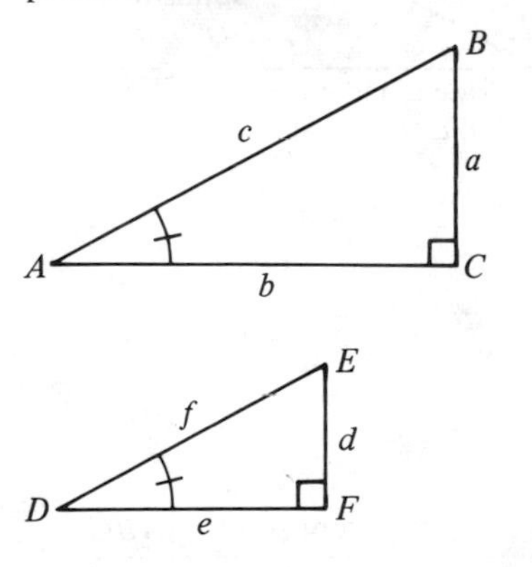

"A" If an acute angle of one right triangle is congruent to an acute angle of another right triangle, then the two right triangles are similar (and so the corresponding sides are proportional and the other two acute angles are congruent).

$$\angle A \cong \angle D \Rightarrow \triangle ABC \sim \triangle DEF \Rightarrow \frac{a}{d} = \frac{b}{e} = \frac{c}{f}$$

In the figure we have followed the standard procedure of denoting the length of the side opposite an angle by the lowercase letter corresponding to the capital letter at the vertex. We have also indulged in the common practice of writing a run-on sentence. It should be kept in mind that the run-on mathematical sentence

$$\frac{a}{d} = \frac{b}{e} = \frac{c}{f}$$

actually represents the compound sentence

$$\frac{a}{d} = \frac{b}{e} \quad \text{and} \quad \frac{b}{e} = \frac{c}{f}$$

from which, of course, it also follows that

$$\frac{a}{d} = \frac{c}{f}$$

(This single common ratio is the ratio of similarity of $\triangle ABC$ to $\triangle DEF$.)

The run-on statement of proportionality among the sides is generally not left in the form

$$\frac{a}{d} = \frac{b}{e} = \frac{c}{f}$$

Instead, first each side of each triangle is named according to its position with respect to the known pair of congruent acute angles:

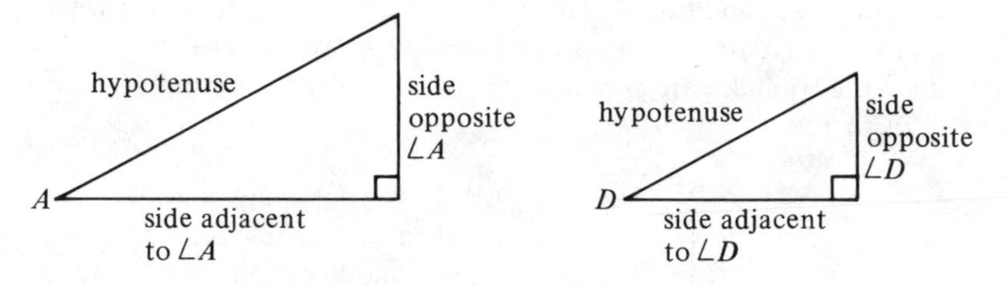

The run-on proportion

$$\frac{a}{d} = \frac{b}{e} = \frac{c}{f}$$

becomes

$$\frac{\text{side opposite } \angle A}{\text{side opposite } \angle D} = \frac{\text{side adjacent to } \angle A}{\text{side adjacent to } \angle D} = \frac{\text{hypotenuse of } \triangle ABC}{\text{hypotenuse of } \triangle DEF}$$

or, using good grammar,

$$\frac{\text{side opposite } \angle A}{\text{side opposite } \angle D} = \frac{\text{side adjacent to } \angle A}{\text{side adjacent to } \angle D} \quad \textit{and}$$

$$\frac{\text{side opposite } \angle A}{\text{side opposite } \angle D} = \frac{\text{hypotenuse of } \triangle ABC}{\text{hypotenuse of } \triangle DEF} \quad \textit{and}$$

$$\frac{\text{side adjacent to } \angle A}{\text{side adjacent to } \angle D} = \frac{\text{hypotenuse of } \triangle ABC}{\text{hypotenuse of } \triangle DEF}$$

Then these three proportions are transformed according to the usual rules for manipulating fractions.

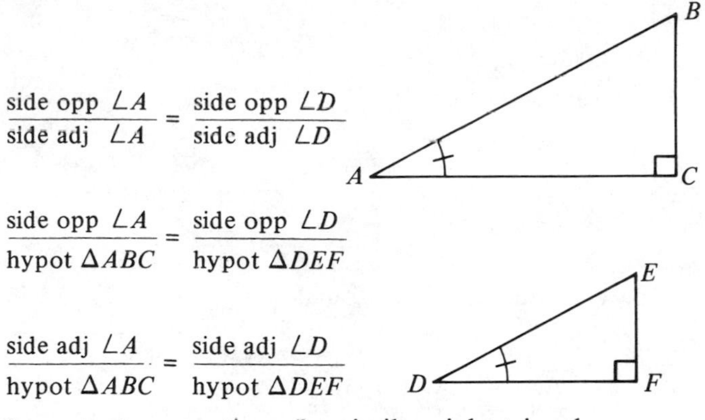

Important proportions for similar right triangles.

These proportions are useful in solving problems involving similar right triangles.

EXAMPLE Find the values of c, e, f.

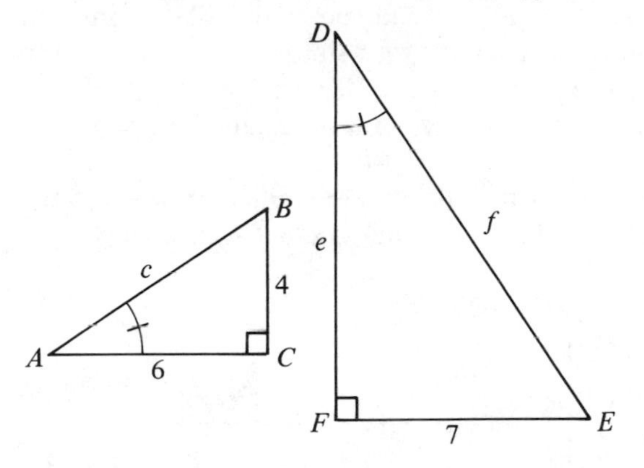

Solution To find c we simply use the Pythagorean theorem:

$$c^2 = 4^2 + 6^2 = 16 + 36 = 52 \Rightarrow c = \sqrt{52}$$

To find e and f we notice that $\triangle ABC \sim \triangle DEF$ by condition "A." Thus

$$\frac{\text{side opposite } \angle A}{\text{side adjacent } \angle A} = \frac{\text{side opposite } \angle D}{\text{side adjacent } \angle D}$$

That is,

$$\frac{4}{6} = \frac{7}{e} \Rightarrow 4 \cdot e = 6 \cdot 7 \Rightarrow e = \frac{42}{4} \Rightarrow e = 10\tfrac{1}{2}$$

Also

$$\frac{\text{side opposite } \angle A}{\text{hypotenuse } \triangle ABC} = \frac{\text{side opposite } \angle D}{\text{hypotenuse } \triangle DEF}$$

That is,

$$\frac{4}{\sqrt{52}} = \frac{7}{f} \Rightarrow 4 \cdot f = 7 \cdot \sqrt{52} \Rightarrow f = 7 \cdot \frac{\sqrt{52}}{4}$$

Having found $e = 10\frac{1}{2}$, we could also have found f by the Pythagorean theorem.

$$f^2 = \left(\frac{21}{2}\right)^2 + 7^2 = \frac{441}{4} + 49 = \frac{(441 + 196)}{4} \Rightarrow f = \sqrt{\frac{637}{4}}$$

(To see that $7 \cdot \sqrt{52}/4 = \sqrt{637/4}$, square both sides and simplify.)

EXERCISES

1. In *any* triangle, right or not, with a vertex A, the expression "side opposite $\angle A$" is unambiguous since there is just one such side. The expression "side adjacent to $\angle A$," however, is ambiguous since there are two sides. What interpretation is given to the expression "side adjacent to $\angle A$" to make it unambiguous in the context where $\angle A$ is an acute angle of a right triangle?
2. Explain why the triangles of each pair below are similar and report the similarity in a way that conveys the correspondence between vertices.

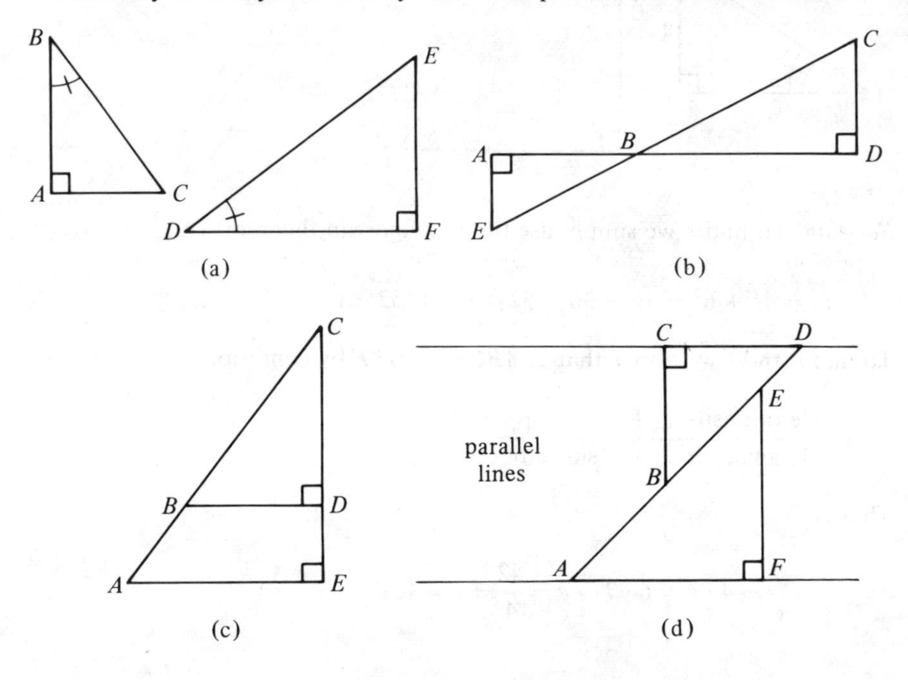

3. Why is $\triangle ABC$ similar to $\triangle DEF$?

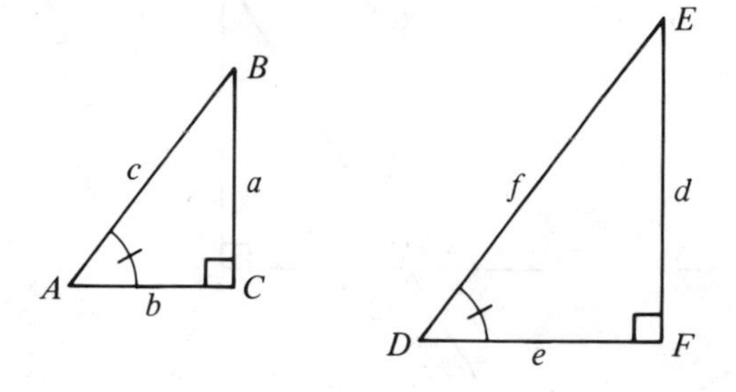

Fill in the boxes.

(a) $\frac{a}{b} = \frac{d}{\square}$

(b) $\frac{b}{\square} = \frac{e}{f}$

(c) $\frac{\square}{b} = \frac{d}{e}$

(d) $\frac{b}{a} = \frac{e}{\square}$

(e) $\frac{c}{\square} = \frac{b}{e}$

(f) $\frac{d}{f} = \frac{\square}{\square}$

(g) $\frac{b}{c} = \frac{\square}{\square}$

(h) $\frac{b}{a} = \frac{\square}{\square}$

(i) $a \cdot \square = b \cdot d$

(j) $c \cdot e = \square \cdot f$

(k) $\square = \frac{a \cdot e}{b}$

(l) $\frac{e}{f} \cdot \square = b$

4. (a) Why is $\triangle ABC \sim \triangle FDE$?

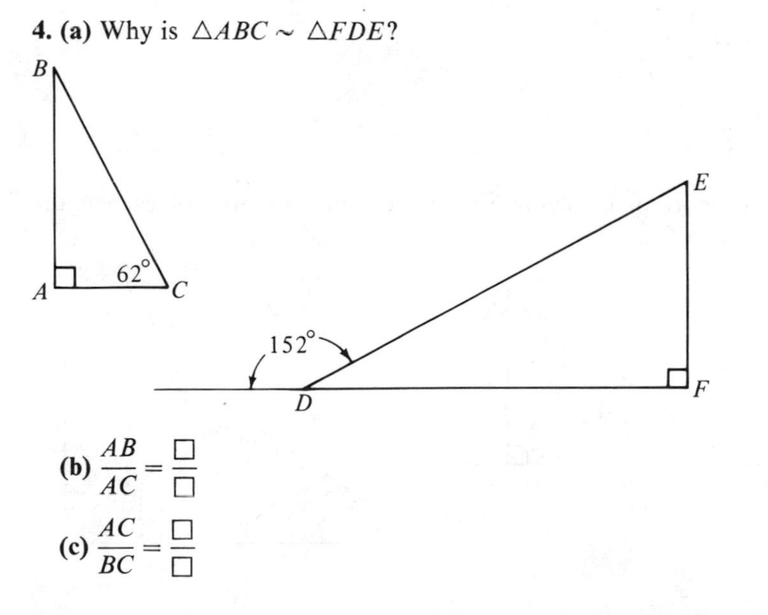

(b) $\frac{AB}{AC} = \frac{\square}{\square}$

(c) $\frac{AC}{BC} = \frac{\square}{\square}$

5. (a) Why is $\triangle BAC \sim \triangle DFE$?

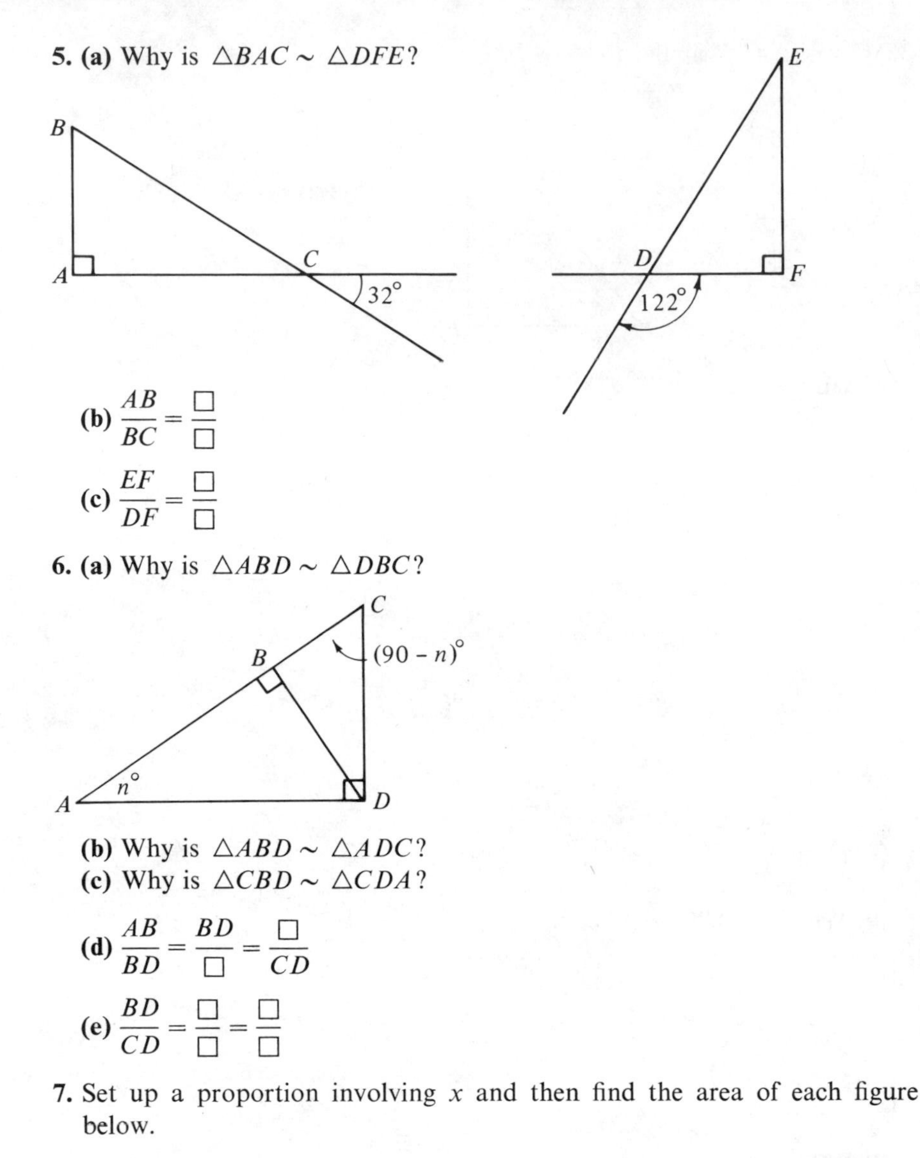

(b) $\dfrac{AB}{BC} = \dfrac{\square}{\square}$

(c) $\dfrac{EF}{DF} = \dfrac{\square}{\square}$

6. (a) Why is $\triangle ABD \sim \triangle DBC$?

(b) Why is $\triangle ABD \sim \triangle ADC$?

(c) Why is $\triangle CBD \sim \triangle CDA$?

(d) $\dfrac{AB}{BD} = \dfrac{BD}{\square} = \dfrac{\square}{CD}$

(e) $\dfrac{BD}{CD} = \dfrac{\square}{\square} = \dfrac{\square}{\square}$

7. Set up a proportion involving x and then find the area of each figure below.

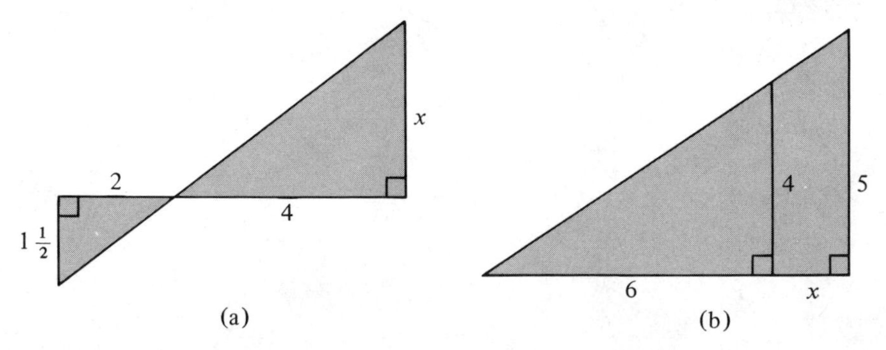

(a) (b)

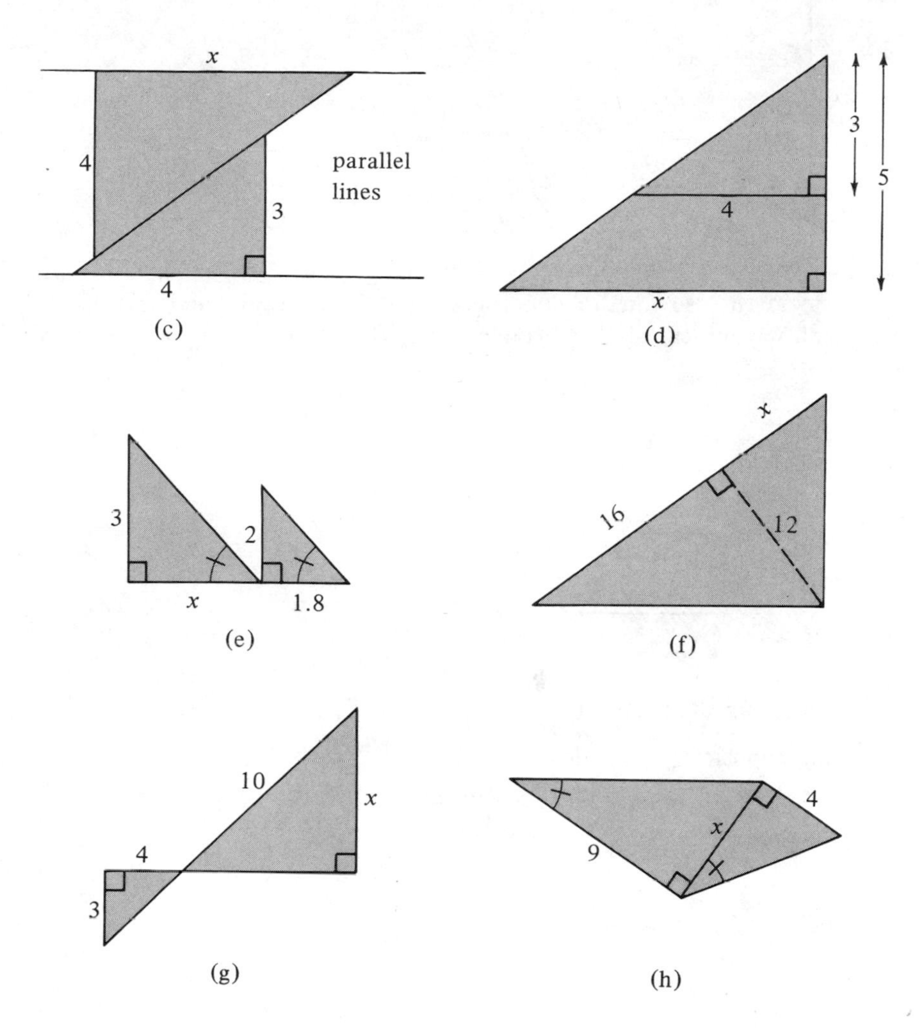

8.7 TRIGONOMETRY

The three important ratios associated with an acute angle in a right triangle,

$$\frac{\text{opposite}}{\text{adjacent}}, \quad \frac{\text{opposite}}{\text{hypotenuse}}, \quad \frac{\text{adjacent}}{\text{hypotenuse}}$$

can be used to solve hard problems.

A HARD PROBLEM A straight cliff rises 30 ft from the edge of a river. You measure the angle between the cliff and a line to the opposite bank of the river and find that it is 55°. You would like to know how wide the river is.

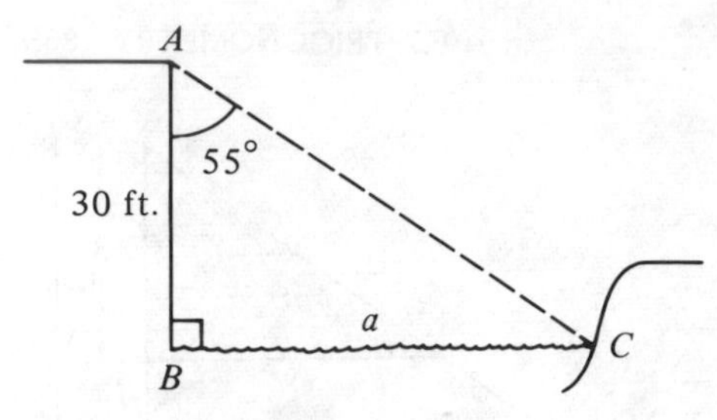

Solution You know that any other right triangle having a 55° angle will be similar to $\triangle ABC$, so you draw one, very carefully, with ruler and protractor. To simplify your calculations you make $A'B' = 1$ in. Then you measure $\overline{B'C'}$. $B'C' \doteq 1\frac{7}{16}$ in.

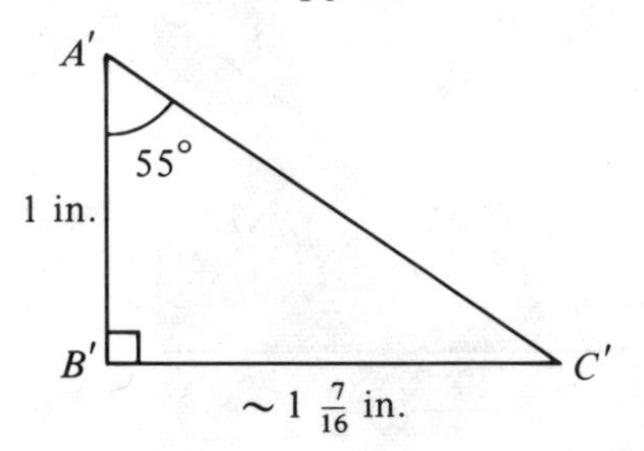

Now you set up the proportion

$$\frac{\text{side opposite } \angle A}{\text{side adjacent } \angle A} = \frac{\text{side opposite } \angle A'}{\text{side adjacent } \angle A'}$$

substitute,

$$\frac{a}{30} \doteq \frac{1\frac{7}{16}}{1}$$

and solve for a,

$$\frac{a}{30} \doteq \frac{23}{16} \Rightarrow a \doteq 30 \cdot \frac{23}{16} = \frac{690}{16} \Rightarrow a \doteq 43 \text{ ft}$$

The river is about 43 ft wide.

This problem would have been easier to solve if you had had a little 89-page handbook of right triangles with the triangle below on page 55.

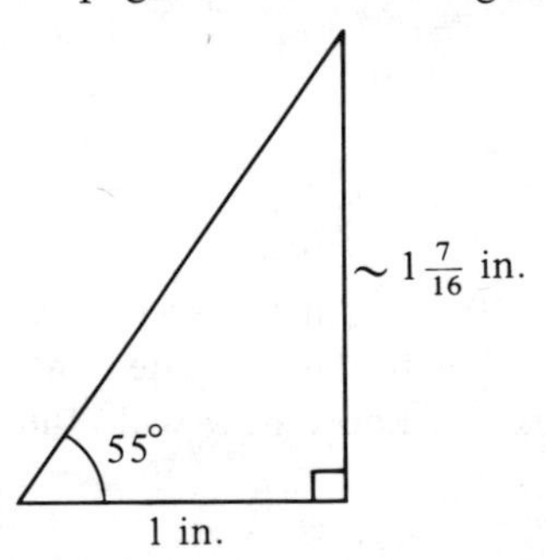

No such little handbooks are available, but something much better is. A table has been compiled of the three important ratios (opp/adj, opp/hyp, adj/hyp) associated with every angle. And it is the *ratio*, side opp 55°/side adj 55°, that you really need to solve the problem about the river. You would find in the table that

$$\frac{\text{side opposite } 55^\circ}{\text{side adjacent } 55^\circ} \doteq 1.428$$

and so the equation you need to solve to find the width of the river becomes

$$\frac{a}{30} \doteq 1.428$$

which easily yields

$$a \doteq 30 \times 1.428 = 42.84$$

or

$$a \doteq 43 \text{ ft}$$

The table we have been talking about is the table of **trigonometric ratios**. (What do you think a "trigon" is?) A "trig table" is given on p. 486. The columns are labeled "sin," "cos," "tan." These are abbreviations for the elegant names—"sine," cosine," and "tangent"—given to the three important ratios associated with an acute angle of a right triangle.

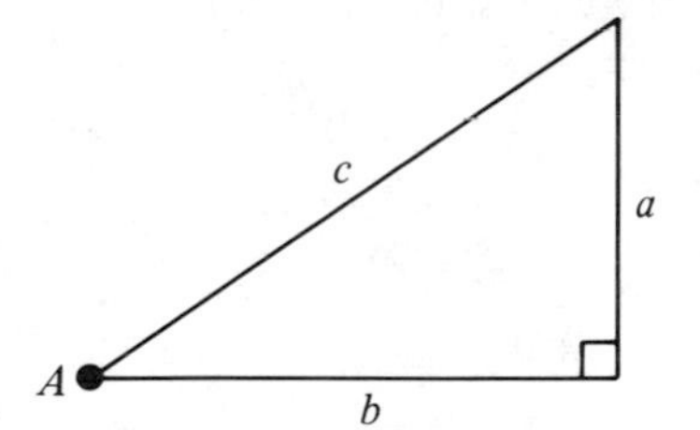

$\sin \angle A = a/c = \text{opposite/hypotenuse}$
$\cos \angle A = b/c = \text{adjacent/hypotenuse}$
$\tan \angle A = a/b = \text{opposite/adjacent}$

Look up tan 55° and check that its value, to three decimal places, is 1.428.

EXAMPLE What are the trig ratios associated with the larger acute angle in a 3–4–5 right triangle?

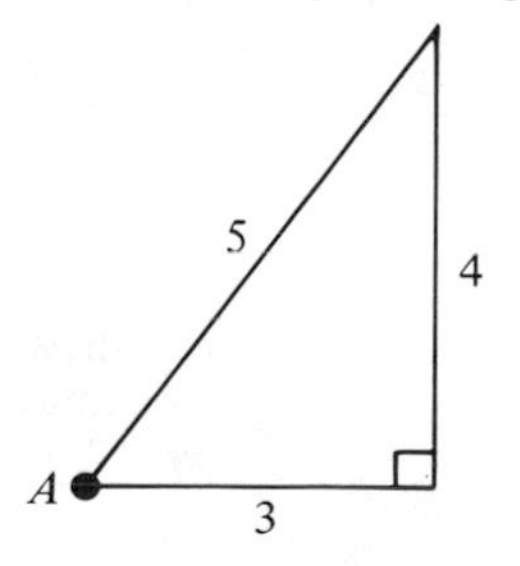

Solution

$$\sin \angle A = \tfrac{4}{5} = .8$$

$$\cos \angle\ = \tfrac{3}{5} = .6$$

$$\tan \angle A = \tfrac{4}{3} = 1.333\ldots$$

EXAMPLE Find the unknown lengths x and y.

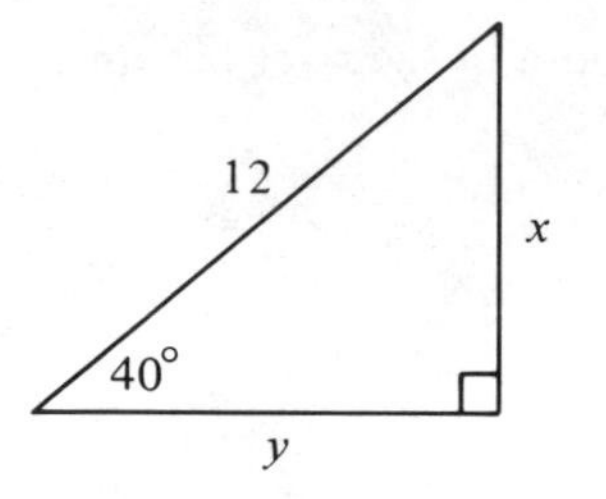

Solution From the table, $\sin 40° \doteq .643$. But also $\sin 40° =$ opposite/hypotenuse $= x/12$. So $x/12 \doteq .643$ or $x \doteq 12 \times .643 = 7.716$.* From the table, $\cos 40° \doteq .766$. But also $\cos 40° =$ adjacent/hypotenuse $= y/12$. So $y/12 \doteq .766$ or $y \doteq 12 \times .766 \doteq 9.192$. Having found x, we could have found y in two other ways:

(i) $7.716/y \doteq x/y = \tan 40° \doteq .839 \Rightarrow y \doteq 7.716/.839 = \ldots$
(ii) $x^2 + y^2 = 12^2 \Rightarrow y^2 \doteq 12^2 - (7.716)^2 \Rightarrow \ldots$

Which of the three ways of finding y involves the easiest computation?

EXERCISES

1. In the figure on the top of p. 367, name:
(a) the hypotenuse
(b) the side opposite $\angle A$
(c) the side adjacent to $\angle A$

* Actually, since trig tables give only approximate values, we can only be certain that

$$.6425 \leq \sin 40° \leq .6435$$

and thus

$$.6425 \times 12 \leq x \leq .6435 \times 12$$

that is,

$$7.710 \leq x \leq 7.722$$

One can always carry out such an error analysis in a trigonometric problem, but we have chosen not to do so in order to keep the exposition simple. Thus we are guilty of "spurious accuracy" in several places in this chapter.

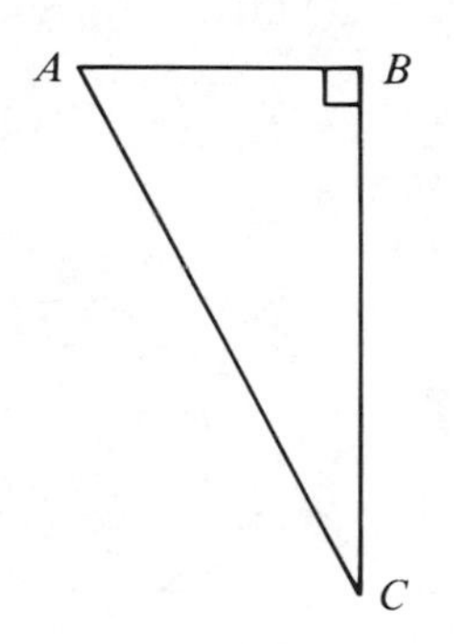

2. Referring to the figure above, match each ratio in the left column with zero, one, or two trig ratios in the right column.

(a) AB/AC $\sin \angle A$
(b) AB/BC $\cos \angle A$
(c) AC/AB $\tan \angle A$
(d) AC/BC $\sin \angle C$
(e) BC/AB $\cos \angle C$
(f) BC/AC $\tan \angle C$

3. Using the figure below, find and write as fractions:

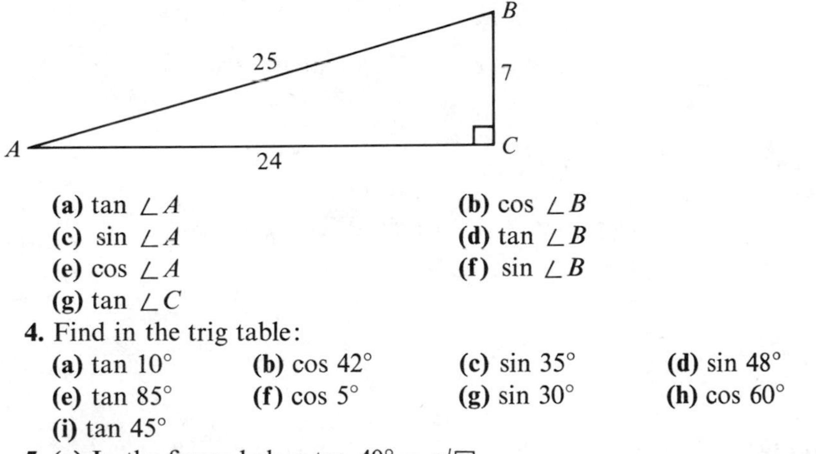

(a) $\tan \angle A$ **(b)** $\cos \angle B$
(c) $\sin \angle A$ **(d)** $\tan \angle B$
(e) $\cos \angle A$ **(f)** $\sin \angle B$
(g) $\tan \angle C$

4. Find in the trig table:

(a) $\tan 10°$ **(b)** $\cos 42°$ **(c)** $\sin 35°$ **(d)** $\sin 48°$
(e) $\tan 85°$ **(f)** $\cos 5°$ **(g)** $\sin 30°$ **(h)** $\cos 60°$
(i) $\tan 45°$

5. (a) In the figure below $\tan 40° = x/\square$.

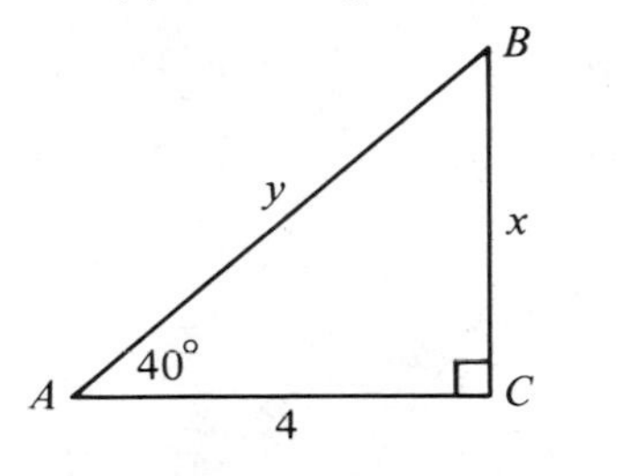

(b) From the trig table, $\tan 40° \doteq$ ▭.
(c) Therefore $x = \doteq \square \times$ ▭ $=$ ▭.

(d) In the previous figure $\cos 40° = \square/y$.
(e) From the trig table, $\cos 40° \doteq$ ▭.
(f) Therefore $y \doteq \square/$ ▭ $\doteq$ ▭.

6. Find x (to the nearest tenth):

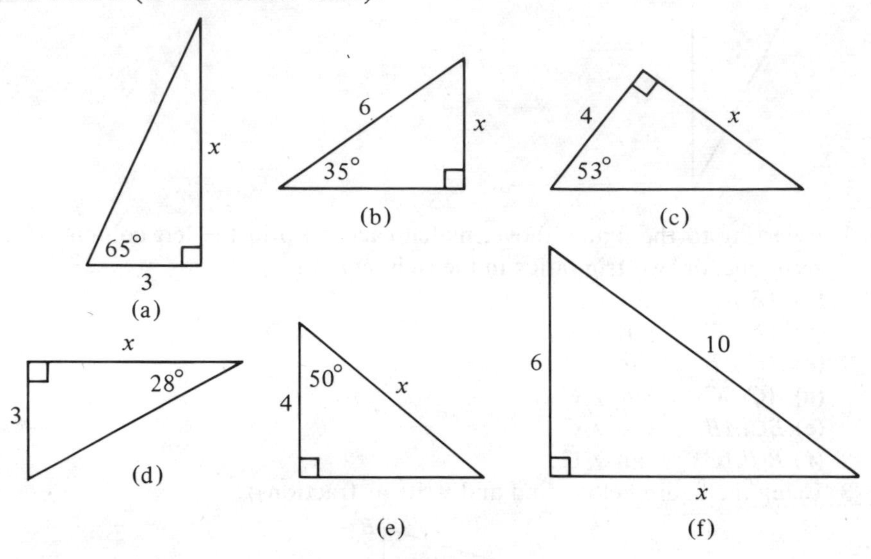

7. Find, to the nearest degree, the measure of an angle whose
(a) tan is .424
(b) sin is .695
(c) cos is .730
(d) tan is 1.8
(e) sin is .330
(f) cos is .5
(g) sin = cos
(h) sin + cos = 1.2

8. In the figure below **(a)** $\tan \angle A = \square/\square =$ ▭.

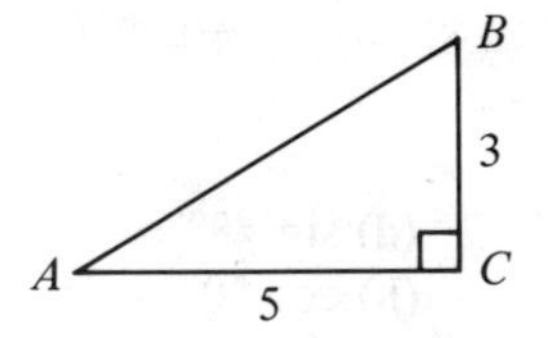

(b) From the trig table, tan ▭° $\doteq .577$ and tan ▭° $\doteq .601$.
(c) Therefore $m \angle A \doteq$ ▭°.

9. Find the measure of $\angle A$ to the nearest degree:

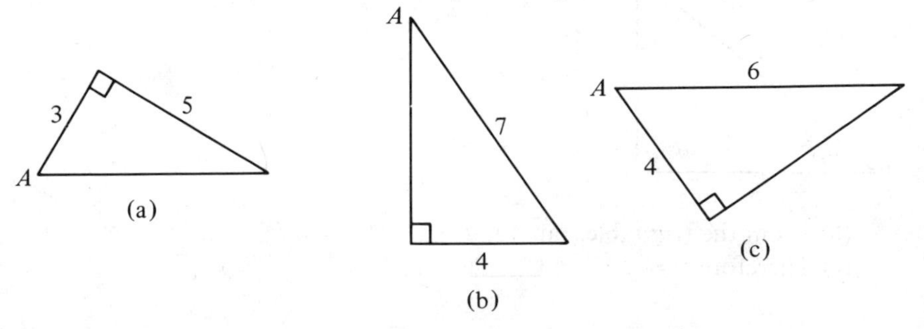

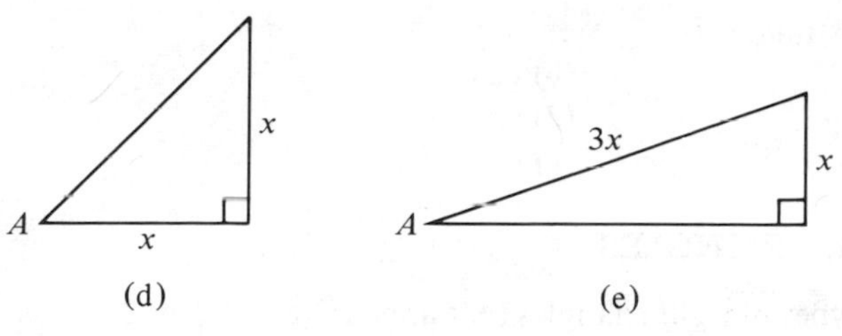

10. How high is the airplane

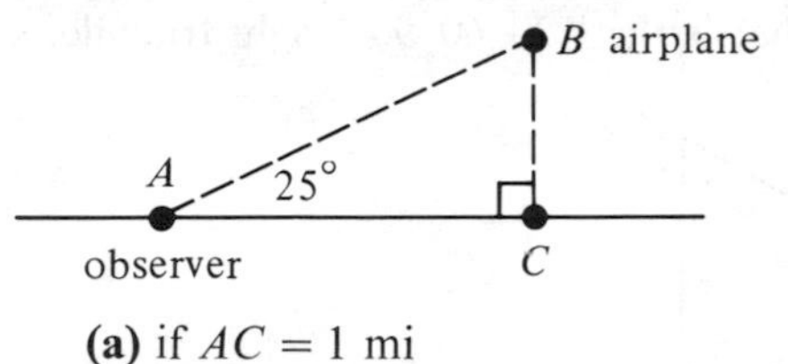

(a) if $AC = 1$ mi
(b) if $AB = 5$ mi

11. The length of $\overline{BC}$ is either sin $\angle A$ or cos $\angle A$ or tan $\angle A$. Pick the right one and explain how this diagram is related to the etymological meaning of the trig ratio you chose.

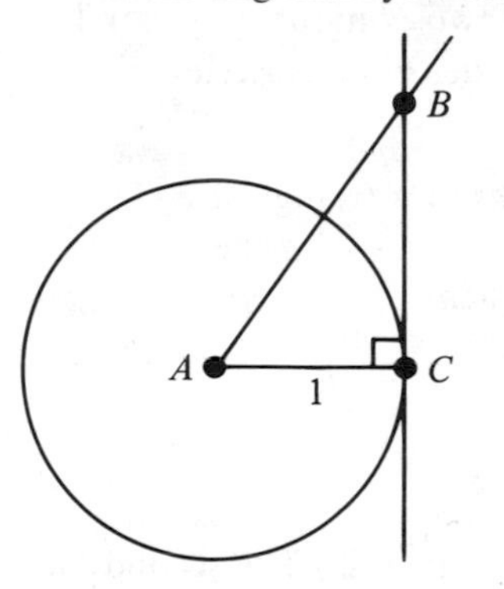

12. In the triangle below, one side has length sin $\angle A$ and one side has length cos $\angle A$.

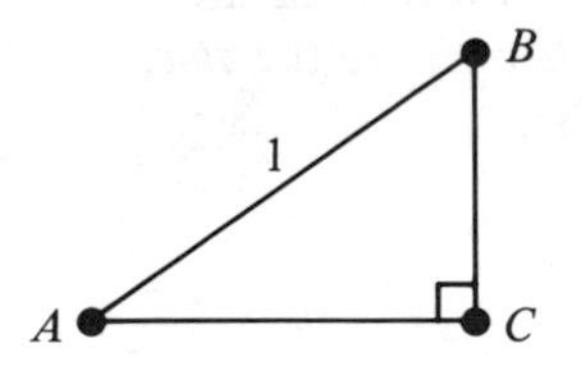

(a) Decide which has which length and explain why.
(b) Show that $(\sin \angle A)^2 + (\cos \angle A)^2 = 1$.
(c) Look up sin 39° and cos 39°, square each number, and add.

13. (a) Look up sin 39° and cos 51°.
(b) Look up sin 63° and cos 27°.
(c) Generalize and explain the pattern appearing in parts (a) and (b).

14. Make reasonable estimates of

(a) sin 19°30′
(b) cos 27°30′
(c) tan 40°30′
(d) sin 19°15′
(e) cos 66°45′
(f) cos 19°10′

8.8 SPECIAL RIGHT TRIANGLES

There are two special types of right triangles that appear quite often. Each has trig ratios that are so easy to find that a trig table is unnecessary. One type is the "45–45–90" right triangle; the other is the "30–60–90" right triangle.

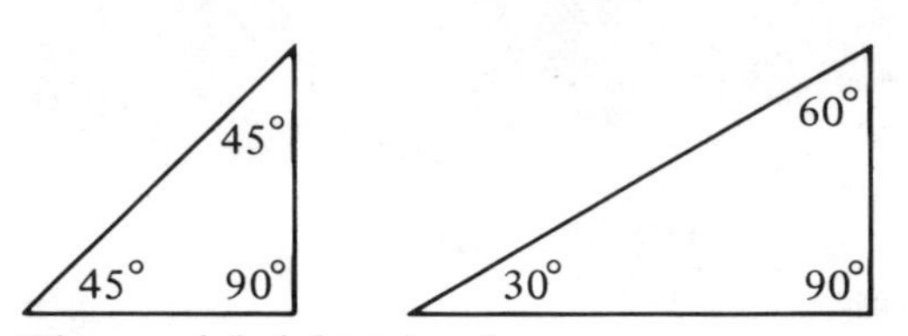

The special right triangles.

To study their properties we shall use two facts about isosceles triangles. (Remember that an isosceles triangle is one with two congruent sides.) The first fact is that the angles opposite the congruent sides of an isosceles triangle are themselves congruent.

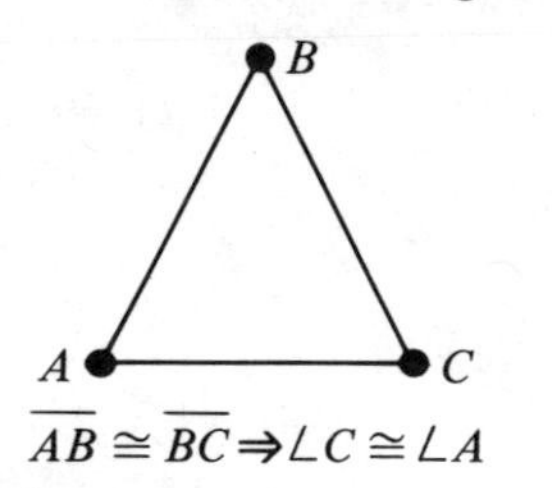

$\overline{AB} \cong \overline{BC} \Rightarrow \angle C \cong \angle A$

If two *sides* of a triangle are congruent, then so are the *angles* opposite them.

(You were asked to establish this fact in Exercise 3, p. 312.) The second fact is very much like the first.

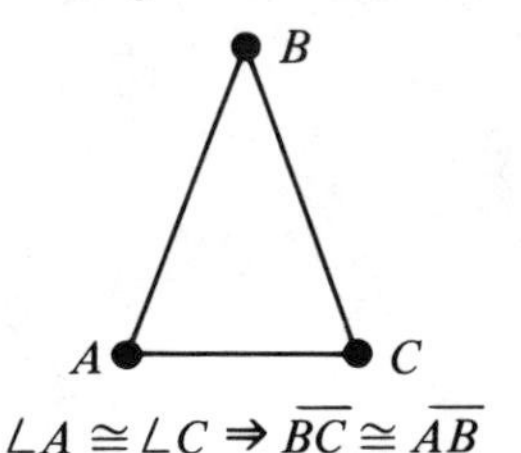

$\angle A \cong \angle C \Rightarrow \overline{BC} \cong \overline{AB}$

If two *angles* of a triangle are congruent, then so are the *sides* opposite them.

So any triangle having a pair of congruent angles must be isosceles. (You are asked to establish this fact in Exercise 13 below.)

The 45–45–90 right triangle is the only *isosceles right* triangle. Its legs are congruent. If each leg has length 1, then, by the Pythagorean theorem, the hypotenuse has length $\sqrt{2}$. Thus

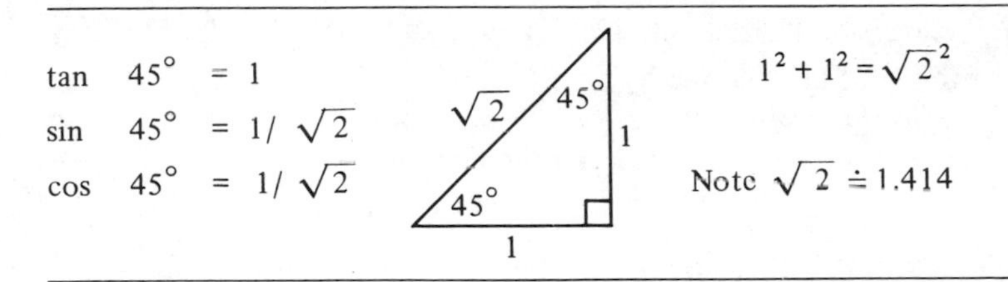

EXAMPLE Find x.

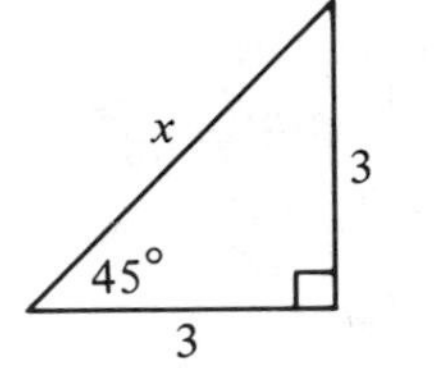

Solution This is easy using the Pythagorean theorem:

$$x^2 = 3^2 + 3^2 = 18 \Rightarrow x = \sqrt{18}$$

It could also be done using trig ratios:

$$\frac{3}{x} = \frac{\text{opposite}}{\text{hypotenuse}} = \sin 45° = \frac{1}{\sqrt{2}} \Rightarrow x = 3\sqrt{2}$$

(Check that $3\sqrt{2} = \sqrt{18}$.)

EXAMPLE Find x.

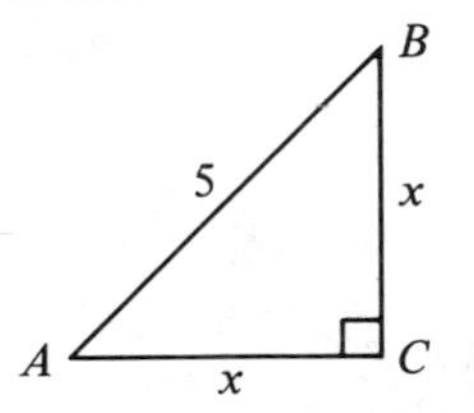

Solution First of all, $\angle A \cong \angle B$ since the sides opposite them are congruent. So $m\angle A = m\angle B$. But $m\angle A + m\angle B = 90°$. Why? Thus $m\angle A = 45° = m\angle B$ and $\triangle ABC$ is a 45–45–90 triangle. Now

$$x/5 = \text{side opposite } \angle A/\text{hypotenuse} = \sin 45° = 1/\sqrt{2} \Rightarrow x = 5/\sqrt{2}.$$

Most people find it easier to remember the triangle

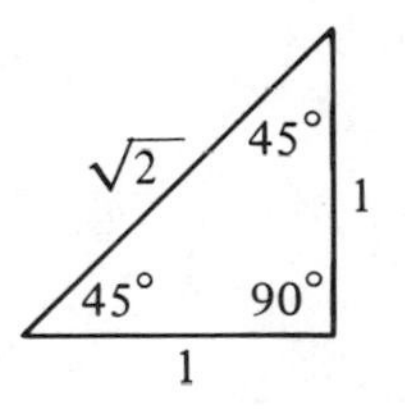

than the trig ratios, $\tan 45° = 1$, $\sin 45° = 1/\sqrt{2}$, and $\cos 45° = 1/\sqrt{2}$. They then solve a problem like this one by sketching the 1–1–$\sqrt{2}$ triangle, noting that it is similar to the given triangle, setting up a proportion without using the language of trigonometry, and solving the proportion.

$$\frac{1}{\sqrt{2}} = \frac{x}{5} \Rightarrow 5 \cdot 1 = \sqrt{2} \cdot x \Rightarrow x = \frac{5}{\sqrt{2}}$$

If the shortest side of a 30–60–90 right triangle has length 1, then the hypotenuse must have length 2 and the other leg must have length $\sqrt{3}$. Here is a derivation of that fact.

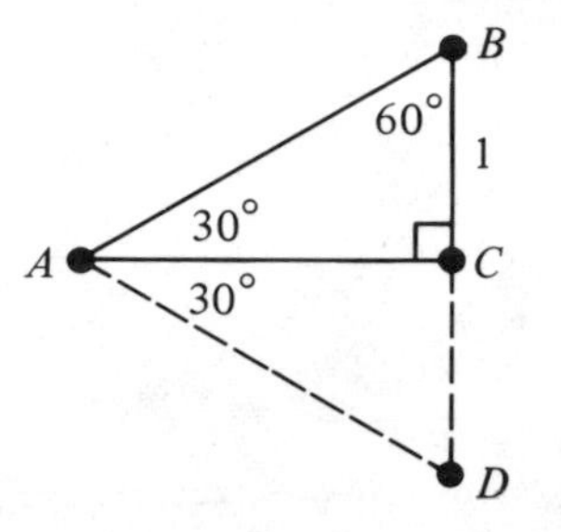

1. $\triangle ACB \cong \triangle ACD$ (Why?)
2. $CD = 1$ (Why?) so $BD = 2$
3. $m\angle D = 60°$ (Why?)
4. $AB = BD\ (= 2)$ (Why?)
5. $AC^2 = AB^2 - BC^2 (= 4 - 1 = 3)$ (Why?)
6. $AC = \sqrt{3}$ (Why?)

Thus

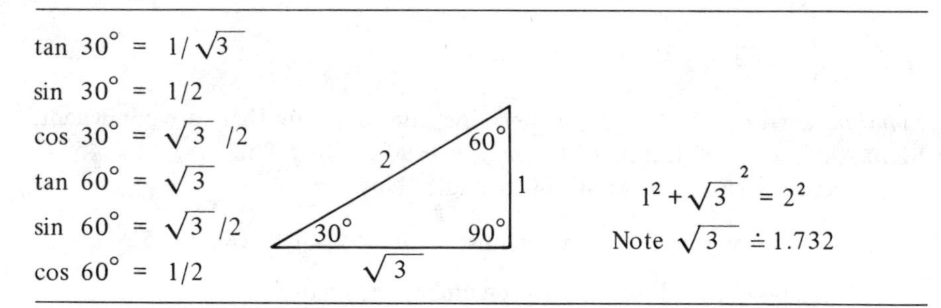

EXAMPLE Find x and y.

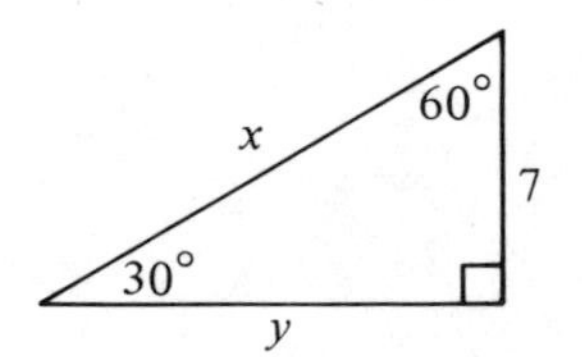

Solution 1 Sketch the similar triangle,

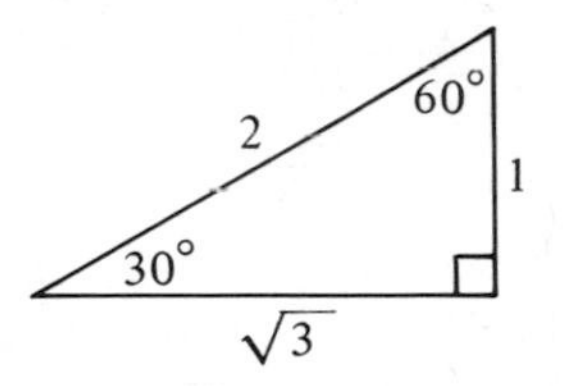

set up proportions,

$$\frac{1}{2} = \frac{\text{opposite } 30^\circ}{\text{hypotenuse}} = \frac{7}{x}$$

$$\frac{1}{\sqrt{3}} = \frac{\text{opposite } 30^\circ}{\text{adjacent } 30^\circ} = \frac{7}{y}$$

and solve for x and y,

$$\frac{1}{2} = \frac{7}{x} \Rightarrow 1 \cdot x = 2 \cdot 7 \Rightarrow x = 14$$

$$\frac{1}{\sqrt{3}} = \frac{7}{y} \Rightarrow 1 \cdot y = 7 \cdot \sqrt{3} \Rightarrow y = 7 \cdot \sqrt{3}$$

Solution 2

$$\frac{7}{x} = \frac{\text{opposite } 30^\circ}{\text{hypotenuse}} = \sin 30^\circ = \frac{1}{2} \Rightarrow 7 \cdot 2 = x \cdot 1 \Rightarrow x = 14$$

$$\frac{7}{y} = \frac{\text{opposite } 30^\circ}{\text{adjacent } 30^\circ} = \tan 30^\circ = \frac{1}{\sqrt{3}} \Rightarrow y \cdot 1 = 7 \cdot \sqrt{3} \Rightarrow y = 7 \cdot \sqrt{3}$$

EXAMPLE Find x and y.

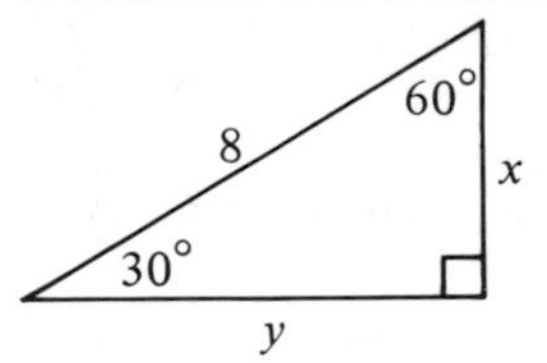

Solution Sketch the similar triangle,

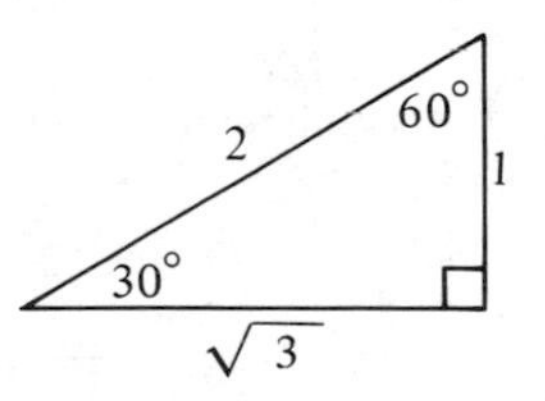

set up proportions, and solve:

$$\frac{1}{2} = \frac{\text{opposite } 30^\circ}{\text{hypotenuse}} = \frac{x}{8} \Rightarrow 1 \cdot 8 = 2 \cdot x \Rightarrow x = 4$$

$$\frac{\sqrt{3}}{2} = \frac{\text{adjacent } 30^\circ}{\text{hypotenuse}} = \frac{y}{8} \Rightarrow \sqrt{3} \cdot 8 = 2 \cdot y \Rightarrow y = 4\sqrt{3}$$

EXERCISES

1. Fill in the missing side and angle measurements.

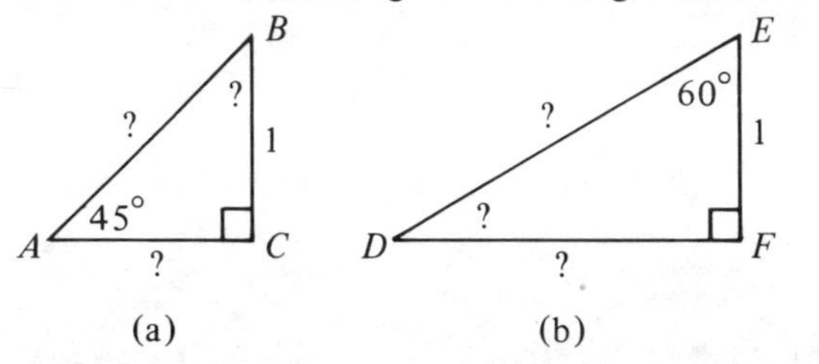

(a) (b)

2. Refer to the above figures to fill in the blanks:

(a) tan 30° = ________ **(b)** cos 60° = ________
(c) sin 30° = ________ **(d)** sin 60° = ________
(e) tan 60° = ________ **(f)** cos 30° = ________
(g) sin 45° = ________ **(h)** cos 45° = ________
(i) tan 45° = ________

3. Refer to the figures in Exercise 1 to fill in the blanks:

(a) If sin $\angle A = 1/\sqrt{2}$, then $m \angle A =$ ________.
(b) If tan $\angle A = 1/\sqrt{3}$, then $m \angle A =$ ________.
(c) If cos $\angle A = \frac{1}{2}$, then $m \angle A =$ ________.
(d) If cos $\angle A = 1/\sqrt{2}$, then $m \angle A =$ ________.
(e) If tan $\angle A = 1$, then $m \angle A =$ ________.
(f) If sin $\angle A = \frac{1}{2}$, then $m \angle A =$ ________.
(g) If cos $\angle A = \sqrt{3}/2$, then $m \angle A =$ ________.
(h) If tan $\angle A = \sqrt{3}$, then $m \angle A =$ ________.
(i) If sin $\angle A = \sqrt{3}/2$, then $m \angle A =$ ________.

4. Name as many "special" right triangles as you can see in each figure below.

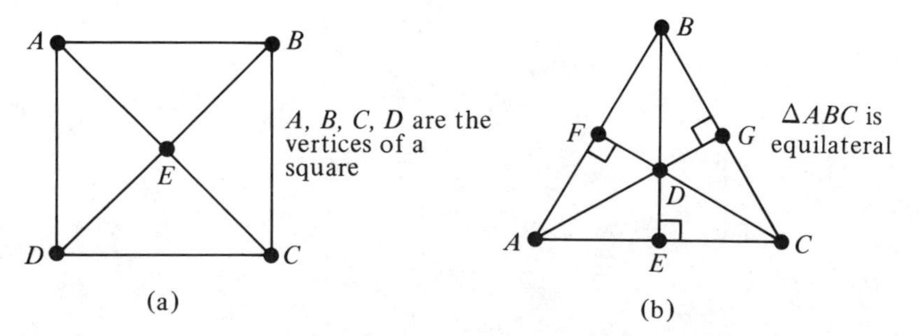

(a) (b)

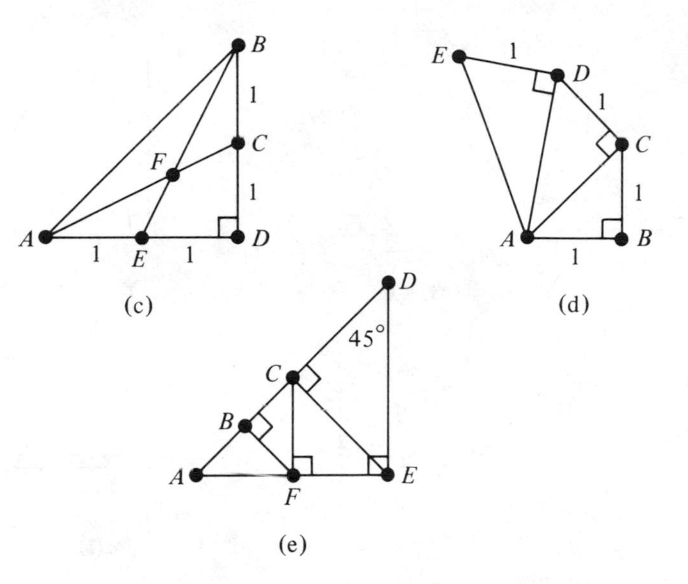

5. Find x and y:

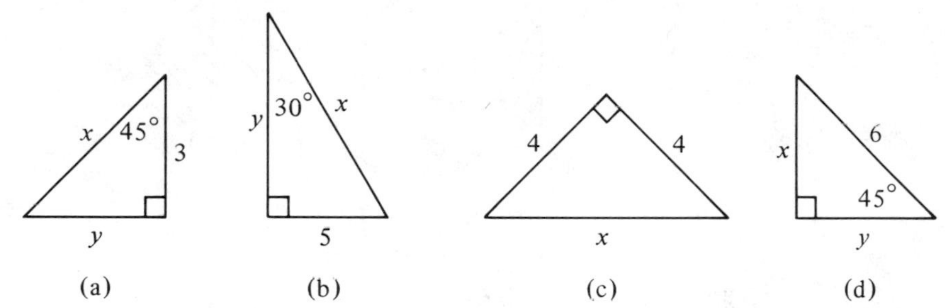

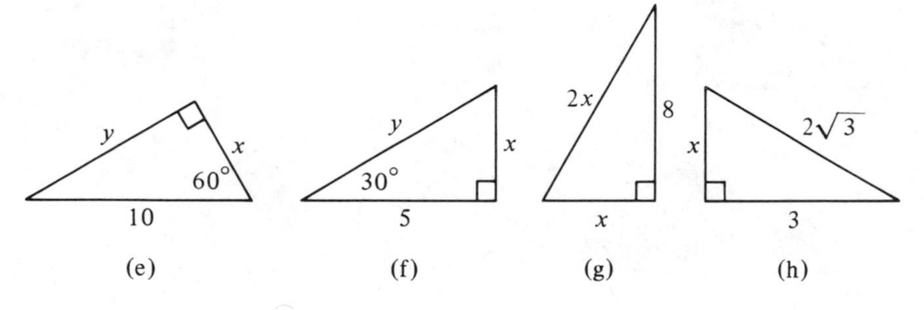

6. Fill in the missing side and angle measurements:

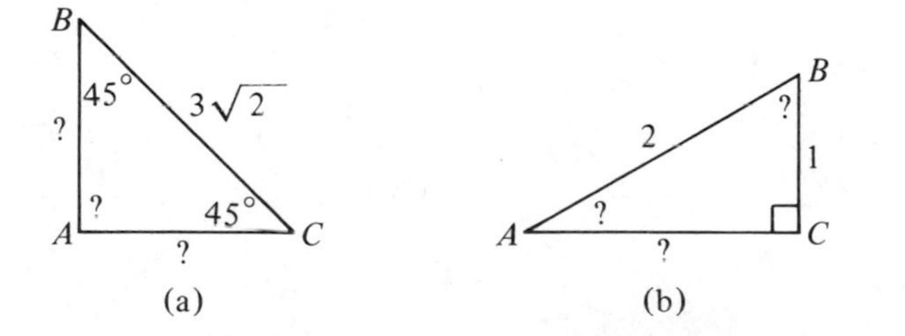

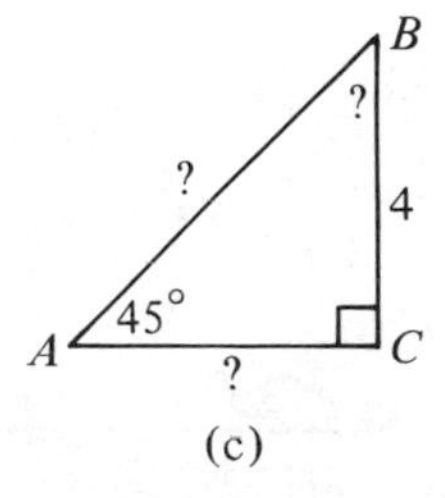

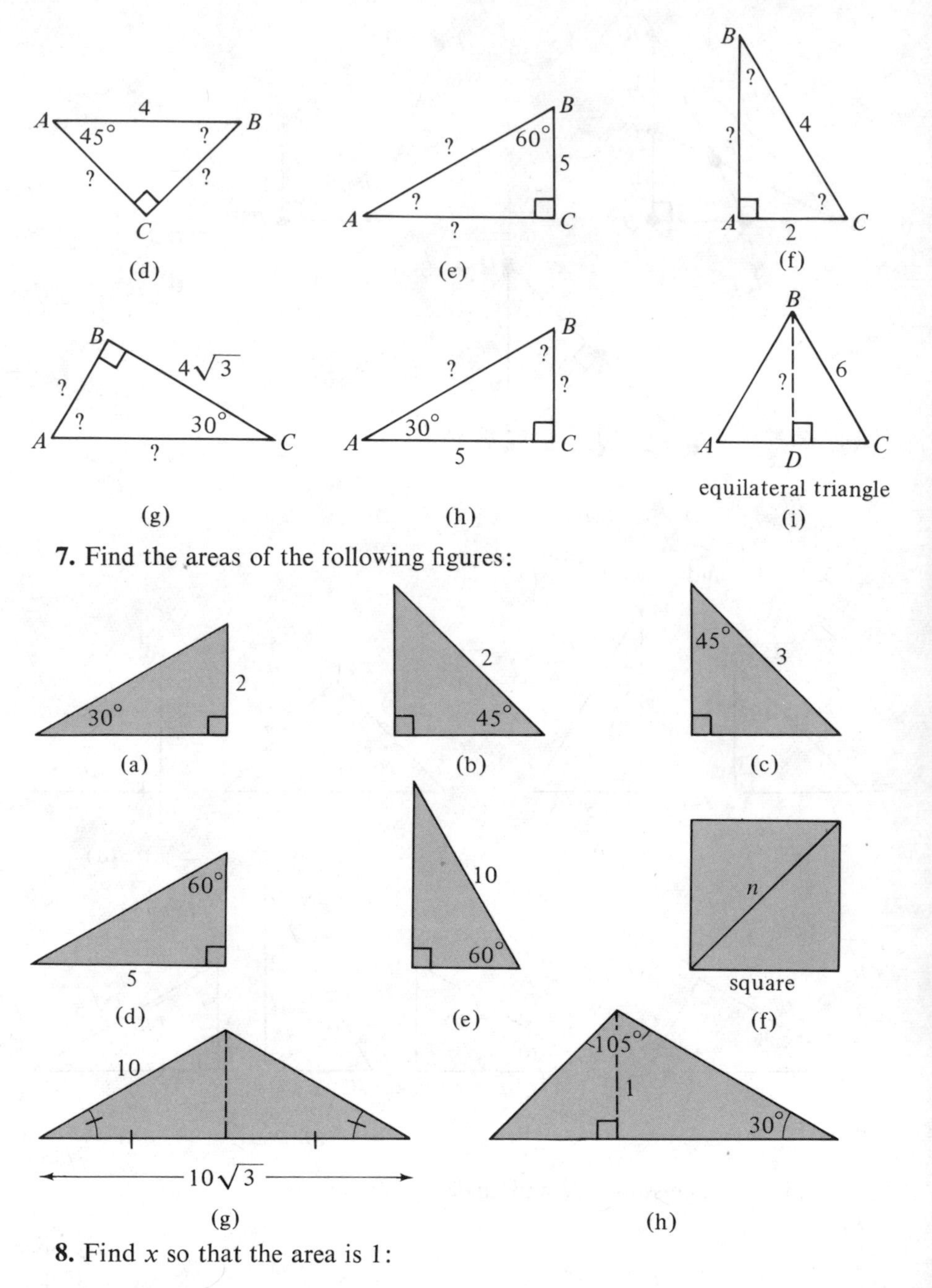

7. Find the areas of the following figures:

8. Find x so that the area is 1:

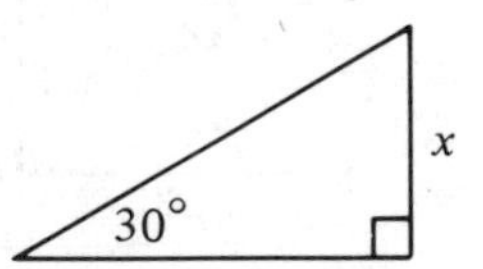

9. Find x so that the area is 1:

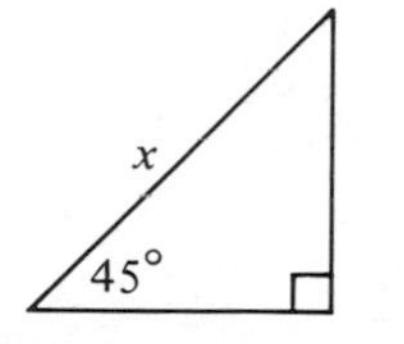

10. Show that the area of square $ACEF$ is double the area of square $ABCD$.

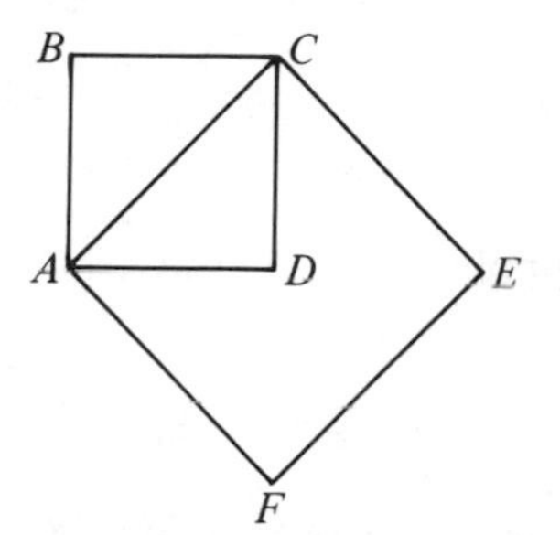

11. (a) Fold a sheet of paper into a 45–45–90 triangle.
(b) Fold a sheet of paper into a 30–60–90 triangle.

12. Joe is lost in the woods without his trusty trig table. He does have a sheet of paper, though, and he knows how to fold it into a 45–45–90 or a 30–60–90 triangle. The pictures suggest some techniques of "indirect measurement" that Joe uses.

(a) How wide is the river?

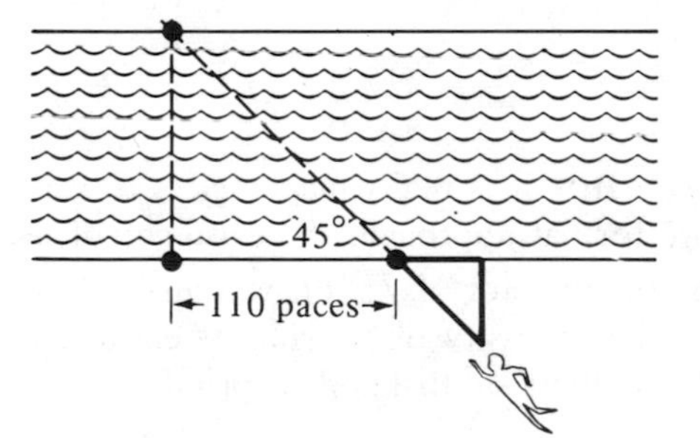

(b) How tall is the tree?

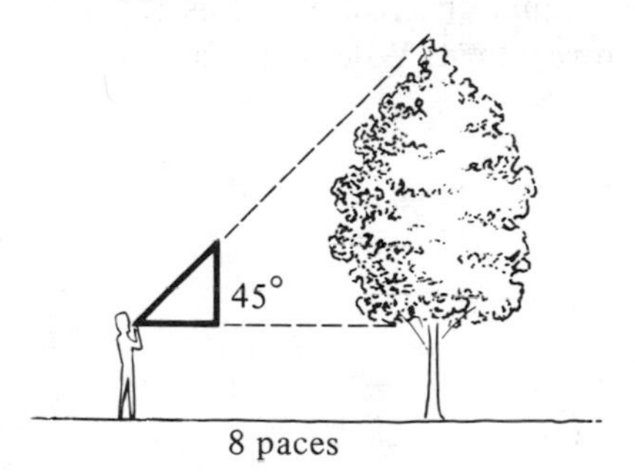

(c) How far is it across the swamp from point A to point B?

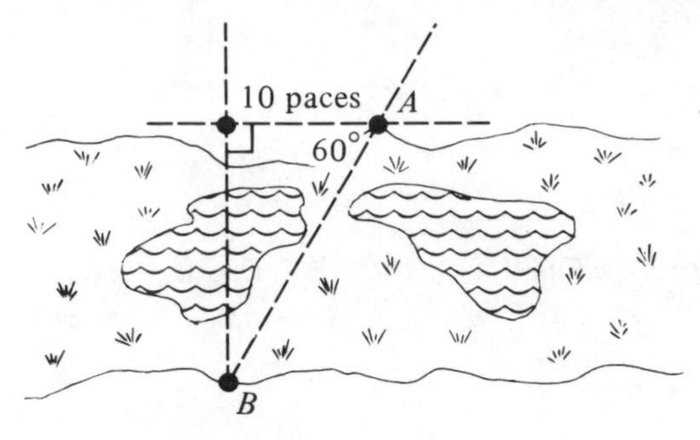

13. Suppose that $\angle A \cong \angle C$ in the figure below. Supply reasons in this argument that $\overline{AB} \cong \overline{BC}$.

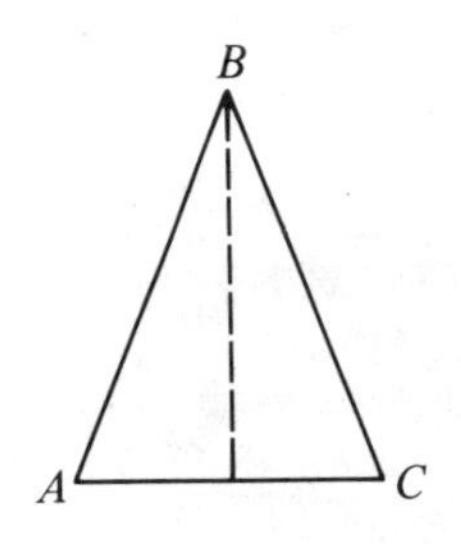

(a) Draw the bisector of $\angle B$ and let D be the point where it intersects $\overline{AC}$.

(b) $\triangle ABD \cong \triangle CBD$.

(c) $\overline{AB} \cong \overline{CB}$.

NOTE In our informal approach to geometry, it is really meaningless to ask for a reason for (a). In Euclid's more formal approach, the existence of a point D on $\overline{AC}$ and simultaneously on the bisector of $\angle ABC$ was also taken for granted. Only in recent years has the necessity of proving its existence been recognized. Using the axioms of Hilbert or Birkhoff, a proof can be given.

***14.** The figure below shows a square of side length 1. Each of the four corners of the square is the center of a quarter circle of radius 1. These four (solid) quarter-circles intersect in region A. What is the area of A?

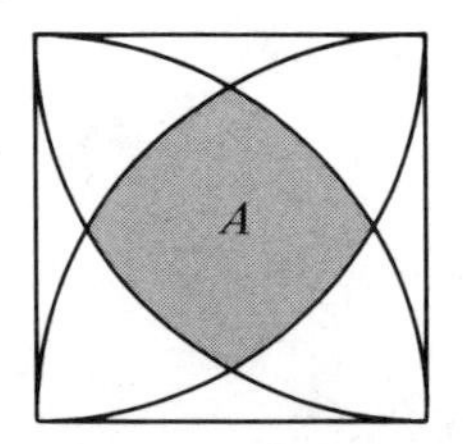

VOCABULARY

8.1 Ratio and Proportion

ratio
proportion

8.2 Similar Triangles

ratio of similarity
"SAS"
"SSS"
"AA"
similarity of triangles, ~

8.3 Similar Geometric Figures; Scale Drawings

similarity of geometric figures
scale drawing
scale
projection point

8.4 Right Triangles and the Pythagorean Theorem

right triangle
hypotenuse
LL congruence condition
Pythagorean triple
acute angle
leg
HL congruence condition

8.5 Using the Pythagorean Theorem; Square Roots

square root of a, $\sqrt{a}$

8.6 Similar Right Triangles

"A" similarity condition
side adjacent to $\angle A$
side opposite $\angle A$

8.7 Trigonometry

trigonometric ratios
sine
cosine
tangent

8.8 Special Right Triangles

45–45–90 triangle
30–60–90 triangle

FUNCTIONS

OVERVIEW

The function concept pervades most of mathematics. While a good deal of mathematical work can be done without making explicit mention of functions, still they are usually present if you look for them. Once the student grasps the idea of function, it can help him to view mathematics as one unified subject rather than as a collection of unrelated special topics. For this reason the concept of function is beginning to receive explicit attention in modern elementary mathematics curricula.

In the first section we recall examples of functions that were encountered earlier in the course. A minimal amount of terminology is introduced, and some schematic techniques for describing a function are illustrated. The first inklings of composition of functions appear in the exercises.

In the next section linear functions are studied. The slides, stretches, shrinks, and flips of the number line, which were first encountered in Chapter 2, are looked at again from a function, or mapping, point of view. Composition of functions is dealt with more explicitly in this geometric context. Such apparently diverse topics as equation solving and similarity of geometric figures on a line are seen to be parts of the study of linear functions.

The investigation of linear mappings of a line is extended, in Section 9.3, to the analogous mappings of a plane. These are seen to be just the similarity mappings of the plane. Congruence mappings (special kinds of similarity mappings) are described as compositions of flips, and then the full set of similarity mappings is obtained by allowing dilations as well as flips into the compositions. In Section 9.4, after the concept of group has been abstracted from the familiar number systems, the sets of similarities and congruences are observed to be groups (with respect to the operation of composition). The section concludes with some work with symmetry groups.

The investigation in Sections 9.3 and 9.4 is carried out in a coordinate-free setting so that the important concepts will stand out clearly, uncluttered by numbers. In Section 9.5 the plane is coordinatized, and algebraic formulas are given for many of the mappings of the plane. The discussion of these mappings now becomes very concrete and down to earth. Two other uses of the coordinate plane are described in Section 9.6: (1) its use as a site for graphing functions from R to R (carefully), and (2) its use as a replacement set for open sentences in two variables (lightly). This section can be expected to be quite useful in teaching.

One goal in this chapter is to illustrate the unifying power of the function concept by actually using it to unify, organize, and summarize some of the mathematics of the first eight chapters. Another is to give a flavor of the increasingly fashionable transformational approach to geometry.

9.1 FUNCTIONS

The concept of function is one of the most basic and important ideas in all of mathematics. It should also be a fairly simple notion to grasp.

Any time you assign to each element of a first set an element of a second set, you have formed a **function.**

EXAMPLE 1 When you assign names to people, you have formed a function.

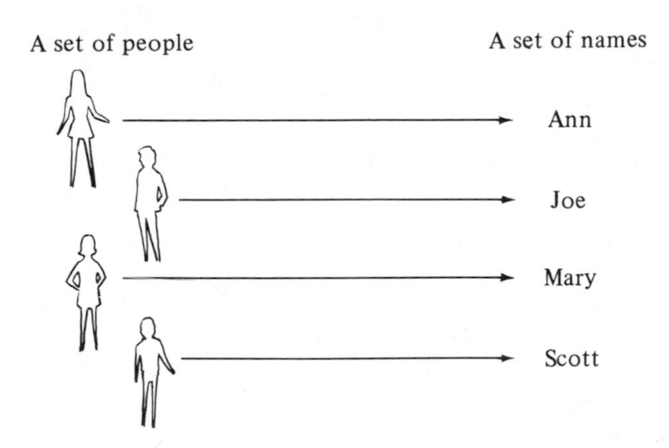

EXAMPLE 2 When you assign a number to each outcome of an experiment, you have formed a function.

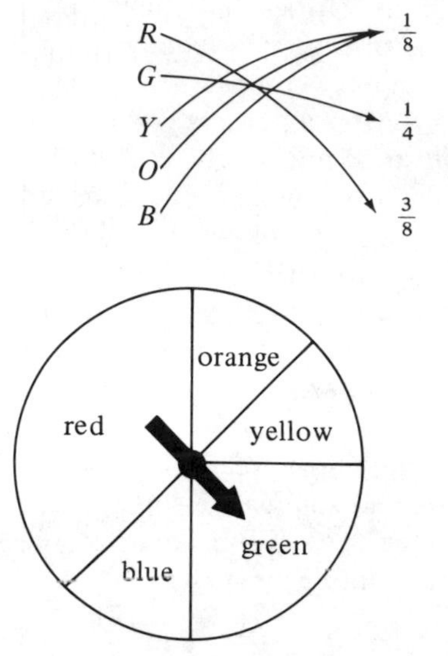

What do we call a function like this one?

EXAMPLE 3 When you assign numbers to geometric figures, as below, you have formed a function.

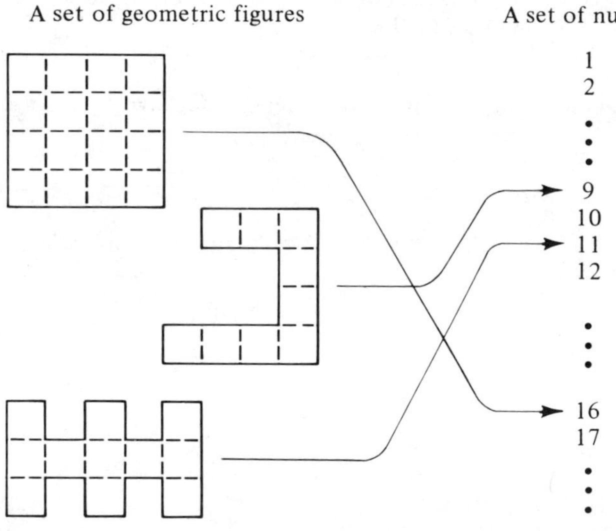

What name do we give to this function?

EXAMPLE 4 What name do we give to the function below?

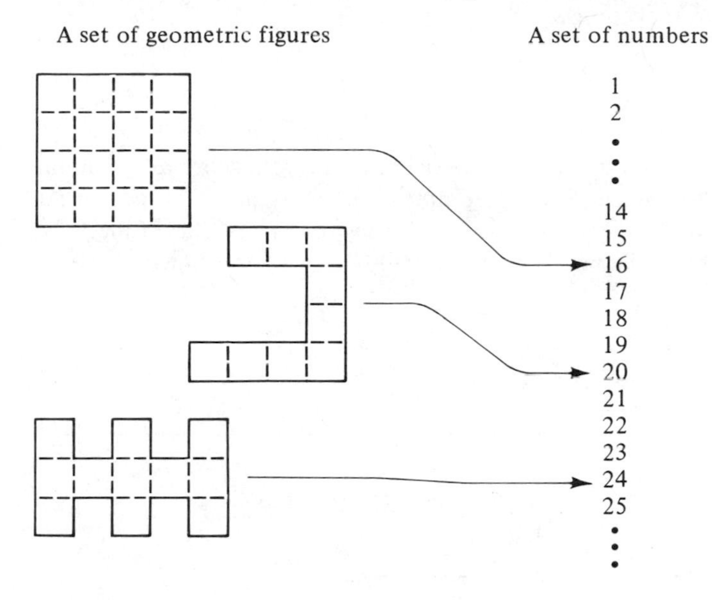

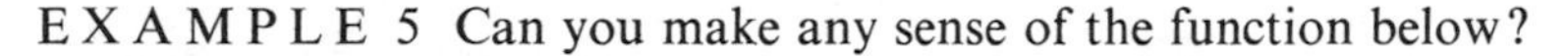

EXAMPLE 5 Can you make any sense of the function below?

A set of geometric figures

A set of numbers

1 2 3 4 5 6 7 8 9 ... 19 20 21 ...

Examples 3, 4, and 5 illustrate that many possible functions, or rules of assignment, can exist between a single pair of sets. For this reason it is very important not only to specify the two sets involved, but also to describe a completely unambiguous **rule** for assigning to each element in the first set one element of the second set. Drawing **arrow diagrams**, as in Examples 1–5, is one way to tell what is assigned to what. Ordinarily, though, it is impractical to draw an arrow diagram. Instead the rule is usually given either in words or in symbols.

EXERCISES

Most of the functions you encounter in arithmetic and algebra assign numbers, or **values**, to things. Several common functions of this type appear in the chart below. If you know ordinary names for any of these functions, give them. Also, fill in the blanks and "simplify" any of the function values that you can.

	THE FUNCTION (RULE) NAMED BY THE SYMBOL	ASSIGNS TO THE "THING" (INPUT)	THE VALUE (OUTPUT)
1. (a)	$\sqrt{\ }$	9	$\sqrt{9}$
(b)	$\sqrt{\ }$	$\frac{25}{16}$	________
(c)	$\sqrt{\ }$	________	$\sqrt{2}$
2. (a)	$\square^2$	5	5^2
(b)	$\square^2$	$-\frac{3}{4}$	________
(c)	$\square^2$	________	2
3. (a)	x^3	4	4^3
(b)	x^3	.5	________
(c)	x^3	________	-8
4. (a)	$+$	(4, 3)	$4 + 3$
(b)	$+$	$(-4, \frac{1}{2})$	________
(c)	$+$	________	1.2
5. (a)	$\cdot$	(4, 3)	$4 \cdot 3$
(b)	$\cdot$	$(-4, \frac{1}{2})$	________
(c)	$\cdot$	________	1.2
6. (a)	$x + 7$	-2	5
(b)	$x + 7$	4	________
(c)	$x + 7$	________	0
7. (a)	$-$	(2, 5)	$2 - 5$
(b)	$-$	(6, 5)	________
(c)	$-$	________	4
8. (a)	$-$	6	-6
(b)	$-$	-2	________
(c)	$-$	________	-1

	THE FUNCTION (RULE) NAMED BY THE SYMBOL	ASSIGNS TO THE "THING" (INPUT)	THE VALUE (OUTPUT)
9. (a)	$\lvert\ \rvert$	-3	$\lvert -3 \rvert$
(b)	$\lvert\ \rvert$	3.14	________
(c)	$\lvert\ \rvert$	________	5
10. (a)	$1/x$	5	$\frac{1}{5}$
(b)	$1/x$	$\frac{2}{3}$	________
(c)	$1/x$	________	4
11. (a)	$\div$	(5, 2)	$5 \div 2$
(b)	$\div$	$(\frac{3}{4}, 2)$	________
(c)	$\div$	________	2
12. (a)	$\div$	(23, 5)	(4, 3)
(b)	$\div$	(16, 3)	________
(c)	$\div$	________	(6, 0)

13. The set of all **inputs** to which a function assigns outputs is called the **domain** of the function. The set of all **outputs** is called the **range** of the function. For example, the function + of Exercise 4 has as its domain the set of all ordered pairs of real numbers (i.e., domain $= R \times R$) and as its range the set of all real numbers (i.e., range $= R$). Describe a (largest possible) domain and range for each of the functions in Exercises 1–11. If the domain in Exercise 12 is $W \times N$, what is the range?

14. Function language is used in ordinary discourse. Complete the following chart and give a reasonable domain and range for each function.

THE FUNCTION	ASSIGNS TO THE INPUT	THE OUTPUT
(a) the capital of	Arkansas	________
the capital of	________	Albany
(b) the birthday of	George Washington	________
the birthday of	"you"	________
(c) the author of	*Moby Dick*	________
the author of	*Don Quixote*	________

15. Functions sometimes appear in tabular form:

STUDENT	AGE	HEIGHT	WEIGHT	IQ
Ann	13	60	96	110
Betty	14	64	110	100

STUDENT	AGE	HEIGHT	WEIGHT	IQ
Charles	13	63	108	115
Don	15	68	130	105
Edward	14	64	125	105

For example, from this table we could make up an arrow diagram of the age function a.

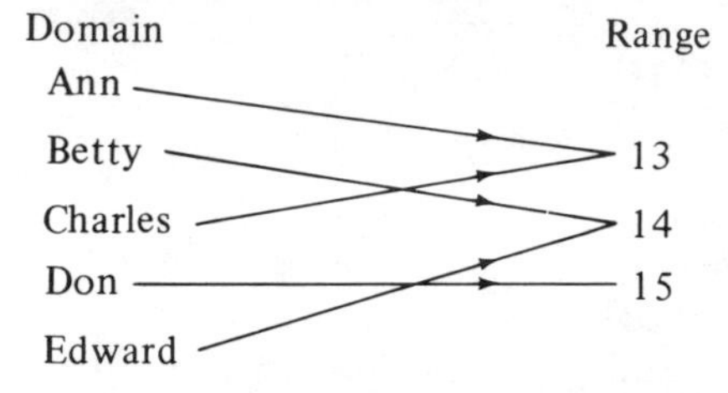

Draw an arrow diagram for the height function h.

16. To keep straight the four functions given by the table above, we name them with letters, a, h, w, i. We also write things such as

"$a(\text{Don}) = 15$" for "the age of Don is 15"
"$w(\text{Charles}) = 108$" for "the weight of Charles is 108"

Translate each of the following into ordinary English and then find all replacements x or y that make it a true statement.

(a) $h(\text{Betty}) = y$ **(b)** $i(\text{Edward}) = y$
(c) $w(\text{Ann}) = y$ **(d)** $h(x) = 63$
(e) $i(x) = 110$ **(f)** $a(x) = 14$
(g) $w(x) = 108$

17. Letters are frequently used to name functions. A favorite letter is f. A function f is defined below by a **formula** that tells what the function does to each domain element x.

$$f(x) = 2x + 3 \qquad (\text{“}f \text{ of } x \text{ equals } 2x \text{ plus } 3\text{”})$$

(a) Find the range element assigned by f to the domain element 6.
(b) Find $f(6)$.
(c) Find the range element assigned by f to the domain element -4.
(d) Find $f(-4)$.
(e) Find the domain element that is assigned 7 by f.
(f) Find x such that $f(x) = 7$.
(g) Find the solution to the equation $7 = 2x + 3$.
(h) Find the domain element that is assigned 0 by f.
(i) Find x such that $f(x) = 0$.

(j) Solve the equation, $2x + 3 = 0$.
(k) Find x such that $f(x) = x$.

18. The function $g(x)$ is defined by the formula $g(x) = 4 - 3x$. Find

(a) $g(-2)$ **(b)** $g(-1)$
(c) $g(0)$ **(d)** $g(1)$
(e) $g(n)$ **(f)** $g(n+1)$
(g) $g(n+1) - g(n)$ **(h)** $g(g(0))$
(i) $g(g(2))$ **(j)** $g(g(x))$

19. The functions h and k are defined by the formulas

$$h(x) = 3x - 4 \qquad k(x) = 5 - 2x$$

Find

(a) $h(0)$ **(b)** $h(1)$
(c) $h(2)$ **(d)** $k(0)$
(e) $k(1)$ **(f)** $k(2)$
(g) x such that $h(x) = k(x)$ **(h)** $h(k(2))$
(i) $k(h(2))$ **(j)** $h(k(h(1)))$

20. Write a formula for

(a) the function f that first doubles and then adds 5
(b) the function g that first adds 5 and then doubles

Are f and g the same function?

21. In the figure,

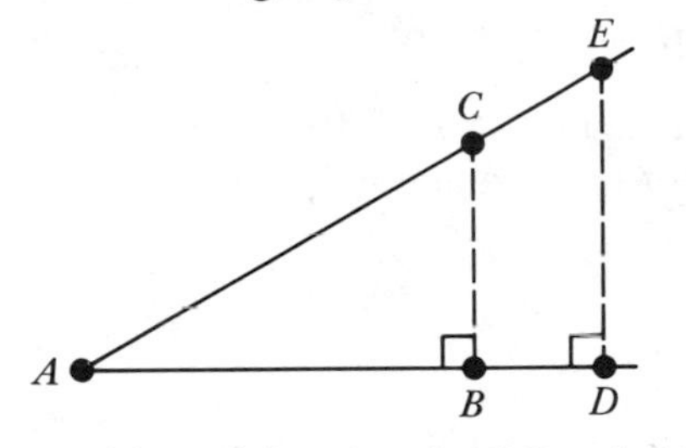

(a) explain why $\triangle ABC \sim \triangle ADE$
(b) explain why $BC/AC = DE/AE$

22. "Sine" can be thought of as a function whose domain is the set of all acute angles. Use your ruler to find, approximately, the function values below.

(a) sine ($\angle A$)

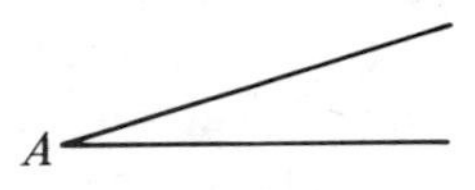

(b) sine ($\angle B$)

B

(c) sine ($\angle C$)

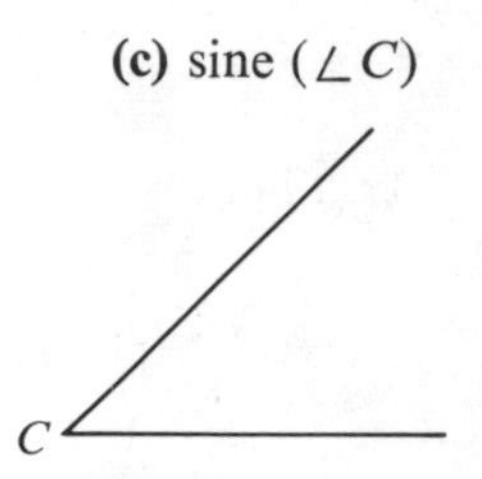

(d) Describe the range of this sine function.

23. Consider "cosine" as a function whose domain is the set of all acute angles.

(a) Use a ruler to find approximate values of cosine ($\angle A$), cosine ($\angle B$), cosine ($\angle C$) in the previous figures.

(b) Describe the range of this cosine function.

24. Describe a possible domain and range for "tangent."

NOTE If you have had an advanced trigonometry or beginning calculus course, you may recall how the function, sine, can be interpreted so that its domain is the set R of all real numbers. At the elementary level it is probably most natural to view sine (and cosine and tangent) as a function whose domain is a set of acute angles.

9.2 LINEAR FUNCTIONS ON THE NUMBER LINE

The simplest kind of functions that assign numbers to numbers are the so-called "linear functions." Some examples of linear functions are

$$x + 4,\ 3x,\ x - 2,\ 3x - 2,\ -x,\ -2x,\ -2x + 5$$

In general,

Any function defined by a formula of the type $f(x) = a \cdot x + b$, where a and b are real numbers and $a \neq 0$, is called a **linear function.**

Ordinarily, we take the domain of a linear function to be the full set R of real numbers. Then the range is also the full set R.

EXAMPLE The function $f(x) = 3x - 2$ is a linear function because it is of the form $a \cdot x + b$, where $a = 3$ and $b = -2$. Its range is all of R, because every real number is an output of the function. To show, for example, that $\frac{13}{47}$ is an output of the function, we must find an x such that $x \xrightarrow{f} \frac{13}{47}$. That is, we must find an x such that $3x - 2 = \frac{13}{47}$. Solving, we get $x = (\frac{13}{47} + 2)/3$. More generally, given any real number r, we can find an x such that $f(x) = r$ as follows:

$$3x - 2 = f(x) = r \Leftrightarrow 3x = r + 2 \Leftrightarrow x = \tfrac{1}{3}(r + 2)$$

If we represent the set R of real numbers by a number line, then the linear functions can be given geometric interpretation. Imagine two copies of the number line

−6 −5 −4 −3 −2 −1 0 1 2 3 4 5 6 7

−6 −5 −4 −3 −2 −1 0 1 2 3 4 5 6 7

The light copy never budges; the dark one can. The linear function $g(x) = x + 4$ can be thought of as a **slide** (or translation) of the dark line four units to the right.

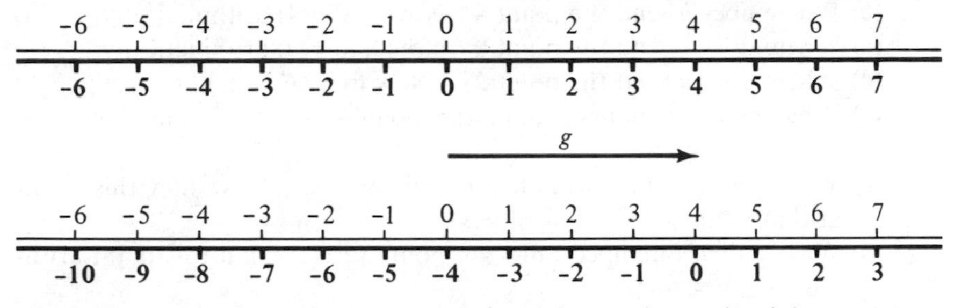

The heavy number line has been translated four units to the right.

The geometric slide sends

(dark) $0 \to 4$ (light)

(dark) $-3 \to 1$ (light)

(dark) $2 \to 6$ (light)

This is precisely the assignment of numbers to numbers that we would have obtained by purely arithmetic substitution in the formula $g(x) = x + 4$.

$$0 \xrightarrow{g} 0 + 4 = 4$$

$$-3 \xrightarrow{g} -3 + 4 = 1$$

$$2 \xrightarrow{g} 2 + 4 = 6$$

Thus we have two equivalent ways of thinking about g: as an arithmetic rule that assigns numbers to numbers; as a geometric motion that "maps" points onto points. Is it clear, geometrically, why the range of g is all of R? The word **mapping** is a synonym for "function," and is generally used in the context where both domain and range are sets of points.

EXERCISES

1. The linear function h is defined arithmetically by the formula $h(x) = x - 3$.

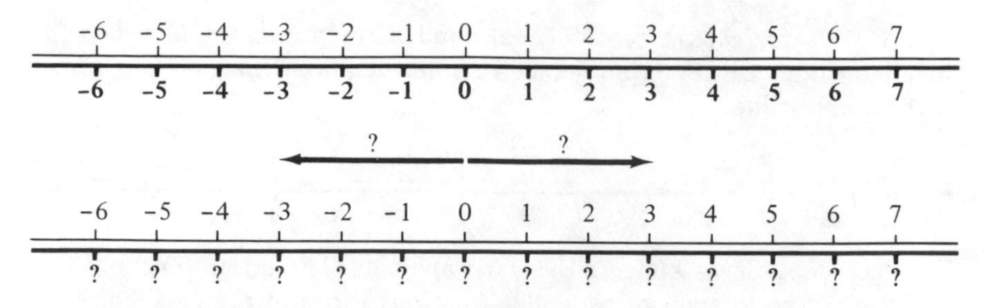

(a) Describe the geometric motion that represents h.
(b) Where does h send the point 4? Now find $h(4)$ arithmetically.
(c) Where does h send the point 0? Now find $h(0)$ arithmetically.
(d) Where does h send the point $2\frac{1}{3}$? Now find $h(2\frac{1}{3})$ arithmetically.
(e) What point is mapped onto the point -5? Now find this point arithmetically.
(f) What point is mapped onto the point -2? Now find this point arithmetically.
(g) What point is mapped onto the point 3? Now find this point arithmetically.
(h) What point is mapped onto the point $-\frac{4}{3}$? Now find this point arithmetically.

2. The linear function k defined arithmetically by $k(x) = 3x$ can be represented geometrically as shown below.

-12 -11 -10 -9 -8 -7 -6 -5 -4 -3 -2 -1 0 1 2 3 4 5 6 7 8 9 10 11 12 13
-12 -11 -10 -9 -8 -7 -6 -5 -4 -3 -2 -1 0 1 2 3 4 5 6 7 8 9 10 11 12 13

-12 -11 -10 -9 -8 -7 -6 -5 -4 -3 -2 -1 0 1 2 3 4 5 6 7 8 9 10 11 12 13
-4 -3 -2 -1 0 1 2 3 4

(a) What might you think of k as "doing" to the dark number line?
(b) Where does k send the point 2? Now find $k(2)$ arithmetically.
(c) Where does k send the point -3? Now find $k(-3)$ arithmetically.
(d) Where is the point $\frac{1}{3}$ sent? Now find $k(\frac{1}{3})$ arithmetically.
(e) Where is the point $\sqrt{2}$ sent?
(f) What point is mapped onto -3? Now find this point arithmetically.
(g) What point is mapped onto 0? Now find this point arithmetically.
(h) What point is mapped onto 2? Now find this point arithmetically.

3. Interpret, geometrically, the linear function $\ell(x) = \frac{1}{2}x$.

Any two linear functions can be **composed**, or "hooked up" or "performed in succession." For example, if $g(x) = x + 4$ (g adds 4) and $k(x) = 3x$ (k multiplies by 3), then arithmetically

$$g(k(2)) = g(3 \cdot 2) = g(6) = 6 + 4 = 10$$

$$g(k(-3)) = g(3 \cdot (-3)) = g(-9) = -9 + 4 = -5$$

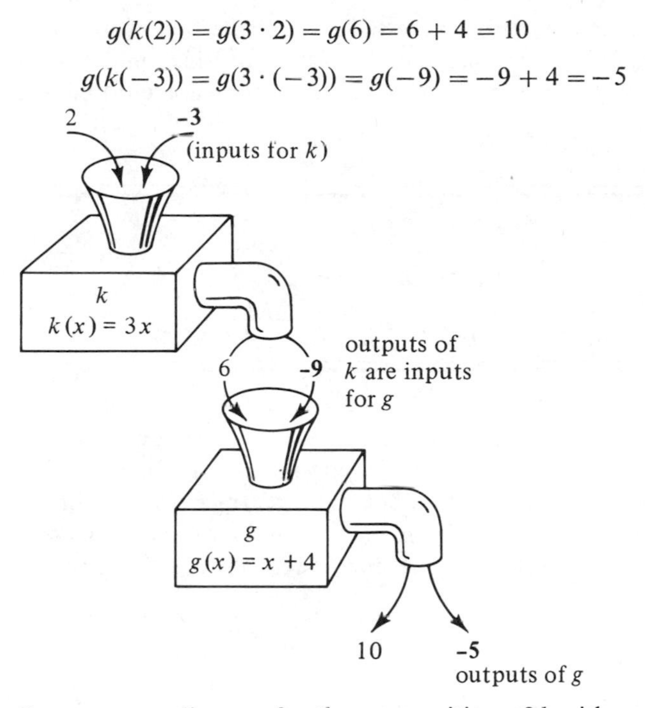

Input-output diagram for the composition of k with g.

More generally, $g(k(x)) = g(3x) = 3x + 4$. (Notice that in computing $g(k(2))$ the function k was applied first and then the function g was applied. The order in which the functions are applied is the *opposite* of the order in which they are read.) We have just seen that the linear function $3x + 4$ is the composition of a 3-**stretch** with a 4-slide-right in that order: first the 3-stretch k, and then the 4-slide-right g.

4. Complete the table:

THE LINEAR FUNCTION	IS THE COMPOSITION OF	FIRST	AND	THEN
(a) $3x + 4$		3-stretch		4-slide-right
(b) $2x - 5$		________		________
(c) $3x + 2$		________		________
(d) $3(x + 2)$		________		________
(e) $3x + 6$		________		________
(f) $\frac{1}{4}x + 2$		$\frac{1}{4}$-**shrink**		________
(g) $\frac{1}{3}x - 5$		________		________
(h) $\frac{2}{3}x + \frac{1}{4}$		________		________
(i) $6 + 2x$		________		________
(j) $3 \cdot 5x$		________		________

5. Not every linear function can be expressed as a composition of stretches, shrinks, and slides. For example, consider the linear function r defined by $r(x) = -x$. (Why is r linear?) Describe the geometric motion that represents r.

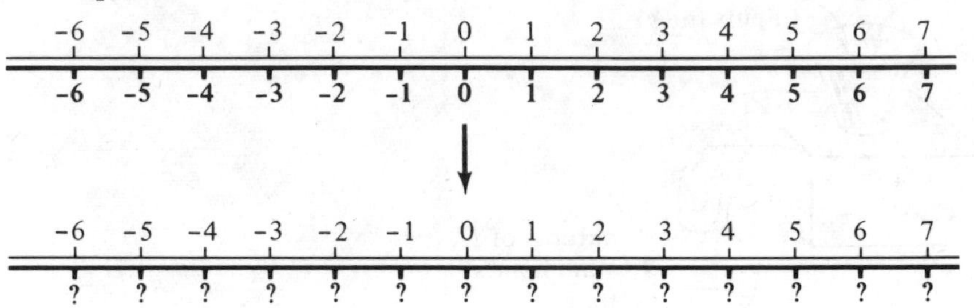

6. Every linear function can be expressed as the composition of slides, stretches, shrinks, and **flips**. For each function below: (i) Express it as such a composition, (ii) draw a number line and trace the motion of the points -1 and 2 under that composition, (iii) verify your work in (ii) arithmetically by substituting -1 and 2, respectively, for x in the formula.

EXAMPLE $-2x + 5$

(i) 2-stretch followed by flip followed by 5-slide-right

(ii)

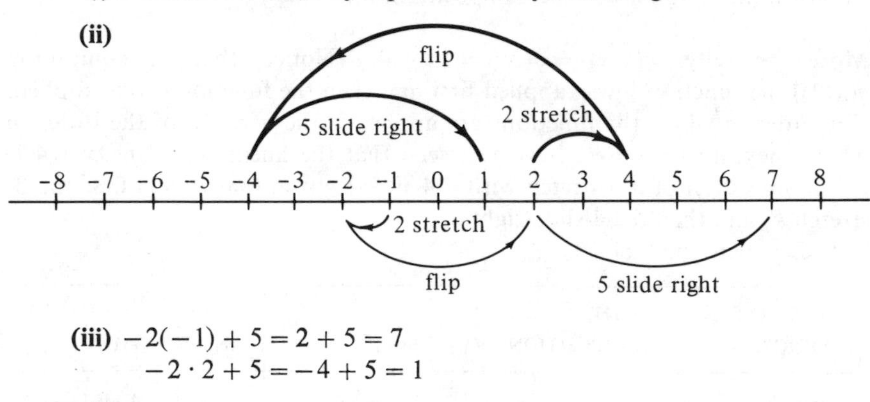

(iii) $-2(-1) + 5 = 2 + 5 = 7$
$-2 \cdot 2 + 5 = -4 + 5 = 1$

(a) $3x - 2$
(b) $\frac{1}{2}x + 4$
(c) $\frac{1}{2} + 4x$
(d) $-x + 3$
(e) $5 - x$
(f) $-(\frac{1}{2}x)$
(g) $-3x + 2$
(h) $2(4 - x)$
(i) $(-3x + 4)/2$

7. Given the linear functions

$$f(x) = 2x - 3 \qquad g(x) = 4 - x \qquad h(x) = \tfrac{2}{3}x + 1$$

compute the following as in the example.

EXAMPLE $g(f(5)) = g(7) = -3$.

(a) $h(g(3))$ **(b)** $g(h(3))$
(c) $f(h(6))$ **(d)** $f(g(h(9)))$
(e) $h(g(f(8)))$ **(f)** $g(f(g(0)))$
(g) $f(f(h(3)))$ **(h)** $f(h(g(f(1))))$
Simplify the following:
(i) $f(g(x))$ **(j)** $g(f(x))$
(k) $h(f(x))$ **(l)** $f(g(h(x)))$

8. Write a formula for a function f that is
(a) a $\frac{1}{3}$-shrink followed by a 6-slide-left
(b) a 2-slide-right followed by a flip
(c) a flip followed by a 5-slide-right followed by a $\frac{1}{2}$-shrink
(d) a 3-stretch followed by a flip followed by a $\frac{1}{2}$-slide-right
(e) a 2-slide left followed by a 7-slide right
(f) a flip followed by a $\frac{2}{3}$-slide-left followed by a flip

9. Describe more simply the function in Exercise 8(e).

10. Describe more simply the function in Exercise 8(f).

Besides keeping track of what happens to individual points under a linear function, we can keep track of what happens to entire sets of points.

EXAMPLE What is the "image" of the right-directed ray from 2 under the linear function, $4 - \frac{1}{2}x$?

Solution The linear function $4 - \frac{1}{2}x$ can be viewed as the composition of a $\frac{1}{2}$-shrink followed by a flip followed by a 4-slide right. The effect of these moves on the given ray is shown below.

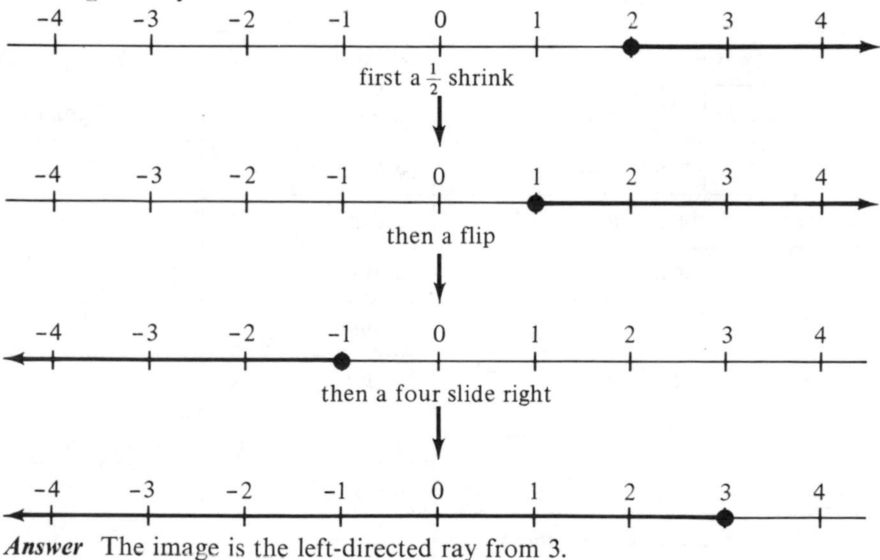

Answer The image is the left-directed ray from 3.

11. Sketch the geometric figure, then sketch and describe its image as in the example. (An explicit definition of the **image** of a figure under a given function may be useful: If f is a function and $\mathscr{A}$ is a figure in the domain of f, then the image of $\mathscr{A}$ under f is the set of all outputs that arise from inputs chosen from $\mathscr{A}$.)

Geometric Figure	Linear Function	Image of Geometric Figure
Example segment from −1 to 2 −3 −2 −1 0 1 2 3	$3x$	segment from −3 to 6 −4 −3 −2 −1 0 1 2 3 4 5 6
(a) segment from −4 to 1 0	$\frac{1}{2}x + 3$	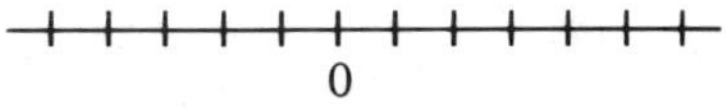 0
(b) left-directed ray from 0 0	$(x + 4)$	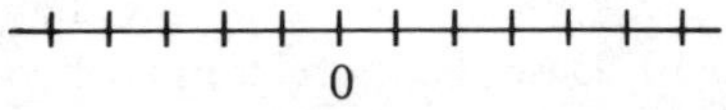 0
(c) segment from −4 to 0 0	$\frac{3}{2} - x$	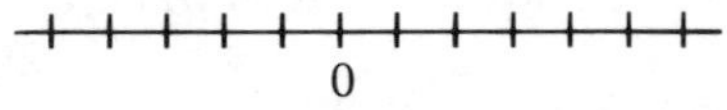 0
(d) right-directed ray from −6 0	$\frac{1}{3}x + 2$	0

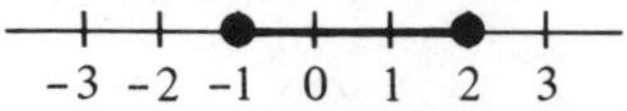

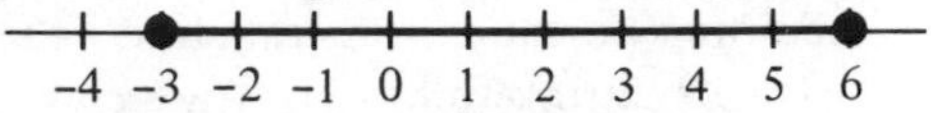

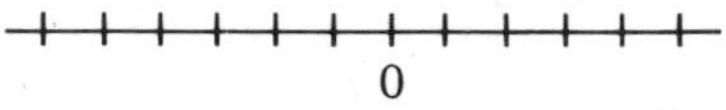

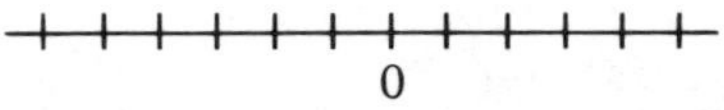

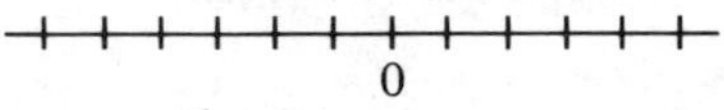

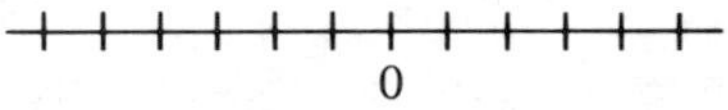

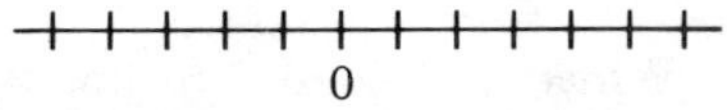

12. Complete the following chart:

THE IMAGE OF		UNDER	IS
Example	a segment	a slide	a segment
(a)	a segment	a stretch	
(b)	a segment	a flip	
(c)	a ray	a shrink	
(d)	a ray	a slide	
(e)	the line	a slide	
(f)	the line	a stretch	
(g)	a half-line (open ray)	a flip	
(h)	an open segment	a shrink	

THE IMAGE OF		UNDER	IS
Example	a segment	a slide	a segment
(i)	a ray	the composition of a slide and a flip	
(j)	a point	the composition of a stretch and a slide	
(k)	a half-open segment	the composition of a flip and a shrink	
(l)	a point	a linear function	
(m)	a ray	a linear function	
(n)	a segment	a linear function	
(o)	an open segment	a linear function	
(p)	the line	a linear function	

13. Would you approve of calling linear functions "similarity mappings"? Why?

***14.** Suppose that $f(x) = ax + b$, where a and b are real numbers and $a \neq 0$.
(a) Find x so that $f(x) = 7$.
(b) Show that the range of f is the entire set R.
(c) Show that different inputs for f lead to different outputs [if $x_1 \neq x_2$, then $f(x_1) \neq f(x_2)$], and conclude that f is a "one-to-one correspondence" between R and R.

9.3 CONGRUENCE AND SIMILARITY MAPPINGS OF THE PLANE

Among the many possible functions from R to R, we singled out only the simplest kind to study in the previous section—the so-called linear functions. Algebraically, a linear function is one given by a formula of the type

$$f(x) = ax + b \qquad \text{where } a, b \in R \text{ and } a \neq 0$$

[Several of the exercises of Section 9.2 illustrated that, although not mentioned explicitly, these functions were the objects of investigation back in Chapter 2 when we studied the solution of "linear equations."] Geometrically, we observed that the linear functions are mappings from a line to itself that are either slides, flips, shrinks, stretches, or compositions of such mappings. We concluded the section by referring to the linear mappings as "similarity mappings" because the image of a figure on a line—such as a segment, a ray, a point, or a half-line—is always a figure of the same kind, a "similar" figure.

The geometric interpretation, with its language of slides, flips, and similarity mappings, suggests rather strongly that the work we did on congruence (Chapter 4) and similarity (Chapter 8) for plane figures might also be summarized in terms of functions. It appears that by using the general concept of function, strong analogies might emerge between the apparently unrelated topics of linear functions from R to R and congruence-similarity. We shall look for geometric analogies in this section. In Section 9.5 we shall look briefly at congruence and similarity from an algebraic point of view.

Congruence Mappings

We decided back in Chapter 4 that two figures in a plane would be considered congruent if and only if one could be made to coincide with the other by means of some sequence of flips, turns, and slides. Any such sequence we referred to as a rigid motion. You might also recall that flips were the most basic type of rigid motion in the sense that any slide or turn, and hence any rigid motion at all, could be accomplished by a sequence of flips. Since flips are so fundamental, we establish special notation for them and give a careful definition.

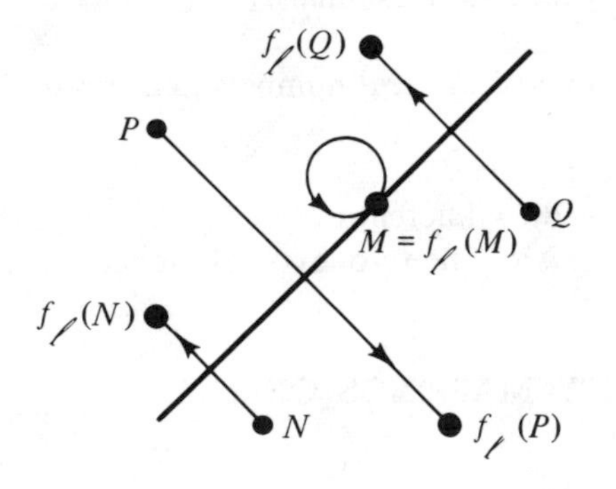

The **flip** across line ℓ, written f_ℓ, is the function, having the plane as both domain and range, with the properties:

1. For any point P off ℓ, ℓ is the perpendicular bisector of segment, $\overline{Pf_\ell(P)}$.
2. For any point M on ℓ, $f_\ell(M) = M$.

The accompanying partial arrow diagram for f_ℓ probably gives a better idea of how f_ℓ behaves than does the careful verbal definition.

EXAMPLE Given the diagram, sketch $f_\ell(A)$, $f_\ell(B)$, and the image of $\mathscr{A}$ under f_ℓ.

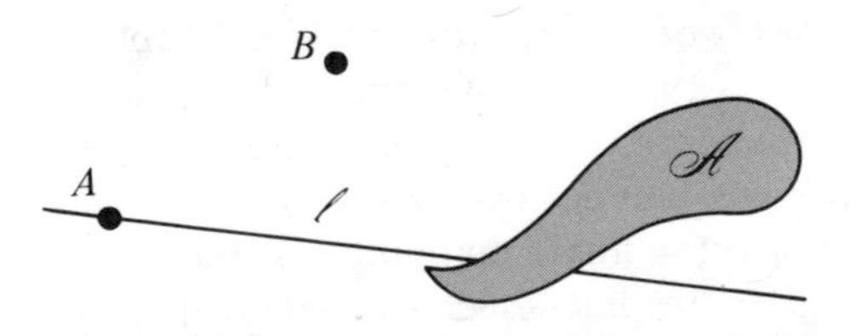

Solution

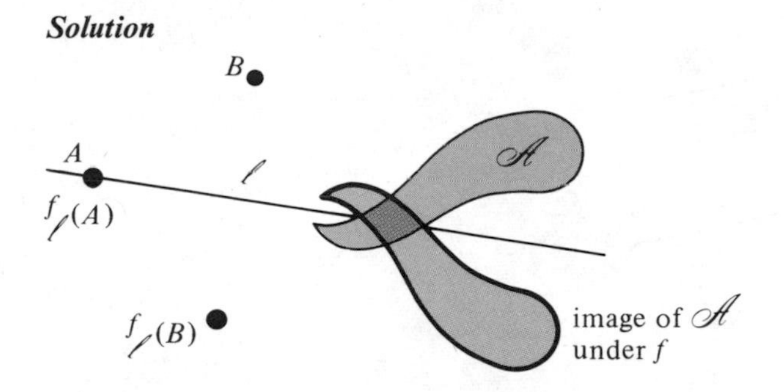

EXAMPLE Sketch the images of A, B, C under the composition, first f_ℓ then f_m. That is, locate $f_m(f_\ell(A))$, $f_m(f_\ell(B))$, and $f_m(f_\ell(C))$.

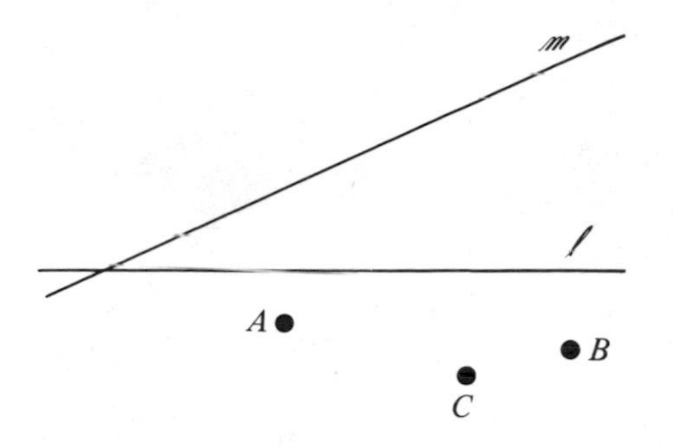

Solution

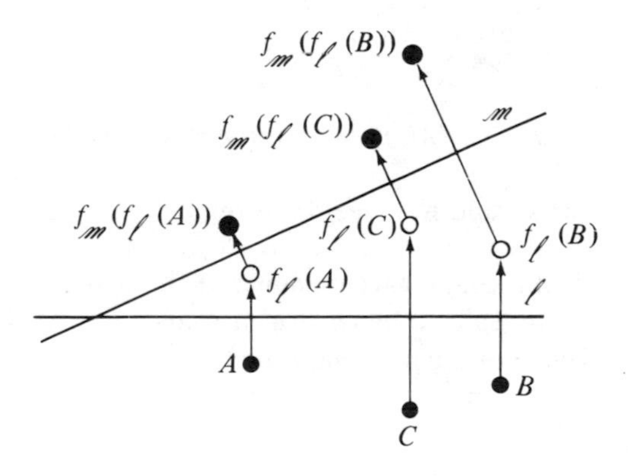

Does it appear that $\Delta f_m(f_\ell(A))f_m(f_\ell(B))f_m(f_\ell(C))$ is the image of ΔABC under some turn (rotation)?

EXAMPLE Sketch the images of A, B, C under the composition, first f_ℓ then f_m. That is, locate $f_m(f_\ell(A))$, $f_m(f_\ell(B))$, and $f_m(f_\ell(C))$.

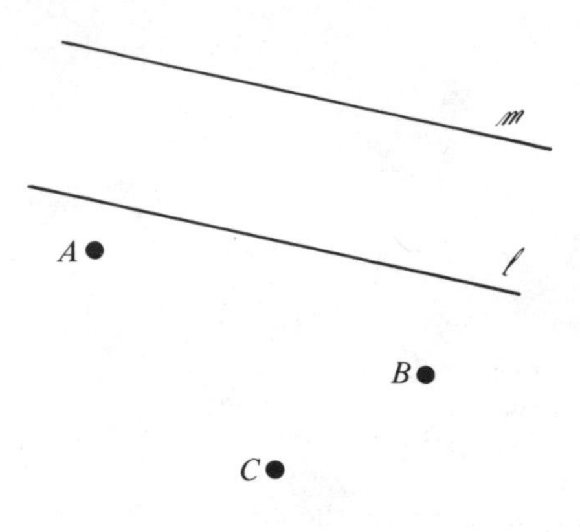

Solution

$f_m(f_\ell(A))$

$f_\ell(C)$ $f_m(f_\ell(B))$

m

$f_\ell(A)$ $f_m(f_\ell(C))$ $f_\ell(B)$

A ℓ

B

C

Does it appear that $\triangle f_m(f_\ell(A))f_m(f_\ell(B))f_m(f_\ell(C))$ is the image of $\triangle ABC$ under some slide (translation)?

The following pair of definitions should now seem reasonable:

Any composition of flips is called a **congruence mapping** of the plane. Two figures A and B in a plane are said to be **congruent** if and only if B is the image of A under some congruence mapping.

Similarity Mappings

We decided back in Chapter 8 that two figures $\mathscr{A}$ and $\mathscr{B}$ in a plane would be considered similar if and only if one could be made to coincide with the other by means of a rigid motion and a "projection drawing."

EXAMPLE Show that $\mathscr{A}$ is similar to $\mathscr{B}$.

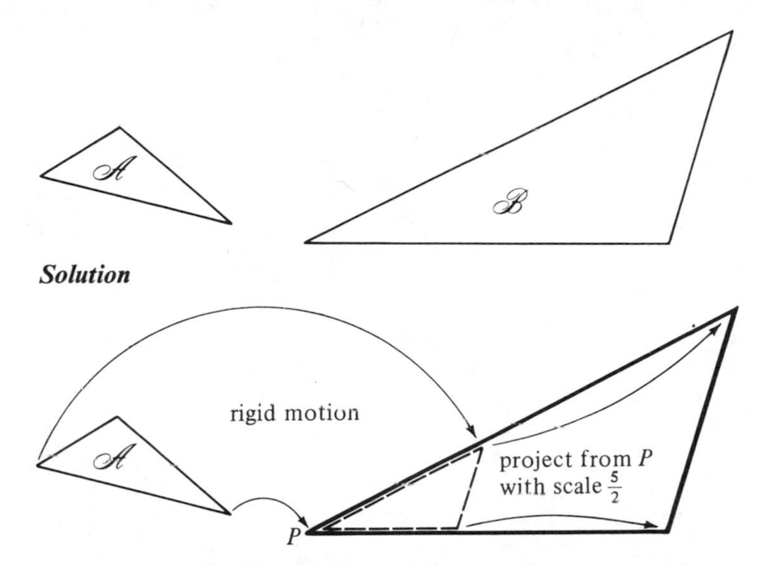

The mappings of the plane that correspond to "projection drawings" (or to stretches and shrinks if one wants to stress the analogy with linear functions) are called dilations. More precisely,

---

The **dilation** with center P and scale r $(r > 0)$, written $d_{P,r}$, is the function, having the plane as both domain and range, with the properties:

1. For any point Q different from P, $d_{P,r}(Q)$ is collinear with P and Q, P is not between Q and $d_{P,r}(Q)$, and the distance from P to $d_{P,r}(Q)$ is r times the distance from P to Q.
2. $d_{P,r}(P) = P$.

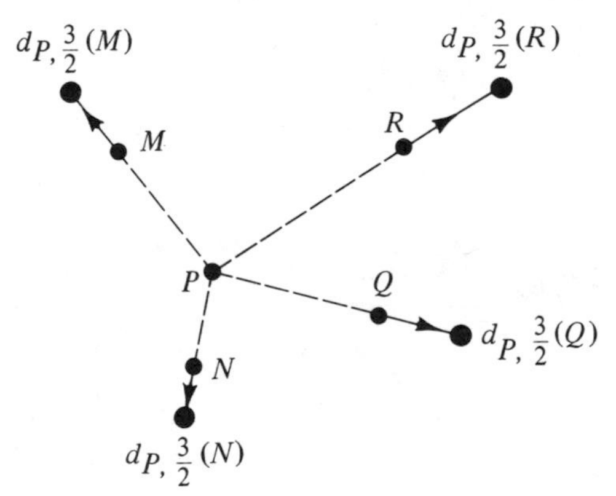

Again the partial arrow diagram, in which r was taken to be $\frac{3}{2}$ for the sake of concreteness, is probably more suggestive than the verbal definition.

EXAMPLE Given the diagram, sketch $d_{P,1/2}(A)$, $d_{P,1/2}(B)$, and the image of $\mathscr{A}$ under $d_{P,1/2}$.

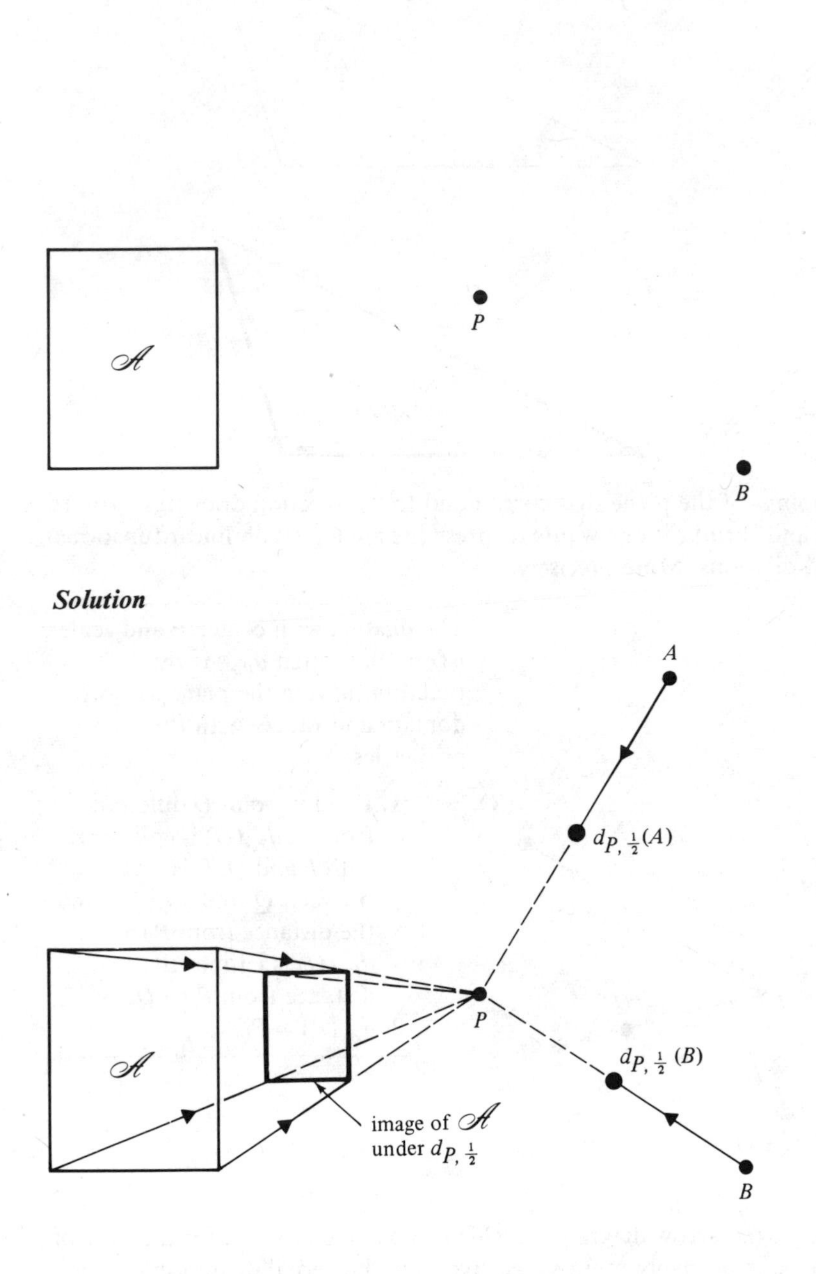

EXAMPLE Given the diagram, sketch the image of $\mathscr{A}$ under the composition, first f_ℓ then $d_{P,2}$, then f_m.

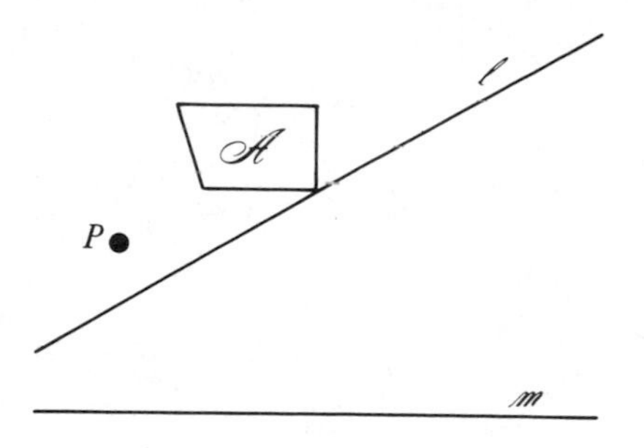

Solution

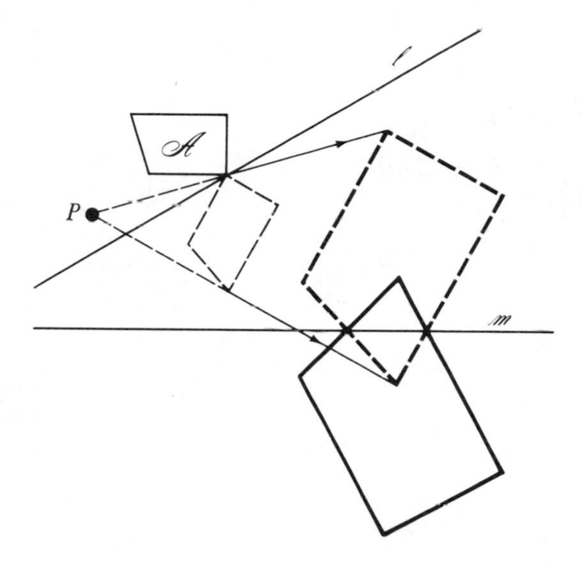

The following pair of definitions should now seem reasonable:

Any composition of flips and dilations is called a **similarity mapping** of the plane.
Two figures $\mathscr{A}$ and $\mathscr{B}$ in a plane are said to be **similar** if and only if $\mathscr{B}$ is the image of $\mathscr{A}$ under some similarity mapping.

EXERCISES

1. **(a)** Is every congruence mapping also a similarity mapping?
   (b) Is every similarity mapping also a congruence mapping?
2. For each figure below: (i) Sketch the image of $\mathscr{A}$ under the given similarity mapping, (ii) decide if the mapping is in fact a congruence mapping.

(a) mapping: f_ℓ

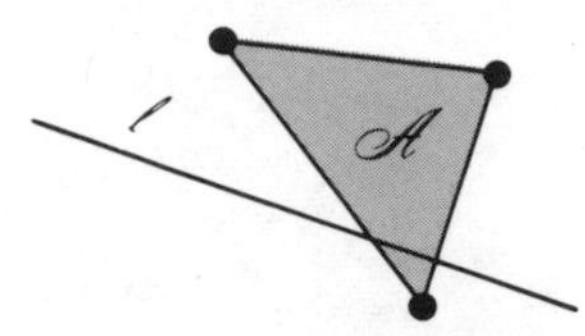

(b) mapping: first f_ℓ, then f_m

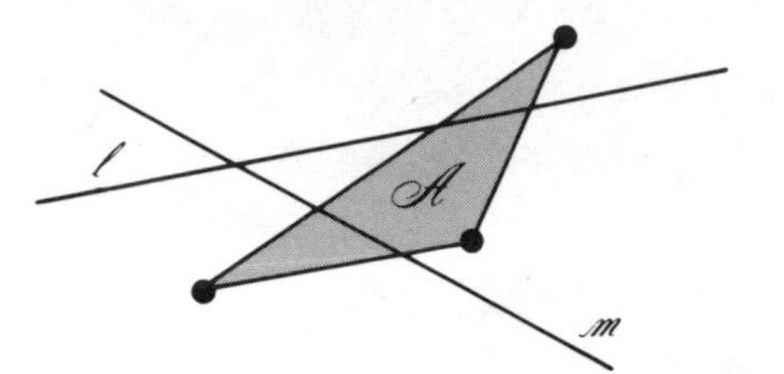

(c) mapping: first f_m, then f_ℓ for the figure in (b)

(d) mapping: first f_ℓ, then f_m

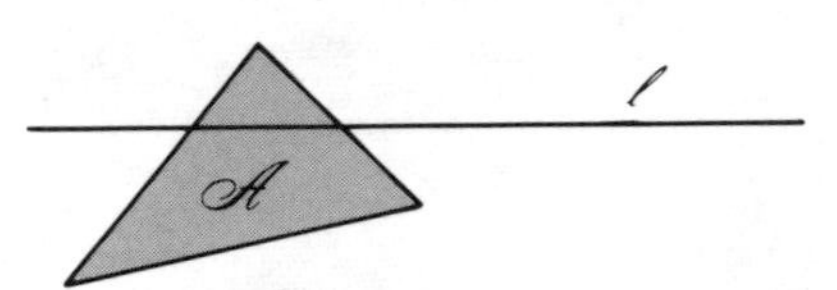

(e) mapping: first f_m, then f_ℓ for the figure in (d)

(f) mapping: $d_{P,3/2}$

(g) mapping: $d_{P,1/2}$ for the figure in (f)

(h) mapping: first $d_{P,2}$, then f_ℓ

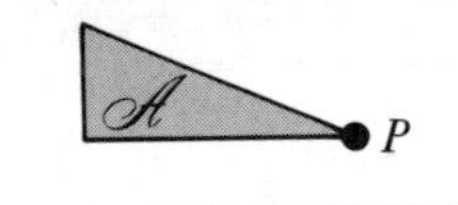

(i) mapping: first f_ℓ, then $d_{P,2}$ for the figure in (h)

(j) mapping: first $d_{P,2}$, then f_ℓ, then $d_{P,1/2}$

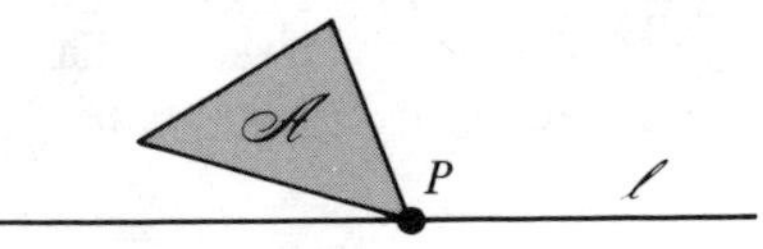

3. In the diagram below you are given a point P and its image $f_\ell(P)$ under a flip across a line ℓ. Construct the line ℓ.

P

$f_\ell(P)$

4. Is $\overline{CD}$ the image of $\overline{AB}$ under a flip across some line ℓ?

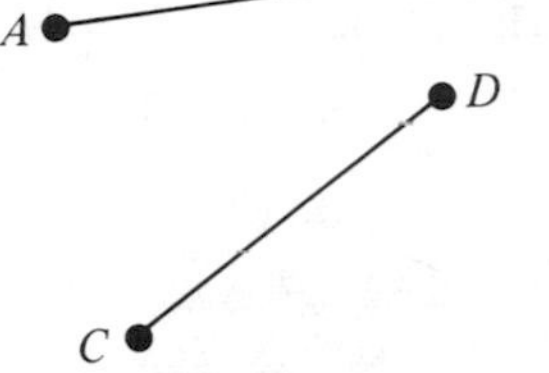

5. Locate lines ℓ_1 and ℓ_2 so that $\mathscr{B}$ is the image of $\mathscr{A}$ under the composition, first f_{ℓ_1} then f_{ℓ_2}.

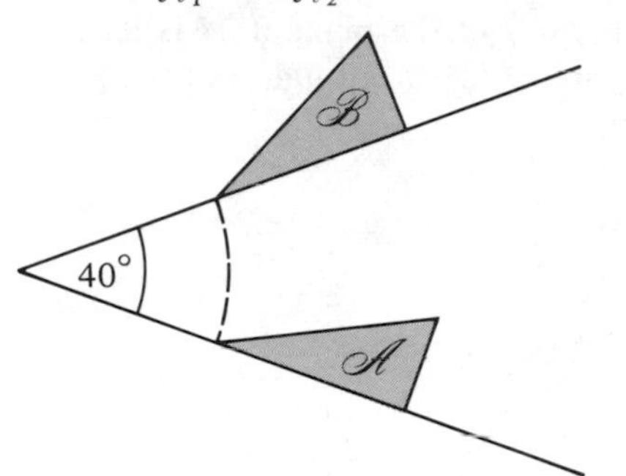

6. Locate lines ℓ_1 and ℓ_2 so that $\mathscr{B}$ is the image of $\mathscr{A}$ under the composition, first f_{ℓ_1} then f_{ℓ_2}.

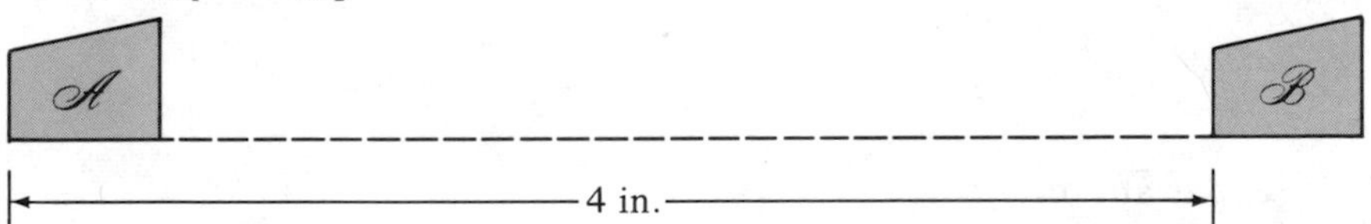

7. Find P and r (approximately) such that $\mathscr{B}$ is the image of $\mathscr{A}$ under $d_{P,r}$.

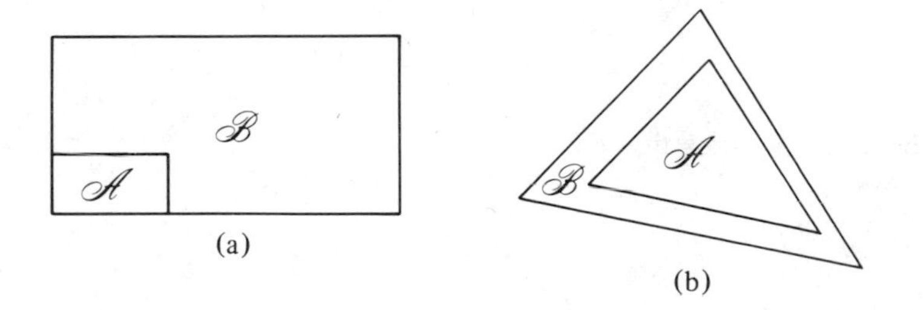

(a)

(b)

(c)

8. For what range of values of r would it be reasonable to refer to $d_{P,r}$ as

(a) a stretch

(b) a shrink

What would you call $d_{P,1}$?

9. A point M is a **fixed point** of a mapping f of the plane if $f(M) = M$. Describe all fixed points of the following mappings.

(a) clockwise rotation about a point P through 30°

(b) a slide right by 2 units

(c) f_ℓ

(d) f_ℓ and then f_ℓ again

(e) $d_{P,1/10}$

(f) $d_{P,1}$

10. A figure $\mathscr{A}$ is a **fixed figure** of a mapping f of the plane if $\mathscr{A}$ is its own image under f. Decide if the given figure $\mathscr{A}$ is a fixed figure of the given mapping f.

(a) f is clockwise rotation about P through 60°

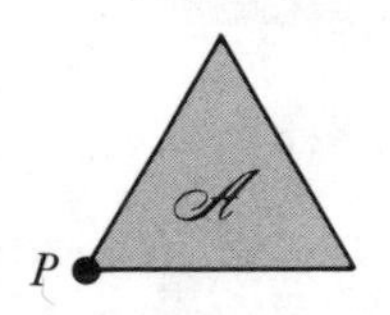

(b) f as above

(c) f as above

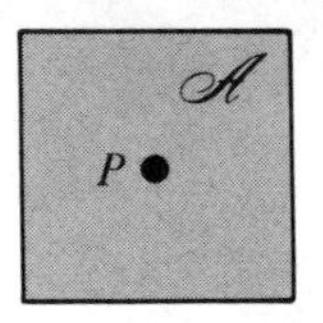

(d) f as above, $\mathscr{A}$ the stick figure

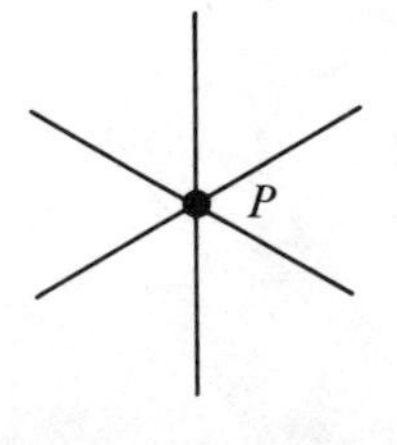

(e) f is still a clockwise rotation about P through 60°

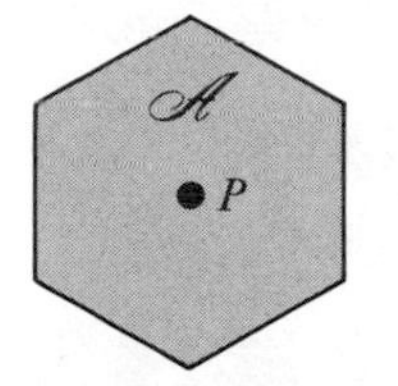

(f) f is a slide 2 in. to the right

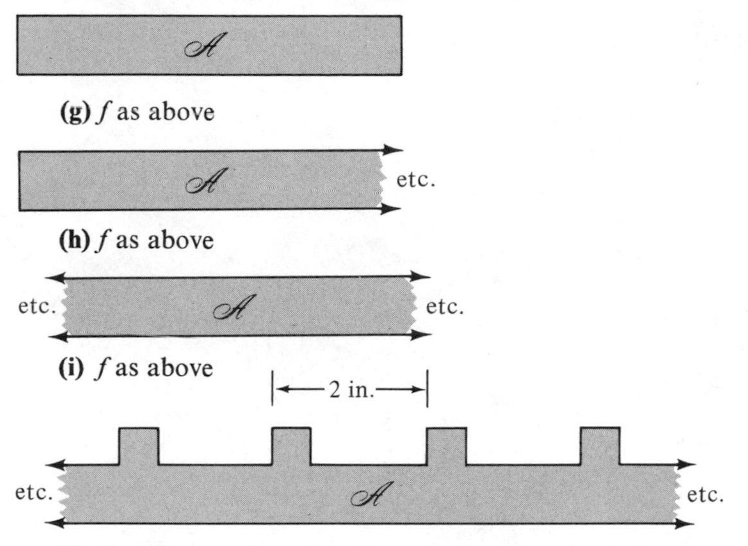

(g) f as above

(h) f as above

(i) f as above

(j) f as above; $\mathscr{A}$ as above except that the protuberances are 1, not 2 in. apart.
(k) f as above; $\mathscr{A}$ as above except that the protuberances are 3 in. apart.
(l) f is f_ℓ

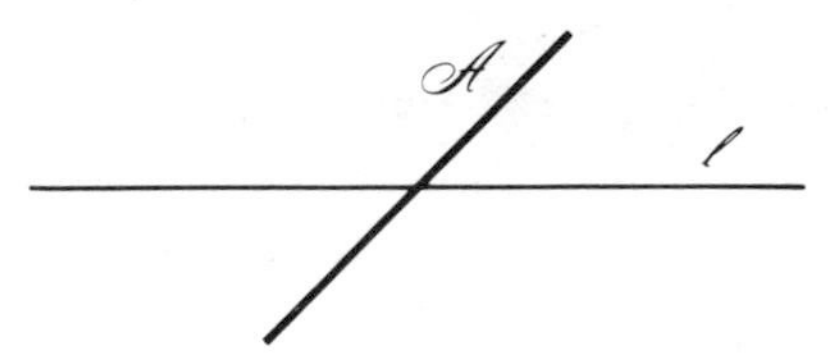

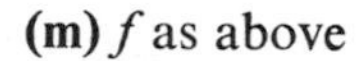

(m) f as above

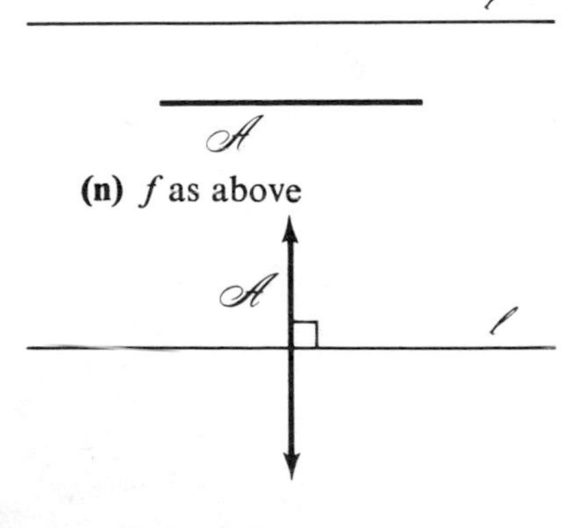

(n) f as above

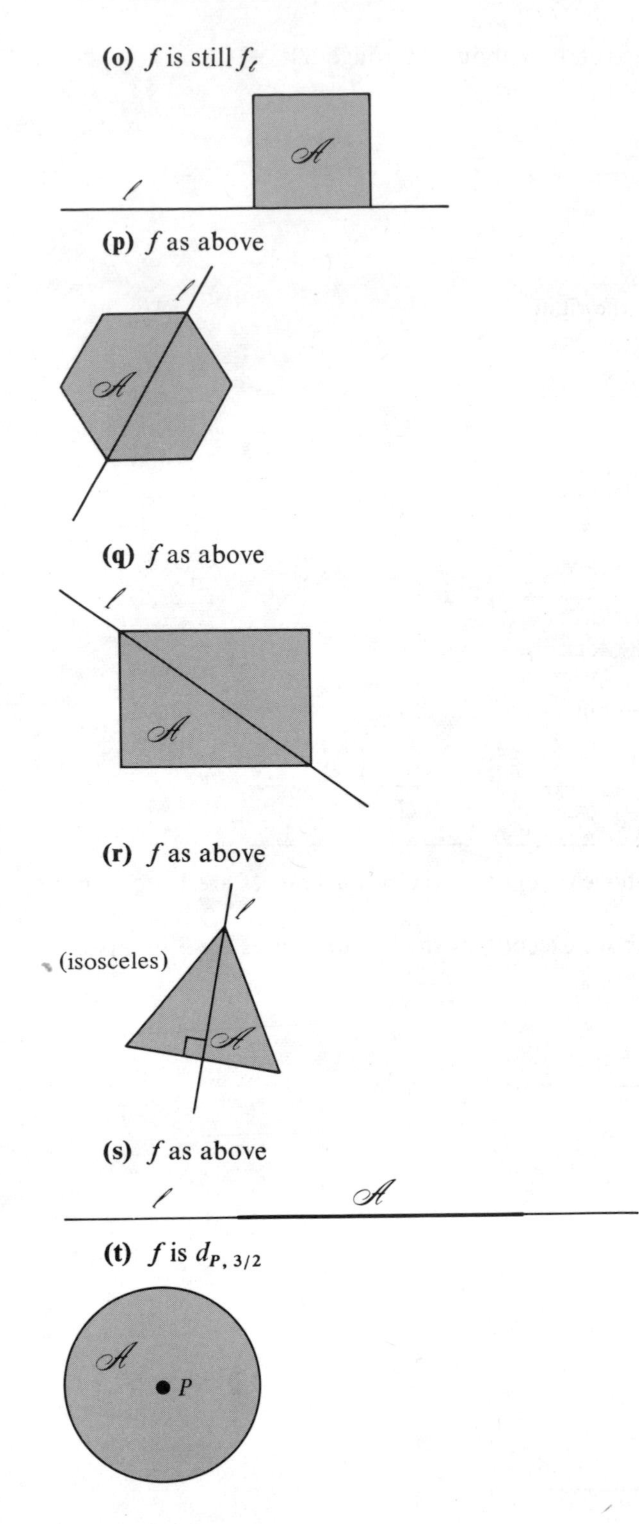
(o) f is still f_ℓ
ℓ
$\mathscr{A}$
(p) f as above
ℓ
$\mathscr{A}$
(q) f as above
ℓ
$\mathscr{A}$
(r) f as above
ℓ
(isosceles)
$\mathscr{A}$
(s) f as above
ℓ
$\mathscr{A}$
(t) f is $d_{P,\,3/2}$
$\mathscr{A}$
P

(u) f is still $d_{P,\,3/2}$

(v) f as above

(w) f as above

(x) f as above

(y) f as above

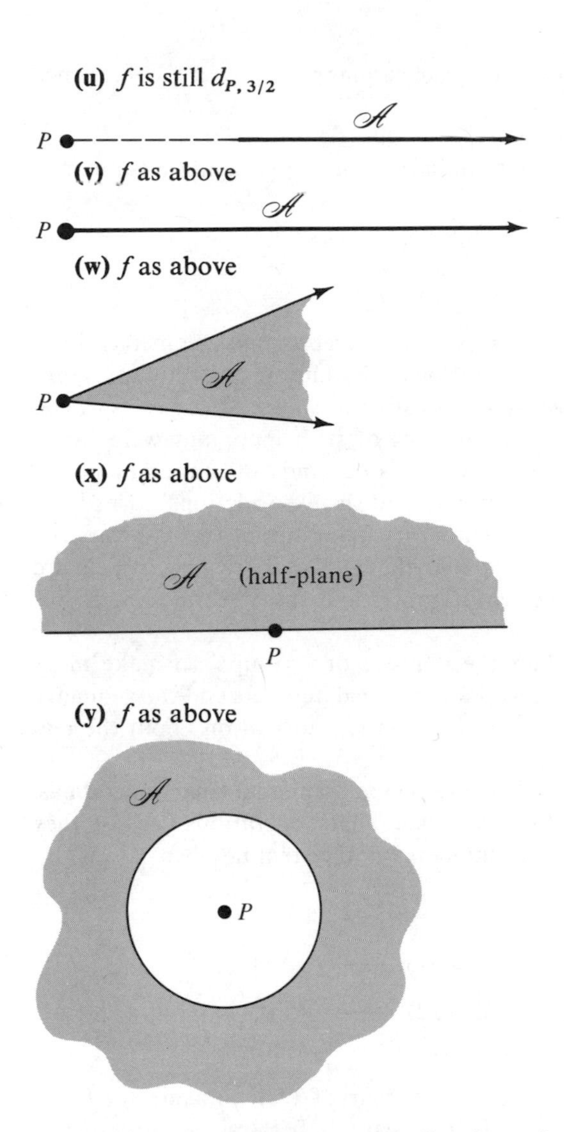

(z) f as above; $\mathscr{A}$ the set of all points of the plane except P

11. (a) Is the set of all fixed points of a mapping a fixed figure of the mapping?

(b) If $\mathscr{A}$ is a fixed figure of f and if $P \in \mathscr{A}$, must P be a fixed point of f? Give examples.

12. It turns out to be true that any congruence mapping can be expressed as the composition of three or less flips. Try to make $\triangle ABC$ below coincide with $\triangle A'B'C'$ using at most three flips. (*Hint*: Choose the first flip so that A is sent to A'.)

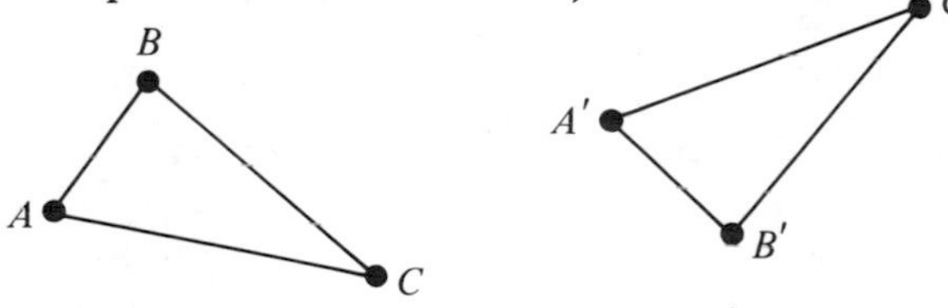

13. (a) Is the composition of two congruence mappings again a congruence mapping?

(b) Is it true that if $\mathscr{A} \cong \mathscr{B}$ and $\mathscr{B} \cong \mathscr{C}$, then $\mathscr{A} \cong \mathscr{C}$?

(c) Is the composition of two similarity mappings again a similarity mapping?

(d) Is it true that if $\mathscr{A} \sim \mathscr{B}$ and $\mathscr{B} \sim \mathscr{C}$, then $\mathscr{A} \sim \mathscr{C}$?

9.4 GROUPS; SYMMETRY

We have been exploring analogies between algebra and geometry. First we observed that the algebraic operations of addition and multiplication of numbers were representable geometrically by slides, stretches, shrinks, and flips of the number line. Compositions of such mappings were called linear functions. Next we investigated the same kinds of mappings of the plane. Compositions of flips with stretches and shrinks (dilations) turned out to be similarity mappings. Thus the similarity mappings of the plane are the exact analogs of the linear functions of the real number line, which are themselves closely related to the real numbers and the algebraic operations of addition and multiplication.

In this section we introduce the concept of "group" to make more explicit the analogy between the algebra of the real numbers and the geometry of the plane. We begin by reviewing some basic information about the real numbers.

On the set of real numbers there are two fundamental **binary operations**, addition and multiplication. They are called *bi*nary operations because they assign to each ordered *pair* of real numbers another real number.

$$(5, 3) \xrightarrow{+} 8 \qquad (5, 3) \xrightarrow{\cdot} 15$$
$$(4, -3) \xrightarrow{+} 1 \qquad (4, -3) \xrightarrow{\cdot} -12$$
$$(0, -12) \xrightarrow{+} -12 \qquad (0, -12) \xrightarrow{\cdot} 0$$
$$\cdots \qquad \cdots$$

The real number assigned to an arbitrary pair of real numbers (a, b) by the addition operation is called $a + b$. The real number assigned to (a, b) by the multiplication operation is called $a \cdot b$.

An obvious, but nevertheless fundamental, property of addition and multiplication is the **associative property**:

$$(a + b) + c = a + (b + c)$$
$$(a \cdot b) \cdot c = a \cdot (b \cdot c)$$

For example,

$$(5 + 6) + 4 = 5 + (6 + 4)$$
$$(5 \cdot 6) \cdot 4 = 5 \cdot (6 \cdot 4)$$

(Which sum do you find easier to compute, the one on the left or the one on the right? Which product?)

A second basic property of addition and multiplication is that each of these binary operations has a **neutral element**. The neutral element with respect to addition is 0, because adding 0 to any number does not change that number. For example,

$$0 + 13 = 13 \qquad -5 + 0 = -5 \qquad 0 + 1 = 1$$

The neutral element with respect to multiplication is 1, because multiplying any number by 1 does not change that number. For example,

$$1 \cdot 13 = 13 \qquad -5 \cdot 1 = -5 \qquad 0 \cdot 1 = 0$$

A further property of addition is that every real number has an **additive inverse**. The additive inverse of 5 is -5, because

$5 + -5 = 0$ the neutral element with respect to $+$

The additive inverse of -13 is 13, because

$-13 + 13 = 0$ the neutral element with respect to $+$

Zero is its own additive inverse.

Every *nonzero* real number has a **multiplicative inverse**. The multiplicative inverse of 5 is $\frac{1}{5}$, because

$5 \cdot \frac{1}{5} = 1$ the neutral element with respect to $\cdot$

The multiplicative inverse of $-\frac{13}{4}$ is $-\frac{4}{13}$, because

$(-\frac{13}{4}) \cdot (-\frac{4}{13}) = 1$ the neutral element with respect to $\cdot$

Zero has no multiplicative inverse. One is its own multiplicative inverse.

We say that the set of real numbers is a **group** with respect to addition to summarize the four properties:

1. *Closure* (The sum of two real numbers is a real number.) For all real numbers a and b, $a + b$ is also a real number.
2. *Associativity.* For all real numbers a, b, and c: $(a + b) + c = a + (b + c)$.
3. There is a *neutral element.* The real number 0 has the property that for any real number a, $a + 0 = a = 0 + a$.
4. Each element has an *inverse.* Corresponding to each real number a is another real number, called $-a$, such that $a + (-a) = 0 = -a + a$.

The set of all real numbers is *not* a group with respect to multiplication because, although properties 1, 2, and 3 are satisfied, property 4 is violated: The real number 0 does not have a multiplicative inverse. The set of all *nonzero* real numbers, however, is a group with respect to multiplication.

1. *Closure*: If a and b are nonzero real numbers, then so is $a \cdot b$.
2. *Associativity*: For all nonzero real numbers a, b, and c: $(a \cdot b) \cdot c = a \cdot (b \cdot c)$.
3. The *neutral element* 1 belongs to the set of nonzero real numbers.
4. Each nonzero number has a multiplicative *inverse* in the set of nonzero real numbers.

The set of all similarity mappings of the plane is also a group, but now the binary operation is composition. To establish this assertion, we consider first the question of closure. Observe that composition does assign to each ordered pair of similarity mappings another similarity mapping. For example,

$(d_{P,1/2}, f_\ell) \xrightarrow{\text{composition}}$ first $d_{P,1/2}$ then f_ℓ (a similarity mapping)

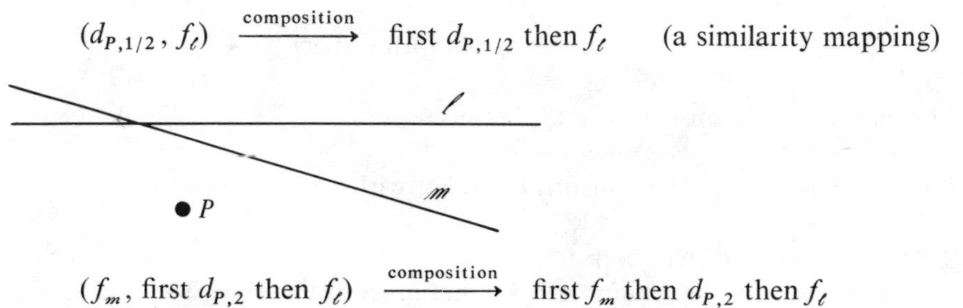

$(f_m,$ first $d_{P,2}$ then $f_\ell) \xrightarrow{\text{composition}}$ first f_m then $d_{P,2}$ then f_ℓ (a similarity mapping)

The similarity mapping assigned to an arbitrary pair of similarity mappings (f, g) by the composition operation is called $f \circ g$ (read this "first f, then g").*

The second thing to check in establishing that the similarity mappings are a group with respect to composition is associativity. The fact that composition is associative is very simple. Given three mappings f, g, h, both $(f \circ g) \circ h$ and $f \circ (g \circ h)$ mean "first f, then g, then h."

The third condition, that there be a similarity mapping that is neutral with respect to composition, is a little tougher. We are looking for a similarity mapping i such that for any similarity mapping f,

$$i \circ f = f \quad \text{and} \quad f \circ i = f$$

That is, preceding or following f by i should effect no change in the mapping

* We have chosen to express the composite function, first f then g, in the natural left-to-right symbolism, $f \circ g$. Thus if x is an input,

$$f \circ g(x) = g(f(x))$$

f. The mapping i with this property is the mapping that leaves every point fixed. We refer to this mapping as the **identity mapping**. The identity mapping is a similarity mapping, in fact a congruence mapping, because

$$i = f_\ell \circ f_\ell \qquad \text{for any line } \ell$$

The final condition for a group is that every similarity mapping have an inverse that is also a similarity mapping. We begin with the two simplest kinds of similarity mappings, flips and dilations. A single flip, f_ℓ is its own inverse by the observation just made:

$$f_\ell \circ f_\ell = i \qquad \text{the neutral mapping with respect to } \circ$$

A single dilation, $d_{P,r}$, has an inverse that is also a dilation, namely, $d_{P,1/r}$:

$$d_{P,r} \circ d_{P,1/r} = i$$

For example, a 2-stretch from P followed by a $\frac{1}{2}$-shrink toward P leaves every point fixed. A more general sort of similarity mapping—for example, $d_{P,3} \circ f_\ell \circ d_{Q,1/2} \circ f_m$, has as its inverse the similarity mapping $f_m \circ d_{Q,2} \circ f_\ell \circ d_{P,1/3}$ because

$$\begin{aligned}
& d_{P,3} \circ f_\ell \circ d_{Q,1/2} \circ \underbrace{f_m \circ f_m} \circ d_{Q,2} \circ f_\ell \circ d_{P,1/3} \\
&= d_{P,3} \circ f_\ell \circ d_{Q,1/2} \circ i \circ d_{Q,2} \circ f_\ell \circ d_{P,1/3} \\
&= d_{P,3} \circ f_\ell \circ \underbrace{d_{Q,1/2} \circ d_{Q,2}} \circ f_\ell \circ d_{P,1/3} \\
&= d_{P,3} \circ f_\ell \circ i \circ f_\ell \circ d_{P,1/3} \\
&= d_{P,3} \circ \underbrace{f_\ell \circ f_\ell} \circ d_{P,1/3} \\
&= d_{P,3} \circ i \circ d_{P,1/3} \\
&= \underbrace{d_{P,3} \circ d_{P,1/3}} \\
&= i
\end{aligned}$$

One merely "undoes" the simple flips and dilations, one by one, in the reverse order in which they were performed in the original similarity mapping.

EXERCISES

1. By checking through the four conditions, decide if the given subset of the real numbers is a group with respect to addition. If it is not a group, give one condition that is violated.
(a) the set of whole numbers, $\{0, 1, 2, 3, \ldots\}$
(b) the set of integers, $\{\ldots -3, -2, -1, 0, 1, 2, 3, \ldots\}$
(c) $\{-1, 0, 1\}$

(d) the set of all rational numbers
(e) the set of all even integers
(f) the set of all odd integers
(g) the set of all integral multiples of 7

2. By checking the four conditions, decide if the given subset of the real numbers is a group with respect to multiplication. If it is not a group, give one condition that is violated.
(a) the set of all integers
(b) the set of positive rational numbers
(c) the set of all rational numbers
(d) $\{-1, 0, 1\}$
(e) $\{-1, 1\}$
(f) the set of all integral powers of 7
(g) the set of all irrational numbers

3. A binary operation $*$ is defined as follows:

$$a * b = \frac{(a+b)}{2}$$

(a) Is the set of all real numbers closed with respect to $*$?
(b) Is $*$ associative?
(c) Is there a neutral element with respect to $*$?
(d) Does each real number have an inverse with respect to $*$?
(e) Is the set of all real numbers a group with respect to $*$?

4. Sketch the image of point A under each of the mappings below.

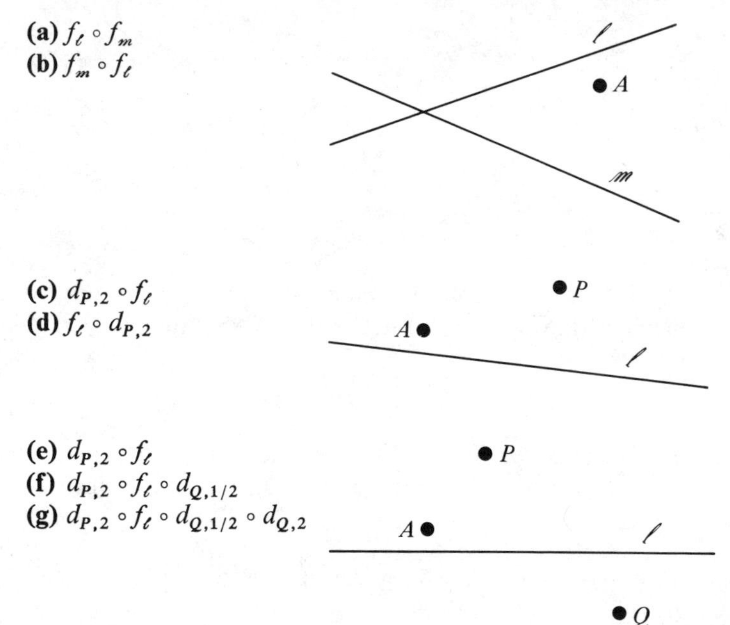

(h) $f_\ell \circ f_m$
(i) $f_m \circ f_\ell$

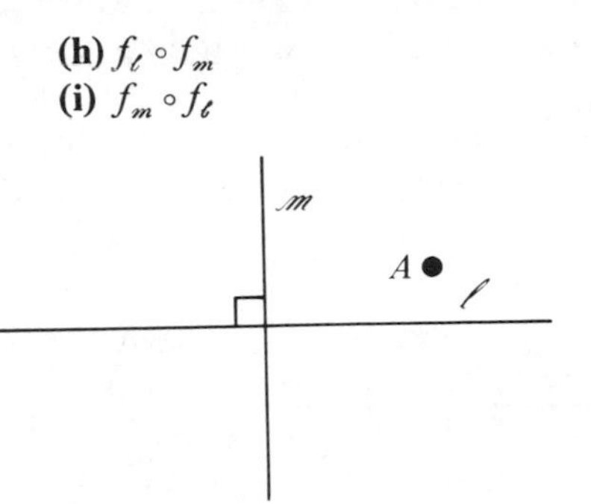

5. Classify each similarity mapping below as a congruence mapping or not a congruence mapping. Then give its inverse and classify that inverse in the same way.
(a) f_ℓ
(b) $d_{P,3}$
(c) $f_\ell \circ f_m$
(d) $d_{P,3} \circ f_\ell$
(e) $f_\ell \circ d_{P,2} \circ f_m$
(f) $f_\ell \circ f_m \circ f_n$
(g) clockwise rotation about P through 45°
(h) counterclockwise rotation about P through 110°
(i) clockwise rotation about Q through 180°
(j) slide right 2 in.
(k) slide up $3\frac{1}{2}$ in.
6. Decide if the set of all congruence mappings is a group with respect to composition. Be prepared to explain your answer.

In the remaining exercises we deal exclusively with congruence mappings.

7. Sketch the image of the rectangle $\mathscr{A}$ under each of the following congruence mappings:

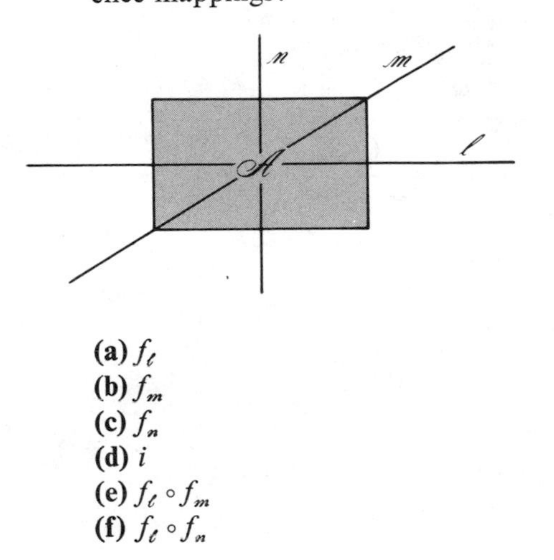

(a) f_ℓ
(b) f_m
(c) f_n
(d) i
(e) $f_\ell \circ f_m$
(f) $f_\ell \circ f_n$

8. From the above congruence mappings, pick out those that map $\mathscr{A}$ onto itself.

9. Any congruence mapping that maps a figure $\mathscr{A}$ onto itself is called a **symmetry** of $\mathscr{A}$. For each figure below, decide if f_ℓ is a symmetry of $\mathscr{A}$.

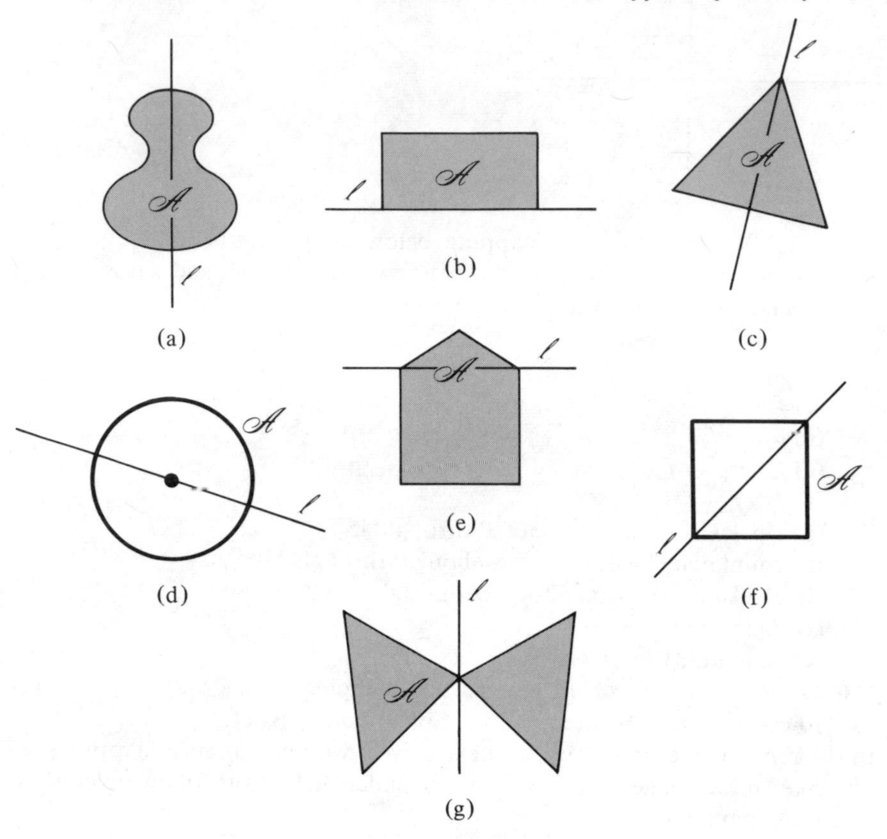

10. For each figure $\mathscr{A}$ below, find a "line of symmetry." That is, find a line ℓ such that f_ℓ is a symmetry of $\mathscr{A}$.

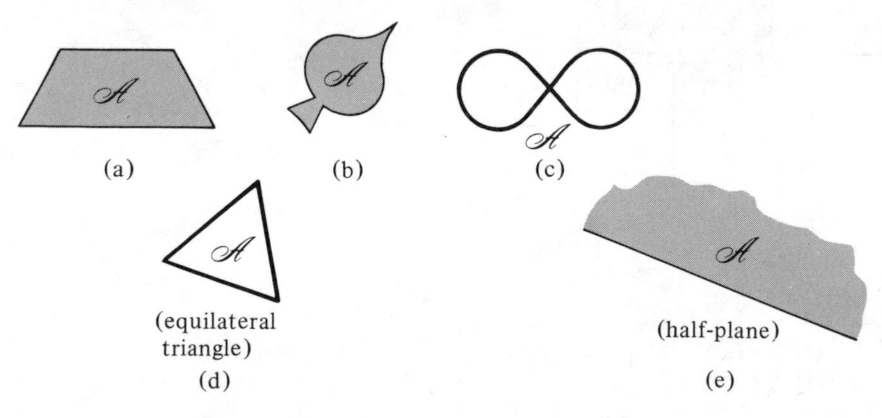

(f) $\mathscr{A}$ is the entire plane

(g) $\mathscr{A}$ is this two-point set:

• •

(h)

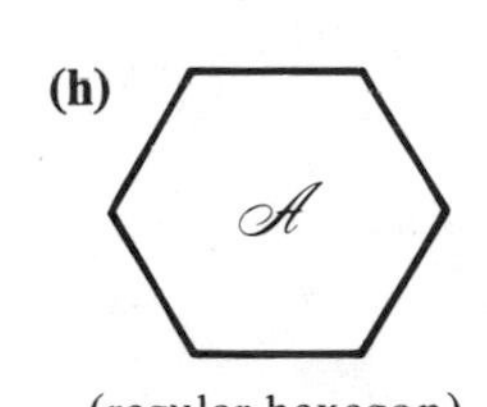

(regular hexagon)

11. Decide which of the figures in Exercise 10 have a rotation (other than a 360° rotation) among their symmetries. Describe the rotation.

12. $\mathscr{A}$ is a figure in the plane, for example, a square.

(a) If f is a symmetry of $\mathscr{A}$, must the inverse of f also be a symmetry of $\mathscr{A}$?

(b) If f and g are symmetries of $\mathscr{A}$, must $f \circ g$ be a symmetry of $\mathscr{A}$?

(c) Is the identity map i a symmetry of $\mathscr{A}$?

(d) Is the set of all symmetries of $\mathscr{A}$ a group (with respect to composition)?

13. Children spend much time in the first few grades working with a portion of the **group table** for the (infinite) group of integers with respect to addition. In the row headed by 3 and the column headed by 4 they are to enter $3 + 4$, that is, 7.

+	...	−2	−1	0	1	2	3	**4**	5	...
⋮										
−2										
−1										
0				0	1	2	3	4	5	
1				1	2	3	4	5	6	
2				2	3	4	5	6	7	
3				3	4	5	6	**7**	8	
4				4	5	6	7	8	9	
5				5	6	7	8	9	10	
⋮										

Fill in some more boxes in this table.

14. Group tables can also be made for symmetry groups. The rectangle $\mathscr{A}$ has just four symmetries:

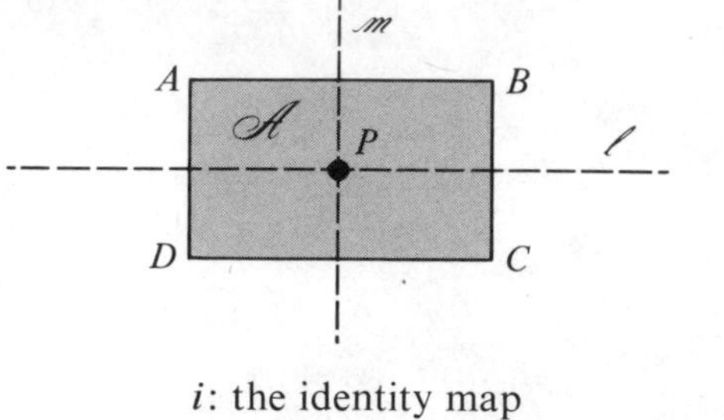

i: the identity map
f_ℓ: flip across ℓ
f_m: flip across m
$r_{P,180}$: rotation (clockwise) about P through 180°

Fill in the missing entries in the following group table for the **group of symmetries** of $\mathscr{A}$. To see that $f_m \circ f_\ell = r_{P,180}$ observe that

$$A \xrightarrow{f_m} B \xrightarrow{f_\ell} C \qquad A \xrightarrow{r_{P,180}} C$$

$$B \xrightarrow{f_m} A \xrightarrow{f_\ell} D \qquad B \xrightarrow{r_{P,180}} D$$

$$C \xrightarrow{f_m} D \xrightarrow{f_\ell} A \qquad C \xrightarrow{r_{P,180}} A$$

$$D \xrightarrow{f_m} C \xrightarrow{f_\ell} B \qquad D \xrightarrow{r_{P,180}} B$$

$\circ$	i	f_ℓ	f_m	$r_{P,180}$
i			f_m	
f_ℓ				
f_m		$r_{P,180}$		
$r_{P,180}$				i

***15.** The equilateral triangle $\mathscr{A}$ has six symmetries:

$$i, f_\ell, f_m, f_n, r_{P,120}, r_{P,240}$$

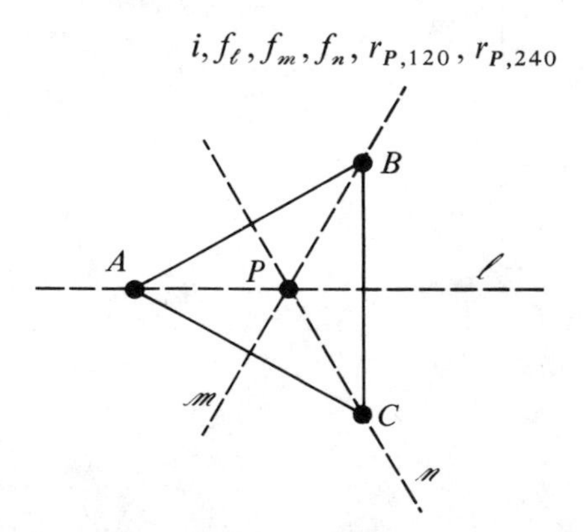

Make a group table for the group of symmetries of $\mathscr{A}$.

***16.** Decide if the set of all linear functions,

$$\{ax + b \mid a \in R, b \in R, a \neq 0\}$$

is a group with respect to composition. *Hint*: Let $f_1(x) = a_1x + b_1$ and $f_2(x) = a_2x + b_2$. Now calculate a formula for

$$f_1 \circ f_2(x) = f_2(f_1(x))$$

17. Which of the following sets of transformations are groups with respect to composition?
(a) all translations
(b) all rotations
(c) all reflections
(d) all rotations with center P
(e) all dilations with center P
(f) the identity mapping all by itself

9.5 MAPPINGS OF THE EUCLIDEAN PLANE

When we worked with mappings from a line to itself, in Section 9.2, we had the advantage of working with a "number line." The fact that each point on the line had a numerical address allowed us to describe, *with formulas*, the various linear mappings.

-2 -1 0 1 2

EXAMPLE

"2-slide right": $f(x) = x + 2$
"flip across origin": $f(x) = -x$
"$\frac{1}{3}$-shrink toward origin": $f(x) = \frac{1}{3}x$

In the last two sections, when we worked with mappings of the plane, each point did not have a numerical address. We had to resort to verbal descriptions of our mappings: f_ℓ, i, $d_{P,r}$, and so on. In this section we remedy that shortcoming.

Where before we "coordinatized" the *line* by matching each point with *a* real number (in such a way that absolute values of number differences corresponded to distances between points), now we coordinatize the *plane* by matching each point with an *ordered pair* of real numbers. This assignment of points to ordered pairs of real numbers (a function) requires fairly lengthy description.

To begin with, a line is drawn, usually horizontally, in the plane.

This line is called the **x axis** (or first axis) and is coordinatized with real numbers in the usual way.

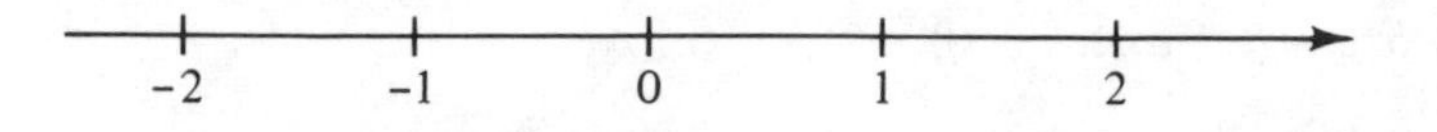

(The arrow on the x axis indicates the direction in which the numbers increase.) Then another line, called the **y axis** (or second axis), is drawn, perpendicular to the x axis, and coordinatized as shown.

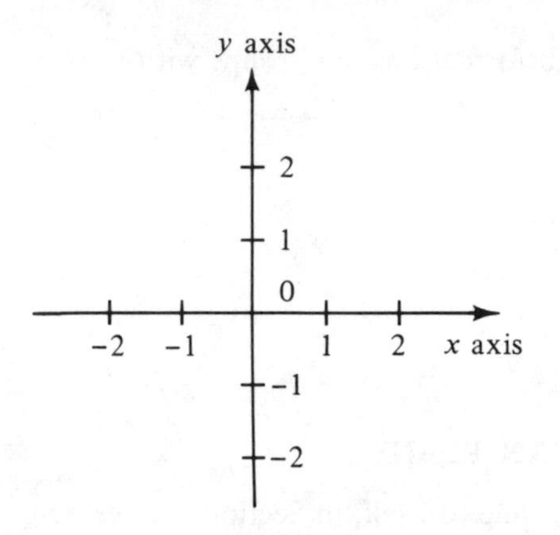

Both axes use the same unit of length. The point of intersection of the two axes is the origin for both and is called the **origin** of the plane.

Now each point in the plane is assigned an ordered pair of real numbers, which you can think of as that point's address. For example, to locate the point with address (3, 2), drive along the x axis until you reach third street, then drive up third street until you reach second avenue.

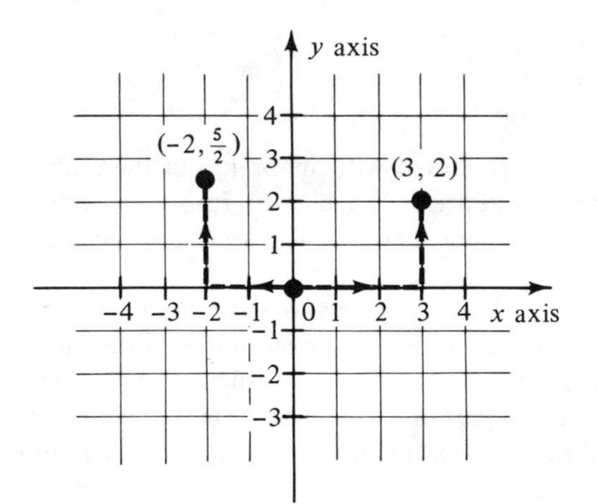

To locate the point corresponding to $(-2, \frac{5}{2})$, drive along the x axis until you get to -2 street, then drive along -2 street until you get to $\frac{5}{2}$ avenue. Mathematicians usually refer to a point's coordinates rather than its address. For example, the point with coordinates $(\frac{4}{3}, -2)$ is $\frac{4}{3}$ units to the right of and 2 units below the origin. The point with coordinates $(-2.6, -1.7)$ is 2.6 units to the left of and 1.7 units below the origin. We say that -2.6 is the **first coordinate** and -1.7 the **second coordinate** of this point.

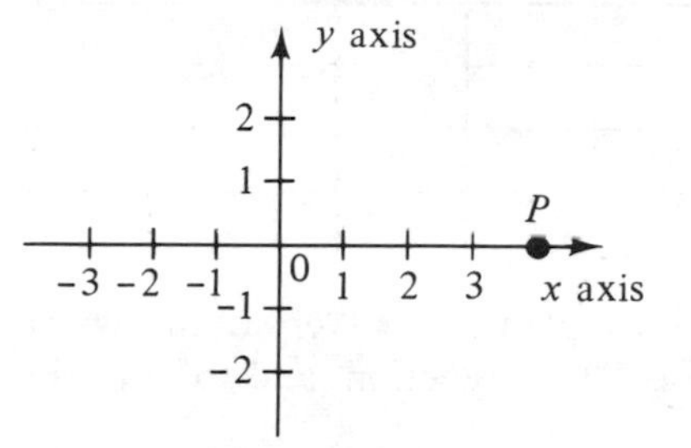

There is one possible source of confusion. Certain points, such as the point P in the figure, qualify for two addresses. As a resident of the x axis, P has address (coordinate) 3. As a resident of the plane, the address (coordinates) of P is $(3, 0)$. The fact that more coordinates are needed to locate a point in the plane than on a line is not surprising. A similar situation is this. As a resident of Lincoln School, Joe has address, room 228. At the same time Joe is a member of a larger community, the community of all students in his city. To specify his location in this larger community, he has to give both school and room number. His address in the larger community is (room 228, Lincoln School).

When ordered pairs of real numbers have been assigned to the points of a plane in the manner just described, it would seem natural to refer to the plane as "the number plane." It is customary, however, to call it instead the **Euclidean plane** (or the cartesian plane).

It is possible to think of mappings of the Euclidean plane in the same way that we thought of mappings of the number line. That is, we can imagine two copies of the Euclidean plane, a transparent dark one on top of a light one. The light copy never budges. The dark one can. For example, a slide T of the dark copy 2 units right and 1 unit up sends

$$\text{dark } (0, 0) \xrightarrow{T} (2, 1) \text{ light}$$

$$\text{dark } (3, 4) \xrightarrow{T} (5, 5) \text{ light}$$

$$\text{dark } (-2, 2) \xrightarrow{T} (0, 3) \text{ light}$$

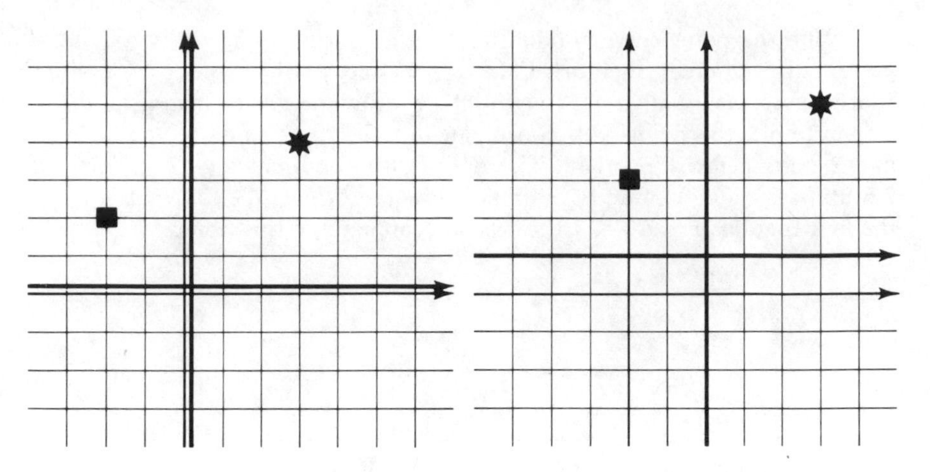

Drawing two copies of the Euclidean plane, however, is messy. It is easier to draw just one, to think of slide T as a function, and to draw an arrow diagram for it.

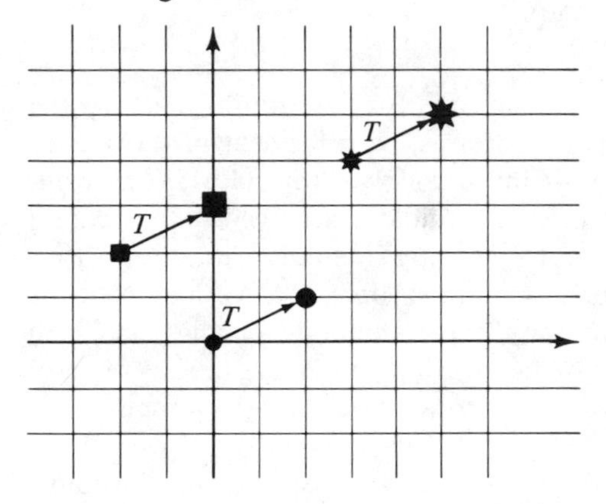

But even this is awkward. If we were to try to draw an arrow from each input to the corresponding output, our Euclidean plane would soon be hidden behind a black curtain of arrows. What we do instead is draw a single arrow (or vector), which stands for the infinitely many arrows that would make up the (undrawable) arrow diagram for T.

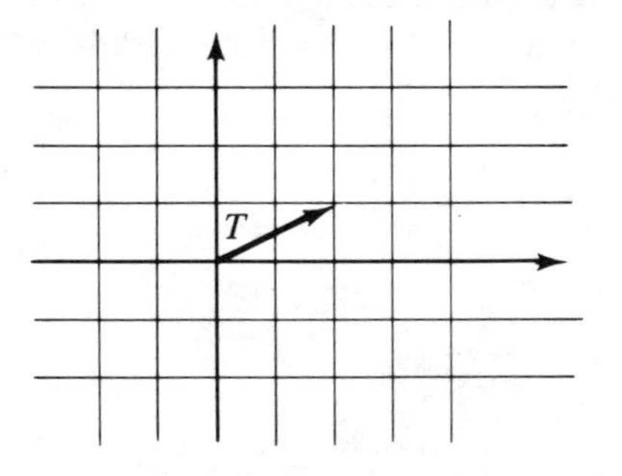

This is a neat geometric description of T, but an even more succinct algebraic description is possible now that the plane has been coordinatized. We can write a *formula* for T that shows how the coordinates of inputs and outputs are related. By studying the data,

$$T(0, 0) = (2, 1) \qquad T(3, 4) = (5, 5) \qquad T(-2, 2) = (0, 3)$$

we see that the effect of T is to increase the first coordinate by 2 and the second coordinate by 1. Thus the general formula is

$$T(x, y) = (x + 2, y + 1) \qquad \text{(for all values of } x \text{ and } y\text{)}$$

EXERCISES

1. On the Euclidean plane given, locate the points A through J with the given coordinates.

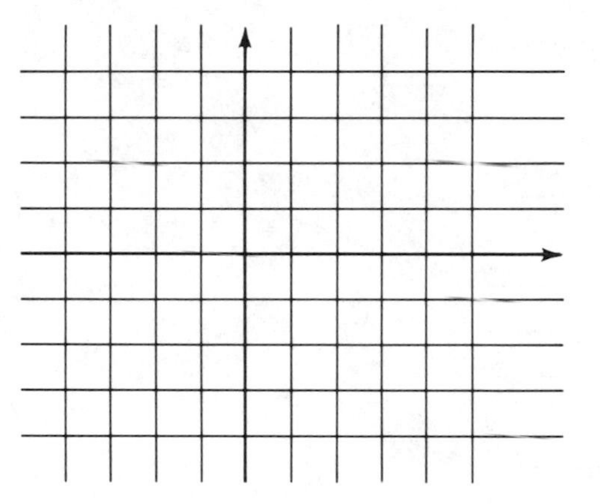

A: (2, 4) B: (4, 2) C: (3, −2) D: $(\frac{3}{2}, -1)$
E: $(-2, \frac{5}{3})$ F: $(-\frac{7}{2}, 0)$ G: (−3, −3) H: (−4, 1)
I: (1, −4) J: (0, −2)

2. Give coordinates of points A through L:

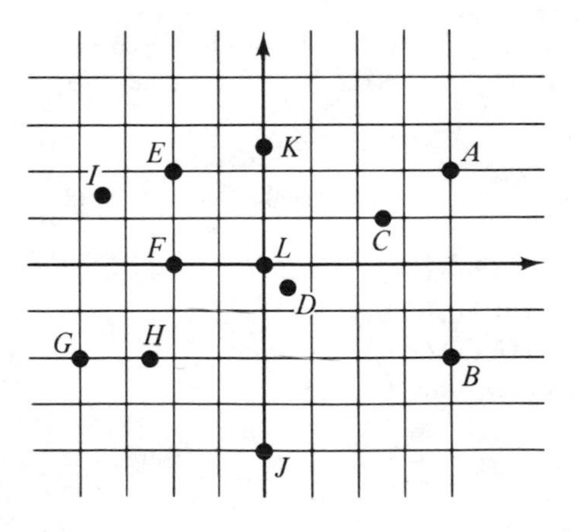

3. Sketch Euclidean planes and shade in the set of all points with
(a) second coordinate 0
(b) first coordinate 2
(c) second coordinate equal to first coordinate
(d) first coordinate positive
(e) first coordinate negative and second coordinate positive
(f) both coordinates between 0 and 1
(g) coordinates that add up to 3

4. Give coordinates of the midpoint of the segment joining the points
(a) $(2, 1)$ and $(4, 3)$
(b) $(1, 2)$ and $(-3, -2)$
(c) $(-1, -2)$ and $(3, 0)$
(d) $(-1, -3)$ and $(1, 2)$
(e) $(-2, 1)$ and $(4, 2)$
(f) (a, b) and (c, d)

5. A slide S is suggested by the figure below. Find the following and sketch appropriate input-output arrows.

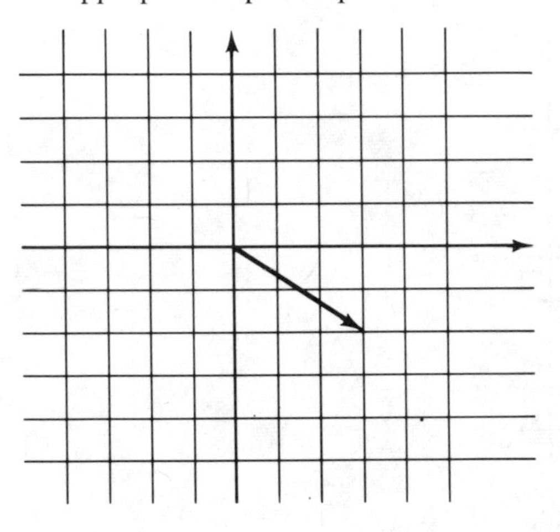

(a) $S(0, 0)$
(b) $S(1, 3)$
(c) $S(-2, 3)$
(d) $S(-3, 0)$
(e) $S(0, -1)$
(f) $S(x, y)$
(g) (a, b) so that $S(a, b) = (0, 0)$

6. On a coordinatized plane, draw a vector to describe the slide T for which $T(0, 0) = (-1, 4)$. Find the following and sketch appropriate input-output arrows:
(a) $T(1, -3)$ **(b)** $T(-2, 0)$ **(c)** $T(3, 2)$ **(d)** $T(x, y)$

7. S and T are the slides of Exercises 5 and 6. Find the following and sketch appropriate input-output arrows.
(a) $S(T(0, 0))$
(b) $S(T(-3, -2))$
(c) $S(T(x, y))$
(d) $T(S(0, 0))$
(e) $T(S(-3, -2))$
(f) $T(S(x, y))$

8. S and T are still the slides of Exercises 5 and 6.
(a) Is $S \circ T$ a slide?
(b) Is $T \circ S$ a slide?

(c) Does $S \circ T = T \circ S$?
(d) On a coordinatized plane, sketch vectors, all emanating from the origin, for the three slides S, T, and $S \circ T$. Do you see any geometric relationship among the vectors? (*Hint*: Look for a parallelogram.)

9. (a) Is the composition of any two slides again a slide?
(b) Is composition a "commutative" operation on the set of all slides?

10. Let M (for magnification) be the dilation with center at the origin and scale 2. Find the following and sketch appropriate input-output arrows on a coordinatized plane.
(a) $M(3, 0)$ (b) $M(1, 2)$
(c) $M(-1, 1)$ (d) $M(-\frac{5}{2}, -1)$
(e) $M(0, -2)$
(f) The general formula is $M(x, y) =$ ________

11. Let C (for contraction) be the dilation with center at the origin and scale $\frac{1}{3}$. Find the following and sketch appropriate input-output arrows on a coordinatized plane.
(a) $C(3, 0)$ (b) $C(3, 6)$
(c) $C(-1, 3)$ (d) $C(0, -4)$
(e) $C(3, -1)$
(f) The general formula is $C(x, y) =$ ________

12. M and C are the dilations of Exercises 10 and 11. Find:
(a) $C(M(2, 3))$ (b) $C(M(-3, -1))$
(c) $C(M(x, y))$ (d) $M(C(2, 3))$
(e) $M(C(-3, -1))$ (f) $M(C(x, y))$

13. M and C are still the dilations of Exercises 10 and 11.
(a) Is $M \circ C$ a dilation?
(b) Is $C \circ M$ a dilation?
(c) Does $M \circ C = C \circ M$?
(d) Is the composition of any two dilations with center at the origin again a dilation with center at the origin?
(e) Is composition a commutative operation on the set of all dilations with center at the origin?

14. Let F be the flip-across-the-y-axis mapping. Find the following and sketch appropriate input-output arrows on a coordinatized plane.
(a) $F(2, 4)$ (b) $F(-1, 2)$ (c) $F(-2, 0)$ (d) $F(-\frac{1}{2}, -\frac{3}{2})$
(e) The general formula is $F(x, y) =$ ________

15. Let G be the flip-across-the-vertical-line-through-(3, 0) mapping. Find the following and sketch appropriate input-output arrows on a coordinatized plane.
(a) $G(1, 2)$ (b) $G(2, -4)$ (c) $G(4, -3)$ (d) $G(\frac{7}{2}, 1)$
(e) The general formula is $G(x, y) =$ ________
(*Hint*: $(3, y)$ is the midpoint of the segment joining (x, y) to $G(x, y)$.)

16. F and G are the flips of Exercises 14 and 15. Find
(a) $G(F(1, 2))$ (b) $G(F(-1, -3))$
(c) $G(F(-3, 0))$ (d) $G(F(x, y))$

(e) $F(G(1, 2))$ **(f)** $F(G(-1, -3))$
(g) $F(G(-3, 0))$ **(h)** $F(G(x, y))$

17. F and G are the flips of Exercises 14 and 15.
(a) What kind of mapping is $F \circ G$?
(b) What kind of mapping is $G \circ F$?
(c) Does $F \circ G = G \circ F$?

(Exercises 15, 16, and 17 suggest an algebraic proof of the fact that any slide can be expressed as a composition of two flips. The procedure is this. A slide is described by a vector in the plane. Coordinatize the plane so that (1) the vector points along the x axis in the direction of increasing x coordinates, (2) the unit of length is one-sixth the length of the vector. Then the given slide is equal to $F \circ G$, where F and G are the flips of Exercises 14 and 15.)

18. Let H be the flip-across-the-x-axis mapping and F again be the flip-across-the-y-axis mapping.
(a) $H(F(1, 2)) =$ ________
(b) $H(F(-3, -1)) =$ ________
(c) $H(F(4, 0)) =$ ________
(d) What kind of mapping is $F \circ H$?

19. For each of the following mappings, A through F, sketch input-output arrows until you can decide what kind of mapping it is (slide, magnification, etc.).
(a) $A(x, y) = (x - 3, y + 2)$ **(b)** $B(x, y) = (\frac{1}{2}x, \frac{1}{2}y)$
(c) $C(x, y) = (-x, y)$ **(d)** $D(x, y) = (2 + x, y - 2)$
(e) $E(x, y) = (y, x)$ **(f)** $F(x, y) = (-y, x)$

NOTE In more advanced courses, where linear transformations and matrices are studied, one learns how to write explicit algebraic formulas for all sorts of mappings, not just the few special ones we have considered here. For example, a counterclockwise turn through 17° about the point (3, 2), call it J, is given by the formula

$$J(x, y) = ([\cos 17°]x - [\sin 17°]y + [3 + 2 \sin 17° - 3 \cos 17°],$$
$$[\sin 17°]x + [\cos 17°]y + [2 - 3 \sin 17° - 2 \cos 17°])$$

20. Use the Pythagorean theorem to find the distance between the following pairs of points. (First locate each pair of points in the Euclidean plane.)
(a) (0, 0) and (3, 4) **(b)** (1, 1) and (3, 2)
(c) (1, −2) and (3, 3) **(d)** (2, −1) and (0, −4)
(e) (−2, −4) and (−4, 1) **(f)** (−4, −1) and (2, 3)
(g) (2, 1) and (x, y) **(h)** (x_1, y_1) and (x_2, y_2)

21. It is geometrically obvious that a flip "preserves distance." That is, the distance from $f_\ell(A)$ to $f_\ell(B)$ is the same as the distance from A to B,

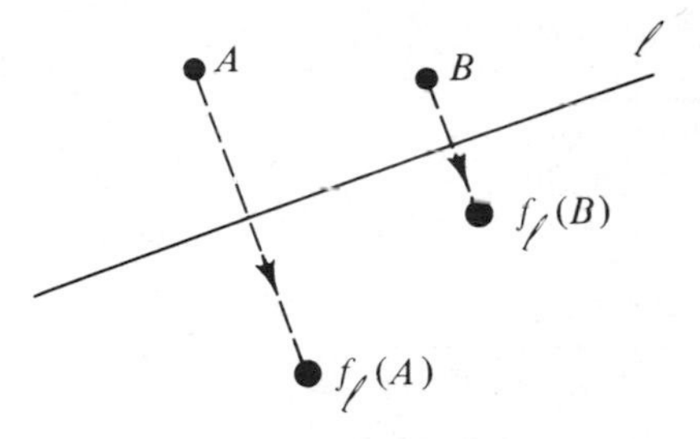

for any pair of points A and B. We can also establish this fact algebraically. We first coordinatize the plane so that ℓ is the x axis. Thus f_ℓ can be described by a formula,

$$f_\ell(x, y) = (x, -y)$$

Next we assign coordinates to A and B, say (x_A, y_A) and (x_B, y_B), respectively.

(a) Use the distance formula you derived in Exercise 20(h) to write an expression for the distance from A to B.

(b) Find the coordinates of $f_\ell(A)$ and $f_\ell(B)$ and then write an expression for the distance from $f_\ell(A)$ to $f_\ell(B)$.

(c) Show that the distances in (a) and (b) are the same.

NOTE In the previous exercise we proved that flips preserve distance. And, of course, it follows immediately that any composition of flips will also preserve distance. Thus all congruence mappings preserve distance, since—by our definition—a congruence mapping is some composition of flips. What is more remarkable is that congruence mappings are the *only* distance-preserving mappings. [A proof of this fact is fairly involved and we shall not go into it.] In some approaches to geometry, congruence mappings (rigid motions) are defined to be those that preserve distance. In others, as in ours, they are defined to be compositions of flips. Whichever approach is taken, it is still the same set of mappings that is called the set of congruence mappings.

9.6 GRAPHING IN THE EUCLIDEAN PLANE

We began our work on functions by looking at ones that assign numbers to numbers. Those of the form $f(x) = ax + b$ (with $a \neq 0$) were called linear functions. The (common) domain and range of such functions is the set R of all real numbers. By matching the real numbers with the points of a line (coordinatizing the line and calling it "the number line"), we were able to interpret the linear functions as compositions of slides, stretches, shrinks, and flips of a line. This led us to consider analogous mappings of a plane. In the last section we coordinatized the plane (and called it "the Euclidean

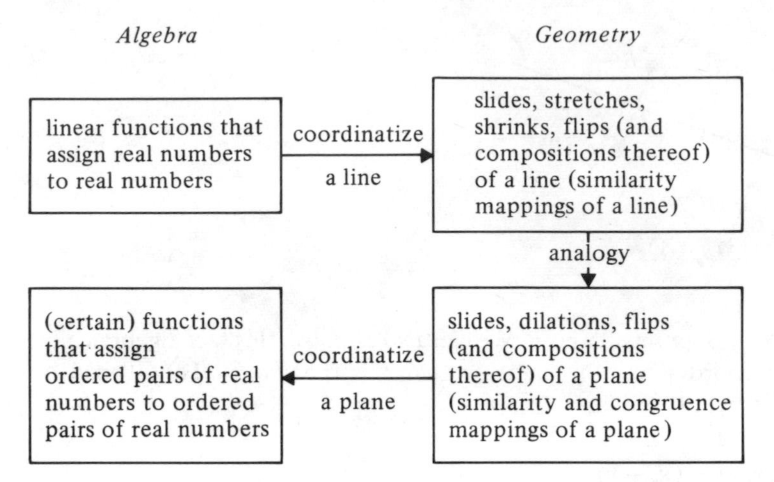

plane") so that these mappings could also be looked at algebraically. Schematically, our course in this chapter has been from algebra to geometry to algebra.

In this section we tie together a new and an old topic. We investigate the use of the coordinatized (Euclidean) plane as a place for representing the functions that got us into all of this in the first place: the functions from R to R. Of course, we shall be studying the ordinary kind of "graphing" that you did in your high school mathematics courses. We have two motives in taking up this topic here. One is that some of this graphing is already in elementary texts (and more is clearly on the way). The other is that the choice of the name "linear" for functions of the form $f(x) = ax + b$ (with $a \neq 0$) is finally vindicated by graphing in the Euclidean plane.

To illustrate what is meant by the **graph** of a function from R to R, consider the function f defined by $f(x) = 2x - 1$. The domain, or input set, of f is the set R of real numbers, which can, of course, be represented by the points on the x axis. The range, or output set, of f is also R, and is represented by the y axis. A partial arrow diagram for f is given below.

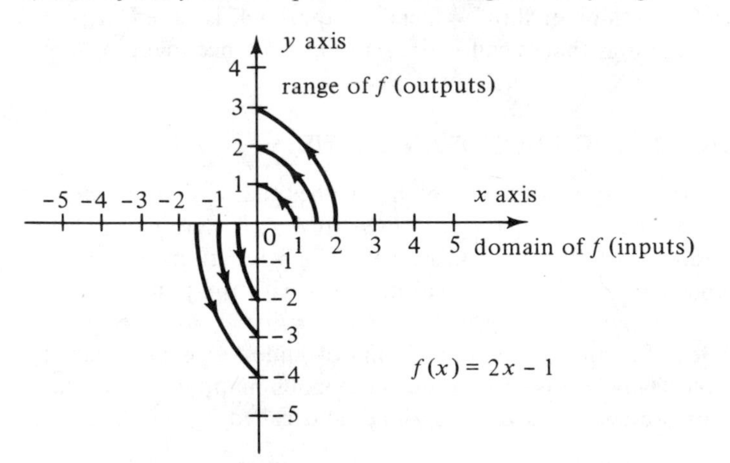

A somewhat neater partial arrow diagram is this:

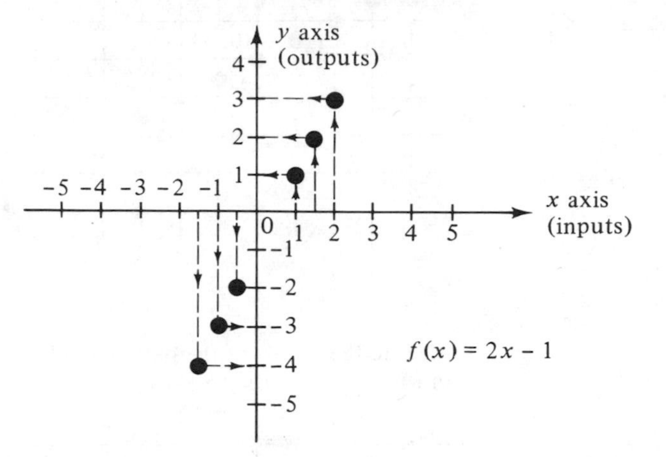

In practice, the arrows themselves are omitted from the arrow diagram. Only the turning points remain.

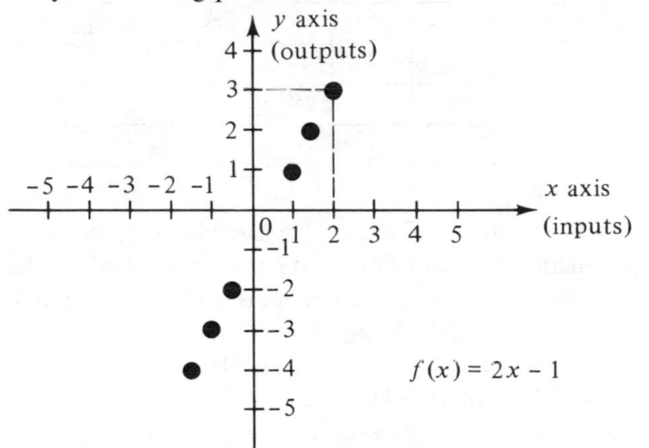

To draw such a graph then, place a heavy dot at the point where the vertical line through an input meets the horizontal line through the corresponding output. For example, the dark dot is placed where the vertical line through the input, 2, meets the horizontal line through the output, $f(2) = 3$.

To read such a graph, that is, to learn from the graph what output the function assigns to each input, begin on the x axis at the input in question, for example at -1, move vertically until you hit the graph, then move horizontally to the y axis. The point you reach, -3, is the output assigned to -1 by f.

We illustrate with a different function,

$$g(x) = 3 - \tfrac{1}{2}x$$

the customary four-step procedure for graphing a function from R to R.

1. Make a partial input-output table

x	$g(x)$
-3	$4\frac{1}{2}$
-2	4
-1	$3\frac{1}{2}$
0	3
1	$2\frac{1}{2}$
2	2
3	$1\frac{1}{2}$

→

2. Make a partial set of ordered pairs

$(-3, 4\frac{1}{2})$
$(-2, 4)$
$(-1, 3\frac{1}{2})$
$(0, 3)$
$(1, 2\frac{1}{2})$
$(2, 2)$
$(3, 1\frac{1}{2})$

→

3. Make a partial graph.

$(-3, 4\frac{1}{2})$ $(-2, 4)$ $(-1, 3\frac{1}{2})$ $(0, 3)$ $(1, 2\frac{1}{2})$ $(2, 2)$ $(3, 1\frac{1}{2})$

↓

4. Join the points of the partial graph in the "most natural" way.

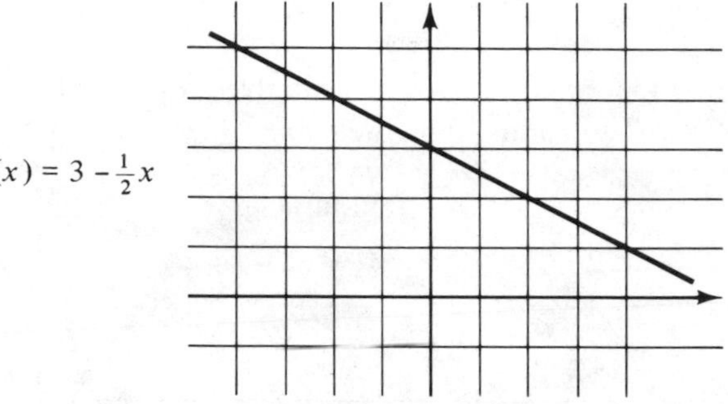

graph of $g(x) = 3 - \frac{1}{2}x$

NOTE In graphing a function (from R to R) in the Euclidean plane, it is very useful to think of the function as a set of ordered pairs. In most advanced mathematics courses, functions are *defined* to be special sets of ordered pairs. There are times, however, when the ordered-pair view is rather cumbersome. For example, one of the "ordered pairs" that makes up the slide $T(x, y) = (x + 2, y + 1)$ of the previous section is $((3, 6), (5, 7))$. At times like that, the mapping view of functions seems more natural.

EXERCISES

1. The function h is defined by the formula $h(x) = \frac{2}{3}x - 1$.

x	$h(x)$
-6	-5
-3	—
-1	—
0	—
1	—
—	0
3	—
6	—

(a) Fill in the blanks in the preceding partial input-output table.
(b) Fill in the blanks in this partial listing of points that belong to the graph of h: $(-6, -5)$, $(-3, \_\_)$, $(-1, \_\_)$, $(0, \_\_)$, $(\_\_, 0)$, $(\_\_, 3)$.
(c) Plot the points from part (b).
(d) Now sketch the complete graph of h:

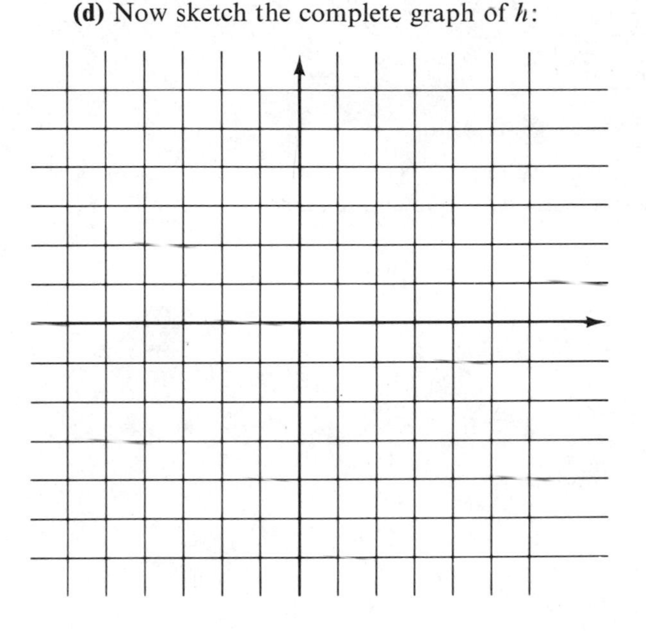

2. The function $k(x)$ is defined by $k(x) = -2x + 3$. Fill in the blanks and then draw the complete graph of k.

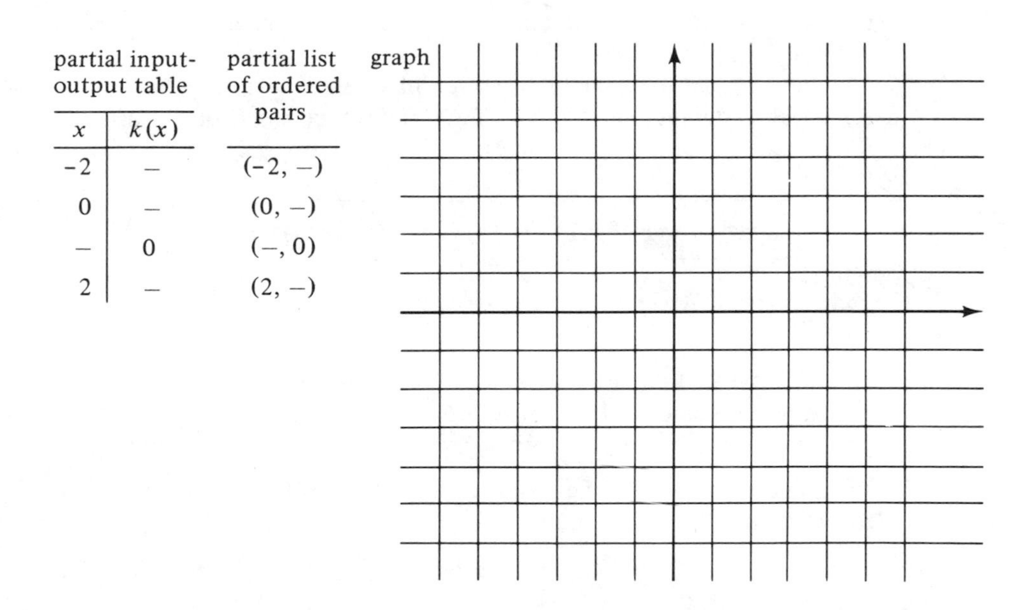

partial input-output table

x	$k(x)$
-2	–
0	–
–	0
2	–

partial list of ordered pairs

(-2, –)
(0, –)
(–, 0)
(2, –)

graph

3. The function $\ell(x)$ is defined by $\ell(x) = -\frac{3}{4}x - 2$. Fill in the blanks and then draw the complete graph of ℓ.

partial input-output table

x	$\ell(x)$
0	–
4	–

partial list of ordered pairs

pairs
(0, –)
(4, –)

4. The function m is defined by $m(x) = x(x - 2)$. Fill in the blanks in the partial input-output table and then sketch the complete graph of m.

x	$m(x)$
4	–
3	–
2	–
$1\frac{1}{2}$	–
1	–
$\frac{1}{2}$	–
0	–
-1	–
-2	–

5. **(a)** Of the functions h, k, ℓ, m of Exercises 1–4, which are linear and which are not linear?

(b) How do the graphs of linear functions differ from the graphs of nonlinear functions?

***6.** You have probably decided that the graph of any linear function, $f(x) = ax + b$, is a straight line. A proof of this fact can be based on the **distance formula** [cf. Exercise 20(h), p. 424],

distance from (x_1, y_1) to $(x_2, y_2) = \sqrt{(x_1 - x_2)^2 + (y_1 - y_2)^2}$

and this result from Chapter 7 (Exercise 8, p. 314):

The set of all points equidistant from two points A and B is the perpendicular bisector of $\overline{AB}$.

We give a proof now for the special case of a linear function $f(x) = ax + b$ for which $b \neq 0$.

(i) Let G be the graph of f; that is,

$$G = \{(x, ax + b) \mid x \in R\}$$

(ii) Let E be the set of all points equidistant from the two points $(ab, 0)$ and $(-ab, 2b)$. [Note that if b were 0, we would have just one point here, not two.]

(iii) Claim $G = E$ (and hence G is a line, namely, the perpendicular bisector of the segment joining $(ab, 0)$ to $(-ab, 2b)$. We establish this claim by showing that $G \subset E$ and $E \subset G$ in steps (iv) and (v).

(iv) $G \subset E$ because every point of form $(x, ax + b)$ is equidistant from $(ab, 0)$ and $(-ab, 2b)$, as use of the distance formula shows:

$$[x - ab]^2 + [(ax + b) - 0]^2 \stackrel{?}{=} [x - (-ab)]^2 + [(ax + b) - 2b]^2$$

$$x^2 - 2abx + a^2b^2 + a^2x^2 + 2abx + b^2 \stackrel{?}{=} x^2 + 2abx + a^2b^2 + a^2x^2 - 2abx + b^2$$

Yes!

(v) $E \subset G$ because if (x, y) is equidistant from $(ab, 0)$ and $(-ab, 2b)$, then $y = ax + b$, as use of the distance formula again shows:

$$[x - ab]^2 + [y - 0]^2 = [x - (-ab)]^2 + [y - 2b]^2$$

$$\Rightarrow x^2 - 2abx + a^2b^2 + y^2 = x^2 + 2abx + a^2b^2 + y^2 - 4by + 4b^2$$

$$\Rightarrow -2abx = 2abx - 4by + 4b^2$$

$$\Rightarrow 0 = 4b(ax - y + b)$$

$$\Rightarrow 0 = ax - y + b \qquad \text{(since } b \neq 0\text{)}$$

$$\Rightarrow y = ax + b$$

Give a similar proof for the case in which $b = 0$. That is, prove that the graph of a linear function $f(x) = ax$ is a straight line. (*Hint*: Use $(1, -1/a)$ and $(-1, 1/a)$ as the pair of points in step (ii).)

7. The graph of a linear function t is given below. Use it to find (approximately)

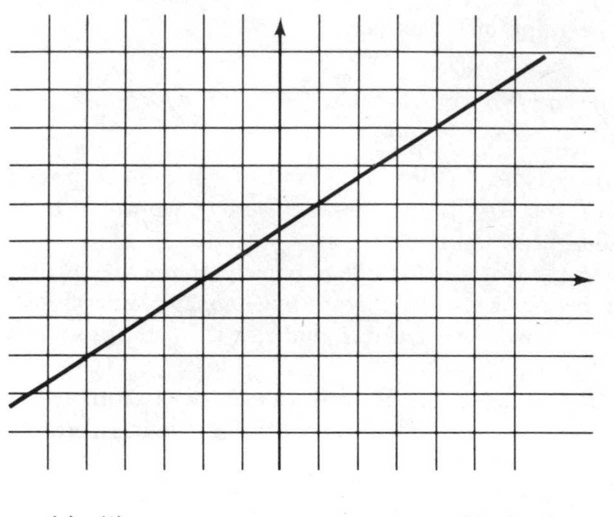

(a) $t(1)$ **(b)** $t(-4)$
(c) $t(4)$ **(d)** $t(0)$
(e) x so that $t(x) = 0$

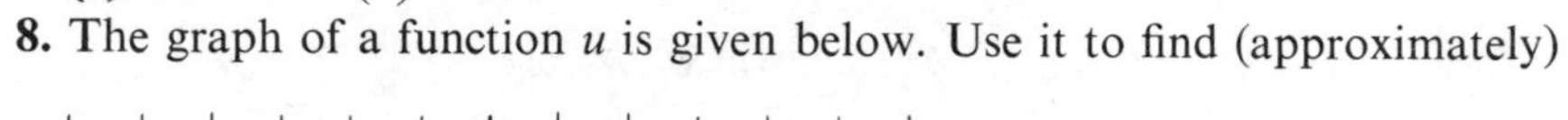

8. The graph of a function u is given below. Use it to find (approximately)

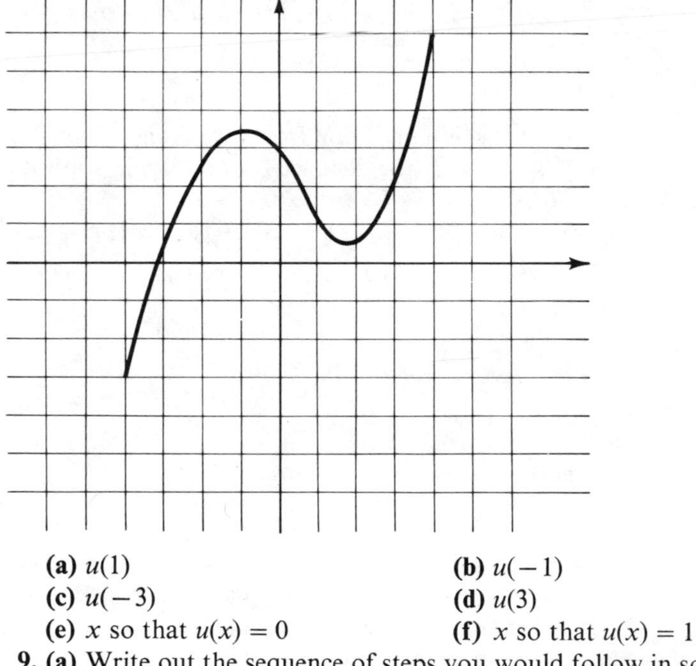

(a) $u(1)$ **(b)** $u(-1)$
(c) $u(-3)$ **(d)** $u(3)$
(e) x so that $u(x) = 0$ **(f)** x so that $u(x) = 1$

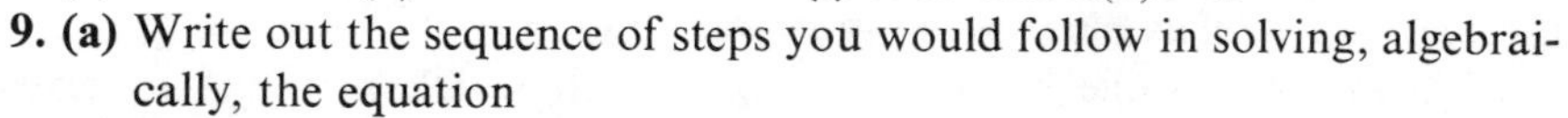

9. (a) Write out the sequence of steps you would follow in solving, algebraically, the equation

$$\tfrac{1}{2}x - 3 = -1$$

(b) Carry out this geometric solution procedure:

(i) Graph the (linear) function $f(x) = \frac{1}{2}x - 3$.
(ii) From the graph, determine (approximately) the x such that $f(x) = -1$.

10. Use the graph you drew in Exercise 9(b)—a graph in the plane— to graph on the number line:
(a) $\{x \mid \frac{1}{2}x - 3 < 0\}$
(b) $\{x \mid \frac{1}{2}x - 3 \geq -2\}$

11. Use the geometric procedure to find (approximately) solutions to the equations, and number-line graphs for the inequalities, below:
(a) $3 - 2x = 1$
(b) $3 - 2x > 0$
(c) $\frac{1}{4}x = \frac{2}{3}$
(d) $\frac{1}{4}x \geq -\frac{1}{2}$

12. Use the graph of $f(x) = x^2$ given below to find (approximately)

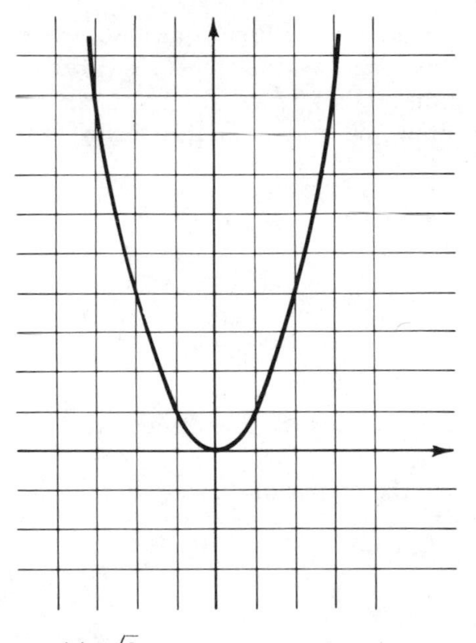

(a) $\sqrt{2}$
(b) $\sqrt{3}$
(c) $\sqrt{4}$
(d) $\sqrt{5}$

13. The U.S. Census Bureau acts as a function p when it assigns population numbers to year numbers. Use the table below to make a partial graph of p. Then complete the graph in a reasonable way. Finally, estimate the population in 1945 and estimate the year in which the population passed 100 million.

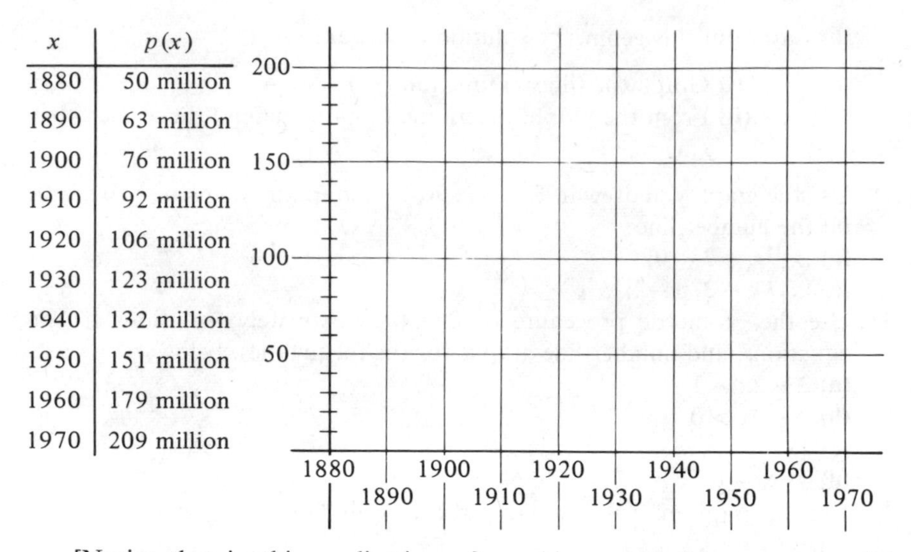

x	$p(x)$
1880	50 million
1890	63 million
1900	76 million
1910	92 million
1920	106 million
1930	123 million
1940	132 million
1950	151 million
1960	179 million
1970	209 million

[Notice that in this application of graphing no useful purpose would be served by choosing the same unit of length on the x and y axes, or by having them intersect at their common origin.]

14. On a single coordinatized plane, draw the graphs of this **family** (set) of linear functions:

$$a(x) = x - 3$$
$$b(x) = x$$
$$c(x) = x + 2$$
$$d(x) = x + 3\tfrac{1}{2}$$

Now find a linear function $r(x)$ whose graph is the line through $(0, -1)$ and $(3, 2)$.

15. On a single coordinatized plane, draw the graphs of this family of linear functions:

$$e(x) = -2x + 1$$
$$f(x) = -2x$$
$$g(x) = -2x - 1$$
$$h(x) = -2x - 2$$

Now find a linear function $s(x)$ whose graph is the line through $(-1, 5)$ and $(2, -1)$.

16. What is the geometric significance of the b in a linear function $ax + b$?

17. On a single coordinatized plane, draw the graphs of this family of linear functions:

$$i(x) = -2x + 3$$
$$j(x) = -x + 3$$
$$k(x) = \tfrac{1}{2}x + 3$$
$$\ell(x) = 3x + 3$$

Now find a linear function $t(x)$ whose graph is the line through (0, 3) and (1, 4).

18. On a single coordinatized plane, draw the graphs of this family of functions:

$$m(x) = 3x - 1$$
$$n(x) = -\tfrac{1}{3}x - 1$$
$$p(x) = 0 \cdot x - 1$$
$$q(x) = -x - 1$$

Now find a linear function $u(x)$ whose graph is the line through (0, -1) and (2, 2).

19. What is the geometric significance of the a in a linear function $ax + b$?

We have been using the Euclidean plane as a place for graphing functions from R to R. We also used it earlier as a place for drawing arrow diagrams of mappings from the plane to itself. Still another use of the Euclidean plane, the one on which the whole subject of plane analytic geometry is built, is as the replacement set for open sentences in *two* variables, x and y. [Recall from Chapter 2 how the number line was used as the replacement set for such open sentences in *one* variable as $2x - 3 < 5$, $|x - 3| \geq \frac{1}{2}$, etc.]

The following exercise gives just a flavor of this use of the Euclidean plane.

20. Graph the truth sets of the following open sentences.

EXAMPLE "$x \geq 0$ and $y \leq 3$"

Solution The graph shows $\{(x, y) \mid x \geq 0 \text{ and } y \leq 3\}$.

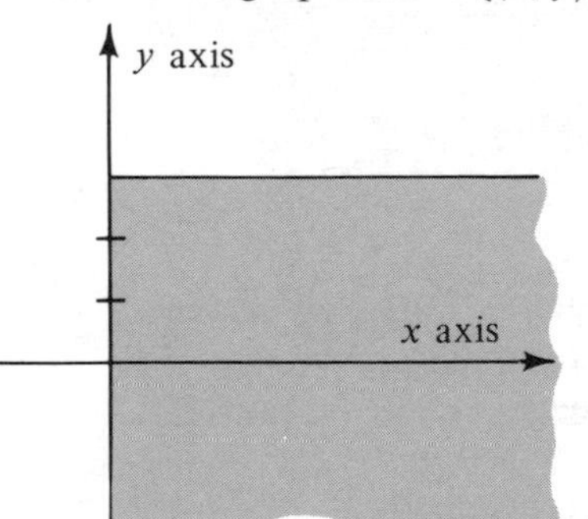

(a) $x \leq 0$ and $y \geq 0$
(b) $0 \leq x \leq 3$ and $-2 \leq y \leq 2$
(c) $0 \leq x \leq 3$ or $-2 \leq y \leq 2$
(d) $x \cdot y \geq 0$
(e) $y = 2x$
(f) $y \neq 2x$
(g) $y > 2x$
(h) $y < 2x$
(i) $y = \frac{1}{3}x - 2$
(j) $y = 2x$ and $y = \frac{1}{3}x - 2$
(k) $y = \frac{1}{2}x - 1$ and $y = -\frac{2}{3}x + 1$
(l) $y < \frac{1}{2}x - 1$
(m) $y < -\frac{2}{3}x + 1$
(n) $y < \frac{1}{2}x - 1$ and $y < -\frac{2}{3}x + 1$
(o) $y < \frac{1}{2}x - 1$ and $y > 2x$
(p) $x \cdot y = 1$
(q) The distance from (x, y) to $(0, 0)$ is 2.
(r) $x^2 + y^2 = 4$
(s) $x^2 + y^2 < 25$

21. Space can be coordinatized by means of *three* mutually perpendicular axes called the x, y, and z axes. The convention for labeling these axes and matching ordered *triples* of real numbers with points of space is illustrated below.

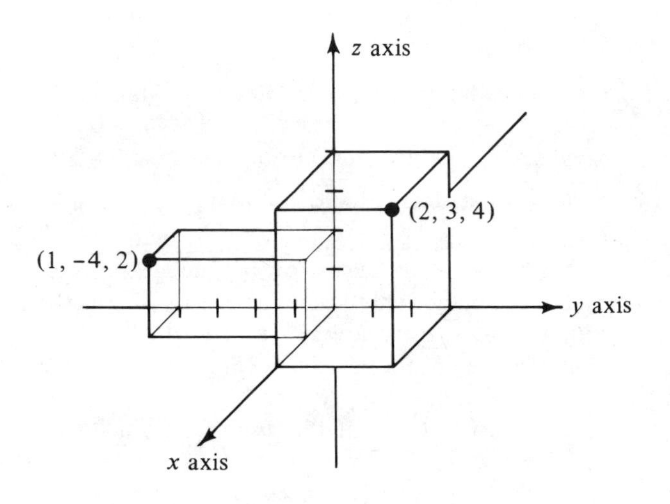

Find the distance from $(1, -4, 2)$ to $(2, 3, 4)$. (*Hint*: First locate $(2, 3, 2)$ and find the distance from $(1, -4, 2)$ to $(2, 3, 2)$.)

VOCABULARY

9.1 Functions

function
rule
arrow diagram
value
input
output
domain
range
formula, $f(x)$
sine function
cosine function
tangent function

9.2 Linear Functions on the Number Line

linear function, $f(x) = ax + b$ $(a \neq 0)$
slide
mapping
compose
stretch
shrink
flip
image

9.3 Congruence and Similarity Mappings of the Plane

flip, f_ℓ
congruent
similarity mapping
fixed point
congruence mapping
dilation, $d_{P,r}$
similar
fixed figure

9.4 Groups; Symmetry

binary operation
neutral element
inverse
identity mapping, i
line of symmetry
group of symmetries
associative property
closure
group
symmetry
group table

9.5 Mappings of the Euclidean Plane

x axis
origin
second coordinate
commutative
y axis
first coordinate
Euclidean plane

9.6 Graphing in the Euclidean Plane

graph
distance formula
input-output table
family (of functions)

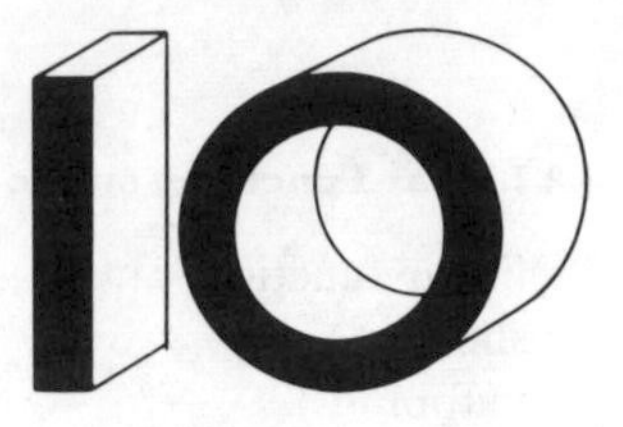

SPACE FIGURES

OVERVIEW

The purpose of this chapter is to introduce the most common types of space figures, and to present surface area and volume formulas for the simpler ones. Currently this material appears mainly at the upper elementary and junior high school levels.

Space figures are introduced in the first section by means of the five regular polyhedra. Many students profit from the experience of actually building physical models of these from patterns. This manipulative work helps them to visualize such space figures as prisms, pyramids, cylinders, cones, and spheres, which are described later on in the chapter via perspective drawings. To balance the manipulative work of this section, there is intellectual work with Euler's formula.

Volume formulas for prisms and pyramids are developed and applied in Section 10.2. While the formula for prisms is transparent and ought to be easily remembered, the one for pyramids is not so clear. For this formula, as for most others in this chapter, we recommend remembering the plausibility argument leading up to the formula (in this case the decomposition of a cube into three congruent pyramids) rather than trying to memorize the formula itself. This is particularly true of the surface area formulas of Section 10.3.

The final three sections deal with the simplest space figures that are not polyhedra; namely, cylinders, cones, and spheres. The surface area and volume formulas for cylinders and right circular cones are completely analogous to those for prisms and regular pyramids. For spheres, new derivations are required. While some of the derivations may seem rather difficult, everyone should be able to *use* the various formulas to calculate surface areas and volumes.

10.1 REGULAR POLYHEDRA AND EULER'S FORMULA

In our work with geometry so far, we have concentrated on figures in a plane. We did less with figures on a line because there is less to do. For example, the whole theory of similarity of figures on a line is subsumed by the study of linear functions. We did less with figures in space for several reasons. One reason is that much of space geometry closely parallels plane geometry. For example, flips and dilations of space—the basis for the theory of similarity and congruence—are defined in much the same way as the corresponding mappings of the plane. For another example, space can be coordinatized and then studied algebraically very much as the plane is.

Another, more practical reason why we gave less attention to space geometry than to plane geometry is that it is more difficult and hence plays a smaller role in the elementary school curriculum. To a large degree the difficulty of space geometry can be traced to the fact that representing three-dimensional figures on a two-dimensional piece of paper is not easy. It takes a good deal of practice to make and interpret "perspective drawings." In the early grades the troublesome middle man, the printed page, is omitted. Young children learn about space figures by handling wooden or construction paper models of them. Some mathematically very interesting figures that can be studied at this early stage are the "regular polyhedra."

Loosely speaking, polyhedra are the figures in space that correspond to polygons in the plane, and regular polyhedra correspond to regular polygons. A careful definition of "polyhedron" is not easy to formulate, and an intuitive understanding of the concept will be sufficient for our purposes anyhow. Thus you should think of a **polyhedron** as a figure in space whose boundary

1. is "closed,"
2. is the union of polygons, called the "faces" of the polyhedron.

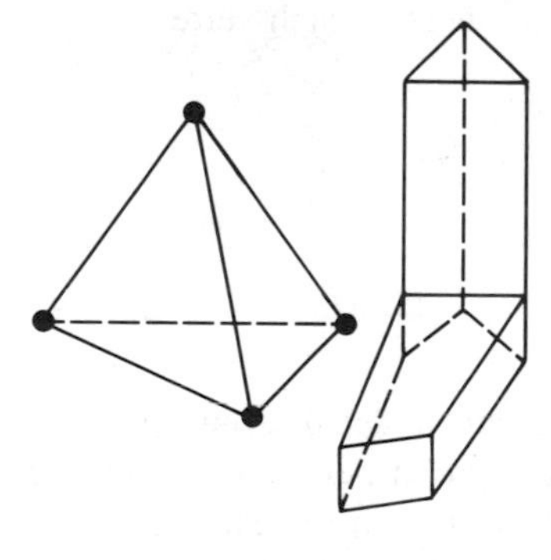

polyhedra

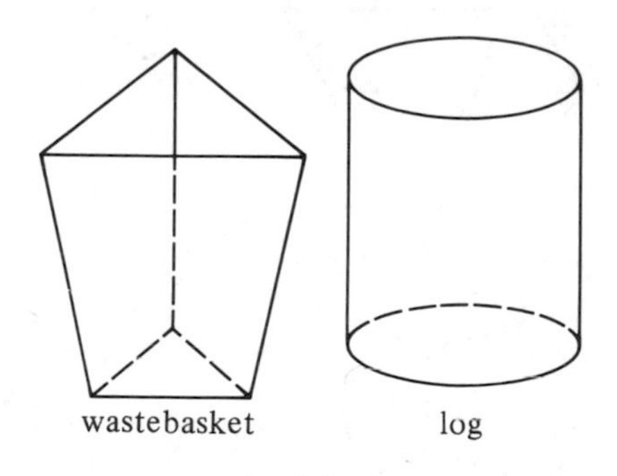

not polyhedra

NOTE The wastebasket is not a polyhedron because its boundary is not closed: It has no lid. The log is not a polyhedron because its "faces" are not polygons.

For the work that follows, we shall require a rather exact definition of a regular polyhedron. Here is one:

A **regular polyhedron** is a polyhedron with the properties:

1. All of its faces are congruent regular polygons.
2. The same number of faces meet at each vertex.

The most familiar example of a regular polyhedron is a cube. Its faces are congruent squares and at each of its vertices three of its faces come together. The ancient Greeks knew about four other regular polyhedra. As we shall see in Exercise 16 below, there are no more. This is in striking contrast with the situation in the plane, where there are regular polygons with any number of sides. The five regular polyhedra are as follows:

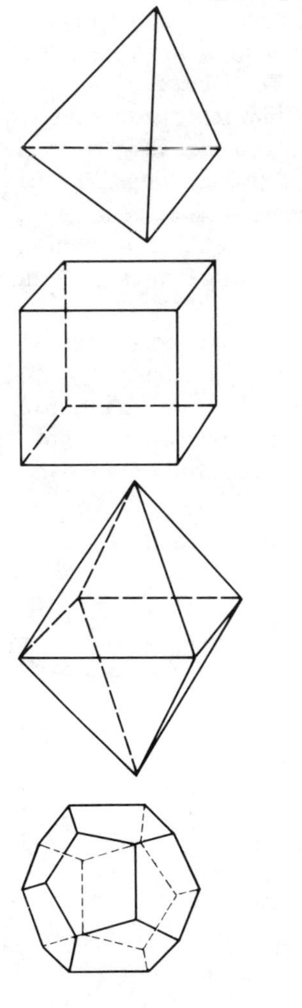

The **regular tetrahedron**: The four faces are equilateral triangles and three faces meet at each vertex.

The **regular hexahedron** (cube): The six faces are squares and three faces meet at each vertex.

The **regular octahedron**: The eight faces are equilateral triangles and four faces meet at each vertex.

The **regular dodecahedron**: The 12 faces are regular pentagons and three faces meet at each vertex.

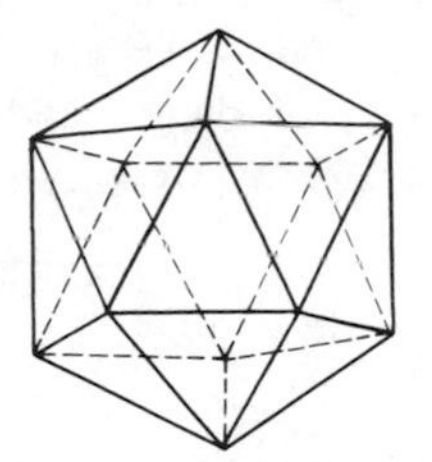

The **regular icosahedron**: The 20 faces are equilateral triangles and five faces meet at each vertex.

The regular dodecahedron, which was discovered by the Pythagoreans, was used as a secret symbol by them. Today it is the emblem of the Mathematical Association of America.

EXERCISES

1. Patterns are provided below for the regular dodecahedron and the regular icosahedron. [You may find it helpful in this section actually to use these patterns to build construction-paper models of a dodecahedron and an icosahedron.] Make (and perhaps use) patterns for a regular **(a)** tetrahedron, **(b)** hexahedron, and **(c)** octahedron.

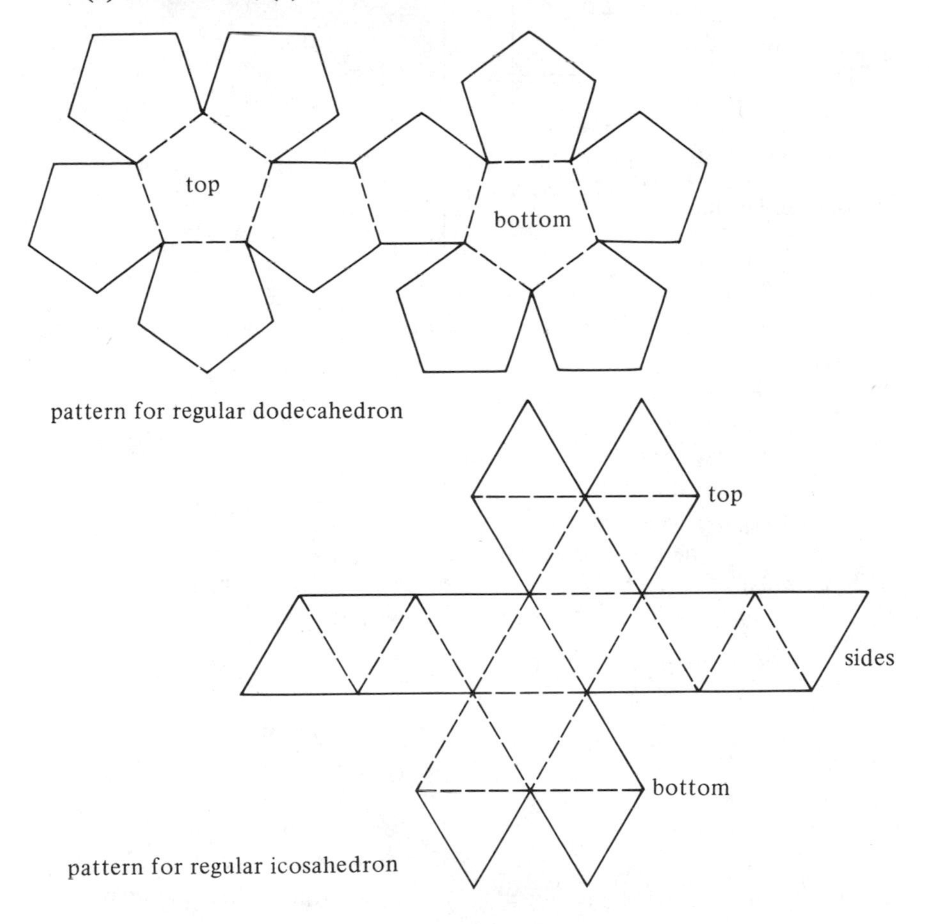

pattern for regular dodecahedron

pattern for regular icosahedron

2. A regular tetrahedron *ADGE* and a regular octahedron *ABCDEF* are held together so that a face of one coincides with a face of the other (*AED*). Is the resulting polyhedron *ABCDFEG* regular? Why or why not?

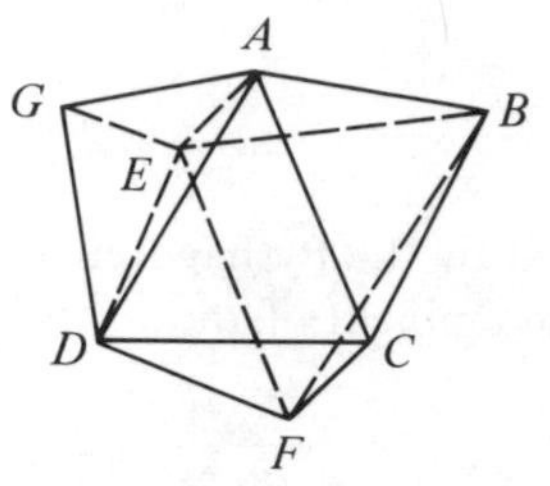

3. Using the sketches of the regular polyhedra provided in the text (or the paper models from Exercise 1 if you made them), fill in the following chart. *V* stands for the number of vertices, *E* for the number of edges, and *F* for the number of faces.

	V	*E*	*F*
regular tetrahedron			
regular hexahedron			
regular octahedron			
regular dodecahedron			
regular icosahedron			

4. Try to guess what sort of polyhedron would have its vertices at the centers of the faces of
(a) a regular tetrahedron
(b) a regular hexahedron
(c) a regular octahedron
(d) a regular dodecahedron
(e) a regular icosahedron
(*Hint*: Use the preceding table.)

5. Calculate $V - E + F$ for each of the five regular polyhedra.

6. For each polyhedron below, find the numbers: $V, E, F, V - E + F$.

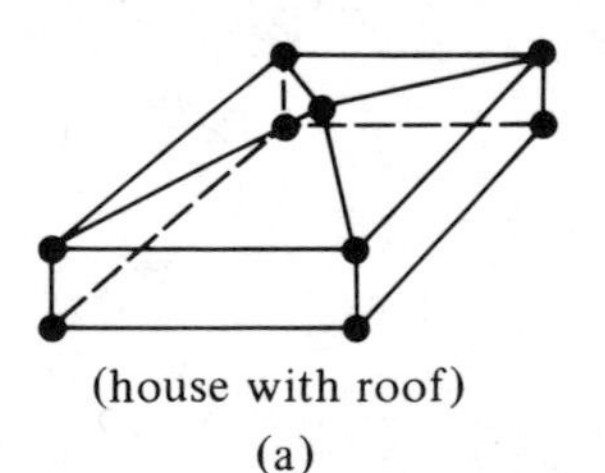

(house with roof)

(a)

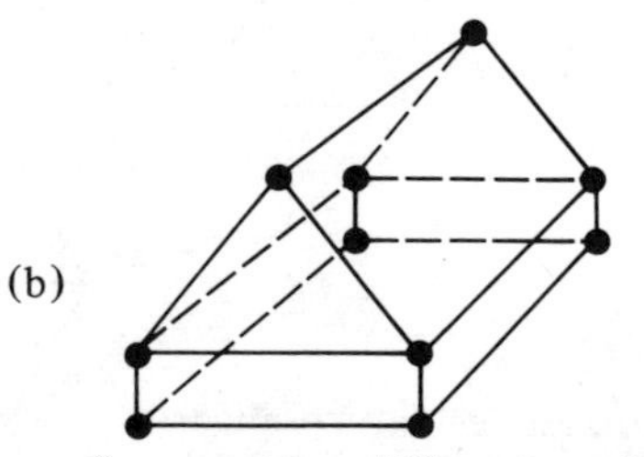

(b)

(house with a different roof)

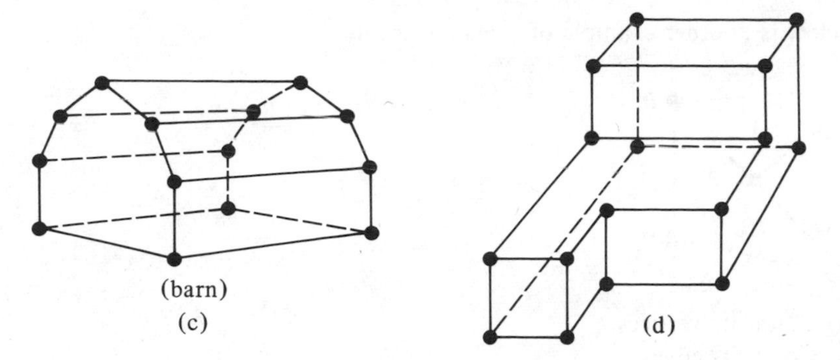

(barn)
(c)

(d)

7. The five regular polyhedra and the four polyhedra of Exercise 6 are examples of "simple" polyhedra. [A polyhedron is said to be **simple** if a rubber model of it could be inflated to be a sphere.] Any simple polyhedron satisfies **Euler's formula**, $V - E + F = 2$. [A proof of this fact is sketched in Exercise 15.] Check that the nonsimple polyhedra below do *not* satisfy Euler's formula.

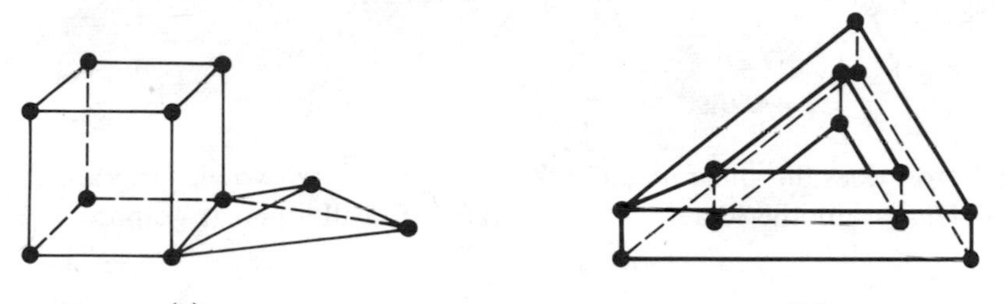

(a)

(b)

8. By using Euler's formula and counting vertices and faces, determine how many edges this polyhedron has:

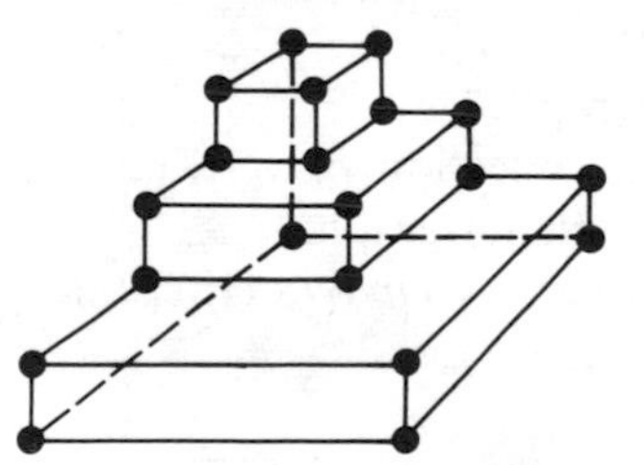

9. Euler's formula is also applicable to "planar graphs." Below is one example of a planar graph. It has five "vertices" (A, B, C, D, E), five "edges" ($\overline{AB}$, $\overline{BC}$, $\overline{CD}$, $\overline{DB}$, $\overline{BE}$), and it partitions the plane into two pieces (I, II).

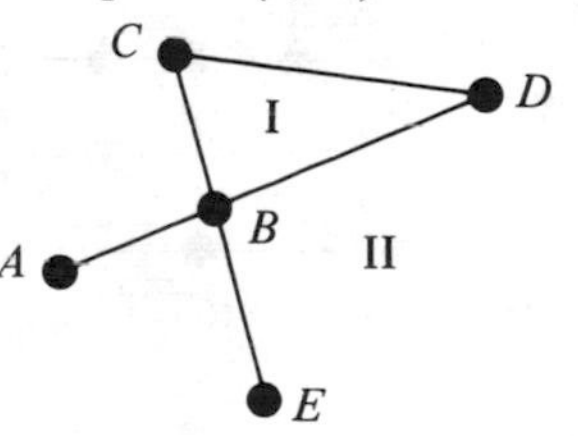

Below is another example of a planar graph.

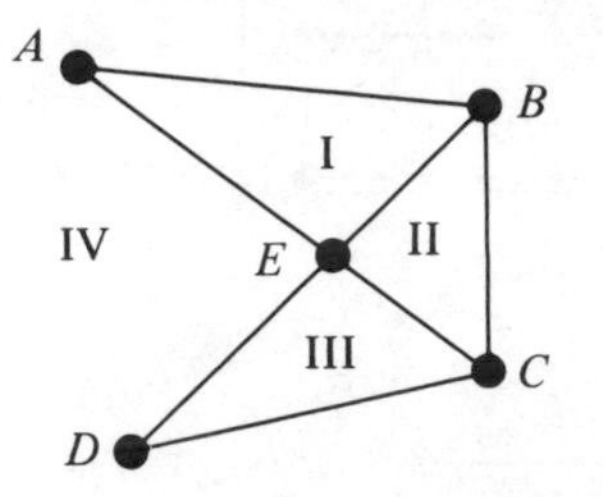

(a) Count its vertices.
(b) Count its edges.
(c) Count the pieces into which it partitions the plane.
(d) For each of the two graphs, compute $V - E + P$ (where P stands for the number of pieces).

10. Not just any collection of dots and segments qualifies as a planar graph. For example, we refuse to call this diagram a planar graph:

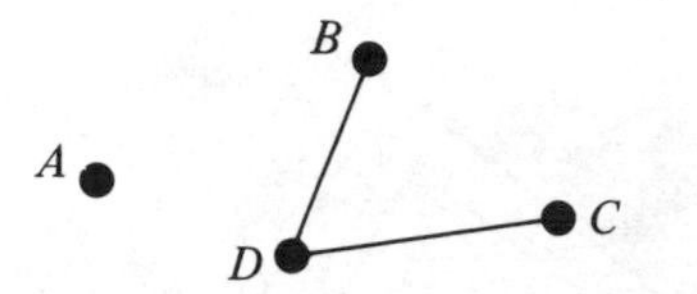

(a) How does this diagram differ from the preceding two planar graphs? Here is another diagram that we refuse to call a planar graph:

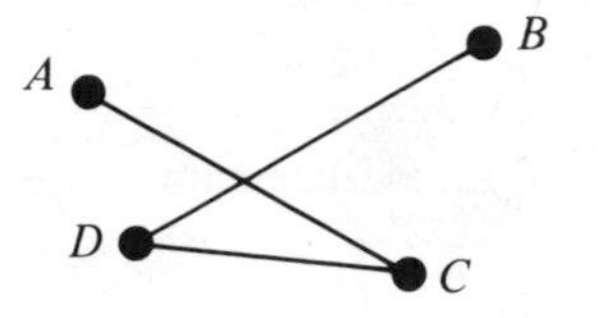

(b) How does it differ from the two planar graphs?
(c) Try to formulate two conditions that a collection of vertices and edges must satisfy in order to be called a planar graph.

11. For each diagram below: (i) Check that it meets your two criteria for a planar graph; (ii) verify that $V - E + P = 2$.

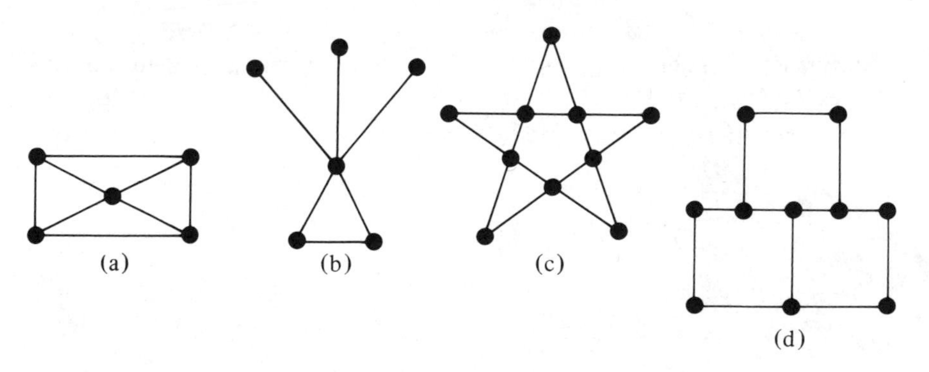

12. Determine the number of edges of the planar graphs below by counting vertices and pieces (a comparatively easy task) and using Euler's formula, $V - E + P = 2$.

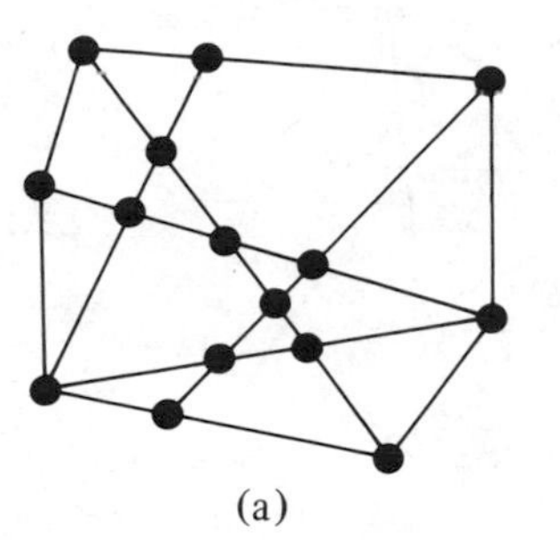

(a)

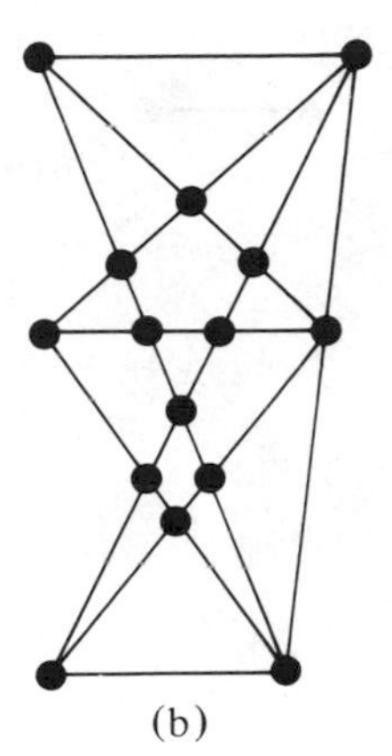

(b)

13. Here is an indication of *why* Euler's formula should hold for planar graphs. Given a complicated graph, for example the one shown, we

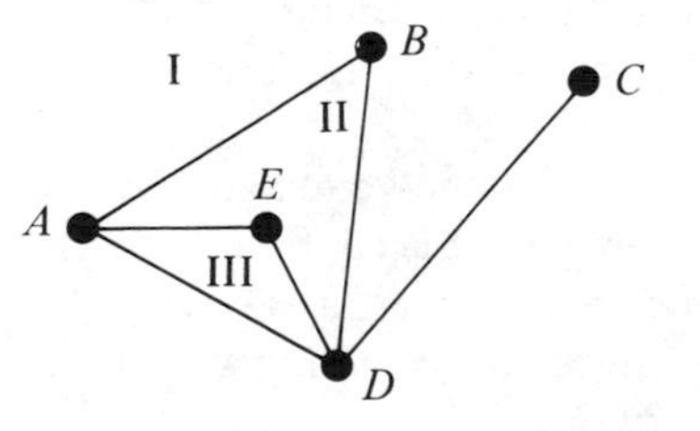

consider two operations for simplifying it. One, which we call "pruning," is to remove jutting edges such as $\overline{CD}$. Notice that in removing $\overline{CD}$ the number of vertices is decreased by one, as is the number of edges. Thus the value of $V - E + P$ is left unchanged when a graph is pruned.

The other operation, which we call "opening," makes one piece out of two by removing an edge that separates them. For example, removing $\overline{BD}$ is an opening operation. Pieces I and II become one on the removal of $\overline{BD}$.

(a) Explain why $V - E + P$ is left unchanged when a graph is opened.

The diagram below suggests that any planar graph can be reduced to the trivial one-vertex graph by repeated pruning and opening.

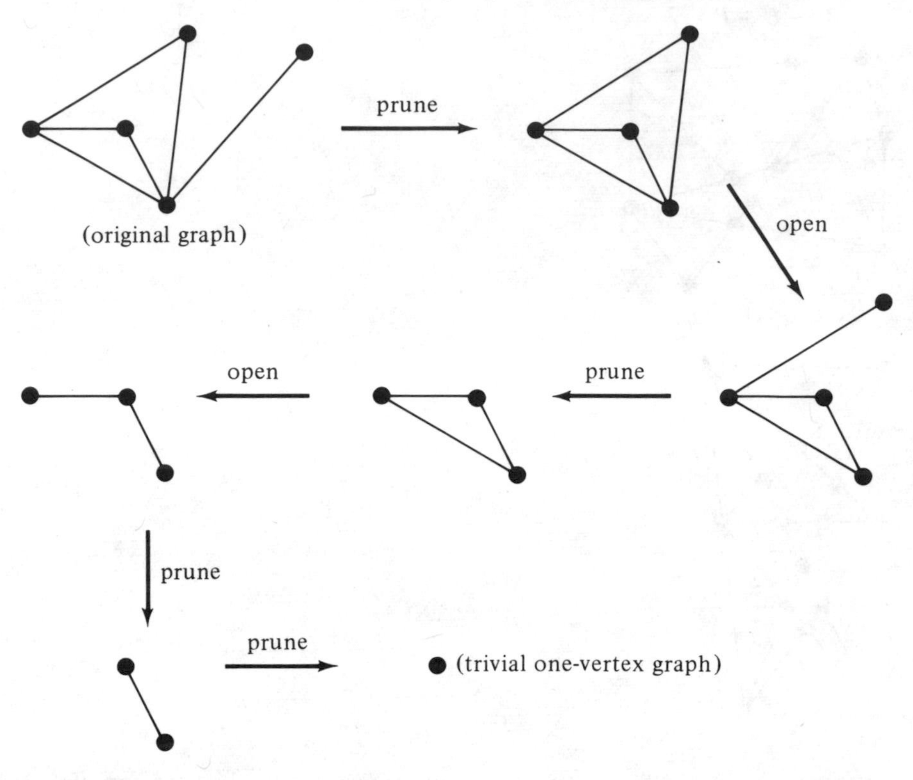

Since $V - E + P$ is unchanged by the pruning and opening operations, the value of $V - E + P$ for the original graph must be the same as for the trivial one-vertex graph.

(b) What is the value of $V - E + P$ for the trivial one-vertex graph?

To convince yourself that *any* planar graph can be reduced to the trivial one-vertex graph by pruning and opening (and hence that Euler's formula, $V - E + P = 2$, holds for any planar graph), reduce the following graphs to the trivial one-vertex graph by pruning and opening.

(c) the graph of Exercise 11(a)

(d) the graph of Exercise 11(b)

14. (a) How many openings will be required to reduce the graph of Exercise 12(a) to the trivial graph? How many prunings?

(b) How many steps, in all, will be required to reduce the graph of Exercise 12(b) to the trivial graph?

(c) How many steps, in all, are required to reduce a graph with V vertices, E edges, and P pieces to the trivial graph?

15. The following sequence of figures suggests why Euler's formula holds for simple polyhedra as well as planar graphs. Study the figures and prepare to explain what happens to each vertex, edge, and face of the original cube. Does $V-E+F$ for the polyhedron equal $V-E+P$ for the planar graph? We derived Euler's formula for the cube. Will the same derivation work for any simple polyhedron? (Recall the definition of simple.)

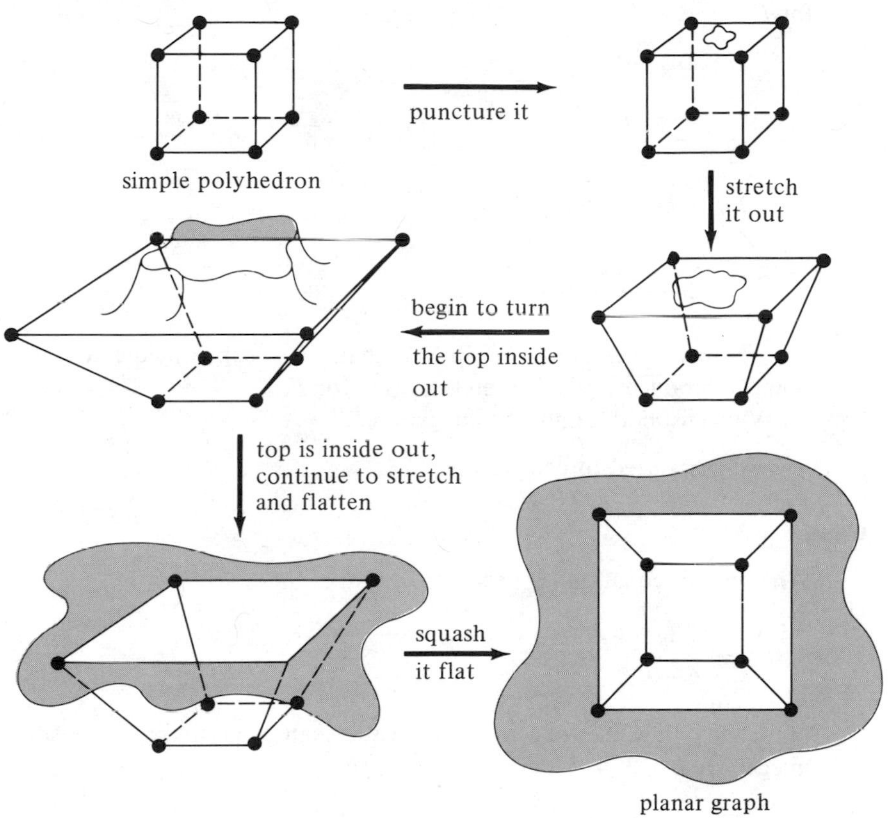

***16.** [A proof, based on Euler's formula, that there are just five regular polyhedra.] Suppose that a regular polyhedron has regular n-gons as its faces and that k of these faces meet at each vertex. (Recall the definition of a regular polyhedron.) As usual let V, E, F stand for the number of vertices, edges, and faces, respectively, of this polyhedron.

We first get an expression for V, the number of vertices. Each face, an n-gon, has n vertices; there are F faces; so there seem to be $n \cdot F$ vertices. But this counts each vertex k times, since each vertex belongs to k faces. Thus

$$V = \frac{n \cdot F}{k}$$

(a) Explain similarly why

$$E = \frac{n \cdot F}{2}$$

Next we substitute these expressions into Euler's formula and solve for F.

$$V - E + F = 2 \Rightarrow$$

$$\frac{n \cdot F}{k} - \frac{n \cdot F}{2} + F = 2 \Rightarrow$$

$$2nF - knF + 2kF = 4k \Rightarrow$$

$$F = \frac{4k}{2n - kn + 2k} \tag{1}$$

Now we seek values of n and k that, when substituted into equation (1), produce a positive integer value for F.

(b) Why must both n and k be integers > 2?

Several cases need to be investigated.

Case $n = 3$

When $n = 3$, equation (1) becomes

$$F = \frac{4k}{6 - k} \tag{2}$$

Now, the only values of k that are greater than 2 and make F a positive integer are $k = 3, 4, 5$.

Subcase $k = 3$

Setting $k = 3$ in equation (2) yields $F = 4$. This is the case of a regular tetrahedron: four faces ($F = 4$), each a regular 3-gon ($n = 3$), three meeting at each vertex ($k = 3$).

(c) Investigate subcase $k = 4$ of case $n = 3$.

(d) Investigate subcase $k = 5$ of case $n = 3$.

Case $n = 4$

When $n = 4$, equation (1) becomes

$$F = \frac{2k}{4 - k} \tag{3}$$

The only value of k that is greater than 2 and makes F a positive integer is $k = 3$, which makes $F = 6$. This is the case of the cube: six faces ($F = 6$), each a regular 4-gon ($n = 4$), three meeting at each vertex ($k = 3$).

(e) Investigate case $n = 5$.
(f) Investigate case $n = 6$.
(g) Show that if $n \geq 6$, then no value of k greater than 2 makes F a positive integer [in equation (1)].

10.2 VOLUMES OF POLYHEDRA

When you speak of the area of a triangle you mean the area of the region it encloses, and when you speak of the volume of a cube you likewise mean the volume of the region it encloses. In this section we learn to compute the volumes of two common types of polyhedra, prisms and pyramids. Much of the work with prisms will be a review of things we did in Chapter 4.

A **prism** is a polyhedron with **bases** that are congruent polygons and lie in parallel planes, and **lateral faces** that are parallelograms.

The prism in the figure below has two bases and five lateral faces.

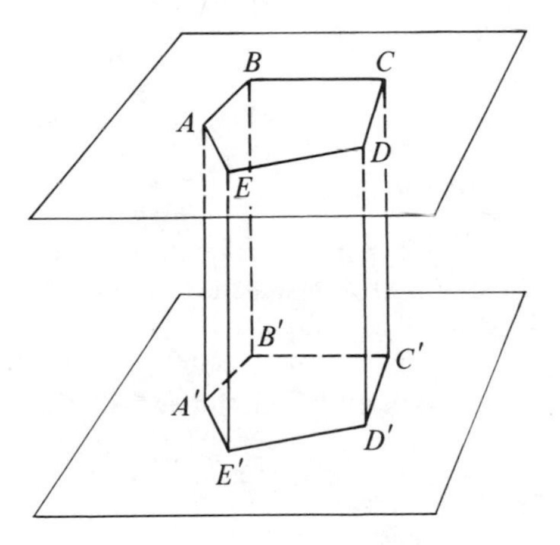

The most familiar prism is the cube:

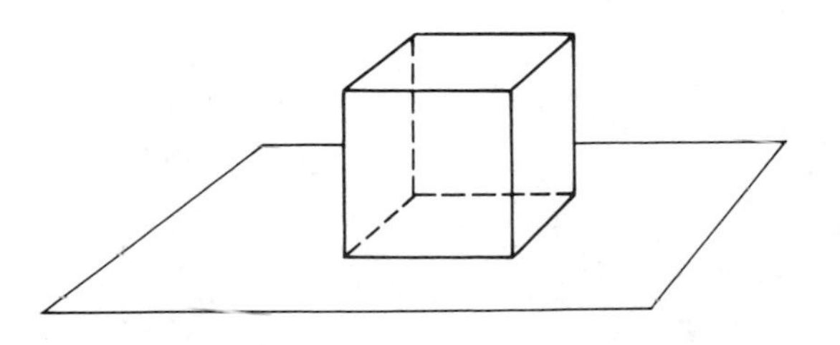

Its bases are squares, as are its lateral faces. The cube is an example of a "right prism" because its edges are perpendicular to its bases. (See Exercise 2.)

A prism is called a **right prism** if its edges are perpendicular to its bases.

The shape of the bases of a prism is also often included in its description. The following examples illustrate how this is done.

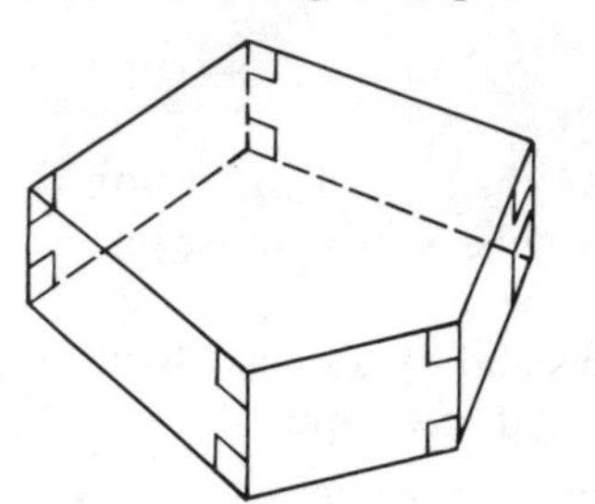

a right pentagonal prism

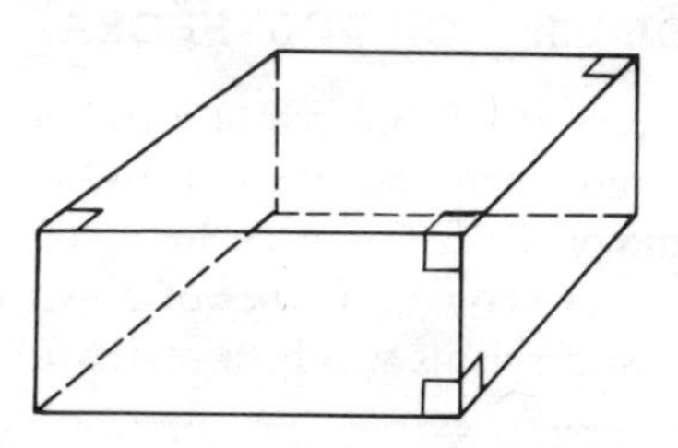

a right rectangular prism

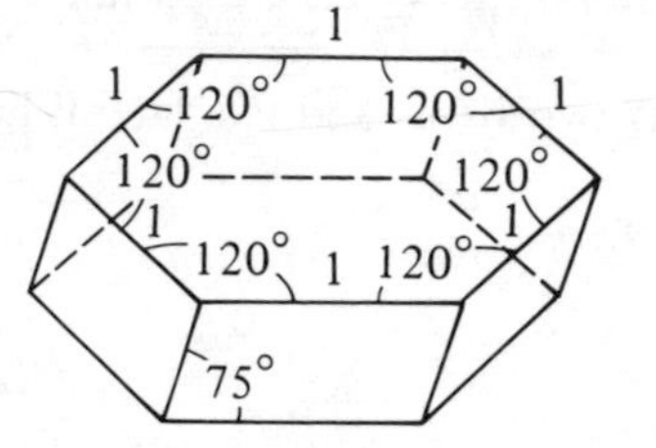

a regular hexagonal prism (not "right")

Notice that there is an unfortunate possibility for misinterpretation in the name given the last prism. We have hyphenated "regular-hexagonal" to stress that "regular" refers to the shape of the bases (regular hexagons) not to the prism as a whole. This prism is certainly *not* a regular polyhedron.

The volume of a right, rectangular prism is given by the simple formula

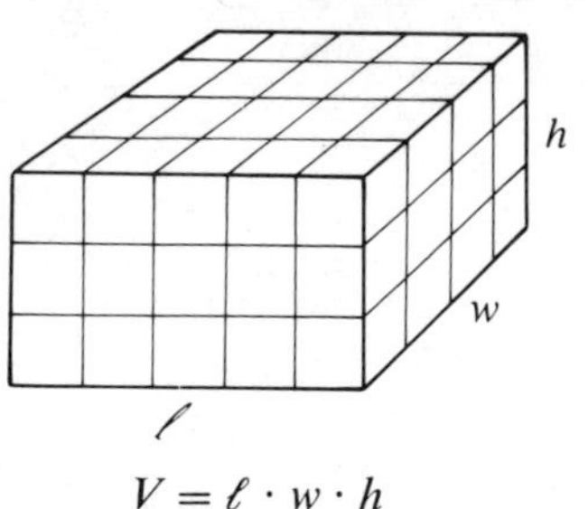

$$V = \ell \cdot w \cdot h$$

or

$$V = B \cdot h$$

where B denotes the area of the base. In the diagram, $\ell = 5$, $w = 4$, $h = 3$ and there are $5 \cdot 4 \cdot 3$ unit cubes in the rectangular prism. You might recall from Chapter 4 that the formula, $V = B \cdot h$, remains true for *any* prism. The prism may or may not be "right," and its bases may have any shape whatsoever.

volume of prism	$V = B \cdot h$	B = area of base h = height

EXAMPLE What is the volume of this pentagonal prism?

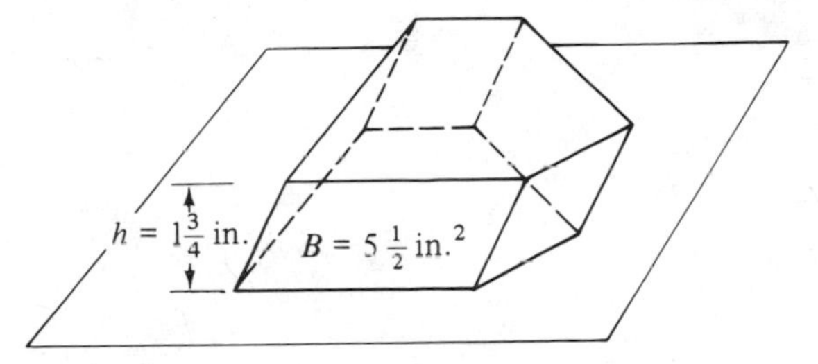

Solution

$$V = B \cdot h = \tfrac{11}{2} \cdot \tfrac{7}{4} = \tfrac{77}{8} = 9\tfrac{5}{8} \text{ in.}^3$$

EXAMPLE What is the volume of this triangular prism?

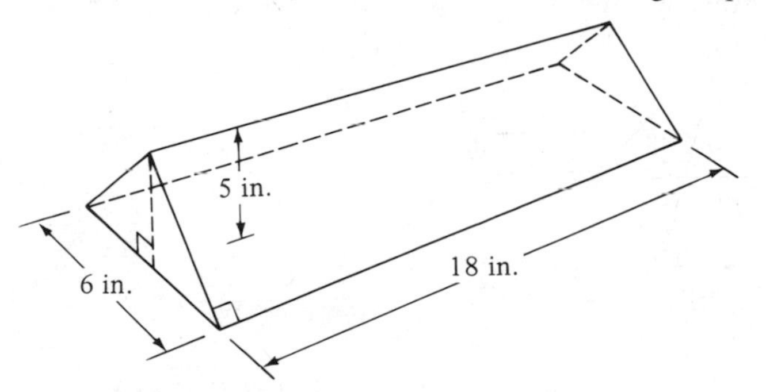

Solution The bases are triangles. Thus $B = \frac{1}{2} \cdot 6 \cdot 5 = 15$ in.^{2} The height is 18 in. Therefore $V = B \cdot h = 15$ in.$^2 \cdot 18$ in. $= 270$ in.^{3}

Another very common type of polyhedron is a pyramid.

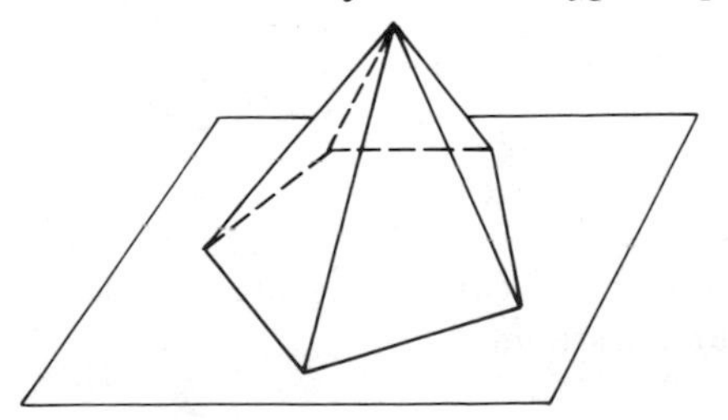

A **pyramid** is a polyhedron with a polygonal **base** lying in a plane and exactly one vertex lying out of that plane.

The volume of any pyramid, no matter what shape its base might have, is given by the formula

volume of pyramid	$V = \frac{1}{3}Bh$	B = area of base h = height

Exercise 7 will suggest why this formula works.

EXAMPLE The volume of this pyramid is

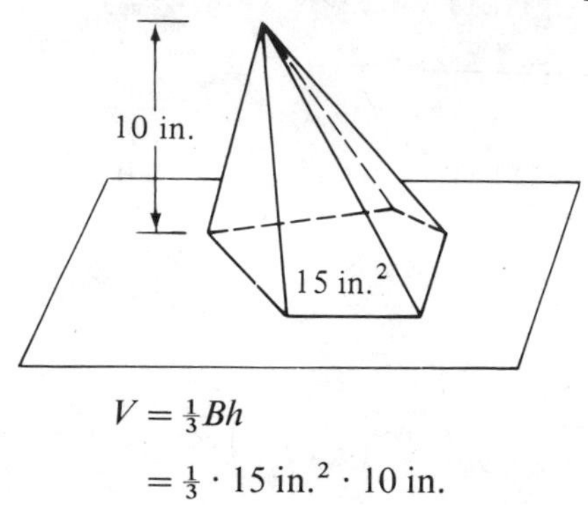

$$V = \tfrac{1}{3}Bh$$
$$= \tfrac{1}{3} \cdot 15 \text{ in.}^2 \cdot 10 \text{ in.}$$
$$= 50 \text{ in.}^3$$

EXAMPLE For this pyramid

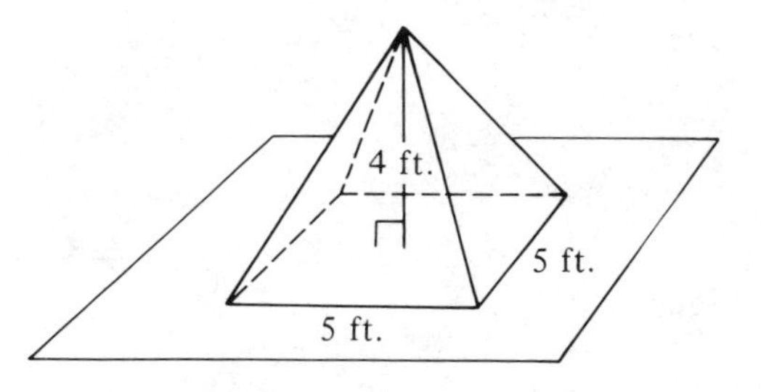

$$V = \tfrac{1}{3}Bh$$
$$h = 4 \text{ ft}$$
$$B = 5 \text{ ft} \cdot 5 \text{ ft} = 25 \text{ ft}^2$$

Thus

$$V = \tfrac{1}{3} \cdot 4 \text{ ft} \cdot 25 \text{ ft}^2 = 33\tfrac{1}{3} \text{ ft}^3$$

EXERCISES

1. **(a)** What do we mean when we say that "two planes are parallel"? Look for parallel planes in your classroom.

(b) Must nonparallel planes intersect?

(c) What does the intersection of two planes look like? Look for intersecting planes in your classroom.

2. Formulate a careful definition of the following: "Line ℓ is perpendicular to plane $\mathscr{P}$ at point A."

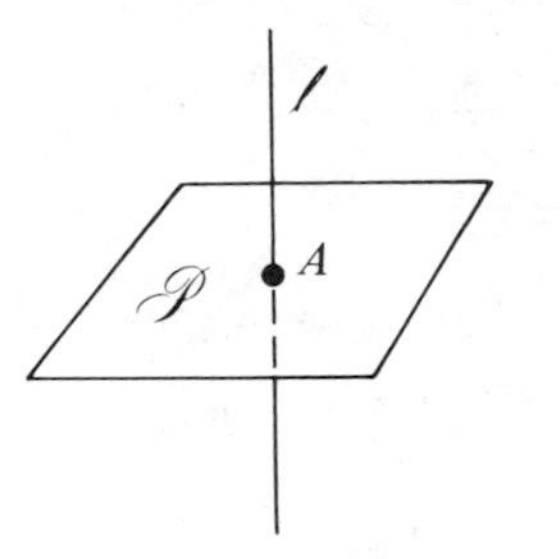

3. Which of the five regular polyhedra are

(a) prisms

(b) pyramids

4. Which of these solid space figures are *convex*? (Do you recall the definition of "convex"?)

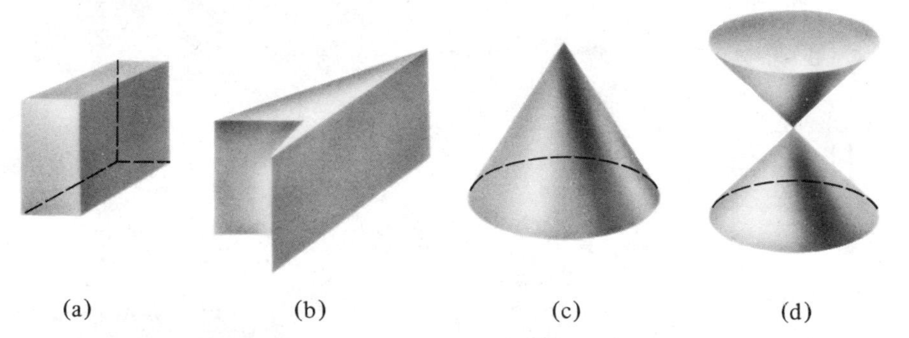

(a) (b) (c) (d)

5. (a) If the bases of a prism are convex plane figures, will the prism be a convex space figure? What if the bases are not convex?

(b) If the base of a pyramid is convex, will the pyramid be convex? What if the base is not convex?

6. Describe the smallest convex space figure that contains each given set of (labeled) points.

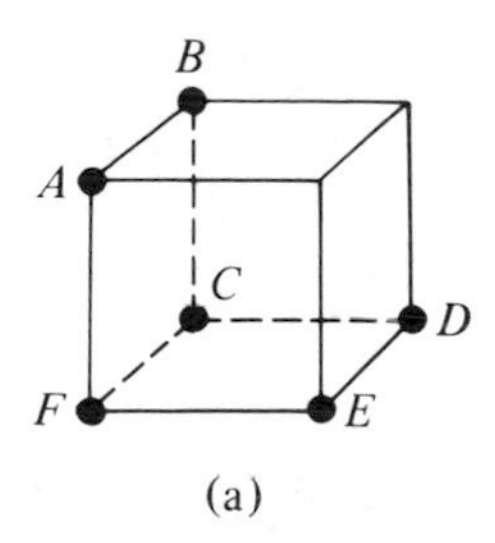

(a)

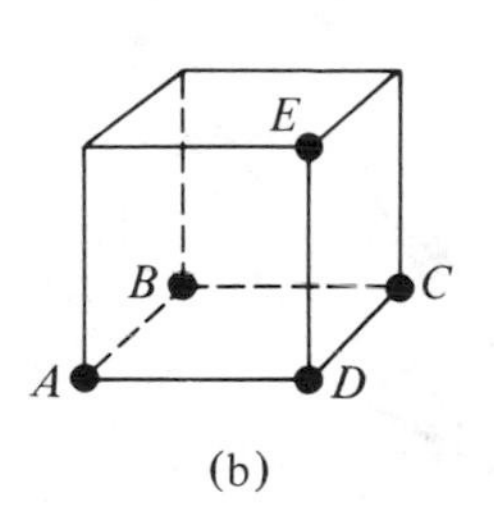

(b)

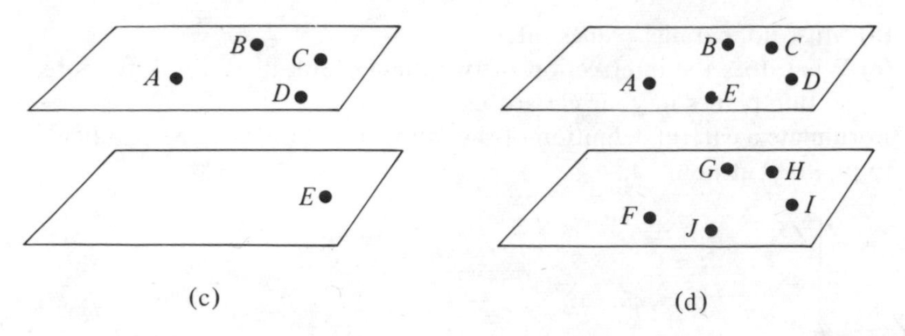

7. **(a)** Make three copies of the pattern below. Cut, fold, and tape each one into a pyramid.

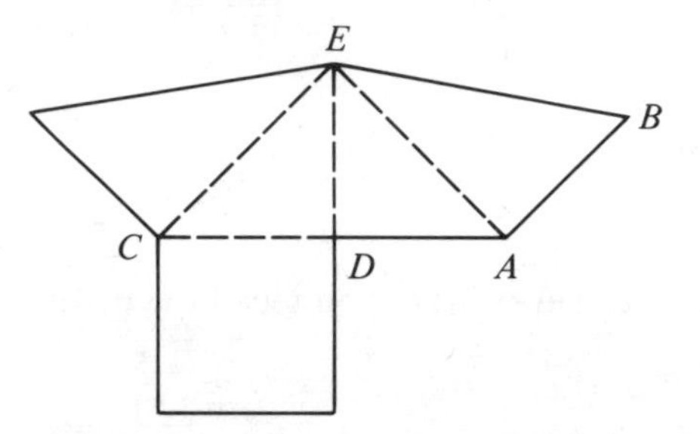

(b) Tape the three pyramids together into a cube.

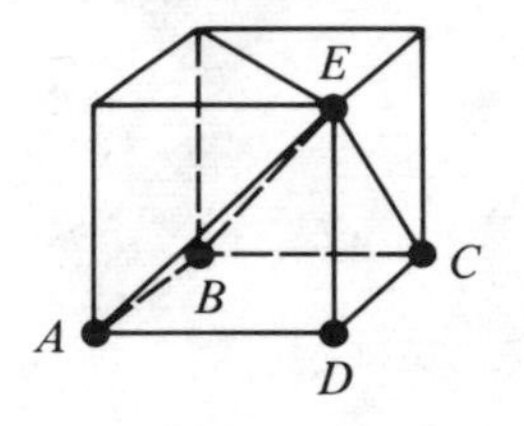

(c) Conclude that for these particular pyramids, volume $= \frac{1}{3} \cdot$ volume of cube $= \frac{1}{3} \cdot$ base $\cdot$ height

8. If the cube in Exercise 7(b) has side length 1

(a) what is the distance from E to A?

(b) what is the distance from E to B?

Check your answers by measuring appropriate segments in the pattern, Exercise 7(a).

9. Which pyramid has the greater volume, $ABCD$ or $EBCD$?

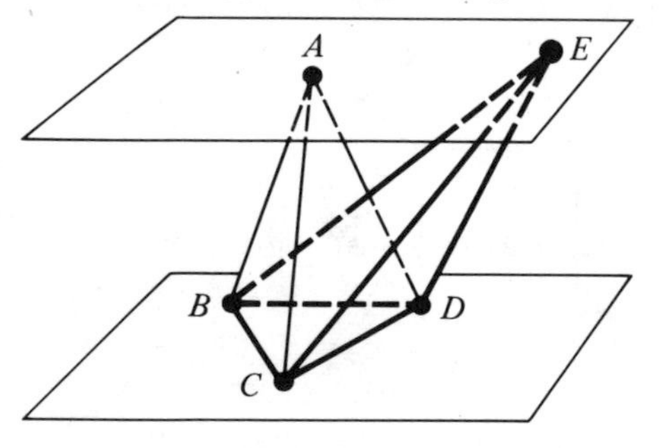

10. Find the volume of each polyhedron shown below.

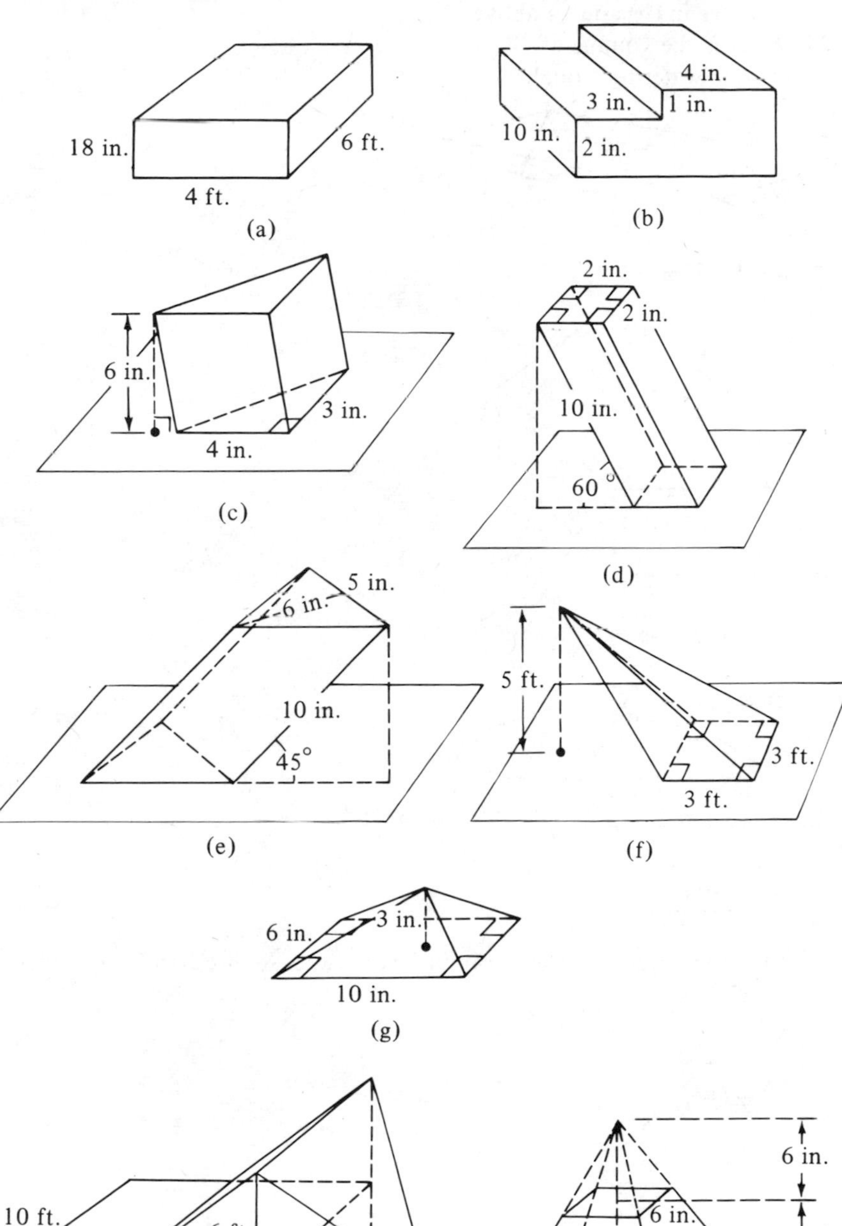

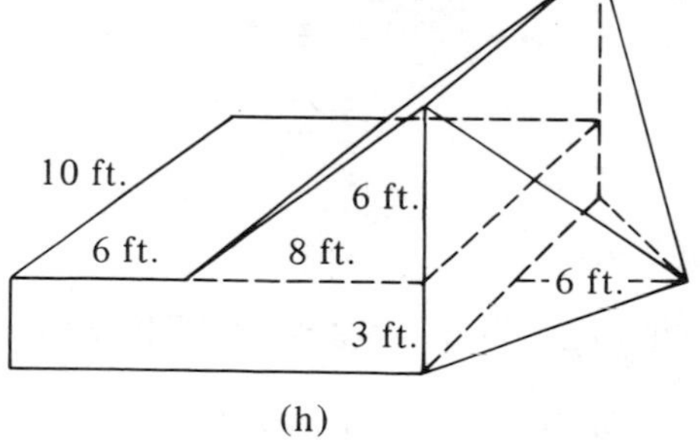

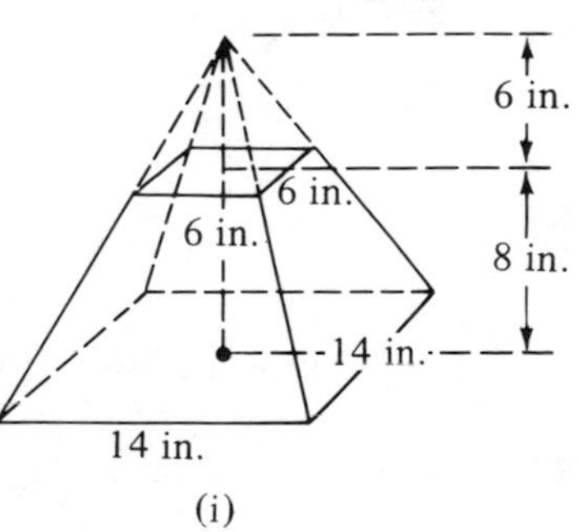

11. On what general principle of measurement did you rely in finding the volumes in (h) and (i) above?

12. What is the volume of

(a) this watering trough:

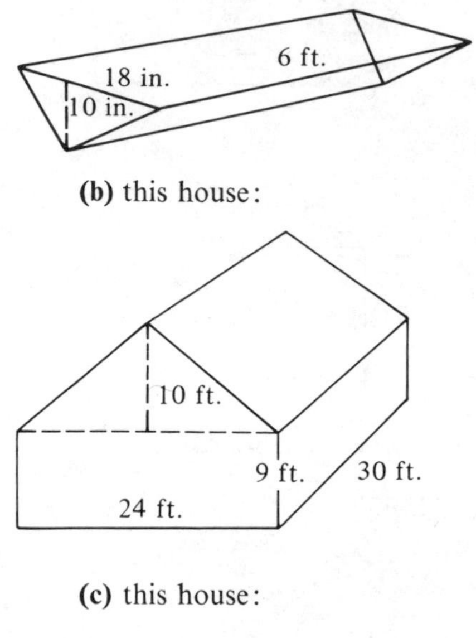

(b) this house:

(c) this house:

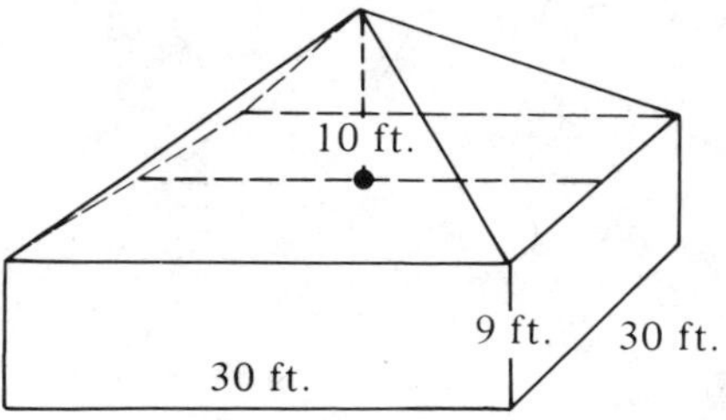

13. Find the volume of the pyramid below (with square base), each edge of which is 5 ft long.

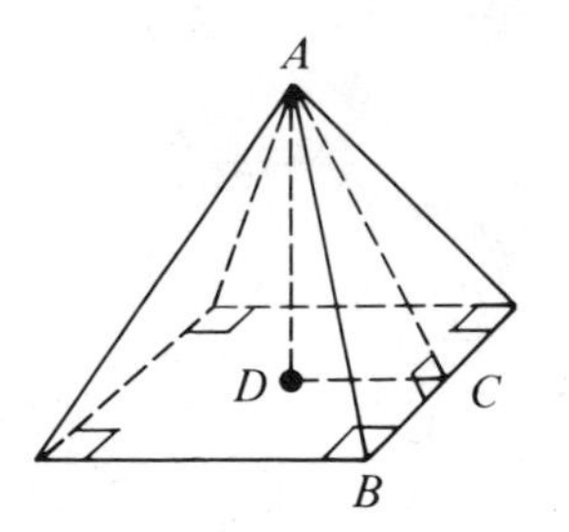

(*Hint*: First work in $\triangle ABC$ to find AC. Then work in $\triangle ACD$ to find AD.)

***14.** Find the volume of the regular tetrahedron of side length 1.

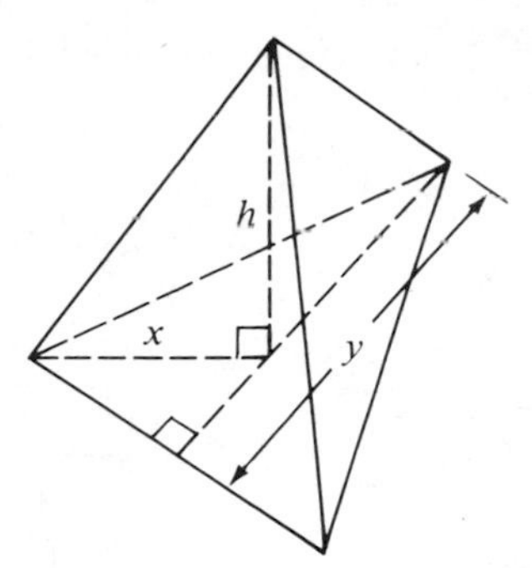

(*Hint*: Find in order: x, h, y, B (area of base), V.)

15. Find the volume of the regular octahedron of side length 1. (*Hint*: Use Exercise 13 above and Exercise 17(b), p. 343.)

***16.** A stick of wood h in. high has the shape shown below. Its "bases" are 1-in. by 2-in. rectangles; its lateral faces are "isosceles trapezoids." What is its volume?

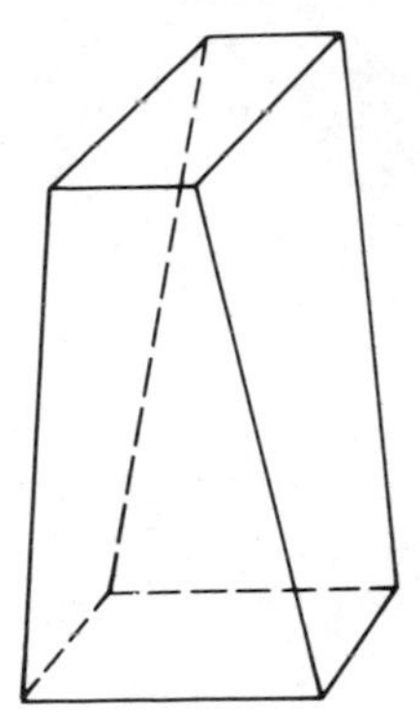

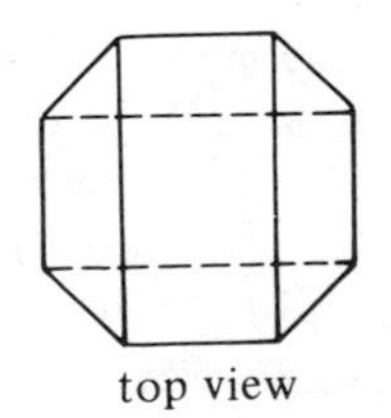

top view

10.3 SURFACE AREAS OF POLYHEDRA

Area formulas for plane figures are based ultimately on the formula for the area of a rectangle. Fill in this basic formula:

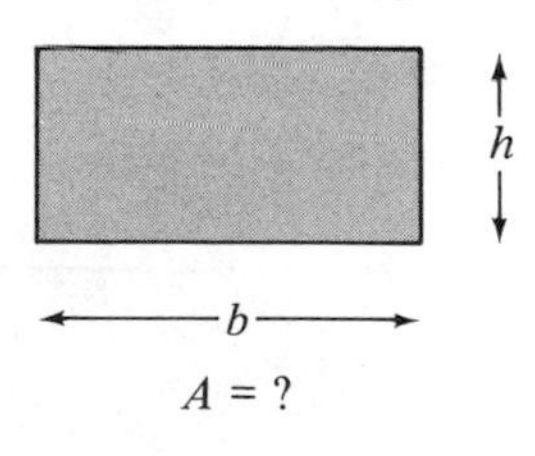

$A = ?$

The figures below suggest formulas for the areas of (plane) parallelograms and triangles. Do you remember those formulas?

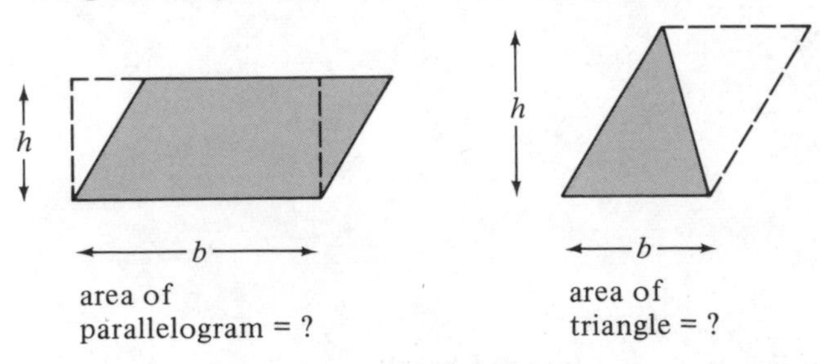

The areas of more general (plane) polygonal regions can be determined by triangulation and the additive property of area measure. What is the area of polygon $ABCDE$?

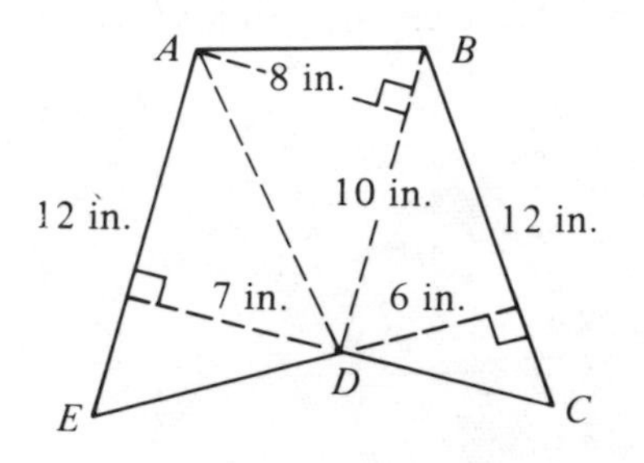

Since a face of a polyhedron is a plane polygon, its area can be determined by one of the above techniques. Then the **total surface area** of the polyhedron, which by definition is the sum of the areas of all of its faces, can be computed. If the polyhedron has many faces, calculating the area of each one and adding them up is tedious. For the "nicest" polyhedra—right prisms and "regular" pyramids—simple surface-area formulas are available.

Recall that the faces of a prism other than the two bases are referred to as its lateral faces. The union of these lateral faces is referred to as the **lateral surface** of the prism. A formula for the **lateral surface area** of a right prism can be found as follows: Think of removing the bases from the right prism, cutting it along a lateral edge, and then flattening out the lateral surface.

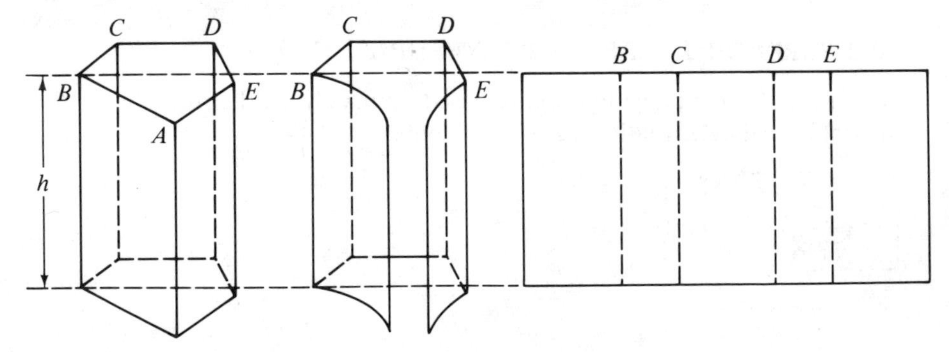

The flattened-out lateral surface is a rectangle with base = perimeter of base of original right prism and height = height of original right prism. Thus

lateral surface area of right prism = perimeter of base · height

And, of course,

surface area of right prism = lateral surface area + 2 · area of base

EXAMPLE Find the surface area of this right rectangular prism.

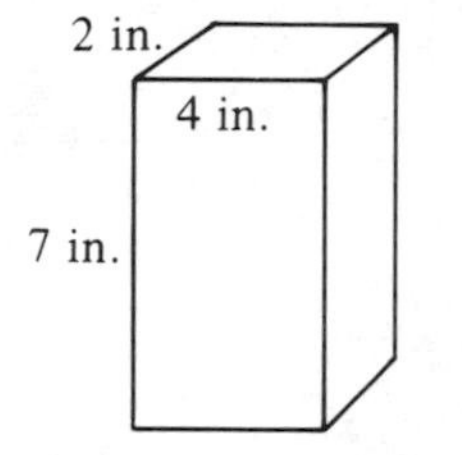

Solution The perimeter of the base is 12 in., so the lateral surface area is 12 in. · 7 in. = 84 in.^{2} The area of each base is 2 in. · 4 in. = 8 in.^{2} Thus the total surface area is 84 in.^{2} + 2 · 8 in.^{2} = 100 in.^{2}

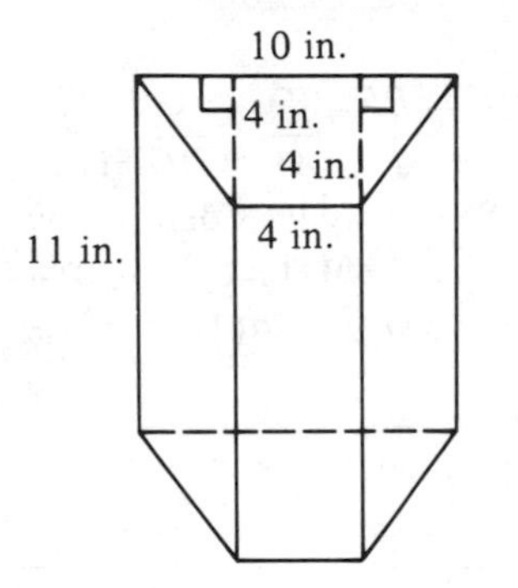

EXAMPLE Find the surface area of this right trapezoidal prism.

Solution The perimeter of the base is 24 in. (Can you see why?), and the height is 11 in., so the lateral surface area is 24 in. · 11 in. = 264 in.^{2} The area of each base is 28 in.^{2} (Can you see why?) Thus the total surface area is 264 in.^{2} + 2 · 28 in.^{2} = 320 in.^{2}

There is no nice, neat formula for the surface area of a general (non-right) prism. Likewise, there is no nice, neat formula for the surface area of a general pyramid. But there is one for "regular" pyramids. By definition,

A regular pyramid is a pyramid in which the base is a regular polygon and all of the lateral edges are congruent.

(See the following figure for one example of a regular pyramid.)

CAUTION The same adjective "regular" has been applied to three different nouns now—"polyhedron," "prism," "pyramid"—with three different meanings. For a prism to be regular it is enough that its bases be regular polygons. For a pyramid to be regular its base must be a regular polygon *and* another condition must be satisfied. (What condition?) Most regular prisms are not regular polyhedra. (Can you name one that is?) Most regular pyramids are not regular polyhedra. (Can you name one that is?) And, of course, no prism, regular or not, is a pyramid.

The surface-area formula for all regular pyramids is readily suggested by studying any one particular type. We shall investigate the regular hexagonal pyramid shown below.

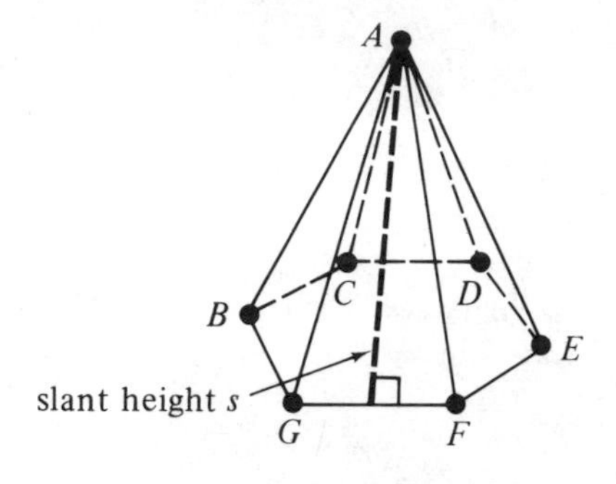

The base is a regular hexagon, so $\overline{BC} \cong \overline{CD} \cong \overline{DE} \cong \overline{EF} \cong \overline{FG} \cong \overline{GB}$. The "lateral edges" are all congruent: $\overline{AB} \cong \overline{AC} \cong \overline{AD} \cong \overline{AE} \cong \overline{AF} \cong \overline{AG}$. Thus the "lateral faces" are all congruent triangles (by SSS). If we let s denote the **slant height** of one of these faces, say face AGF, then that face has area $\frac{1}{2} \cdot GF \cdot s$. The entire lateral surface area is then $6 \cdot \frac{1}{2} \cdot GF \cdot s$. But $6 \cdot GF =$ perimeter of base, so

---

lateral surface area of regular pyramid = $\frac{1}{2}$ · perimeter of base · slant height

---

And, of course

---

surface area of regular pyramid = lateral surface area + area of base

---

EXAMPLE What is the surface area of this regular pyramid?

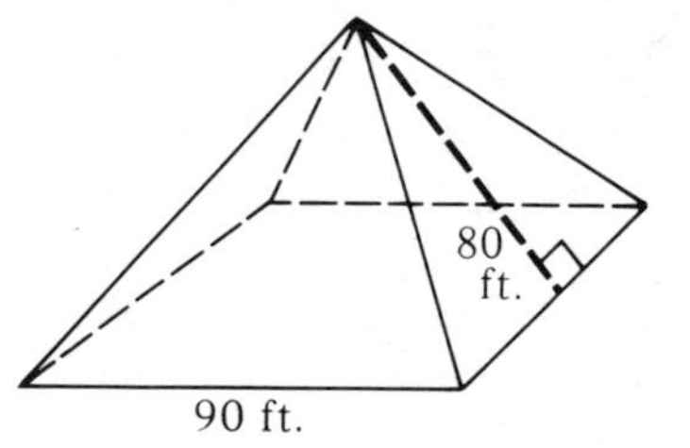

Solution Lateral surface area = $\frac{1}{2}$ · perimeter of base · slant height = $\frac{1}{2} \cdot 360 \cdot$

$80 = 14{,}400\ \text{ft}^2$. Area of base $= 90 \cdot 90 = 8100\ \text{ft}^2$. Total surface area of pyramid $= 14{,}400 + 8100 = 22{,}500\ \text{ft}^2$.

EXAMPLE What is the surface area of this regular pyramid?

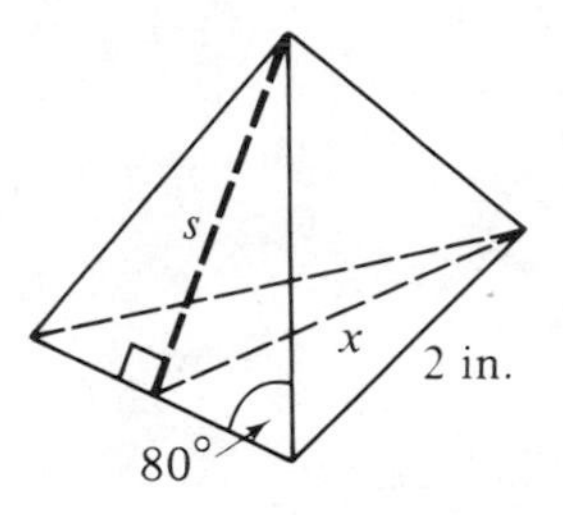

Solution

$$\frac{s}{1} = \tan 80° \Rightarrow s = \tan 80° \doteq 5\tfrac{2}{3}$$

perimeter of base = 6

Thus lateral surface area $\doteq \frac{1}{2} \cdot 6 \cdot \frac{17}{3} \doteq 17\ \text{in.}^2$

$$\frac{x}{2} = \sin 60° \Rightarrow x = 2 \cdot \sin 60° = \sqrt{3}$$

So area of base of pyramid $= \frac{1}{2} \cdot 2 \cdot \sqrt{3} = \sqrt{3} \doteq 1.7\ \text{in.}^2$ Total surface area of pyramid is about $18.7\ \text{in.}^2$

EXERCISES

1. Without looking back in the text, try to fill in this chart with formulas:

	LATERAL SURFACE AREA	TOTAL SURFACE AREA	VOLUME
right prism			
regular pyramid			

2. Formulate definitions for the following:
 (a) lateral face of a pyramid
 (b) lateral surface of a pyramid
 (c) lateral edge of a pyramid
 (d) lateral surface area of a pyramid

3. For each polyhedron below, find (approximately) the total surface area. All the pyramids are regular.

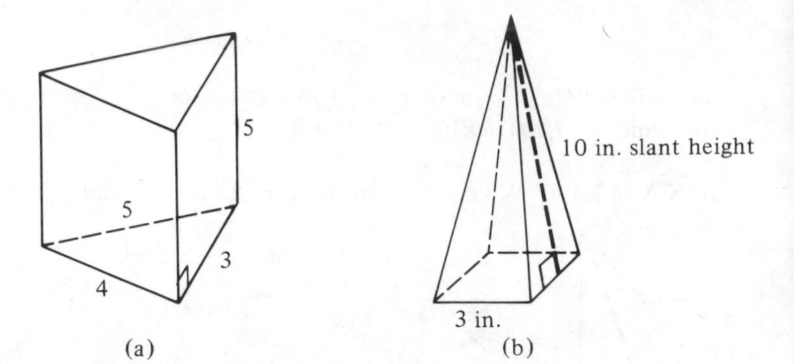

(a) (b)

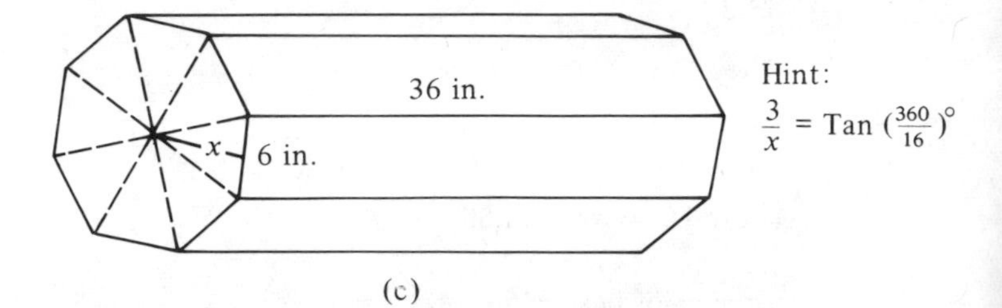

(c)

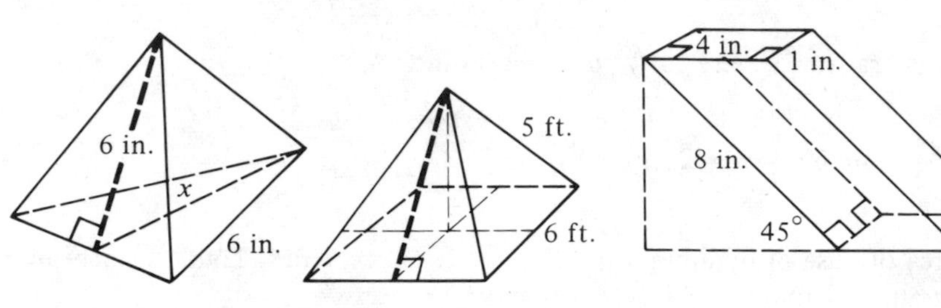

(d) (e) (f)

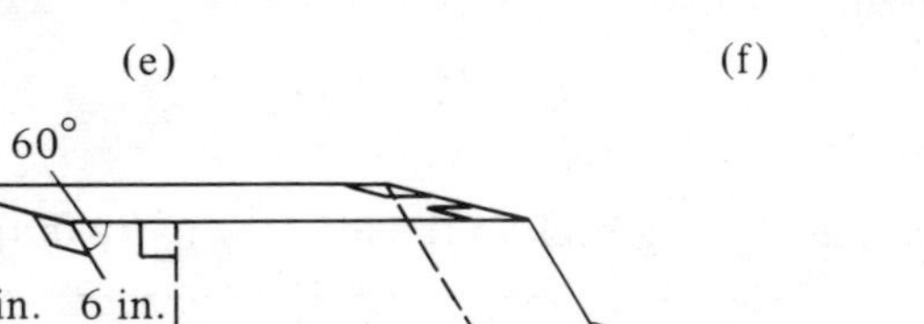

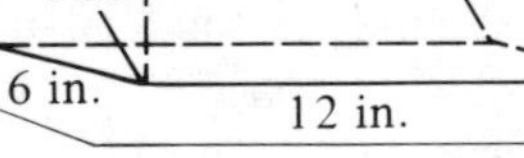

(g)

4. Find the surface areas of the houses on p. 456.

5. (a) Which container below holds more? (Both are portions of "regular" figures.)

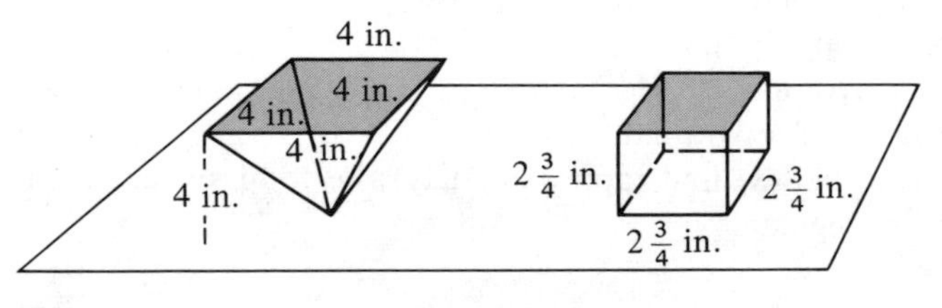

(b) Which has greater surface area?

6. Taking edge length to be 1 in each case, find the total surface area of each of the five regular polyhedra.

10.4 CYLINDERS

Up until now all of the space figures we have studied have been polyhedra; that is, they have had "flat" faces. In this and the next two sections we shall extend our investigation to three common types of space figures that can have "curved surfaces": cylinders, cones, and spheres.

The lateral surface of a prism can be viewed as the union of

1. two congruent (hollow) polygons lying in parallel planes, with
2. the parallel line segments joining corresponding points on these polygons.

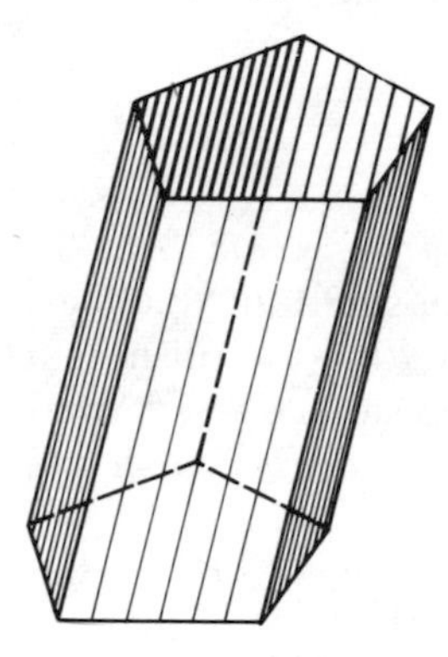

The bases of the prism, which are not shown in the diagram, are of course the plane regions bounded by the two congruent polygons.

There is a more general type of space figure, the **cylinder**, of which prisms are special cases. The **bases** of a cylinder are regions, in parallel planes, that are bounded by congruent curves. The **lateral surface** of a cylinder is the union of (1) the two congruent curves with (2) the parallel line segments joining corresponding points on these curves.

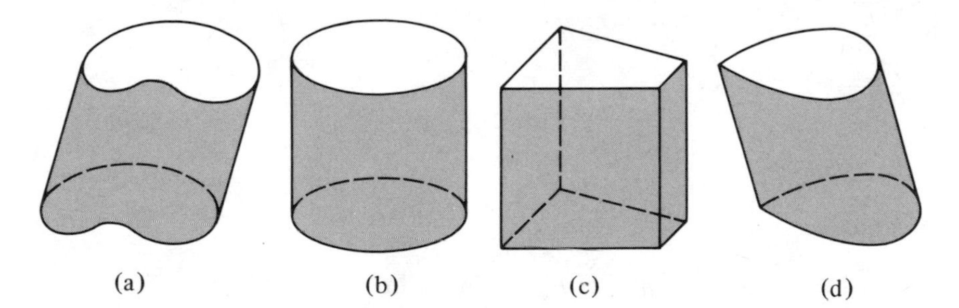

Cylinders, with lateral surfaces shaded

If the segments that make up the lateral surface are perpendicular to the bases, the cylinder is called a **right** cylinder. In the previous figure, (b) and (c) are right cylinders. Cylinder (c) is a prism because its bases are polygons. Cylinder (b) is called a right **circular** cylinder because its bases are circles. This is the figure most people think of when they hear the word "cylinder."

The formula for the volume of a general cylinder and the surface area of a right cylinder are listed below. Before you look at them, study the figure and see if you can guess what they are.

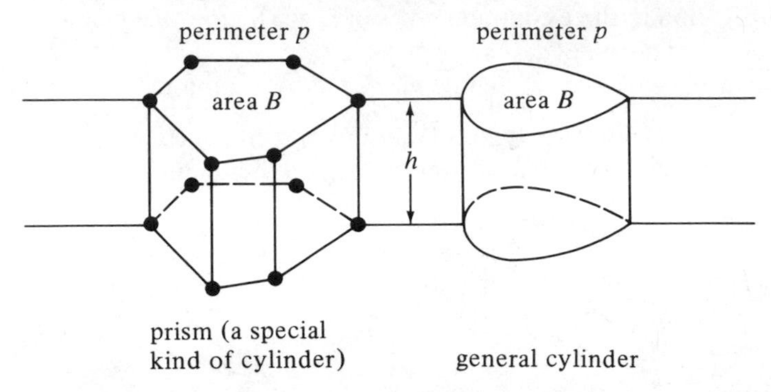

Do you remember the formula for the volume of the prism? Can you guess the formula for the volume of the cylinder? Assuming that the prism is a right prism, do you remember the formula for its lateral surface area? What do you suppose is the lateral surface area of a right cylinder? You should have guessed these formulas:

volume of cylinder = area of base · height = $B \cdot h$
lateral surface area of right cylinder = perimeter of base · height = $p \cdot h$

EXAMPLE Find the volume and total surface area of this right cylinder:

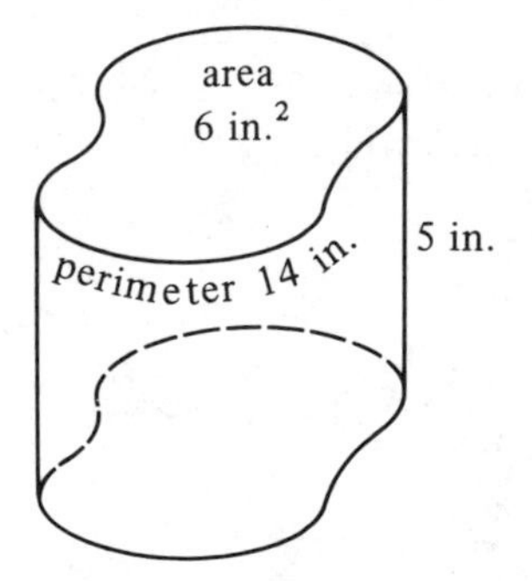

Solution

$$\text{surface area} = p \cdot h + 2 \cdot B = 14 \text{ in.} \cdot 5 \text{ in.} + 2 \cdot 6 \text{ in.}^2$$
$$= 70 \text{ in.}^2 + 12 \text{ in.}^2 = 82 \text{ in.}^2$$

$$\text{volume} = B \cdot h = 6 \text{ in.}^2 \cdot 5 \text{ in.} = 30 \text{ in.}^3$$

EXAMPLE Find the volume and surface area of this right cylinder:

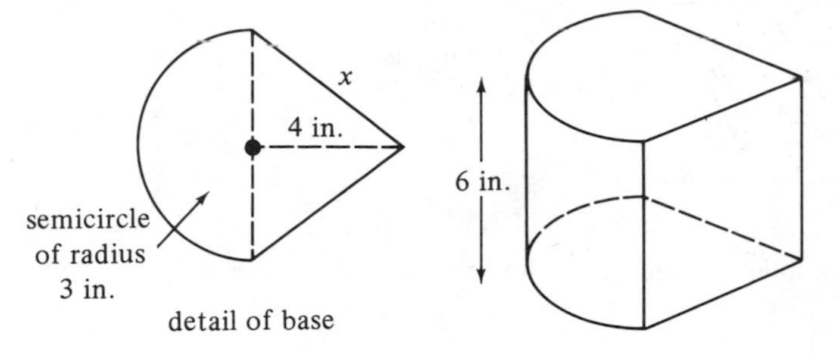

Solution $x = 5$ by the Pythagorean theorem, so the perimeter of the base $=$ $5 + 5 + \pi \cdot 3 = 10 + 3\pi$.

$$\text{area of base} = \pi \cdot \frac{3^2}{2} + \frac{1}{2} \cdot 6 \cdot 4 = 12 + \frac{9}{2}\pi$$

Thus

$$\begin{aligned}\text{surface area} &= p \cdot h + 2 \cdot B = (10 + 3\pi) \cdot 6 + 2 \cdot (12 + \tfrac{9}{2}\pi) \\ &= 84 + 27\pi \doteq 84 + 85 = 169 \text{ in.}^2\end{aligned}$$

and

$$\text{volume} = B \cdot h = (12 + \tfrac{9}{2}\pi) \cdot 6 = 72 + 27\pi \doteq 72 + 85 = 157 \text{ in.}^3$$

EXERCISES

1. Why is each figure below not a cylinder?

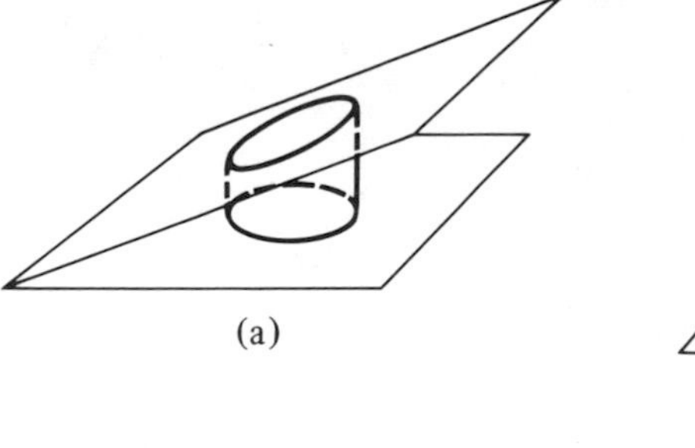

(a)

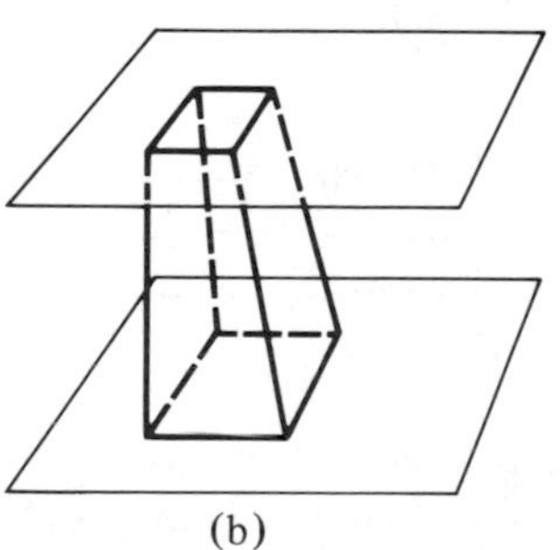

(b)

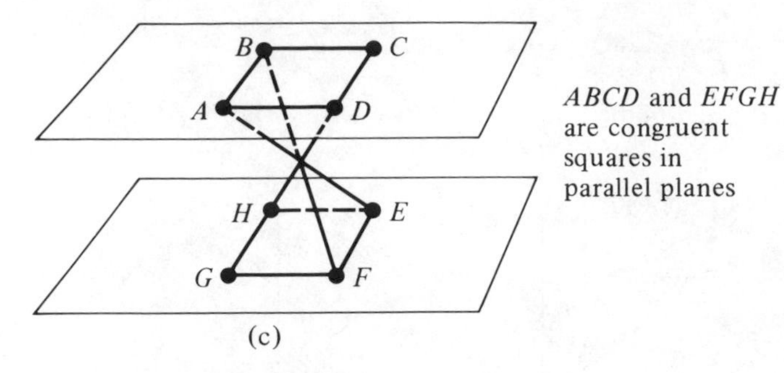

(c)

2. In this right circular cylinder, find the following. (Use $\pi \doteq \frac{22}{7}$.)

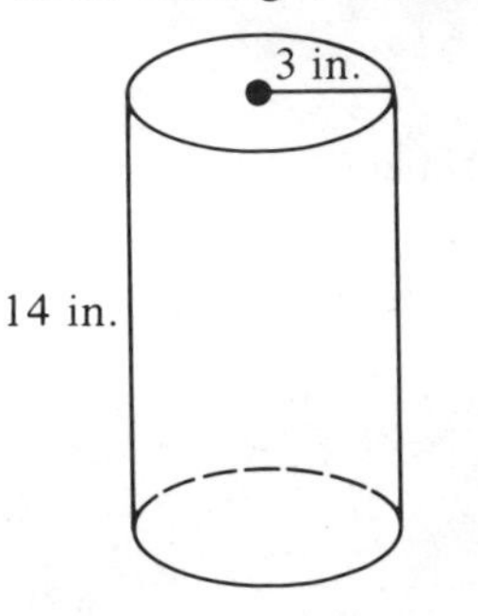

(a) perimeter of base
(b) lateral surface area
(c) area of base
(d) total surface area
(e) volume

3. For right circular cylinders, the formulas for area and volume can be stated completely in terms of the radius r of the base, the height h of the cylinder, and π.

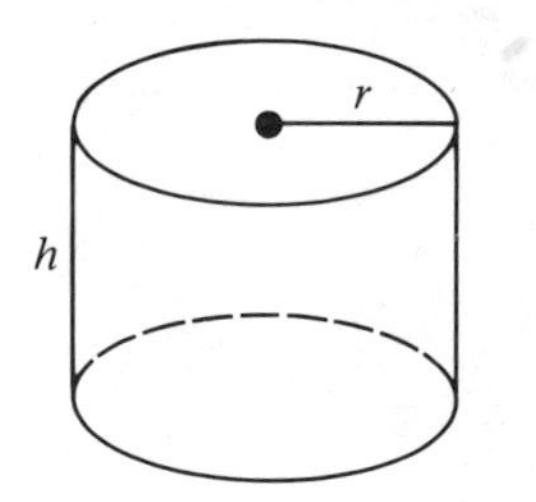

(a) perimeter of base = ________
(b) lateral surface area = ________
(c) area of base = ________
(d) total surface area = ________
(e) volume = ________

4. Find the volume and surface area of each right cylinder below. (All curves are portions of circles. Leave your answers in terms of π.)

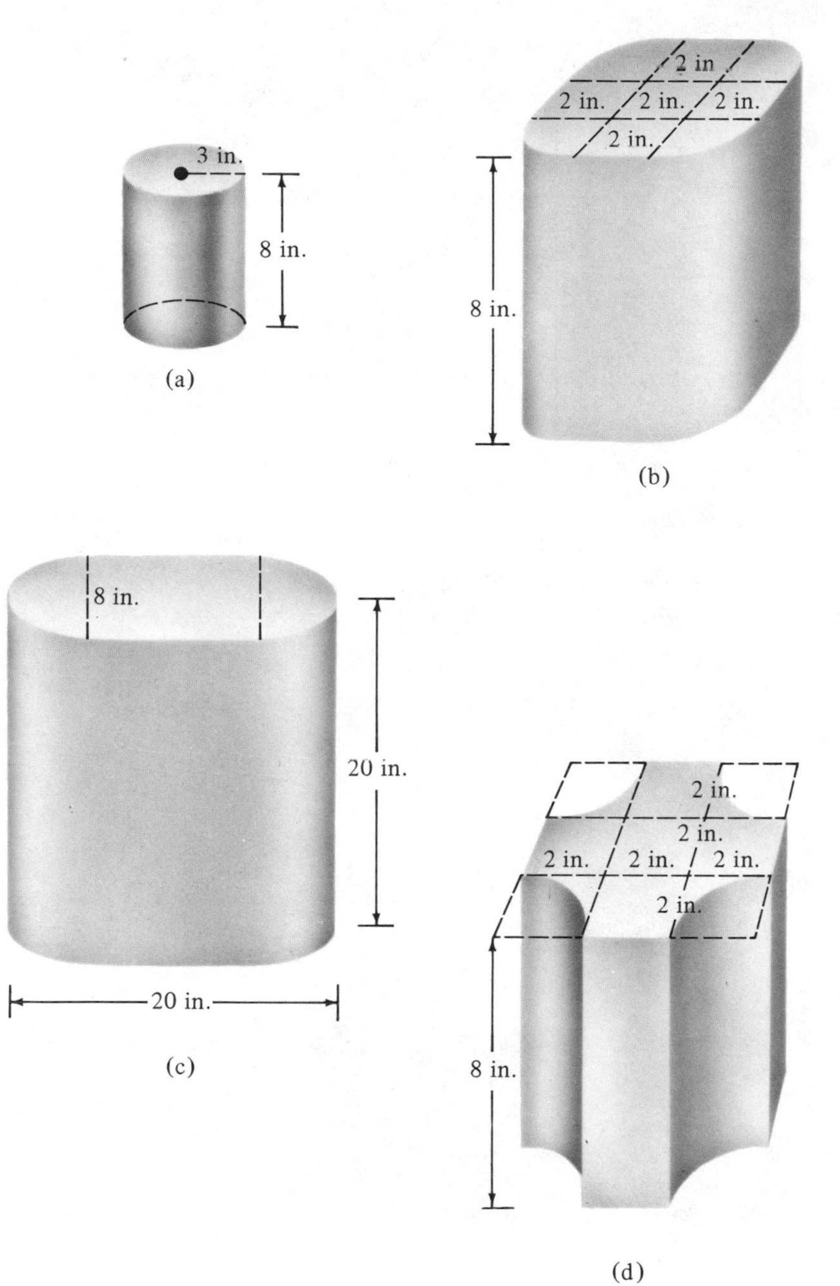

5. (a) Which tin can holds more: can A of radius 2 in. and height 2 in., or can B of radius 1 in. and height 7 in.?

(b) Which requires the most tin?

6. Find the volume of metal in this piece of pipe. (Leave your answer in terms of π.)

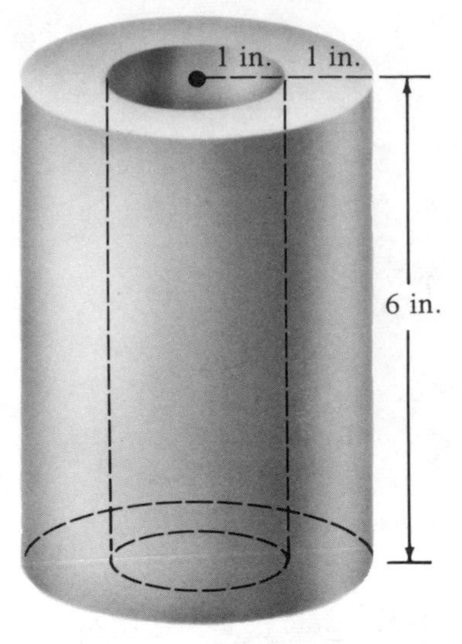

7. Find the surface area of the piece of pipe.
8. What is the volume of metal in this nut?

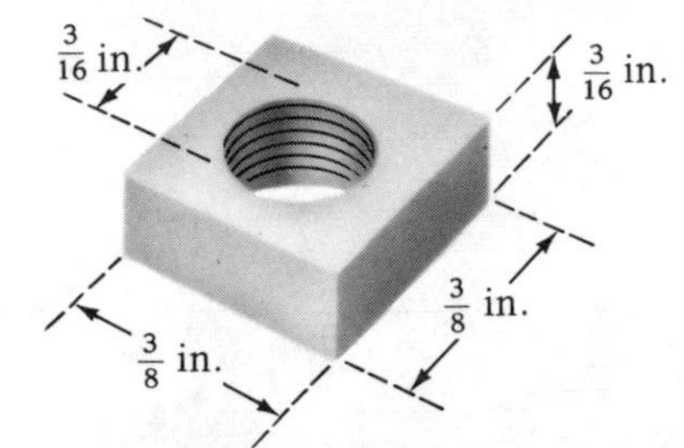

9. (a) Which container holds more?

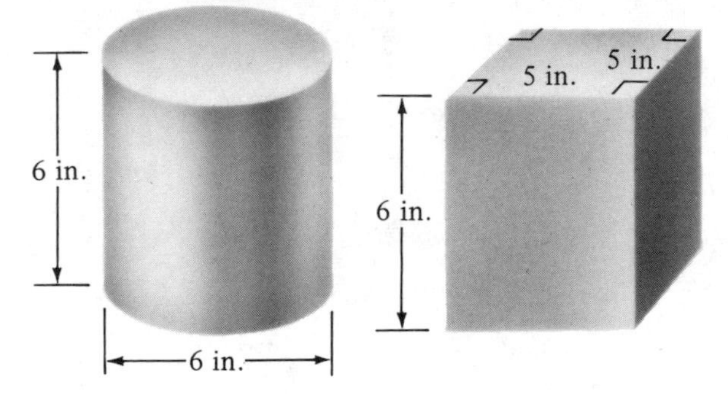

(b) Could the smaller container be placed inside the larger one?

10.5 CONES

The notion of cone, like the notion of cylinder, is a very general one. All of the figures below are cones.

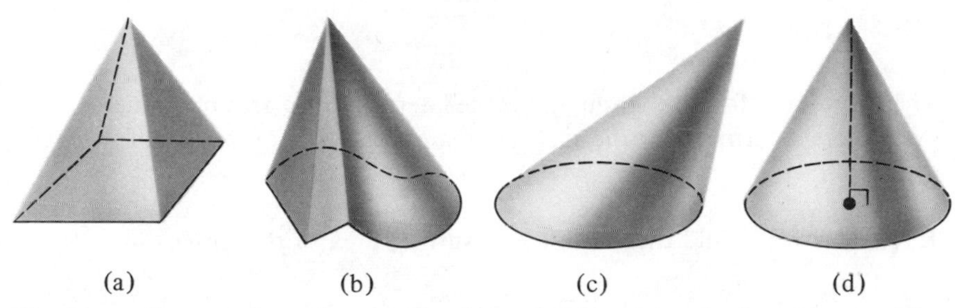

The **base** of a cone is a plane region bounded by a simple closed curve. The **lateral surface** of a cone is made up of all line segments joining the boundary of the base to a single point not in the plane of the base.

Figure (a) shows that a pyramid is a special kind of cone, namely, one with a polygonal base. Figure (d) is what most people think of when they hear the word "cone." It is a **right circular cone**. The segment joining the center of the circular base to the peak of the cone is perpendicular to the base. Right circular cones can be thought of as smoothed-out regular pyramids.

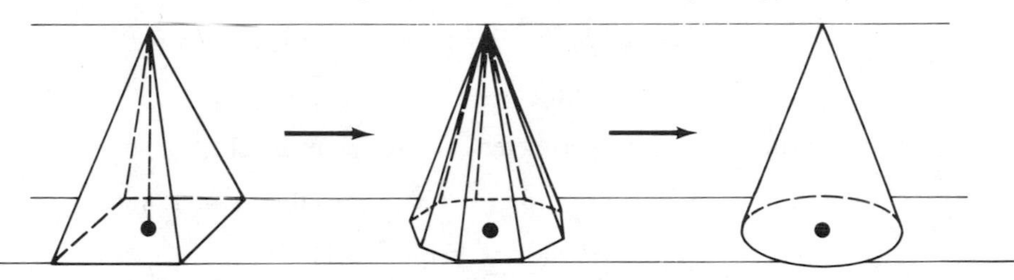

It should not be surprising that the formulas for surface area and volume of right circular cones are the same as for regular pyramids. Before looking ahead, see if you remember formulas for:

lateral surface area of a regular pyramid
total surface area of a regular pyramid
volume of an arbitrary pyramid

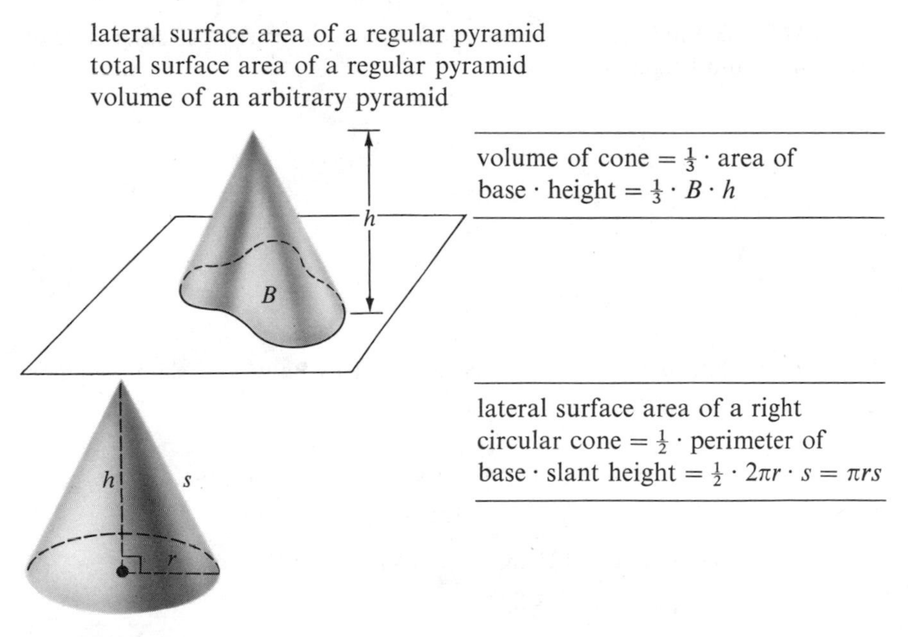

There is no nice formula, though, for the lateral surface area of a general cone. (Was there a formula for the surface area of a general pyramid? of a general cylinder?)

EXAMPLE Find the volume and surface area of this cone.

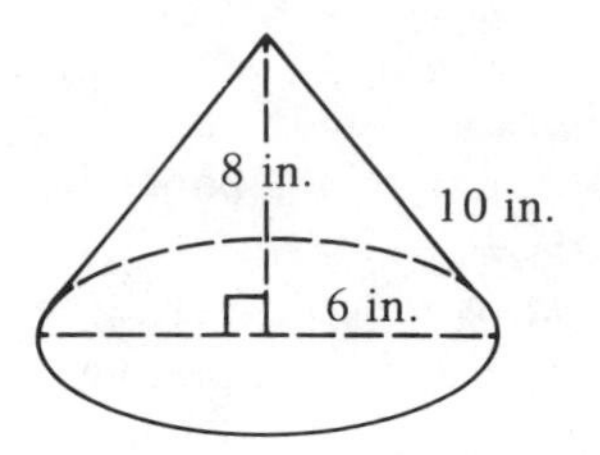

Solution

$$\text{volume} = \tfrac{1}{3} \cdot B \cdot h = \tfrac{1}{3} \cdot \pi \cdot (6 \text{ in.})^2 \cdot 8 \text{ in.} = 96\pi \text{ in.}^3$$
$$\doteq 302 \text{ in.}^3$$

$$\text{lateral surface area} = \tfrac{1}{2} \cdot \text{perimeter of base} \cdot \text{slant height}$$
$$= \tfrac{1}{2} \cdot 2\pi \cdot (6 \text{ in.}) \cdot (10 \text{ in.}) = 60\pi \text{ in.}^2 \doteq 188 \text{ in.}^2$$

$$\text{area of base} = \pi \cdot (6 \text{ in.})^2 = 36\pi \text{ in.}^2 \doteq 113 \text{ in.}^2$$

$$\text{total surface area} = \text{lateral surface area} + \text{area of base}$$
$$\doteq 188 \text{ in.}^2 + 113 \text{ in.}^2 = 301 \text{ in.}^2$$

EXAMPLE Find the volume and surface area of the right circular cone of radius 4 in. and height 10 in.

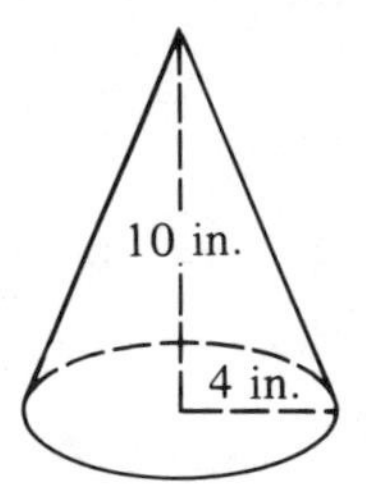

Solution To find lateral surface area we need to know the slant height s. By the Pythagorean theorem, $s = \sqrt{116} \doteq 10.77$ in. Thus

$$\text{lateral surface area} = \tfrac{1}{2} \cdot \text{perimeter of base} \cdot \text{slant height}$$
$$\doteq \tfrac{1}{2} \cdot 2 \cdot \pi \cdot 4 \text{ in.} \cdot 10.77 \text{ in.} \doteq 135 \text{ in.}^2$$

$$\text{total surface area} \doteq 135 \text{ in.}^2 + \pi \cdot (4 \text{ in.})^2 \doteq 135 \text{ in.}^2 + 50 \text{ in.}^2$$
$$\doteq 185 \text{ in.}^2$$

$$\text{volume} = \tfrac{1}{3} \cdot B \cdot h = \tfrac{1}{3} \cdot \pi \cdot (4 \text{ in.})^2 \cdot (10 \text{ in.}) = \frac{160\pi}{3} \text{ in.}^3$$
$$\doteq 167 \text{ in.}^3$$

EXERCISES

1. Stalactites and carrots suggest cones. Can you think of any other natural objects that suggest cones? Any man-made objects?

2. Find the volume and surface area of each figure below. All cones are right circular cones. Leave your answers in terms of π and square roots.

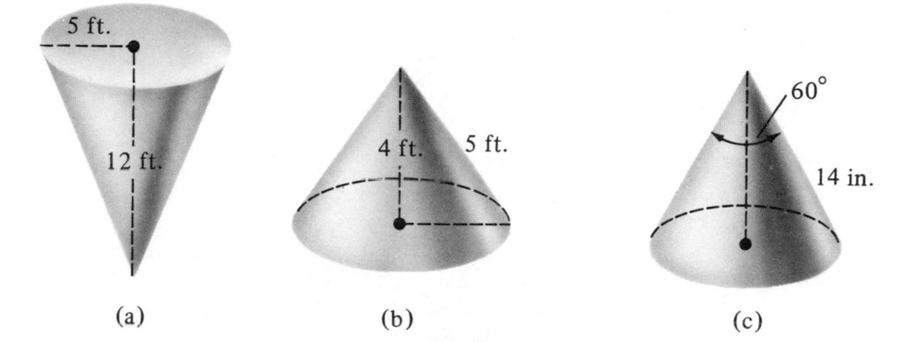

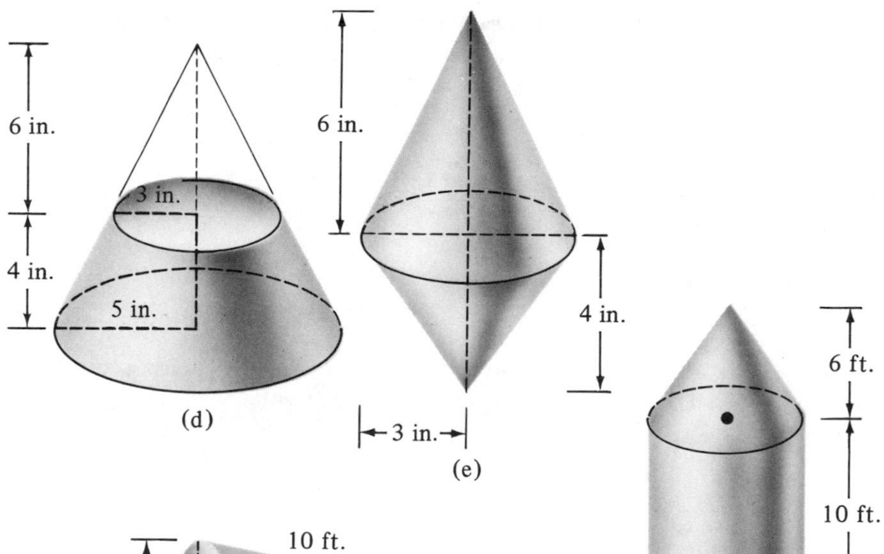

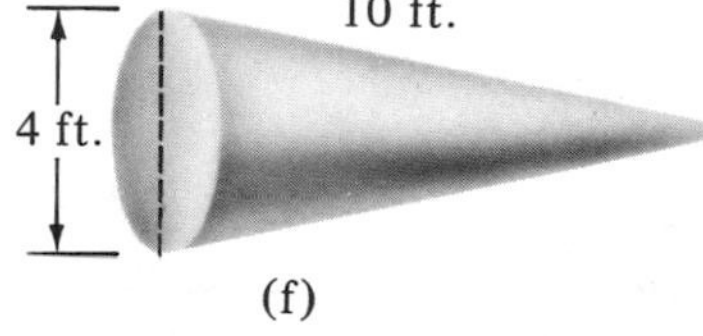

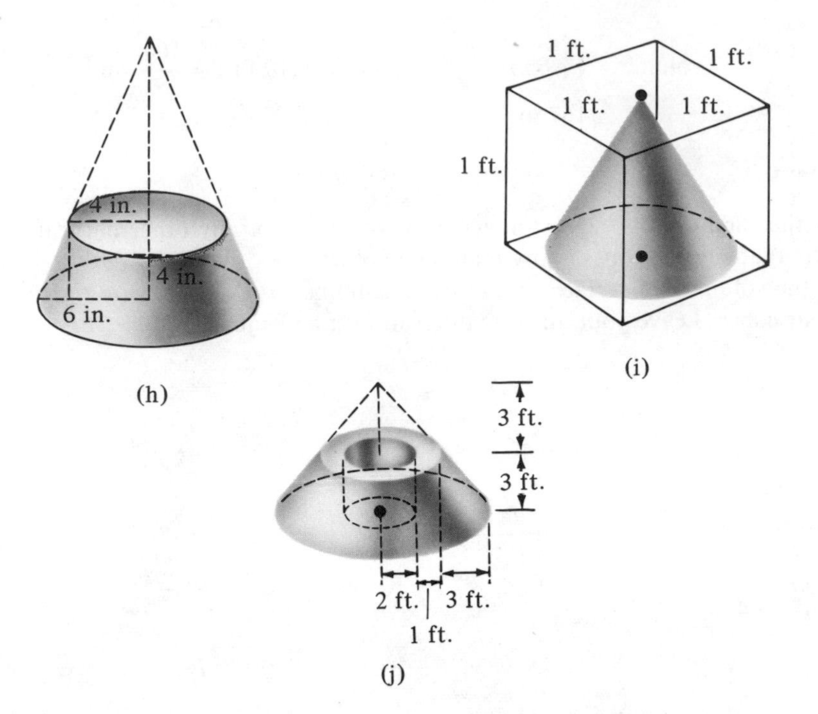

3. A right circular cone is carved out of a right circular cylinder of radius r and height r as shown below. What volume remains? (Simplify your answer. We shall refer to this answer in the next section.)

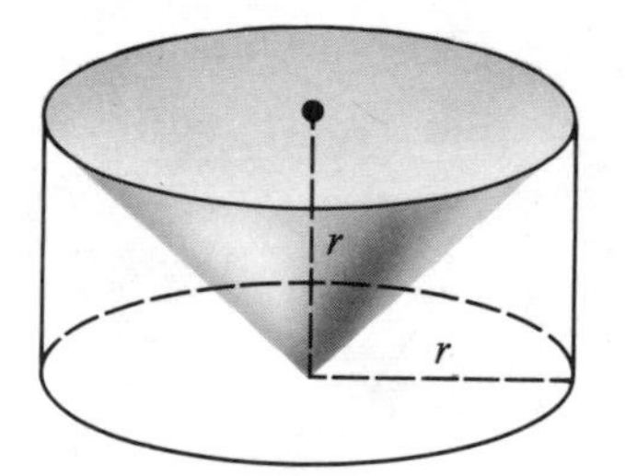

4. After sharpening a (right circular cylindrical) pencil in sharpener A, you decide to sharpen it again in sharpener B. What volume of wood and lead is ground away by sharpener B?

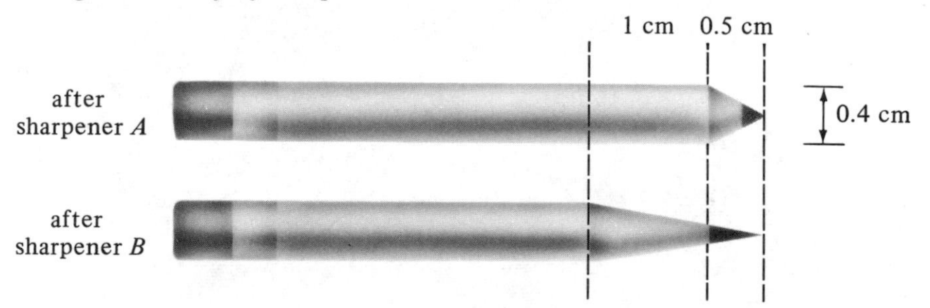

5. How do the volumes of these two cones compare?

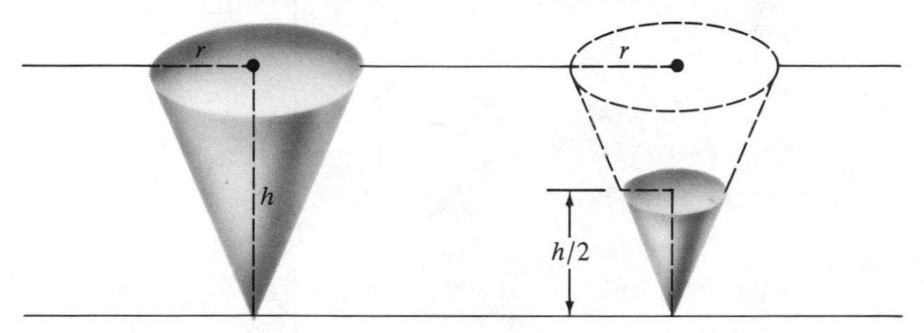

6. Joe claims that his cup is only half full. Is he right?

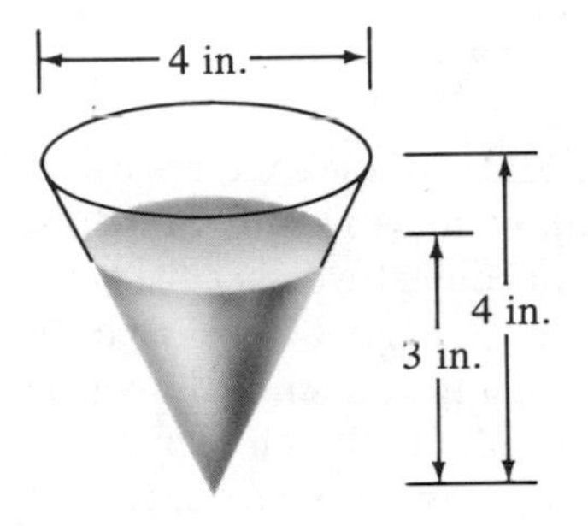

7. Shown below is the pattern for a right circular cone.

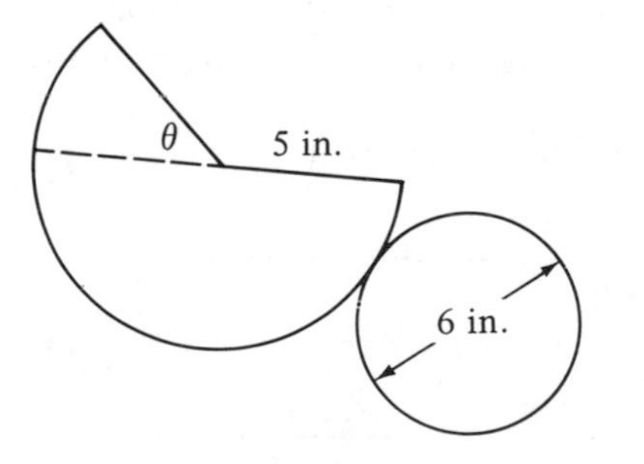

(a) What will be its height?
(b) What is the measure of $\angle\theta$?

10.6 SPHERES

A **sphere** is the geometric figure suggested by a basketball. A sphere can also be defined in purely mathematical terms as the set of all points in space that are equidistant from a given point, called the **center of the sphere**.

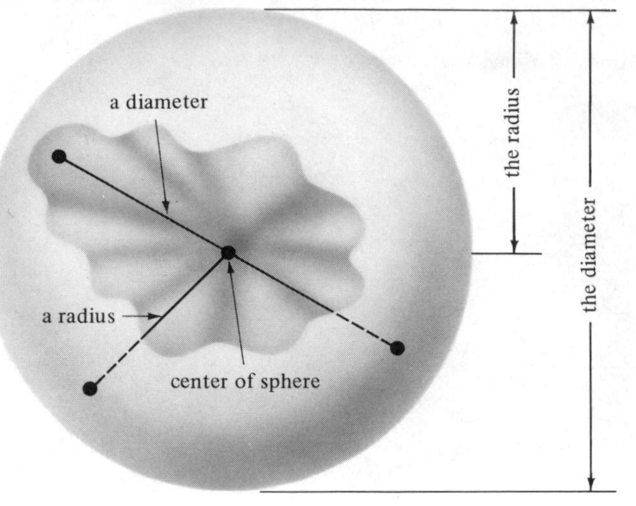

a sphere with a hole torn in it

Each segment joining a point of the sphere to the center of the sphere is called **a radius** of the sphere. The common length of all of these radii is called **the radius** of the sphere. You will have to decide from context whether the word "radius" is being used to refer to a segment or to a number. A segment joining two points on the sphere and passing through the center of the sphere is called **a diameter** of the sphere. The common length of all such segments is called **the diameter** of the sphere. Again you will have to decide from context whether the word "diameter" refers to a segment or to a number.

When we speak of the volume of a sphere we mean, of course, the volume of the chunk of space inside the sphere. The solid figure consisting of a sphere and all points inside it is called a *solid sphere* or a "three-ball," the word "three" referring to the dimension of the solid figure. (What is the dimension of the sphere alone?)

The formulas for the surface area and volume of a sphere are very simple to write down, but are quite a bit harder to derive.

Sphere:
surface area $= 4\pi r^2$ (r — radius)
volume $= \frac{4}{3}\pi r^3$

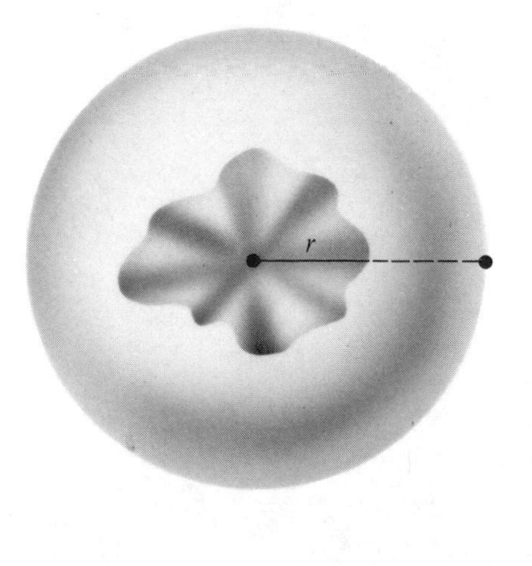

A physical experiment for discovering the volume formula for a sphere is described in the exercises. A plausible mathematical derivation is also outlined. The surface-area formula can be derived from the volume formula, as we now show.

Think of the solid sphere as being made up of lots of little pieces, each of which is very nearly a pyramid with peak at the center of the sphere.

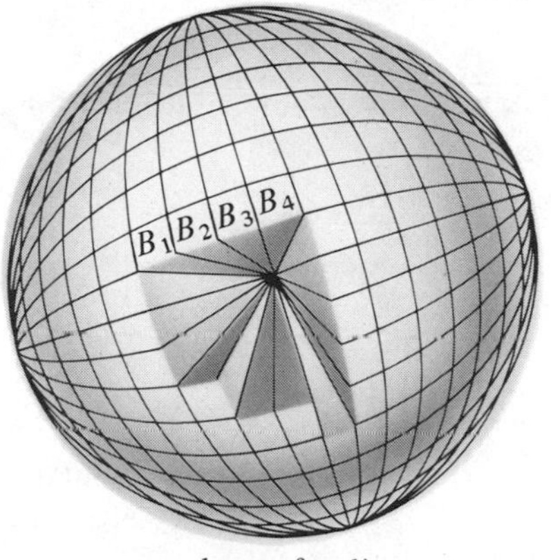

sphere of radius r

By the volume formula for pyramids, each little piece has volume

$$= \tfrac{1}{3} \cdot \text{area of its base} \cdot \text{height*}$$
$$= \tfrac{1}{3} \cdot \text{area of its base} \cdot r.$$

Let us say that the area of the base of the first piece is B_1, of the second piece B_2, and so on. The total volume of the sphere is the sum of the volumes of its pieces.

$$\begin{aligned}\text{volume of sphere} &= \tfrac{1}{3} \cdot B_1 \cdot r + \tfrac{1}{3} \cdot B_2 \cdot r + \tfrac{1}{3} \cdot B_3 \cdot r + \cdots \\ &= \tfrac{1}{3} \cdot r \cdot (B_1 + B_2 + B_3 + \cdots)\end{aligned}$$

But the sum of the areas of the bases of the pieces is the surface area of the sphere; that is,

$$\text{volume (of sphere)} = \tfrac{1}{3} \cdot r \cdot \text{surface area (of sphere)}$$

Thus, if we know that the volume is $\frac{4}{3}\pi r^3$, we can find the surface area:

$$\tfrac{4}{3}\pi r^3 = \tfrac{1}{3} \cdot r \cdot \text{surface area} \Rightarrow 4\pi r^2 = \text{surface area}$$

* Technically, we ought to write "$\doteq$" instead of "$=$" here. We write "$=$" because if the sphere is cut up into tinier and tinier pieces, the approximation here gets better and better.

On the other hand, if we had some independent way of deriving the surface area formula, surface area $= 4\pi r^2$, then we could deduce the volume formula:

$$\begin{aligned} \text{volume} &= \tfrac{1}{3} \cdot r \cdot \text{surface area} \\ &= \tfrac{1}{3} \cdot r \cdot 4\pi r^2 \\ &= \tfrac{4}{3}\pi r^3 \end{aligned}$$

EXAMPLE Find the approximate surface area and volume of a sphere of radius 7 in. (Use $\pi \doteq \frac{22}{7}$.)

Solution

$$\text{surface area} = 4\pi r^2 \doteq 4 \cdot \tfrac{22}{7} \cdot (7 \text{ in.})^2 = 88 \cdot 7 \text{ in.}^2 = 616 \text{ in.}^2$$

$$\text{volume} = \tfrac{4}{3}\pi r^3 \doteq \tfrac{4}{3} \cdot \tfrac{22}{7} \cdot (7 \text{ in.})^3 = 88 \cdot \tfrac{49}{3} \text{ in.}^3 = 1437 \text{ in.}^3$$

EXAMPLE Find the volume and (total) surface area of this solid hemisphere:

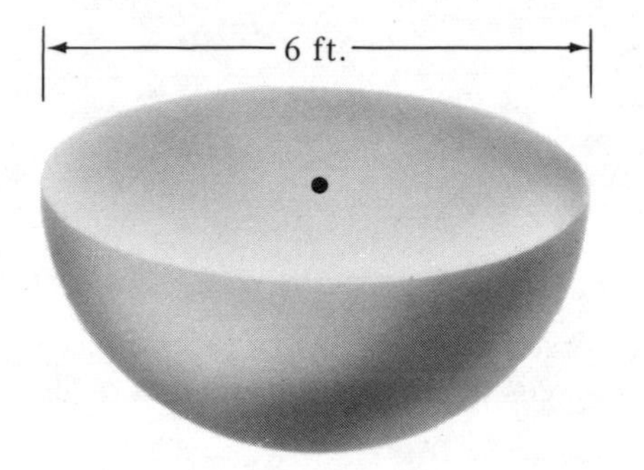

Solution

$$\begin{aligned} \text{diameter} &= 6 \text{ ft} \Rightarrow \text{radius} = 3 \text{ ft} \\ \text{volume of hemisphere} &= \tfrac{1}{2} \cdot \text{volume of sphere} = \tfrac{1}{2} \cdot \tfrac{4}{3} \cdot \pi \cdot (3 \text{ ft})^3 \\ &= 18\pi \text{ ft}^3 \\ \text{curved surface area} &= \tfrac{1}{2} \cdot \text{surface area of sphere} = \tfrac{1}{2} \cdot 4 \cdot \pi \cdot (3 \text{ ft})^2 \\ &= 18\pi \text{ ft}^2 \\ \text{flat surface area} &= \pi r^2 = \pi(3 \text{ ft})^2 = 9\pi \text{ ft}^2 \\ \text{total surface area} &= 18\pi \text{ ft}^2 + 9\pi \text{ ft}^2 = 27\pi \text{ ft}^2 \end{aligned}$$

EXERCISES

1. Write a definition of a circle that is similar to the "purely mathematical" definition of a sphere given in the text.

2. Explain how you would use the apparatus shown below to discover the formula for the volume of a sphere.

hemispherical
bowl of radius r

right circular
conical paper cup
of radius r
and height r

3. Compute the volume and (total) surface area of each figure below. Leave your answers in terms of π.

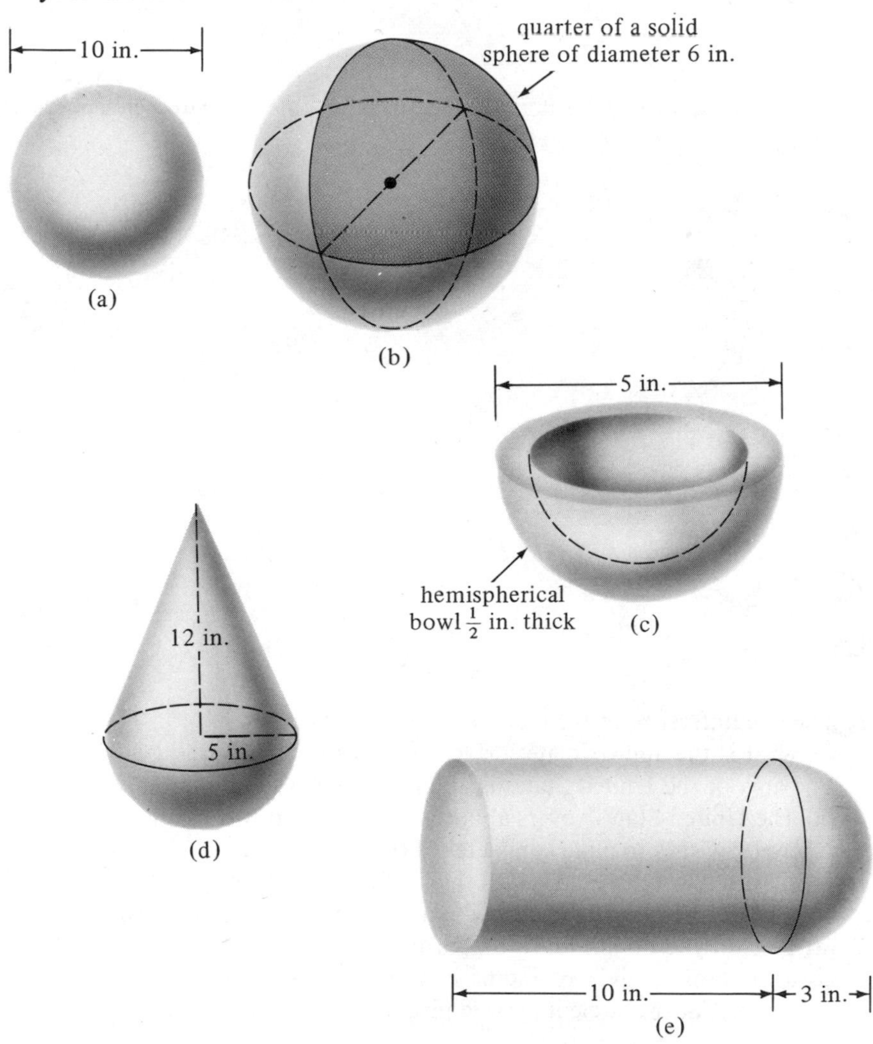

(a) (b) (c) (d) (e)

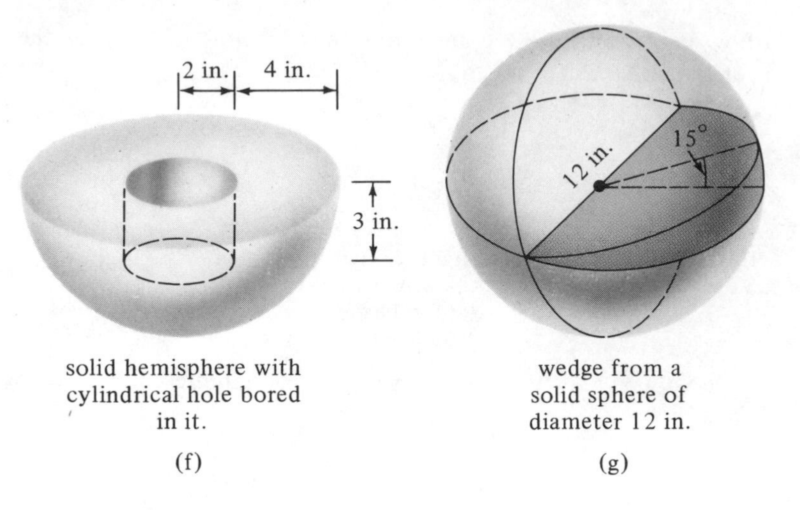

solid hemisphere with cylindrical hole bored in it.

(f)

wedge from a solid sphere of diameter 12 in.

(g)

4. How many quarts of paint are needed to paint this silo with hemispherical cap, if 1 gallon covers 500 ft^2?

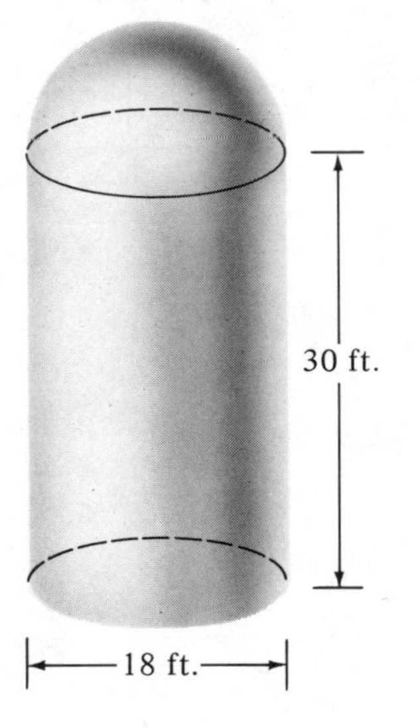

5. The circumference of the earth is about 25,000 miles.
(a) what is the approximate surface area of the earth?
The area of the United States is about 3,600,000 square miles.
(b) The United States covers about what percent of the earth's surface?

6. A jeweler wishes to plate completely a thin hemispherical bowl of diameter 14 cm with a layer of gold 0.1 cm thick. If 1 cc of gold weighs 19.3 g and gold costs \$3.90 per gram, about how many dollars worth of gold will he use?

7. **(a)** Pretend you do not know the formulas for the circumference C and the area A of a circle of radius r. Derive a formula relating C and A. (*Hint*: Cut the circle into little pieces. Each is nearly a triangle.)

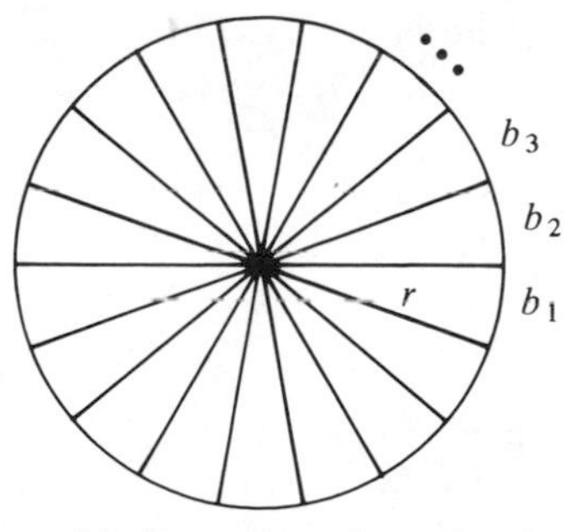

(b) Assuming that $C = 2\pi r$ and using your formula from (a), derive the formula for A in terms of π and r.

(c) Assuming that $A = \pi r^2$ and using your formula from (a), derive the formula for C in terms of π and r.

8. (Derivation of volume formula for sphere without using formula for surface area of sphere) In Exercise 3, p. 472, we considered the figure remaining after a right circular cone of radius r was carved from a right circular cylinder of radius r and height r. A cross section of the figure is sketched here:

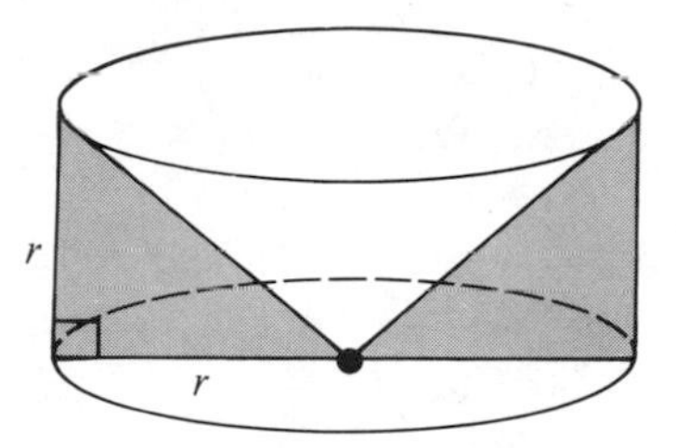

Suppose that we consider the figure as made up of a lot of flat washers piled on top of each other.

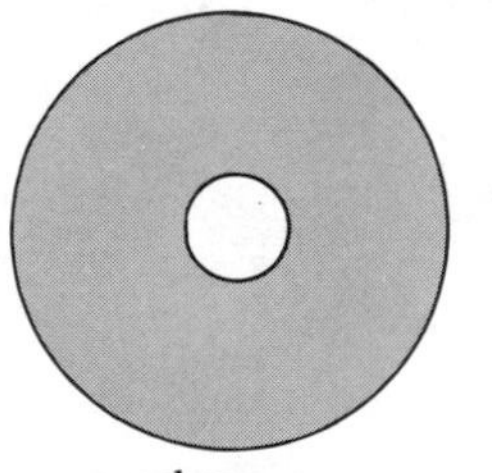

washer near bottom of stack

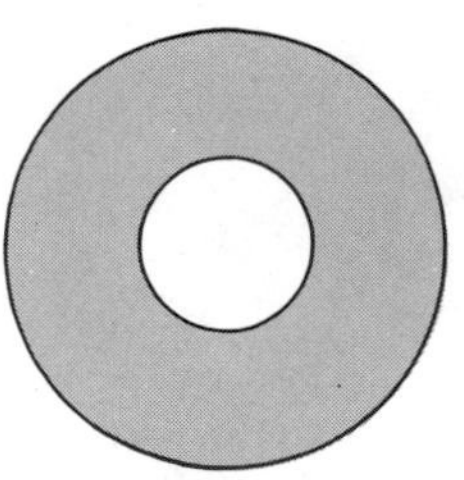

washer near middle of stack

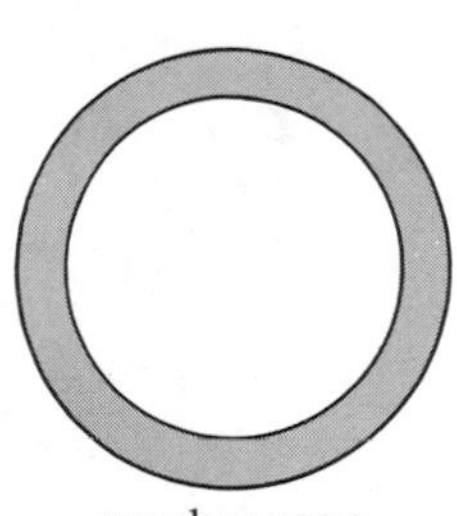

washer near top of stack

(a) Then the washer that lies at height x above the base of the figure has inner radius ________ and outer radius ________.

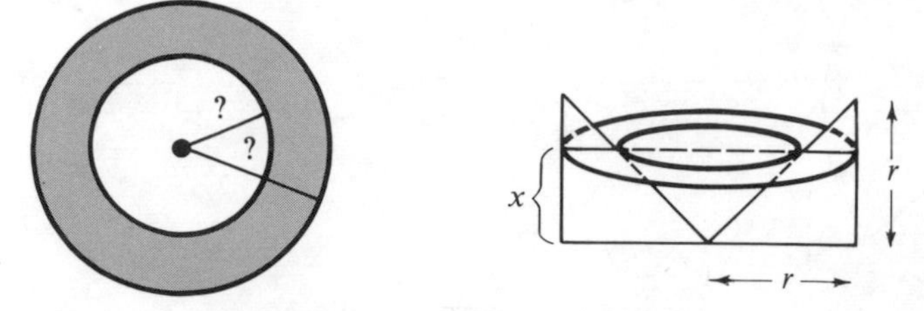

(b) What is the area of one face of this washer?

A hemisphere of radius r can also be thought of as a pile of disks.

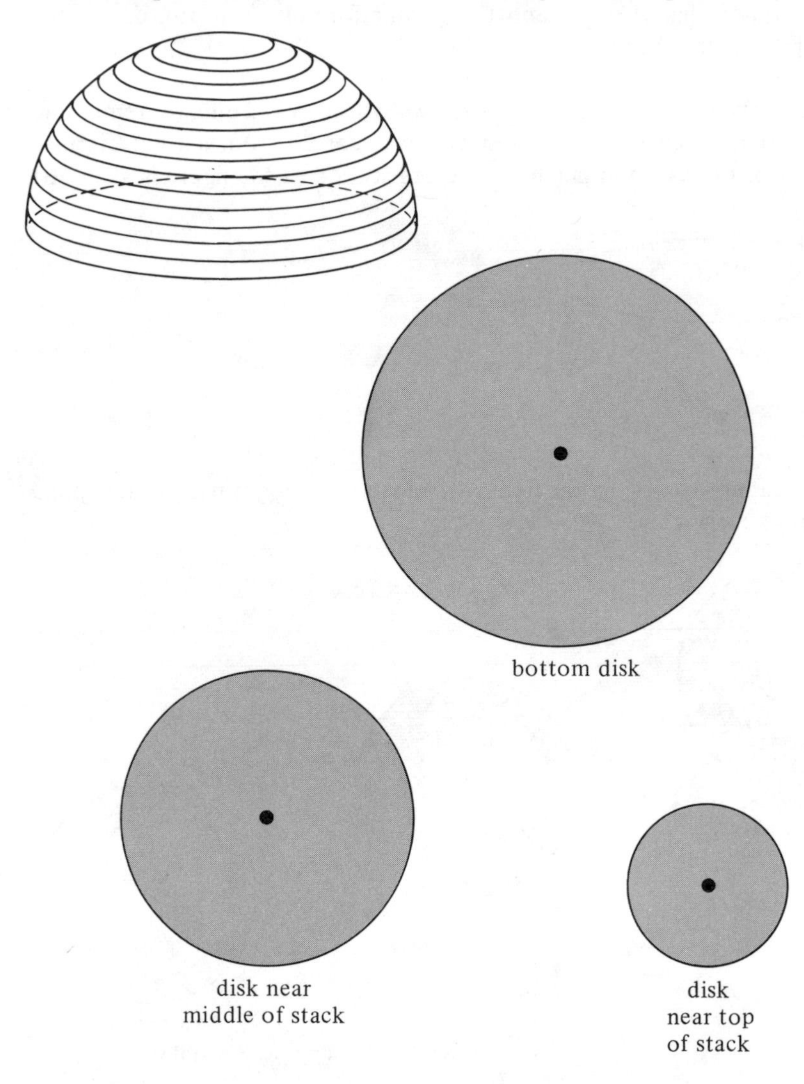

(c) Then the disk that lies at height x above the base of the hemisphere has radius ________.

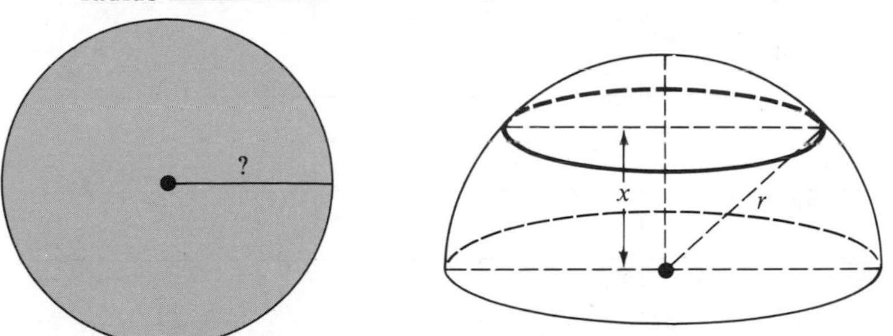

(d) What is the area of one face of this disk?

You should have concluded that the washer at height x above the base of the carved-out cylinder has the same area as the disk at height x above the base of the hemisphere. Thus the volume of the hemisphere (a stack of disks) should be the same as the volume of the carved-out cylinder (a stack of washers). But in Exercise 3 on p. 472 we found the volume of the carved-out cylinder to be $\frac{2}{3}\pi r^3$. Thus

$$\text{volume hemisphere} = \tfrac{2}{3}\pi r^3$$

and

(e) volume of sphere = ________.

TEACHING NOTE Washer and disk sets for the derivation in Exercise 8 are commercially available teaching aids.

9. Fill in as many formulas as you can in this chart. If you do not recall the formula right away, try to remember how it was derived. Shade in those boxes in the chart for which no general formula exists.

SPACE FIGURE	VOLUME	LATERAL SURFACE AREA	TOTAL SURFACE AREA
prism			
right prism			
cylinder			
right cylinder			
right circular cylinder			
pyramid			
cone			

SPACE FIGURE	VOLUME	LATERAL SURFACE AREA	TOTAL SURFACE AREA
regular pyramid			
right circular cone			
sphere			

VOCABULARY

10.1 Regular Polyhedra and Euler's Formula

polyhedron
faces (of a polyhedron)
regular polyhedron
regular tetrahedron
regular hexahedron
regular octahedron
regular dodecahedron
regular icosahedron
simple polyhedron
Euler's formula (for simple polyhedra)
planar graph
Euler's formula (for planar graphs)

10.2 Volumes of Polyhedra

prism
lateral faces (of a prism)
pyramid
base (of a prism)
right prism
base (of a pyramid)

10.3 Surface Areas of Polyhedra

total surface area (of a polyhedron)
lateral surface (of a prism)
lateral surface area (of a right prism)
regular pyramid
lateral edges (of a pyramid)
lateral faces (of a pyramid)
slant height (of a regular pyramid)
lateral surface (of a pyramid)
lateral surface area (of a regular pyramid)

10.4 Cylinders

cylinder
lateral surface (of a cylinder)
right circular cylinder
base (of a cylinder)
right cylinder

10.5 Cones

base (of a cone)
lateral surface (of a cone)
right circular cone

10.6 Spheres

spherc
a radius
a diameter
solid sphere (three-ball)
center
the radius
the diameter

APPROXIMATE SQUARE ROOTS					
NUMBER n	SQUARE ROOT $\sqrt{n}$	SQUARE ROOT OF 10 TIMES n $\sqrt{10 \times n}$	NUMBER n	SQUARE ROOT $\sqrt{n}$	SQUARE ROOT OF 10 TIMES n $\sqrt{10 \times n}$
1	1.000	3.162	40	6.325	20.000
2	1.414	4.472	41	6.403	20.248
3	1.732	5.477	42	6.481	20.494
4	2.000	6.325	43	6.557	20.736
5	2.236	7.071	44	6.633	20.976
6	2.449	7.746	45	6.708	21.213
7	2.646	8.367	46	6.782	21.448
8	2.828	8.944	47	6.856	21.679
9	3.000	9.487	48	6.928	21.909
10	3.162	10.000	49	7.000	22.136
11	3.317	10.488	50	7.071	22.361
12	3.464	10.954	51	7.141	22.583
13	3.606	11.402	52	7.211	22.804
14	3.742	11.832	53	7.280	23.022
15	3.875	12.247	54	7.348	23.238
16	4.000	12.649	55	7.416	23.452
17	4.123	13.038	56	7.483	23.664
18	4.243	13.416	57	7.550	23.875
19	4.359	13.784	58	7.616	24.083
20	4.472	14.142	59	7.681	24.290
21	4.583	14.491	60	7.746	24.495
22	4.690	14.832	61	7.810	24.698
23	4.796	15.166	62	7.874	24.900
24	4.899	15.492	63	7.937	25.100
25	5.000	15.811	64	8.000	25.298
26	5.099	16.125	65	8.062	25.495
27	5.196	16.432	66	8.124	25.690
28	5.292	16.733	67	8.185	25.884
29	5.385	17.029	68	8.246	26.077
30	5.477	17.321	69	8.307	26.268
31	5.568	17.607	70	8.367	26.458
32	5.657	17.889	71	8.426	26.646
33	5.745	18.166	72	8.485	26.833
34	5.831	18.439	73	8.544	27.019
35	5.916	18.708	74	8.602	27.203
36	6.000	18.974	75	8.660	27.386
37	6.083	19.235	76	8.718	27.568
38	6.164	19.494	77	8.775	27.749
39	6.245	19.748	78	8.832	27.928

APPROXIMATE SQUARE ROOTS (*continued*)					
NUMBER n	SQUARE ROOT $\sqrt{n}$	SQUARE ROOT OF 10 TIMES n $\sqrt{10 \times n}$	NUMBER n	SQUARE ROOT $\sqrt{n}$	SQUARE ROOT OF 10 TIMES n $\sqrt{10 \times n}$
79	8.888	28.107	90	9.487	30.000
80	8.944	28.284	91	9.539	30.166
81	9.000	28.461	92	9.592	30.332
82	9.055	28.636	93	9.644	30.496
83	9.110	28.810	94	9.695	30.659
84	9.165	28.983	95	9.747	30.822
85	9.220	29.155	96	9.798	30.984
86	9.274	29.326	97	9.849	31.145
87	9.327	29.496	98	9.899	31.305
88	9.381	29.665	99	9.950	31.464
89	9.434	29.833	100	10.000	31.623

APPROXIMATE TRIGONOMETRIC RATIOS							
ANGLE	SIN	COS	TAN	ANGLE	SIN	COS	TAN
0°	.000	1.000	.000	41°	.656	.755	.869
1°	.017	1.000	.017	42°	.669	.743	.900
2°	.035	.999	.035	43°	.682	.731	.933
3°	.052	.999	.052	44°	.695	.719	.966
4°	.070	.998	.070	45°	.707	.707	1.000
5°	.087	.996	.087	46°	.719	.695	1.036
6°	.105	.995	.105	47°	.731	.682	1.072
7°	.122	.993	.123	48°	.743	.669	1.111
8°	.139	.990	.141	49°	.755	.656	1.150
9°	.156	.988	.158	50°	.766	.643	1.192
10°	.174	.985	.176	51°	.777	.629	1.235
11°	.191	.982	.194	52°	.788	.616	1.280
12°	.208	.978	.213	53°	.799	.602	1.327
13°	.225	.974	.231	54°	.809	.588	1.376
14°	.242	.970	.249	55°	.819	.574	1.428
15°	.259	.966	.268	56°	.829	.559	1.483
16°	.276	.961	.287	57°	.839	.545	1.540
17°	.292	.956	.306	58°	.848	.530	1.600
18°	.309	.951	.325	59°	.857	.515	1.664
19°	.326	.946	.344	60°	.866	.500	1.732
20°	.342	.940	.364	61°	.875	.485	1.804
21°	.358	.934	.384	62°	.883	.469	1.881
22°	.375	.927	.404	63°	.891	.454	1.963
23°	.391	.921	.424	64°	.899	.438	2.050
24°	.407	.914	.445	65°	.906	.423	2.145
25°	.423	.906	.466	66°	.914	.407	2.246
26°	.438	.899	.488	67°	.921	.391	2.356
27°	.454	.891	.510	68°	.927	.375	2.475
28°	.469	.883	.532	69°	.934	.358	2.605
29°	.485	.875	.554	70°	.940	.342	2.747
30°	.500	.866	.577	71°	.946	.326	2.904
31°	.515	.857	.601	72°	.951	.309	3.078
32°	.530	.848	.625	73°	.956	.292	3.271
33°	.545	.839	.649	74°	.961	.276	3.487
34°	.559	.829	.675	75°	.966	.259	3.732
35°	.574	.819	.700	76°	.970	.242	4.011
36°	.588	.809	.727	77°	.974	.225	4.332
37°	.602	.799	.754	78°	.978	.208	4.705
38°	.616	.788	.781	79°	.982	.191	5.145
39°	.629	.777	.810	80°	.985	.174	5.671
40°	.643	.766	.839	81°	.988	.156	6.314

APPROXIMATE TRIGONOMETRIC RATIOS (*continued*)							
ANGLE	SIN	COS	TAN	ANGLE	SIN	COS	TAN
82°	.990	.139	7.115	87°	.999	.052	19.081
83°	.993	.122	8.144	88°	.999	.035	28.636
84°	.995	.105	9.514	89°	1.000	.017	57.290
85°	.996	.087	11.430	90°	1.000	.000	—
86°	.998	.070	14.301				

INDEX